Forensic Chemistry

Published and forthcoming titles in the Forensic Science in Focus series

Published

The Global Practice of Forensic Science
Douglas H. Ubelaker (Editor)

Forensic Chemistry: Fundamentals and Applications
Jay A. Siegel

Forthcoming

The Future of Forensic Science
Daniel A. Martell

Humanitarian Forensics and Human Identification
Paul Emanovsky and Shuala M. Drawdy

Forensic Anthropology: Theoretical Framework and Scientific Basis
Clifford Boyd and Donna Boyd

Forensic Microbiology
David O. Carter, Jeffrey K. Tomberlin, M. Eric Benbow and Jessica L. Metcalf

Forensic Chemistry

Fundamentals and Applications

EDITED BY

Jay A. Siegel

Emeritus Professor of Forensic Science, Michigan State University, USA

WILEY Blackwell

This edition first published 2016 © 2016 by John Wiley & Sons, Ltd

Registered Office
John Wiley & Sons, Ltd, The Atrium, Southern Gate, Chichester, West Sussex, PO19 8SQ, UK

Editorial Offices
9600 Garsington Road, Oxford, OX4 2DQ, UK
The Atrium, Southern Gate, Chichester, West Sussex, PO19 8SQ, UK
111 River Street, Hoboken, NJ 07030-5774, USA

For details of our global editorial offices, for customer services and for information about how to apply for permission to reuse the copyright material in this book please see our website at www.wiley.com/wiley-blackwell.

Library of Congress Cataloging-in-Publication Data
Forensic chemistry : fundamentals and applications / edited by Jay A. Siegel.
    pages   cm. – (Forensic science in focus)
  Includes bibliographical references and index.
  ISBN 978-1-118-89772-0 (cloth)
1. Chemistry, Forensic.   2. Forensic sciences.   I. Siegel, Jay A.
  RA1057.F66 2015
  614′.12–dc23
                                    2015022497
A catalogue record for this book is available from the British Library.

Wiley also publishes its books in a variety of electronic formats. Some content that appears in print may not be available in electronic books.

Set in 10.5/13.5pt Meridien by SPi Global, Pondicherry, India

1   2016

Contents

About the editor

Jay A. Siegel, PhD, is a consultant in forensic science. He retired in 2012 as Chair of the Department of Chemistry and Chemical Biology from Indiana University–Purdue University Indianapolis (IUPUI). He also served as founder and director of the Forensic and Investigative Sciences Program at IUPUI until 2011. Prior to his tenure at IUPUI, Dr Siegel was Director of the Forensic Science Program at Michigan State University for 25 years. He retired from that position in 2003 as Emeritus Professor. Prior to his stint at Michigan State, he was an assistant professor of forensic science at Metropolitan State College (now university), Denver, Colorado. From 1974–1977, Dr Siegel was a forensic chemist with the Virginia Bureau of Forensic Sciences. During his career since becoming a forensic chemist, Dr Siegel has testified more than 100 times as an expert witness in Federal and U.S. Military Courts as well as courts in seven states.

Currently, Dr Siegel serves as an Adjunct Professor with the Forensic and Investigative Sciences program at IUPUI. He is the coauthor of two forensic science textbooks: *Forensic Science: The Basics, 3rd Edition* (2015) with Kathy Mirakovits and *Fundamentals of Forensic Science, 3rd Edition* (2015) with Max Houck. He is also the author of *Forensic Science: A Beginner's Guide, 2nd Edition* (2015). In addition, he is coeditor in Chief of *Forensic Science Policy and Management: An International Journal,* with Max Houck. Dr Siegel has served on the editorial board of the *Journal of Forensic Sciences* and was a member of the National Academy of Sciences Forensic Science Committee (2006–2009). He currently serves on the Scientific Advisory Board of the Washington, DC, Forensic Science Laboratory.

Dr Siegel received a Bachelor of Science, Master of Science and PhD in Analytical Chemistry from the George Washington University. He has been a member of the American Academy of Forensic Sciences and was recognized as a Distinguished Member in 2011. He was also given the Distinguished Alumni Scholar Award from the George Washington University in 2011.

Contributors

Jose Almirall
Department of Chemistry and
Biochemistry and
International Forensic Research Institute,
Florida International University, USA

Richard E. Bisbing
Richard E. Bisbing LLC, USA
(Retired from Michigan State Police
Forensic Science Division, and McCrone
Associates, Inc., USA)

Amanda A. Frick
Nanochemistry Research Institute and
Department of Chemistry, Curtin University,
Australia

Patrick Fritz,
Nanochemistry Research Institute and
Department of Chemistry, Curtin University,
Australia

John Goodpaster
Department of Chemistry and Chemical
Biology
Forensic and Investigative Sciences Program
Indiana University–Purdue University
Indianapolis (IUPUI), USA

Max Houck
Forensic & Intelligence Services, LLC, USA

Paul Kirkbride
School of Chemical and Physical Sciences,
Flinders University,
Australia

Gerald M. LaPorte
Office of Investigative and Forensic Sciences,
National Institute of Justice, Department of
Justice, USA

Simon W. Lewis
Nanochemistry Research Institute and
Department of Chemistry, Curtin University,
Australia

Reta Newman
Pinellas County Forensic Laboratory, USA

Niamh Nic Daéid
Centre for Anatomy and Human Identification,
University of Dundee, UK

Christopher S. Palenik
Microtrace LLC, USA

Walter F. Rowe
Department of Forensic Sciences, The George
Washington University, USA

Ruth Smith
Forensic Science Program, School of Criminal
Justice, Michigan State University,
USA

Tatiana Trejos
Department of Chemistry and Biochemistry
and International Forensic Research Institute,
Florida International University, USA

Series preface

Forensic Science in Focus

The forensic sciences represent diverse, dynamic fields that seek to utilize the very best techniques available to address legal issues. Fueled by advances in technology, research and methodology, as well as new case applications, the forensic sciences continue to evolve. Forensic scientists strive to improve their analyses and interpretations of evidence and to remain cognizant of the latest advancements. This series results from a collaborative effort between the American Academy of Forensic Sciences and Wiley Blackwell to publish a select number of books that relate closely to the activities and objectives of the American Academy of Forensic Sciences. The book series reflects the goals of the AAFS to encourage quality scholarship and publication in the forensic sciences. Proposals for publication in the series are reviewed by a committee established for that purpose by the AAFS and also reviewed by Wiley Blackwell.

The American Academy of Forensic Sciences was founded in 1948 and represents a multidisciplinary professional organization that provides leadership to advance science and its application to the legal system. The eleven sections of the AAFS consist of Criminalistics, Digital and Multimedia Sciences, Engineering Sciences, General, Pathology/Biology, Questioned Documents, Jurisprudence, Anthropology, Toxicology, Odontology and Psychiatry and Behavioral Science. There are over 6000 members of the AAFS, originating from all 50 States of the United States and many countries beyond. This series reflects global AAFS membership interest in new research, scholarship and publication in the forensic sciences.

Douglas H. Ubelaker
Senior Scientist
Smithsonian Institution
Washington, DC, USA
Series Editor

Preface

The most important question that an author or editor should ask when contemplating a new text or reference book is: Who is the audience? The desired readership guides the depth and breadth of the book and in the case of *Forensic Chemistry: Fundamentals and Applications,* the selection of authors. This is not an introductory text in forensic science. It is not meant for the casual reader who may have an interest in the broad subject of forensic chemistry or for the student beginning a secondary or tertiary (college) level course. This book is about the applications of chemistry, especially analytical chemistry to various aspects of forensic science. When I conceptualized this book, I wanted the readership to come in with a solid foundation in the principles of chemistry. Prior knowledge of forensic chemistry is not a necessary antecedent, but is a helpful one. I asked the authors of the twelve chapters of this book to write for an audience of advanced college students, both chemistry and forensic science majors, faculty who teach forensic and/or analytical chemistry, forensic chemistry practitioners and, finally, scholarly researchers in analytical and forensic chemistry. If you are one of these people, I believe that you will find this book to be the definitive reference and textbook in the field of forensic chemistry.

There are twelve chapters in this book. Ten of them are subject specific; they cover the common areas of forensic chemistry; abused drugs, textile fibers, paints and polymers, fire residues, explosives, glass, and soil; and some areas that are not normally thought of as being part of forensic chemistry, but they surely are. These include chemical analysis of questioned documents, finger marks, and firearms. The other two chapters could be termed "foundational," because they span the entire field of forensic chemistry; these are forensic microscopy and chemometrics.

Having decided the scope of the book, I very carefully chose the authors. All of them are outstanding experts in their fields. Every one of them has been or continues to be a practitioner in forensic chemistry in a private or public forensic science laboratory. Two are currently forensic science laboratory administrators. Seven are currently educators and researchers with major universities worldwide, all of which are known for their education in forensic chemistry.

I have been a forensic chemist since 1974 and have been in forensic science and chemical education since 1977. I am familiar with all of the major literature in forensic chemistry. As the twelve chapters of this book came to me for my reading as the editor, I was constantly blown away by the depth and breadth of knowledge exhibited by every one of the authors. I believe that you will be similarly impressed. I hope you enjoy and learn from this book.

Finally, I wish to thank all twelve of the experts who contributed chapters to this book. They all donated many hours of their precious and valuable time to make this the standard work in forensic chemistry. All of the royalties generated by this book go to the **Young Forensic Scientists Forum** of the American Academy of Forensic Sciences. They are the future of forensic science in the United States and deserve our support.

Jay A. Siegel, PhD
Editor

CHAPTER 1

Drugs of abuse

Niamh Nic Daéid

Centre for Anatomy and Human Identification, University of Dundee, UK

1.1 Introduction

The aim of this chapter is to provide a brief overview of illegal drugs as they are controlled internationally, the specific nature of some of these compounds and how they can be analysed. The United Nations Office on Drugs and Crime (UNODC) provides an annual report (the World Drug Report) that collates international data to develop a picture of global drug markets in terms of production and use (UNODC, 2014a). The UNODC also provides an excellent international resource for information relating to analytical procedure and data interpretation that the reader is directed towards.

Generally speaking, controlled drug production and use tends to concentrate around four drugs or drug types: cannabis, opiates, cocaine and amphetamine-type substances. Each of these is discussed together with a brief overview of the emerging new psychoactive substances (NPS). These substances can also be categorized through their mechanisms of production as naturally occurring materials (cannabis), semi-synthetic materials derived from natural products (opiates and cocaine) and synthetic products (the amphetamine-type substances and the NPS).

Chemical substances controlled under International or National laws were used by between 190 and 324 million people worldwide in 2012 with 16–39 million users categorized as regular drug users (UNODC, 2014a). The prevalence of drug use varies regionally and the main problem drugs at the global level continue to be the opiates (notably heroin), followed by cocaine. For most of Europe and Asia, opiates continue to be the main problem drug; in South America, drug related treatment demand continued to be mainly linked to the abuse of cocaine; and in Africa the bulk of all treatment demand is linked to cannabis (UNODC, 2014a).

Forensic Chemistry: Fundamentals and Applications, First Edition. Edited by Jay A. Siegel.
© 2016 John Wiley & Sons, Ltd. Published 2016 by John Wiley & Sons, Ltd.

1.2 Law and legislation

Over a century ago, the international effects of the "national" Chinese heroin problem were addressed with the creation of The Shanghai Opium Commission in 1909. Out of this was born the first mechanism for the international legislation of drugs, The Hague Opium Convention of 1912. Countries party to this convention were required to implement domestic legislation restricting drugs to medical use. The Shanghai Opium Commission and the Hague Opium Convention were the catalysts for international drug control as it is today.

Half a century later, the United Nations (UN) created the first of three major international drug control treaties that are still in effect: the Single Convention on Narcotic Drugs 1961. As with any international convention, signatories are required to implement domestic legislation in accordance with the recommendations set out therein, and they are obligated to evaluate the policies against the international standard. More than 180 countries are party to the UN treaties. The UN regularly names the countries that have not signed the treaties and continually urges them to do so.

The 1961 Single Convention on Narcotic Drugs was formed out of a need to combine many different nations' legislation on narcotic drugs into one primary instrument. It seeks to limit the manufacture, possession and use of drugs to science and medicine. Countries that are party to this convention are required to provide annual statistical returns and estimates for the use of, and need for, drugs covered by this convention. It also encourages international cooperation as a tool for fighting illicit international drug trafficking (UN, 1961; UNODC, nd).

The second major international treaty on drug control is the Convention on Psychotropic Substances 1971, which was written in response to the widening range of drugs of abuse. This convention establishes control over the use of psychotropic substances, including many synthetic drugs, based on their potential harm and therapeutic value (UNODC, nd; UN, 1971). A psychotropic substance is defined by the convention as "any substance, natural or synthetic, or any natural material in Schedule I, II, III or IV" of the convention (UN, 1971).

In contrast to the 1961 and 1971 treaties that seek to limit the trade in and use of controlled substances to legitimate scientific and medical purposes, the third major treaty was formed in an effort to deal with the growing problem of drug trafficking. The Convention Against the Illicit Traffic in Narcotic Drugs and Psychotropic Substances 1988 provides for the extradition of drug traffickers, and it offers provision against money laundering and the divergence of precursors (UNODC, 1988).

The effectiveness of this international drug control system is difficult to evaluate due to patchy data over the years; a concealed, illegal activity does not lend itself to simple quantitative monitoring. However, reports published annually by the UN now highlight the worldwide drug situation in yearly increments based in part on statistics provided by signatories to the major conventions.

On occasion, enough data are available to allow analysis of a drug trend over the last century. For example, The Shanghai Commission of 1909 was formed when the worldwide production of opium was estimated to be 30,000 metric tons, with 75% originating in China. Over one hundred years later, the global production has fallen by 80%, yet the world's population has tripled. Therefore, in terms of opium, it is reasonable to conclude that the global drug problem is contained and the major international conventions are effective.

Chemicals are essential to the manufacture of narcotic drugs. They are an integral component in the case of synthetic drugs and are required for the processing of coca and opium into heroin and cocaine. Only marijuana, of the major illicit drugs of abuse, is available as a natural, harvested product.

Chemicals used in drug manufacture are divided into two categories, precursor and necessary chemicals, although the term "precursors" is often used to identify both. Precursor chemicals are chemicals that are essential to the production of a controlled substance and for which no substitution can be made. Essential chemicals are used in the refining of the chemical processes, for example the refining of coca and opium into cocaine and heroin. Although some remain in the final product, the basic raw material is the coca or opium. Many essential chemicals required for illicit drug manufacture have extensive commercial applications, are widely traded, and are available from numerous source countries (King, 2003).

One of the major shortcomings of forensic science in general, is a lack of uniform consensus and international standards for the analysis of particular types of evidence. DNA typing has probably come the closest to achieving international consensus; however, in more recent times this has fallen short in the interpretation of mixed DNA profiles. The area of seized (abused) drugs, however, has come a long way in the development of standards and guidelines for analysis. This effort is personified by the work of organizations such as the UNODC, the European Network of Forensic Science Institutes (ENFSI) and the Scientific Working Group for the Analysis of Seized Drugs (SWGDRUG; http:// www.SWGDRUG.org) in the United States. Because of the major advances the latter group has made, its widespread adoption, especially in the United States and its international consensus, it is discussed in some detail.

During the last decade of the 20th century, the US Federal Government sponsored a group of discipline specific Scientific Working Groups (SWGs). With one exception, the SWGs were funded and managed by the Federal Bureau of Investigation. The SWG that was created to set up standards and guidelines for the analysis of seized drugs (SWGDRUG) was set up and managed by the US Drug Enforcement Administration (DEA) in 1997 and still remains very active. It contains approximately twenty members, nearly half of which are international. During the intervening years, the SWGDRUG has been funded by the DEA, the US Office of National Drug Control Policy (ONDCP), and the National Institute of Standards and Technology (NIST).

The SWGDRUG was created to develop standards and guidelines for the analysis of seized drugs. Many of its recommendations have been submitted to the ASTM (American

Society for Testing and Materials, now ASTM International) for adoption as standards for the analysis of drugs.

The SWGDRUG's recommendations fall into four major categories:

- Code of Professional Practice
- Education and Training
- Methods of Analysis
- Quality Assurance.

The code of professional practice includes guidelines for professional conduct, characteristics of casework and standards for reporting the results of drug analysis.

The section on education and training specifies the minimum educational background and training for an analyst as well as provisions for continuing professional development. Initial training requirements are also spelled out.

The methods of analysis section is quite comprehensive. It has subsections on sampling considerations, methods of drug analysis, clandestine laboratory evidence and a new section on drug analogs and structural class determination. The section on methods of analysis has had a major impact on the field of seized drug analysis. It groups the most common analytical techniques into three categories based on the probative value and reviewable data of each one and then assigns analysis protocols to each category.

The last major section of the SWGDRUG's recommendations is on quality assurance. This is very comprehensive, covering general practices of quality assurance, validation of analytical methods and uncertainty.

As the SWGDRUG develops new guidelines, it provides ample opportunities for public comment before they are adopted.

1.3 Sampling

The taking of samples either at a crime scene or thereafter has obvious consequences for any subsequent investigation of that crime. The consequences of incorrect sampling may result in unusable data and potentially incorrect and improper conclusions where the adage of "rubbish in equals rubbish out" may easily apply. There is much published literature regarding sampling theory and the methods of sampling, much of which relates to specific disciplines or problems. Additionally, a specific aspect of sampling relating to the forensic sciences is that of sample integrity (being able to state that the sample gathered is the sample analyzed and the result obtained is from that sample), which must be maintained, as, if not, subsequent analytical results become questionable. Information regarding the sample is paramount, whether it has been gathered at the scene and transferred to the laboratory for analysis or has been generated *in situ* in the laboratory. Correct recording of the circumstances of the sample are essential: Who has handled it? For how long? Why?

Two things generally dictate sampling protocols:
- A legal obligation that may require that all items in a seizure are described and sampled.
- A policy left to the expert that will include a description of the items and the selection of a sample from the items.

The sampling strategy is fully dependent on the question being asked and, thus, the problem that has to be solved. There may be different needs for possession, production, or trafficking:
- Is a drug present? Minimal sampling (this may require just one positive result).
- Is a drug present in (more than) a specified proportion of the items?
- Is a drug present in all the items? (All samples to be examined.)

The criteria for selecting the type of sampling protocol undertaken should include a balance between the loss of completeness in terms of information versus time saving and that each new sample analyzed should produce new information.

The first stage in any procedure where sampling is to occur is an understanding of what it is that needs to beachieved. This will depend upon what type of sampling is being undertaken, whether sampling is occurring at the crime scene and the nature of that scene, or sampling at the laboratory. Whichever scenario is faced there are a few universal objectives:
- The sample should be representative of the material in question.
- The sample should be randomly chosen as appropriate
- Samples should be of adequate size for the analysis to be performed and to allow subsequent analysis to take place.

The basis of sampling is that the composition found in the samples taken reflects – in principle – the composition of the whole. As a consequence, only a fraction of the total packages in a seizure should need to be investigated.

Many laboratories have developed their own sampling protocol for drug samples. In all cases, samples should be taken from items that have the same morphology (also called sampling by attributes). This means that if a seizure is found to contain different types of materials, then each type should be sampled separately. This is particularly true for samples containing tablets with different logos or of different colors, or items that have been packaged separately from each other. Where blocks of resin are being sampled this should be away from the edges of the block, as such edges can potentially be used to make physical fits between the blocks.

In cases where samples of powder are seized, it is necessary to ensure that the powder is homogenous. This is generally accomplished using one of three methods:
- Cone and square method – the sample is ground up first and then poured onto a flat surface and flattened. The material is divided into quarters at right angles and one quarter shaped into a cone. The cone is quartered again at right angles and a quarter chosen and so on. This process is continued until a sample of sufficient size is prepared.

- Shaking in a plastic or nylon bag – the sample is placed in its entirety into the bag and shaken to mix it. This can result in larger particles falling to the bottom of the bag or particles of the sample preferentially adhering to the sides of the container because of charge differences depending on the size of compounds that may be present.
- Blending – the sample is placed into a blender and mixed mechanically. This method is very effective in homogenizing samples but it must be ensured that the blender is completely clean.

Sampling protocols will also be dictated by the nature of the analysis, identification/ quantification, profiling or sampling at clandestine laboratories. An international system of sampling that has been adopted and promoted by the UNODC is:

Powders and tablets and packages Single package of material: The material should be removed from all packaging and weighed to a constant dry weight. The sample is then homogenized and a sample removed.

More than one package: The material in each package should be examined by eye for color differences. The contents of each package should be weighed (cleaning the balance in between each weighing) and each package should be tested using a color spot test or thin layer chromatography. If it is assumed that all packages contain the same material then:

- if there are less than 10 packages, all should be tested;
- if there are between 10 and 100 packages, 10 should be tested at random;
- if there are more than 100 packages, the square root of the total should be tested at random.

Liquids If all of one phase, then a sample of the liquid is removed for testing. If there is more than one phase present, then a sample of each phase should be removed for testing.

Control/comparison samples There are normally two types of control samples, positive controls (comparison samples) and negative controls (sometimes the same as blank samples).

1.3.1 Random sampling and representative sampling

A representative sampling procedure can be performed on a population of units with sufficient similar external characteristics (e.g., size, color). Many analytical methods require that random sampling of the item occur. A random sample is a selection of some members of a population in a manner such that each member of the sample is independently chosen and each member of the population has a non-zero probability of being chosen. Random does not mean haphazard but requires planning to ensure randomness. Random samples can be with or without replacement. Sampling without replacement (SWOR) means that once an item/element is removed from the population it is not put back and can, therefore, only ever be chosen once. Sampling with replacement (SWR) involves replacing items into the population after sampling.

The theoretical way to select a truly random, unbiased representative sample from a population is to individually number each item in the population and then use a random number generator to choose which item to select. This is not possible in practice, especially for large populations containing many thousands of units.

When sampling, two principles must be maintained:

- The properties of the sample are a true reflection of the properties of the population from which the samples were taken.
- Each unit in the population has an equal chance of being selected.

In reality, it is more difficult to adhere to these principles than it first seems because, when the population is high, it is impossible to number all the units and use a protocol based on a random selection of numbers. The practical solution is quite easy: after separating the samples by attributes all of the units in a group can be placed in a "black box" (plastic bag or similar) and samples selected. This kind of solution can be applied to practical cases, such as seizures of a thousand heroin street doses in similar external packages or a thousand tablets. This "black box" sampling method can reduce to a minimum any bias that may be introduced by the person selecting the samples and aims to prevent the sampler from consciously selecting a specific item from the population.

Any samples gathered must accurately reflect the material being tested (this is particularly true of control samples). The control sample and sample to be analyzed must, as far as possible, be adequate in size, so as to ensure that all presumptive and analytical tests can be carried out by forensic scientists acting for both the prosecution and defense. This can be difficult in the cases of trace amounts of illicit substances but there can, however, be no excuse for taking too little of a control sample. It is rare and in some cases impossible that an item can be re-visited in order to correct mistakes of inadequate sampling.

1.3.2 Arbitrary sampling

There are various arbitrary sampling methods that are used in practice across different drug analysis laboratories and work well in many situations. However, many have no statistical foundation and may lead to very large samples in cases where the original seizure is itself large. Not all existing sampling procedures are given here. Some laboratories use variations of these. In each case N = the total number of units and n represents the number of samples chosen (UNODC, 2009a):

1. All ($n = N$)
   Advantage(s): 100% certainty about the composition of the population.
   Disadvantage(s): Excessive sample sizes for larger populations.

2. $n = 0.05N$, $n = 0.1N$, etc.
   Advantage(s): Simple approach.
   Disadvantage(s): Excessive sample sizes for larger populations.

3. $n = \sqrt{N}$, $n = 0.5\sqrt{N}$, $n = \sqrt{\dfrac{N}{2}}$, etc.
   Advantage(s): Widely accepted approach.

Disadvantage(s): The samples may be too small when the population is small.

4. $n = 20 + 10\% (N - 20)$ *(where $N > 20$)*

Advantage(s): Heterogeneous populations likely to be discovered before analysis is complete.

Disadvantage(s): Excessive sample sizes for larger populations.

5.

$$N < x \qquad n = N$$
$$x \leq N \leq y \qquad n = z$$
$$N > y \qquad n = \sqrt{N}$$

(where x, y and z are arbitrary numbers; $x < y$ and $x \leq z < y$)

Advantage(s): UN recommended method ($x = 10$, $y = 100$, $z = 10$).

Disadvantage(s): Excessive sample sizes for larger populations.

6. $n = 1$

Advantage(s): Minimum amount of work.

Disadvantage(s): Least amount of information on the characteristics of the seizure.

1.3.3 Statistical sampling methods

There are two recognized approaches used in statistical sampling methods: a frequentist approach and a likelihood ratio approach (UNODC, 2009a).

1.3.3.1 Frequentist approach

The assumption behind a frequentist approach is that a fixed but unknown proportion of the seizure contains drugs and the proportion of drugs in a sample can estimate the proportion of drugs in the parent seizure. The sample proportion will, however, vary over different samples. One sample will give a higher proportion than another sample. Therefore, the frequentist methods provide a confidence, $(1 - \alpha)100\%$ (for instance 95% if α is selected to be 0.05), that with a given sample proportion the seizure proportion is at least k × 100% (for instance 90% if k is selected to be 0.9). This means that if the sampled proportion is found to be as assumed then it can be stated with 95% confidence that the proportion of drugs in the whole seizure is at least 90%. The hypergeometric distribution can be used to calculate these probabilities (UNODC, 2009a).

1.3.3.2 Likelihood ratio approach

The assumption behind a likelihood ratio approach is that the proportion of drugs within the sample is known and fixed. This proportion is used to calculate probabilities on certain values of the unknown proportion of drug in the seizure, which at that point is still assumed variable. With this approach it is possible to incorporate some knowledge about the seizure that you may possibly have. The seizure proportion is not known but often some ideas about this proportion exist. For instance, if all plants in a cannabis nursery appear similar they probably are all cannabis plants. It is also possible that there is no information about the amount and type of drugs in a seizure. These various forms of prior information will result in different mathematical models to estimate a desired sample size in the Bayesian approach (UNODC; 2009a).

1.4 Specific drug types

1.4.1 Cannabis
1.4.1.1 Introduction

Cannabis is the most consumed illicit drug in the world and, therefore, the subject of the most illegal trafficking in one form or another. The botanical name of cannabis is *Cannabis Sativa L.* and it is referred to as marijuana (herbal cannabis) and hashish (cannabis resin), terms associated with cannabis grown for its illicit use, or as hemp, a term usually associated with cannabis plants grown for their fiber content (UNODC, 2014a).

Cannabis is native to the mountainous areas of central and south Asia and man has used the plant for over 6,000 years. Cannabis grows over a wide variety of geographic terrains, altitudes and latitudes. It is grown in many countries and on all continents. Although it prefers the higher temperatures and longer growing seasons of the equatorial areas of the world, it has been cultivated as far north as 60° latitude. With the more recent prevalence for hydroponic growth, cannabis is known to be grown in most countries and remains one of the most used controlled substances globally (UNODC, 2014a).

A number of forms of the drug may be encountered, including plant material, resin and "hash oil." The active ingredient in all of these is Δ^9-tetrahydrocannabinol (Δ^9-THC). Also found is Δ^9-tetrahydrocannabinolic acid (THCOOH), which is converted to Δ^9-THC through smoking. Also present are the compounds Δ^8-tetrahydrocannabinol, cannabidiol (CBD), and cannabinol (CBN), as well as upwards of 50 minor cannabinols.

Plant material

Cannabis plant material can occur as live plants or as dried plant material. It is important when examining plant material that it should be in the dried state, as water content will affect any weight measurements, a crucial factor in determining the seriousness of any charges. Male and female cannabis plants are separate, with the male plant flowering before the female to ensure cross-pollination. The plant grows between 30 cm and 6 m in height; it grows from seed to maturity in about three months, although harvesting can occur after two months. The leaves are palmate with serrated edges and are opposite and alternate around a hollow four-cornered stem. The leaves are coated with upward pointing unicellular hairs called trichomes. Non-glandular trichomes are found on the stems and leaves together with a few glandular trichomes, which contain the Δ^9-THC. The greatest concentration of glandular trichomes is in the flowering tops of the plants and those of the female plant have a greater concentration of Δ^9-THC. Material prepared from the flowering tops or leaves is commonly called marijuana and usually contains 0.5–5% Δ^9-THC. Cannabis plants grown under controlled hydroponic conditions, which are predominantly female plants (skunk), generally have a higher quantity of Δ^9-THC. (9–25%) and can reach full maturity in about 13 weeks.

Resin

This is also produced directly from the plant material. The glandular trichomes produce a resin that is scraped from the surface of the plant and pressed into blocks. Resin contains about 2–10 wt-% of the active constituent, Δ^9-THC. Usually between 100 and 400 mg of resin is used in a joint (cigarette), though this varies from user to user (UNODC, 2014a). The high increase in hydroponic growth, particularly in Europe, has seen a significant increase in the usage of highly potent plant material and a reduction in cannabis resin usage.

When examining cannabis resin several facts are recorded, including whether or not the blocks seized fit together, any striation marks (cutting marks) that may be present and whether these can they be linked to a cutting instrument. The color and different layers within the resin block should also be examined and recorded (Figure 1.1).

Hash oil

This is a manufactured product of cannabis produced by extracting the whole plant material using an organic solvent (usually alcohol, ether or benzene). On extraction the resulting oil can contain between 10 and 30 wt-% Δ^9-THC. The oil is used in various ways, including smoking in special pipes with tobacco.

1.4.1.2 The cannabinoids

There is a large number of known cannabinoids that can be divided into major and minor cannabinoids. The main cannabinoids (Figure 1.2) are cannabinol (CBN), cannabidiol (CBD), two isomers of tetrahydrocannabinol (Δ^8 and Δ^9-THC) and tetrahyrdocannininolic acid (THCOOH). Δ^9-THC is the cannabinoid that has the greatest pharmacological activity and it is this cannabinoid that needs to be detected in order to demonstrate that a sample is legally cannabis. CBD is the precursor of THC and CBN is the decomposition product of THC. THC is converted to THCOOH on smoking.

Other minor cannabinoids will exhibit similar oxidations; for example, THV (tetrahydrocannabivarin) will oxidize to CBV (cannabivarin). Both CBN and CBV are often absent from fresh plant materials, while the presence of CBD and its ratio to Δ^9-THC can often be a geographical indicator. The presence of THCOOH, THVA (tetrahydrocannabivarinic acid) and CBCh (cannabichromene) as well as others are also used in profiling, and it should also be noted that in gas chromatography (GC) analysis the THCOOH is converted thermally to Δ^9-THC (Lewis *et al.*, 2005; Le Vu *et al.*, 2006; Cadola *et al.*, 2013).

The major cannabinoids will decompose over time and, as such, the potential to link, for example, resin samples together using chemical composition is severely hampered.

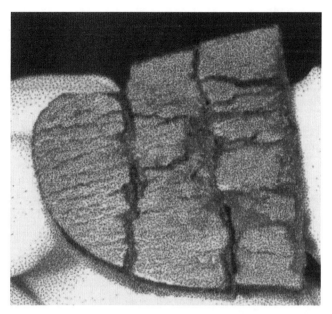

Figure 1.1 Three blocks of cannabis resin that physically fit together

Furthermore, studies have also exposed the issues associated with long-term storage of resin and plant extracts and demonstrated a steady decay in the concentration of Δ^9-THC and a rise in CBN levels (Trofin *et al.,* 2012; Lindholst, 2010).

1.4.1.3 Analysis of cannabis

A comprehensive text on the analysis of cannabis in all of its formulations is available from the UNODC Botanical analysis of Cannabis (UNODC, 2009b). Because cannabis is presented to the laboratories in plant form, the morphological characteristics of the plant are used to provide an identification of the material. Leaf shape and, in particular, the presence of the various trichomes of the plant are characteristic of *Cannabis Sativa L.* (UNODC, 2009b).

The main presumptive tests for cannabis are the Duquenois Levine test, which is carried out in test tubes and gives a violet layer in the presence of cannabinoids, and the Fast blue BB test, which gives a purple or red color change for cannabinoids. These tests are quite sensitive to the amount of Δ^9-THC present and can give negative results for "old" plant and resin samples, as the Δ^9-THC content may be reduced (UNODC, 2009b).

Thin layer chromatography (TLC) can be used to identify which drug is present within a drug class, as each different compound within the drug group can be separated on the basis of chemical structure, molecular weight, and so on. In many cases, the TLC plates need to be developed using chemical developers that give characteristic colors for the compounds under test.

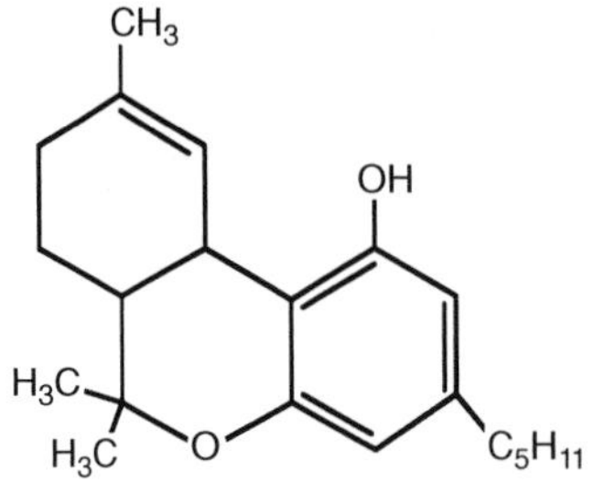

Δ^{9}-tetrahydrocannabinol (THC)

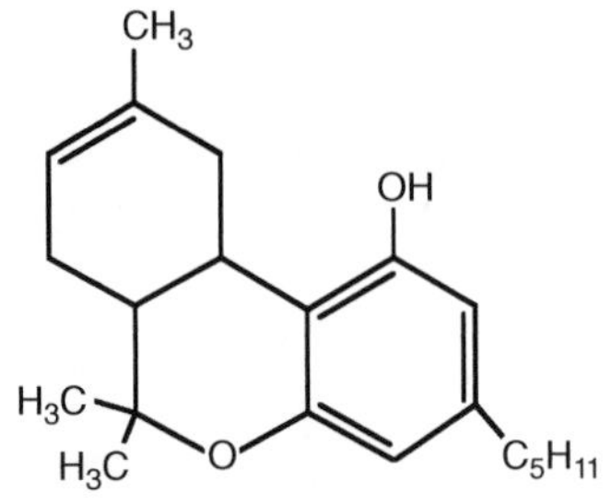

Δ^{8}-tetrahydrocannabiniol (THC)

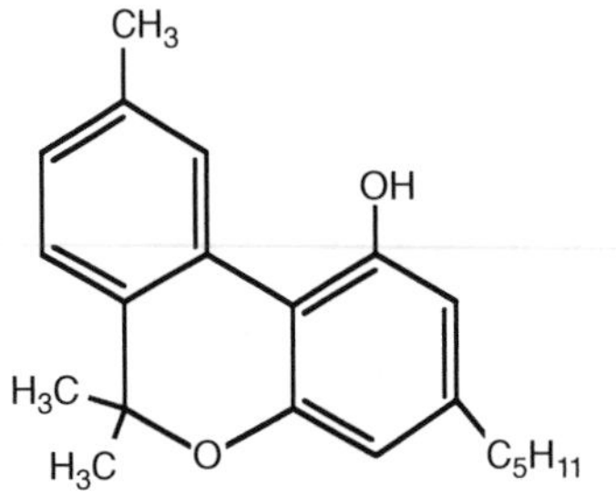

Cannabinol (CBN)

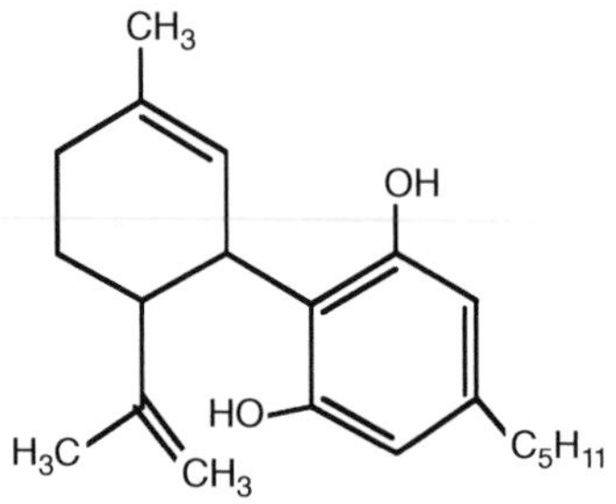

Cannabidiol (CBD)

Figure 1.2 Chemical structures of the main cannabinoids

Both high performance liquid chromatography (HPLC) and gas chromatography or gas chromatography mass spectroscopy (GCMS) are also used as confirmatory analysis. However, in many cases microscopy, presumptive tests and TLC are deemed sufficient for court purposes.

Cannabinoids are weakly acidic compounds because of the phenolic group. This results in the compounds undergoing ionization in aqueous media forming compounds that will strongly interact with the stationary phase in HPLC analysis, causing tailing of the peaks when analyzed using liquid chromatographic methods. Because of this, ionization is suppressed by the presence of acetic acid in the mobile phase.

GC and GCMS analysis of cannabinoids tends to be favored over HPLC. However, the cannabinoids suffer severely from thermal decomposition and, as a result, do not chromatograph well (Lewis *et al.*, 2005). As a consequence, extracts from cannabis samples are often derivatized using, for example, N,O-bis(trimethylsilyl)acetamide (BSA). Figure 1.3 illustrates the main mass spectral fragmentation patterns for derivatized cannabinoids.

In more recent years, there has been a focus on the analysis of cannabis using DNA techniques. In particular, short tandem repeat (STR) DNA analysis has been used very successfully as a potential linkage tool between cannabis plants and does not suffer from the restrictions of chemical analysis (Gilmore *et al.*, 2003; El Alaoui *et al.*, 2013). The use of DNA analysis to explore the potential to link resin blocks together has also shown promise (Nic Daéid *et al.*, 2003).

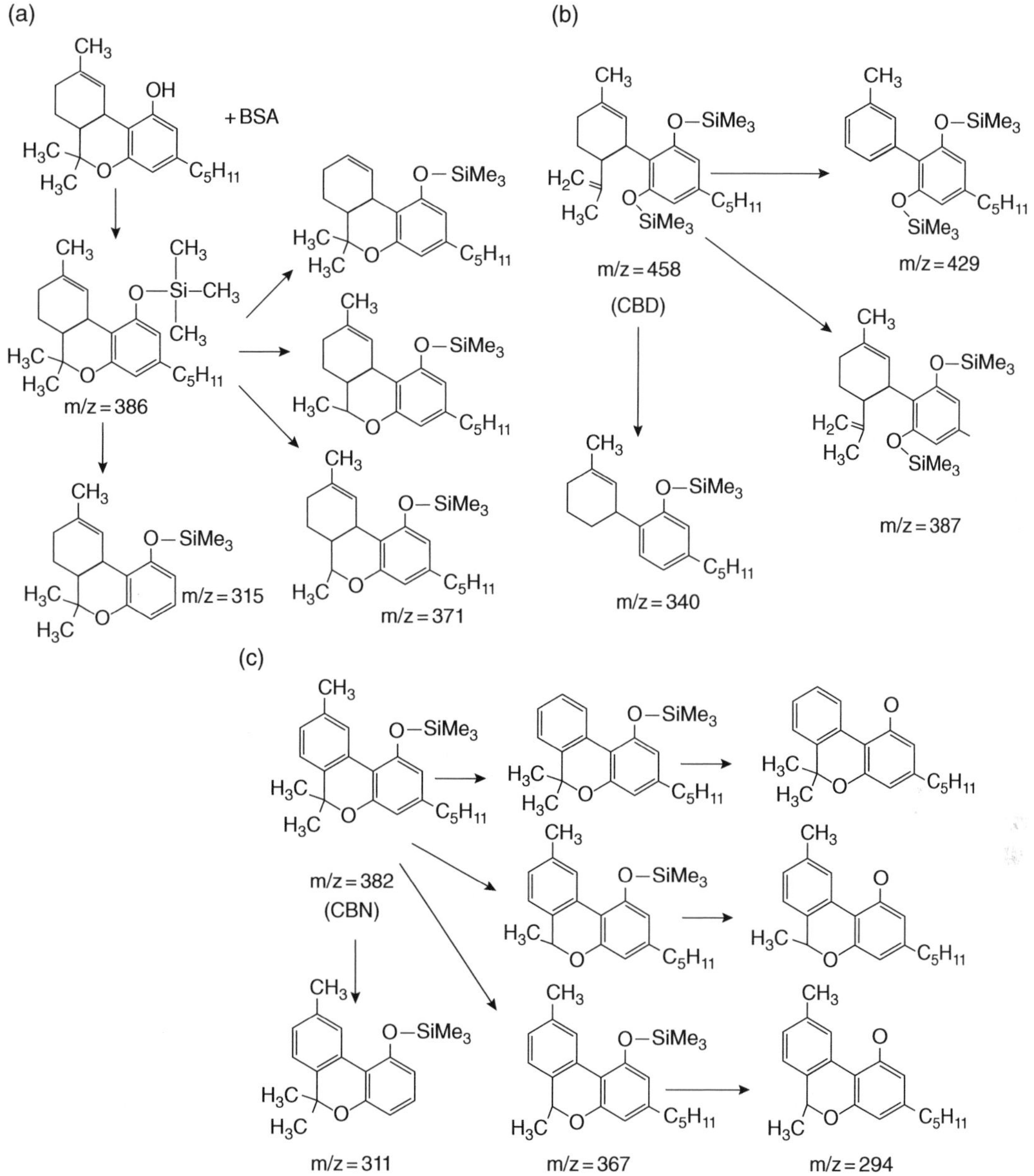

Figure 1.3 Fragmentation patterns for derivitized (a) THC (b) CBD (c) CBN

1.4.1.4 Recent issues in the control of cannabis

There have been a number of recent developments in the control of marihuana, princi-pally in the United States, although some of these are occurring worldwide. The development of forms of so-called synthetic marihuana or synthetic cannabis has intro-duced serious issues of control and safety into the drug using community. Synthetic can-nabis usually consists of substances that are designed to mimic the effects of cannabis that are sprayed onto some type of plant material or herbal substances such as parsley. These formulations are sold as brand names such as Spice and K2. Synthetic cannabis first came on the scene in the United States in around 2000 and was initially thought to be comprised of all natural substances. A few years later it was discovered that the active ingredients were a large variety of synthetic substances. These include cannabicyclo-hexanol and other, more esoterically named, substances such as JWH-XXX or HU-XXX, where X = an integer. These substances were used to circumvent the laws that make cannabis illegal by substituting them for THC.

Another, more promising trend in the use of cannabis is the recognition that certain compounds, chiefly THC, may have significant medical benefits. The use of cannabis for medical purposes dates back thousands of years but little research had been done until recently to determine what diseases it may be used for, what doses and forms are best for medical treatment and what safety issues accrue with short or long-term use. In the United States, research has been hampered by the classification of cannabis in Federal Schedule I, which is reserved for drugs with a high potential for abuse and no accepted medical use, by the US Food and Drug Administration. Some medical agencies have urged the US Congress to move cannabis from Schedule I to Schedule II to facilitate research. A number of potential beneficial medical effects have been identified but most have not been proven. These include appetite stimulation, antiemetics (nausea reduction), control of involuntary spasms and even some pain management effects. It has also been experimented with in the control of some symptoms of AIDS. More than 30 States have adopted medical cannabis laws or are considering them. Bills have been introduced into the US Congress to remove Federal penalties against cannabis possession when it is used for approved medical purposes.

The most surprising development in the United States has been a move by three states (Alaska, Colorado, Oregon) and the District of Columbia to legalize outright the growing, possession and use of small amounts of cannabis. The initiative is most advanced in Colorado, where cannabis was legalized in 2013 and an entire growing and selling enterprise has been set up with the blessing of state government. The US Attorney General has directed the Drug Enforcement Administration (DEA) to suspend enforcement of Federal marihuana laws in states where it has been legalized by legislation. Although a large majority of citizens in DC voted in 2014 to legalize cannabis, elements of the US Congress seem bent on preventing implementation. Other states are now considering similar legislation.

1.4.2 Heroin
1.4.2.1 Introduction

Opiates belong to a family of compounds known as narcotic analgesics, narcotic meaning "tending to induce sleep" and analgesic meaning without pain. The natural opiate, morphine is extracted from the opium poppy; acetylation affords the semi-synthetic opiate diacetylmorphine (diamorphine). Heroin refers to a crude mixture of opium alkaloids obtained from the extraction of morphine from opium and the subsequent acetylation reaction (Figure 1.4).

The psychological effects of opium have been known since 4000 BC and morphine was first isolated in 1805, followed by other opium alkaloids codeine and papaverine in 1832 and 1848. The pure alkaloids were prescribed for the relief of pain, coughs and diarrhea. In the 1860s, morphine was extensively used as a painkiller for wounded soldiers, often resulting in morphine addiction. Diamorphine was first synthesized in 1874 by acetylating morphine in an attempt to produce a new non-addictive painkiller; however, diamorphine was found to have narcotic and addictive properties far exceeding those of morphine. In 1914, the US Congress called for control of each phase of the preparation and distribution of opium, making it illegal to possess or supply these controlled substances. This, in turn, led to the start of illegal smuggling trades with the large scale smuggling of heroin into the United States in 1967 (Besacier and Chaudron-Thozet, 1999).

Papaver somniferum (var. *album* and *Papaver somniferum* var. *glabrum*) are the two varieties of opium poppy plants cultivated for the illicit production of heroin because of their high morphine content (*Papaver setigerum* plants also contain morphine) within the opium latex produced from the poppy seedpods. The raw latex of 1414 poppy seed pods were analyzed by the United Nations International Drug Control Programme (UNIDCP) and in

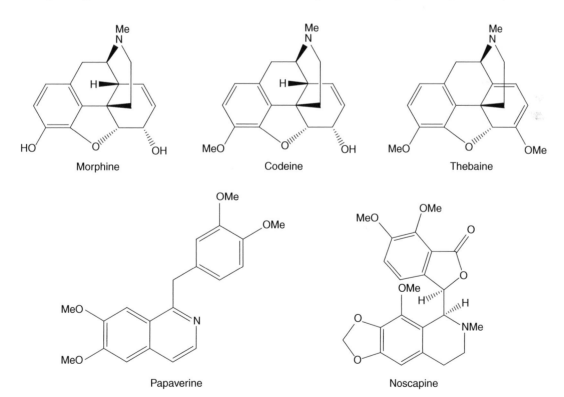

Figure 1.4 Chemical structures of major alkaloids

Table 1.1 Major alkaloid composition of opium

Alkaloid			
Class	**Structure**	**Name**	**Composition (%)**
Phenanthrene		Morphine	3.1–19.2
		Codeine	0.7–6.6
		Thebaine	0.2–10.6
Isoquinoline		Papaverine	<0.1–9.0
		Noscapine	1.4–15.8

total crude opium derived from *Papaver somniferum* was found to contain in the region of 25–30 alkaloids. The major alkaloid compositions are given in Table 1.1 (Drummer, 2001).

1.4.2.2 Heroin production

The raw material for the production of heroin is the opium poppy. The variety and age of the *Papaver somniferum* plant, together with the climate, altitude, soil fertility and moisture levels encountered during growth, affect the level and number of alkaloids found in the opium poppy.

The majority of clandestine laboratories extract morphine from opium using the "lime method." The raw latex from the opium poppy seeds is dissolved in boiling water to remove the insoluble plant material. Lime (calcium hydroxide) is added to the opium solution, converting water-insoluble morphine into water-soluble calcium morphenate; other insoluble alkaloids precipitate on cooling and are removed. The calcium morphenate solution is heated and the pH adjusted to 8–9 by addition of ammonium chloride. Upon cooling the precipitated morphine base is collected by filtration. Laboratories producing morphine as an end product perform further purification to remove traces of codeine, thebaine, papaverine and noscapine (Besacier and Chaudron-Thozet, 1999).

Clandestine manufactures typically acetylate the crude morphine base by addition of acetic anhydride, followed by heating to generate diamorphine via the intermediate 3-monoacetylmorphine (3-MAM). The cooled reaction mixture is typically treated with sodium carbonate to precipitate the heroin base. Dissolving the heroin base in acetone with the addition of hydrochloric acid generates the heroin hydrochloride salt.

Use of acetyl chloride is documented in New Zealand for the "home bake" preparation of heroin, which involves the initial production of morphine by demethylation of codeine with pyridine hydrochloride. Use of a mixture of trifluoroacetic anhydride and acetic acid or, alternatively, ethylene diacetate are reported; each of the different acetylation routes afford route specific markers (UNIDCP, 1998). The raw opium natural product and subsequent synthetic steps involved in the illegal manufacture of heroin provide specific chemical profiles for the material.

1.4.2.3 Heroin impurities

Major opiate impurities include morphine, codeine, papaverine and noscapine. Thebaine is rarely observed in illicit heroin, as it decomposes during acetylation generating acetyl-thebaol (Figure 1.5). Minor opiate impurities from the opium include benzylisoquinolines (laudosine, narceine), tetrahydroisoquinolines, and cryptopine, as well as alkaloids of unknown structure. Non-opiate derived impurities from opium include meconin (Figure 1.5).

Impurities generated from the acetylation of codeine and thebaine are acetylcodeine (Figure 1.6) and acetylthebaol (Figure 1.5), respectively.

In addition, the incomplete acetylation of morphine using acetic anhydride affords 3-MAM while non-quantitative acetylation of morphine with acetyl chloride affords both 3-MAM and 6-MAM (Figure 1.6). The degree of skill of the illicit heroin manufacturer also determines the extent of diamorphine hydrolysis to generate predominantly 6-MAM. Huizer (1983) demonstrated that the reaction conditions used to convert diamorphine base to diamorphine hydrochloride also afford 3-MAM but further deacetylation to morphine occurs at a much faster rate than for 6-MAM. Occluded solvents trapped in the heroin matrix may include the acetone used in the conversion of diamorphine base to diamorphine hydrochloride or acetic acid generated during diamorphine deacetylation.

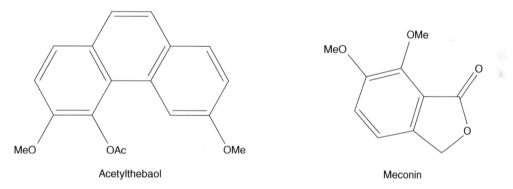

Figure 1.5 Chemical structures of acetylthebaol and meconin

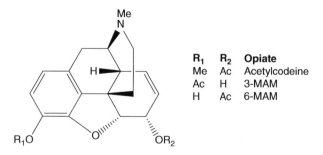

Figure 1.6 Chemical structures of acetylcodeine, 3-MAM and 6-MAM

1.4.2.4 Heroin cutting agents

Pharmacologically active adulterants are added to heroin. Caffeine and paracetamol are most common in Western European countries; paracetamol increases heroin base volatility, thereby increasing the effects from smoking heroin. Other adulterants include phenobarbitone, diphenyhydramine, procaine, and quinine. Inert diluents are also added to expand the heroin bulk and increase profit; the most frequently encountered diluents are sugars (mannitol, lactose and glucose) (Besacier and Chaudron-Thozet, 1999). Common cutting agents are listed in Table 1.2.

Kaa (1994) reported that heroin samples stored in the dark appear to be stable with no significant changes in their concentrations of acetylcodeine, papaverine and noscapine. However, storage for more than five years often resulted in diamorphine decomposition with increased 6-MAM and morphine content. Postprocessing diamorphine hydrolysis can also readily occur if the sample contains water that is not bound or excess acid (UNODC, 2005). Diamorphine decomposition studies varying pH and temperature have been conducted determining the ratio of 3-MAM and 6-MAM isomers obtained (Bernhauer *et al.*, 1983).

Although most opium poppy and opium production cultivation occurs in Afghanistan, there are three distinct production centers for opiates that supply three distinct markets. Trafficking flows tend to be from (UNODC, 2014a):

Table 1.2 Common cutting agents and their presence in heroin samples

			Heroin	
Cutting agent	**Description**	**Everyday use**	**Adulterant**	**Diluent**
Benzocaine	Local anesthetic	Throat lozenges and topical medicines	✓	
Caffeine	Stimulant	Ingredient in beverages and foodstuffs	✓	
Diazepam	Sedative	Medication for anxiety	✓	
Griseofulvin	Anti-fungal Agent	Medication for athletes foot	✓	
Mannitol	Sugar alcohol	Diabetic Sweetener		✓
Paracetamol	Analgesic	Pain relief preparations	✓	
Phenacetin	Analgesic	None	✓	
Phenolphthalein	Laxative and pH indicator	None as medicine	✓	
Procaine	Local anesthetic	Dentistry	✓	

- Afghanistan to neighboring countries and onwards to Europe through the Middle East ("Southern Route") or through the "Balkan route" across Eastern and Central Europe. There is increasing evidence to suggest other markets (Oceania and South East Asia) are also being targeted.
- Myanmar/Laos to neighboring countries of South-East Asia, (notably China) and to the Oceania region (mainly Australia).
- Latin America (Mexico, Colombia and Peru) to North America

There have been a large number of scientific investigations aiming to relate the relative and/or absolute alkaloid content of opium to the geographical source of the opium. Each major geographic source area produces a heroin that can usually be recognized as a chemically distinct type.

Southeast Asia heroin

Southeast Asia heroin is typically a white powder with high diamorphine purity (80%), as the hydrochloride salt, with few other alkaloids or adulterants. Papaverine and noscapine are seldom present, indicating effective purification of the intermediate morphine. The high similarity between chromatograms obtained from very pure samples results in poor resolution between subgroups (different areas of manufacture or processing techniques) in the Southeast Asia group.

Southwest Asia heroin

In contrast, Southwest Asia heroin samples are far more variable than those from Southeast Asia. The most common form is medium brown and of lower diamorphine purity (40–60%) as the base with noscapine (20–30%), papaverine (2–6%), and acetyl-codeine (5–9%), plus many trace level alkaloid-related impurities. The second most common Southwest Asia heroin is light brown with a higher diamorphine purity (60-85%) with proportional decreases in the remaining alkaloids. Huizer (1983) similarly reported that heroin from Southwest Asia characteristically has high levels of both noscapine and papaverine with further discrimination possible based on the levels of acetylthebaol present.

Mexican heroin

Mexican heroin is unique by its appearance as a sticky black tar (30–60% diamorphine purity) or, less commonly, as a dark brown powder. It is often called "Mexican mud" and is sometimes cut with cocoa powder.

South American heroin

South American heroin is characteristically of high purity (>90%) with low acetyl codeine (<3.5%) and a low total content of 3-MAM and 6-MAM isomers (<5%). Other reports indicate low thebaine content with high papaverine levels (Odell *et al.*, 2006).

Indian heroin

Heroin from India contains less acetylcodeine than heroin from other countries suggesting a lower proportion of codeine to morphine (Kaa, 1987).

The United Nations determined the alkaloid ratios for heroin samples of known provenance and applied the ratios to identify the geographical origin of unknown samples as southeast Asia, southwest Asia, Mexican or South American. The variance between the ratios was found to be quite significant and there was a large overlap for each data set across the different heroin producing regions.

1.4.2.5 Analysis of heroin

Various literature reviews have summarized the many different methods employed in the comparative analysis of heroin samples (Chiarotti and Fucci, 1999; Dams *et al.*, 2001; Nic Daéid and Waddell, 2005).

Presumptive color tests provide an indication that heroin or other opiate alkaloids may be present but are not specific, as many other compounds also give similar colors with the test reagents, necessitating an additional confirmatory technique. The Marquis reagent (8–10 drops of 40% formaldehyde added to 10 mL concentrated sulfuric acid) (UNIDCP, 1998) is widely used as a heroin field test and gives a characteristic purple color with diamorphine, morphine, codeine, 6-MAM or acetylcodeine; papaverine gives no color and noscapine gives a bright yellow color (UNIDCP, 1998).

TLC is often used as a simple and rapid screening technique to identify the opiate alkaloids and other components that may be present in heroin samples prior to examination by other methods (Levy *et al.*, 1996). Examples of solvent systems are cyclohexane/toluene/diethylamine (75:15:10) and toluene/acetone/ethanol/ammonia (45:45:7:3). Developed TLC plates can be visualized using UV light at 254 nm or various spray reagents (Iodoplatinate, Dragendorff and Marquis).

Infrared spectroscopy is a useful screening technique that is more appropriate for the analysis of pure samples, such as heroin, rather than multiple component mixtures. However, Fourier transform infrared spectroscopy (FTIR) has successfully been used for heroin profiling (Simonov *et al.*, 2000; Drozdov *et al.*, 2001; Cai and Wu, 2007).

Major IR peaks are listed below in order of magnitude of wave number absorbance (cm^{-1}) for the heroin base and heroin hydrochloride; The unique IR spectra enable their differentiation (UNIDCP, 1998):

Heroin base:	1243	1196	1727	1214	1444	1757	1054	1370
Heroin hydrochloride:	1245	1736	1177	1194	1448	1765	1157	1386

However, determination of the heroin form within a sample is often not practical if the sample contains mixtures of the heroin salt and base forms or if the heroin base is adulterated with different salts.

Gas chromatography coupled with mass spectrometry (GCMS) generates highly specific mass spectral data enabling the definitive identification of both known and unknown components within heroin samples. Brenneisen and Hasler (2002) used GCMS to study heroin pyrolysis; they obtained 72 pyrolysis products. Heating the heroin street samples from 250 to 400°C produced substantial diamorphine degradation, whereas morphine, codeine, acetylcodeine, papaverine and caffeine were heat stable. The fragmentation pattern for heroin is presented in Figure 1.7.

Liquid chromatography (LC) techniques employed to study and/or profile major and minor heroin components include high performance liquid chromatography (HPLC) (Collins *et al.*, 2006; Hibbert *et al.*, 2010), sonic spray ionization for liquid chromatography mass spectrometry (LCMS) (Dams *et al.*, 2002), and ultra-performance liquid chromatography-tandem mass spectrometry (UPLC-MS/MS) (Lurie and Toske, 2008). LC techniques eliminate the adsorption, heat instability and transesterification problems often associated with GC heroin analysis and sample derivatization is not required. LC limitations include the need for component solubility, poorer resolution, high solvent consumption and associated solvent waste disposal.

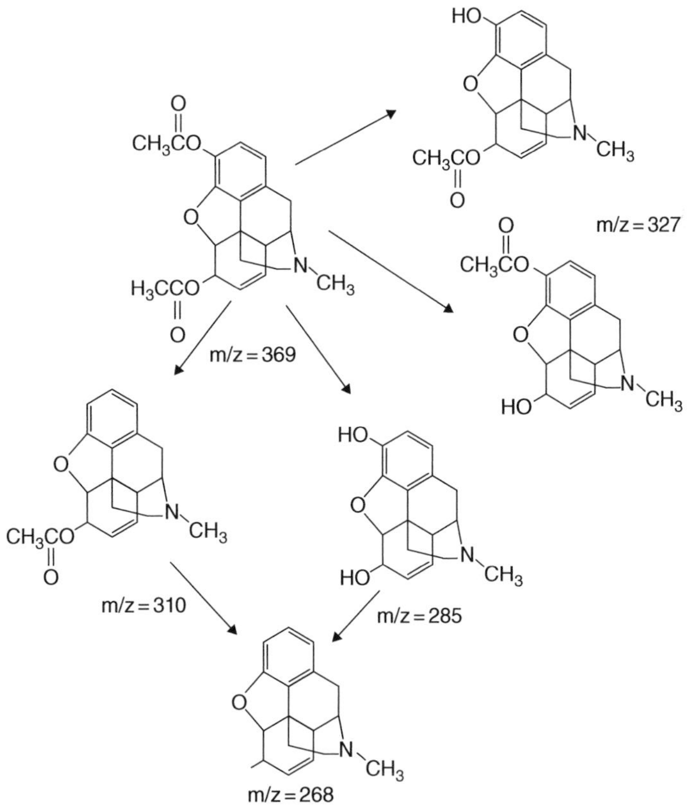

Figure 1.7 The fragmentation pattern for heroin (no derivitization)

Capillary electrophoresis (CE), capillary electrochromatography (CEC) with laser-induced fluorescence (LIF), and capillary electrophoresis-mass spectrometry (CE-MS) have been widely used to analyze heroin samples (Anastos *et al.*, 2005). The separation obtained by CE in narrow bore capillaries under the influence of an electric field is highly efficient, selective, rapid, and may be applied to both charged and neutral species. In addition, CE has been used to determine the concentration of carbohydrates (glucose, sucrose, lactose, mannitol and mannose) found in heroin samples (Lurie *et al.*, 2006).

A range of techniques have been studied to examine the trace inorganic impurities found in heroin samples originating from the elements present in the original opium poppy plus those introduced during the manufacturing process. Methods include elemental analysis (EA) to determine major metal (Ca, Mg, Al, Fe, Zn, Ba) and trace metal (Mn, Cu, Pb, Cd) concentrations (Bora *et al.*, 2003).

Isotopic analysis of heroin samples to determine isotope ratios of $^{13}C/^{12}C$, $^{15}N/^{14}N$ and $^{18}O/^{16}O$ as markers for their geographical origin has been extensively studied using gas chromatography-isotope ratio mass spectrometry (GC-IRMS) (Idoine *et al.*, 2005a, 2005b; Zhang *et al.*, 2005).

1.4.3 Cocaine

1.4.3.1 Introduction

Cocaine is a semi synthetic drug produced from the coca plant, mainly in South America. Extraction from the plant produces a large number of cocaine alkaloids, most of which do not appear in the final product. The main cocaine alkaloids found in street cocaine are shown in Figure 1.8.

No two illicit samples of cocaine will be exactly the same, though the variation from sample to sample is much less than for heroin, for example. Cocaine is commonly encountered in two forms: the salt (cocaine hydrochloride) and the base (cocaine). Certain forms of cocaine base are also known as "crack." Both forms are a white crystalline powder; however, the base form can appear as a more "waxy" solid.

The cocaine content in the plant is usually from 0.3 to 1.5% relative to dry leaf weight but 10–15% *cis*- and *trans*-cinnamoylcocaine is also present. The cocaine purity in coca paste varies from 30 to 80% depending on the extraction technique.

1.4.3.2 Cocaine production

There are over 250 species of the coca plant, two of which are known to produce cocaine: *Erythroxylum coca* and *Erthroxyylum novogranatense* (Nordegren, 2002). *Erthroxyylum novogranatense* is grown mostly in Columbia and other Central American countries, while *Erythroxylum coca* is grown in Boliva and Peru (Freye and Levy, 2009). Columbia, Bolivia and Peru are the three main areas for the production of cocaine (UNODC, 2014a). The coca plant grows in tropical rainforest climates between 100 and 1700 m above sea level.

The appearance of the coca leaves can vary depending on the variety of the coca plant, although there are some commonalities between them. For example, the upper side of the leaf, which can be green or greyish in color, is always darker than the underside, which has two lines running parallel to the midrib (UN, 1986).

The growth cycle of the coca bush is anywhere from six to nine months and can be harvested between three and eight times during the year, but this is dependent on a number of factors, including seasonal variations such as rainfall and temperature. Other factors include disease or spraying of herbicides onto crops in counter narcotics operations, coca species, soil type and horticultural practices, for example, fertilization and irrigation (UNODC, 2014a). However, the main growth period is from December to April. Once picked the coca leaves must be dried so as to preserve the cocaine content (Nordegren, 2002).

The production of most illicit cocaine occurs in very crude and rudimentary settings, often by the coca farmers themselves, and, as a result, there can be a number of variations in the reagents and quantities used. The production process begins with the mixing of the leaves with water and a basic material, for example, lime, followed by the crushing of the leaves to form a pulp. Kerosene is then added and mixed with the pulp to extract the alkaloids from the coca leaves. The kerosene is then removed and separated from the leaves. Acidified water is added to the kerosene, extracting the alkaloids into the aqueous layer. The aqueous layer is removed and used to make the coca paste (UN, 1986). Calcium carbonate (lime) or ammonium is added to make the solution basic, precipitating out the basic alkaloids, which are then collected and allowed to dry, forming the coca paste (UN, 1986).

In order to produce cocaine, the coca paste is added to dilute sulfuric acid, dissolving it, and causing the hydrolysis of the alkaloids, which forms ecognine. This is then reacted with 10% boron trichloride to form ecognine methyl ester. Potassium permanganate is

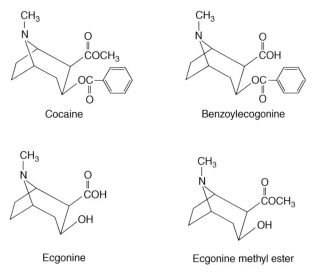

Figure 1.8 The main cocaine alkaloids

also added until the solution turns and remains pink, in order to eliminate any cinnamoylcocaine isomers present. The solution is then filtered and a base, such as ammonia, is added, causing the precipitation of the cocaine free base (UN, 1986).

The cocaine free base is filtered off, washed with water and allowed to dry. It is then dissolved in ether and the mixture is filtered, after which hydrochloric acid and acetone are added to precipitate out the cocaine hydrochloride, which is then filtered off and allowed to dry.

The free base form is more potent than cocaine hydrochloride and, as such, it is often reverted back to its free base form using one of two methods. The less common and more dangerous method is called the ether wash method. The baking soda method is much more popular and involves mixing cocaine hydrochloride and sodium bicarbonate (baking soda) and heating. The result is a hard, crystal like material known as "rock cocaine" or "crack" (Freye and Levy, 2009).

1.4.3.3 Cocaine cutting agents

A range of impurities are present in cocaine, some of which are hydrolysis by-products, formed during the extraction and purification processes, while others are present in the solvents and processing chemicals used. Some may also due to the packaging materials used and the storage conditions under which the packages are kept (Casale and Waggoner, 1991).

Adulterants and diluents are also added to the cocaine before being sold. Common adulterants include lidocaine, procaine and benzocaine, and diluents include sugars such as mannitol, lactose or glucose. A table of common cutting agents is provided in Table 1.3.

1.4.3.4 Analysis of cocaine

Analysis of cocaine follows the normal methodologies of other drug samples where initial colorimetric tests provide a tentative identification, followed by TLC and then more specific chromatographic techniques. The main presumptive test for cocaine is the cobalt thiocynate test. A rapid development of a bright blue color indicates the presence of cocaine; however, a similar color may develop over time for barbiturates as well.

Gas chromatographic analysis of cocaine alkaloids is not straightforward and generally requires derivatization of the molecules in order to be successful. Casale and Waggoner (1991) developed a one-step derivatization method for the identification and quantification of a number of coca related impurities in unadulterated cocaine samples. They concluded that by using this method along with statistical analysis it was possible to identify samples of a common origin or batch, and that samples from the same batch gave "virtually identical chromatographic profiles," and while hundreds of batches were tested no two batches gave similar chromatographic profiles. Cocaine itself, however, does produce reasonably well defined peaks using GC or GCMS. Derivatization is generally accomplished with BSA or BSTFA (N,O-Bis(trimethylsilyl)trifluoro acetamide). The derivatization products and mass fragmentation are given in Figure 1.9.

Table 1.3 Adapted table of common cutting agents and their presence in cocaine samples

			Cocaine	
Cutting agent	**Description**	**Everyday use**	**Adulterant**	**Diluent**
Aspirin	Analgesic	Pain relief preparations	✓	
Benzocaine	Local anesthetic	Throat lozenges and topical medicines	✓	
Boric Acid	Antiseptic and insecticide	Acne medication and insect control preparations		✓
Caffeine	Stimulant	Ingredient in beverages and foodstuffs	✓	
Creatine	Dietary supplement	Body-building		✓
Diazepam	Sedative	Medication for anxiety	✓	
Glucose	Sugar	Ingredient in beverages and foodstuffs		✓
Lactose	Sugar	Filler in tablets and ingredient in foodstuffs		✓
Levamisole	Antibiotic	Medication for worm infections		✓
Lignocaine	Local anesthetic	Dentistry and topical medications	✓	
Mannitol	Sugar alcohol	Diabetic sweetener		✓
Paracetamol	Analgesic	Pain relief preparations	✓	
Phenacetin	Analgesic	None (withdrawn from use in the UK)	✓	
Procaine	Local anesthetic	Dentistry	✓	
Starch	Carbohydrate	Ingredient in foodstuff		✓
Sucrose	Sugar	Ingredient in beverages and foodstuffs		✓

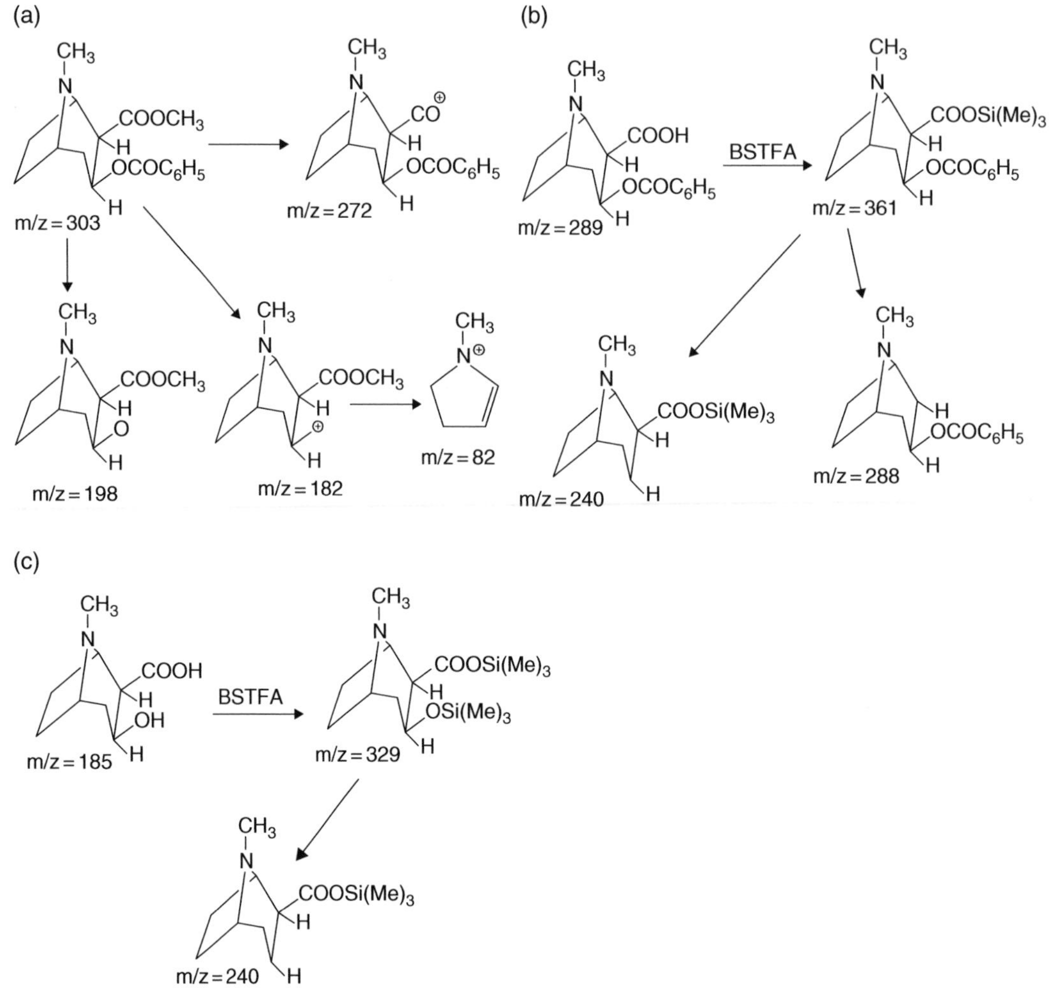

(a)

(b)

(c)

Figure 1.9 Fragmentation patterns for (a) underivitized cocaine (b) derivitized benzoyl ecgonine and (c) derivitized ecgonine

LeBelle *et al.* (1991) found that while the alkaloid composition may be similar between cocaine samples, the solvent residues found can be different, thus differentiating two samples as different solvents were used during manufacturing. They also concluded that from the profiles generated it was possible to differentiate between large production batches or manufacturers. Morello and Meyers (1995) successfully developed a method for the simultaneous identification and quantification of the residual solvents in cocaine a few years later.

A significant body of work has been undertaken in the use of isotope ratio mass spectrometry (IRMS) for cocaine analysis, particularly by Ehleringer and colleagues. In cocaine samples from the four major growing regions they found that the samples "differed by

0.6‰ and 7.1‰ in their $\delta^{13}C$ and for $\delta^{15}N$ values," and, as such, it was concluded that there is the potential for using isotope analysis to identify the geographical origin of cocaine (Ehleringer *et al.*, 1999). Ehleringer *et al.* (2000) also carried out a large study on 200 cocaine samples. They analyzed ^{13}C and ^{15}N and also the truxilline and trimethoxycocaine content. They found that there was a wide range in the $\delta^{13}C$ and $\delta^{15}N$ values obtained; the ^{13}C and ^{15}N values varied from −32.4 to −25.3 for ^{13}C and from 0.1 to 13.0 for ^{15}N. They also noted that "truxilline content and cocaine ^{15}N are positively correlated and that cocaine ^{13}C and trimethoxycocaine are negatively correlated." Combining these factors they correctly identified the country of origin of 192 of the samples. Casale *et al.* (2005) examined the isotopic fractionation of carbon and nitrogen and found that the processing of the cocaine only "caused minimal changes in the $\delta^{13}C$," although fractionation against ^{15}N does occur, during the conversion of the cocaine base to the hydrochloride salt form, but is less prominent the higher the yield of cocaine hydrochloride obtained.

1.4.4 Amphetamine-type stimulants
1.4.4.1 Introduction
Amphetamine-type stimulants (ATS), including amphetamines, methylamphetamine and ecstasy, remain the second most widely consumed group of illicit substances. Amphetamines are β-phenethylamine derivatives and totally synthetic drugs and, as such, are very different in terms of the potential for analysis, as in all cases there are many synthetic by-products carried through each stage of the synthetic process. These synthetic by-products can help to identify the specific synthetic mechanism used in their production. Amphetamine-type stimulants are generally amphetamine sulfate, methyl-amphetamine hydrochloride or free base and 3,4-methylenedioxymethylamphetamine hydrochloride (ecstasy).

ATS are commonly encountered in one of four forms: tablet, powder, free base or as crystaline material. Tablets normally contain 3,4-methylenedioxymethylamphetamine hydrochloride or methylamphetamine hydrochloride and caffeine. In the United States, methylamphatamine tablets are encountered in different colors and shapes that can be flavored, scented and stamped with various logos in a similar fashion to conventional ecstasy tablets.

The powder form is normally either amphetamine or methylamphetamine. It is bitter to taste and water soluble. The color of the powder ranges from off-white to reddish-brown depending on the chemicals used in the manufacturing process.

Crystalline methylamphetamine is usually obtained by recrystallizing the hydrochloride powder using isopropyl alcohol or water. This form of methylamphetamine has the highest purity compared to other forms. The structures of these compounds are provided in Figure 1.10.

Most laboratories that have been dismantled by law enforcement have been manufacturing methylamphetamine and were concentrated in North America, East and Southeast Asia. Approximately half of the methylamphetamine seizures in the United States now

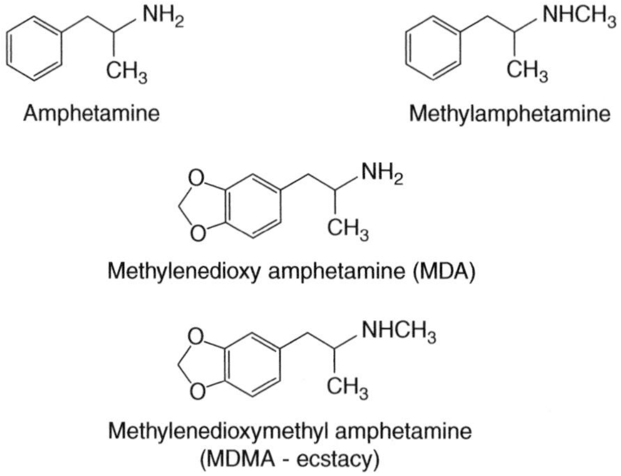

Figure 1.10 Amphetamine-type stimulants

occur at the US–Mexican border and large-scale production is now known to occur in Mexico (UNODC, 2014a). As Mexico respond has responded with strong counter methylamphetamine initiatives, manufacturing activities are tending to move south to Latin America, including Argentina, Guatemala, Honduras, and Peru. Similar shifts may also be occurring in South Asia, where India and Sri Lanka reported their first operational methylamphetamine laboratories in 2008, and reported seized manufacturing equipment and chemicals in 2007.

Large quantities of amphetamine seizures continue to be reported in the Middle East, in particular by Jordan, Saudi Arabia and the Syrian Arab Republic. Major quantities of "ecstasy" were seized in East and Southeast Asia, followed by Europe (Southeastern Europe and Western and Central Europe). All three regions account for nearly three quarters of global "ecstasy" seizures.

1.4.4.2 Synthesis of amphetamine-type stimulants

The most common methods used in the production of ATS usually involve more than one synthetic step and involve simple reflux apparatus or more complicated pressure controlled reaction vessels. Three categories of chemicals are used in the synthesis of amphetamine type stimulants: precursors, reagents and general purpose chemicals. The molecular structure of the precursor is generally quite similar to that of the final product. The purpose of reagents in the synthesis is to chemically modify or combine precursors in a chemical reaction. General purpose chemicals are used to facilitate the reaction and isolate the product. Examples include solvents, alkalis and acids. In many countries, the sale of these chemicals is placed under strict control and regulated by the individual legislation of the countries involved. Besides chemicals, another important tool required in the clandestine manufacture of ATS is appropriate glassware, which can be obtained either from chemical supply companies or made out of household items.

Amphetamine

The most common European method for the production of amphetamine is called the Leuckart synthesis which is a two-step formulation and hydrolysis of phenyl-2-propanone (P2P) (also known as benzylmethylketone [BMK]), phenylacetone and phenylpropanone. Reductive amination of P2P can also be used to produce amphetamine; this reaction is carried out under reduced pressure and homemade pressure reaction vessels are commonly used. The third common method is via a nitrostyrene intermediate to produce amphetamine. P2P is the starting product for each of these methods and as the major precursor chemical is itself under legislative control. Each synthetic route will produce a "family" of different reaction by-products and impurities. Some of these will be specific to the route used while others will be common across routes.

The Leuckart synthesis is one of the most common routes to synthesis of amphetamine as the sulfate salt. It involves a formylation reaction between P2P and formamide followed by a hydrolysis reaction to form the amphetamine.

Reductive amination of P2P is a popular method for the production of amphetamine in parts of Europe. P2P is reacted with ammonia to form an imine. The imine is then reduced under high temperature and pressure to the corresponding amine.

The nitrostyrene synthesis involves the use of lithium aluminum hydride as a reducing agent. The reaction is a two-step reaction involving the formation of nitrostyrene followed by a lithium reduction. The initial reaction involves mixing benzaldehyde with nitroethane and ammonium acetate in acetic acid. The mixture is refluxed to form nitrostyrene, which is purified by recrystallization in ethanol. This is followed by a dilution of the nitrostyrene in a solvent (this can be ether, toluene or THF) and the addition of this solution, drop wise, to a solution of lithium aluminum hydride, keeping the temperature of the reaction mixture cool. The reaction is refluxed, after which the excess lithium aluminum hydride is quenched and removed. Amphetamine free base is extracted into ether and precipitated as the sulfate salt with sulfuric acid.

Methylamphetamine

In general, methylamphetamine is synthesized from one of two different types of known precursors: *l*-ephedrine or *d-pseudo*ephedrine and phenyl-2-propanone (P2P). Ephedrine and *pseudo*ephedrine are used commonly in cold medication and sold as over-the-counter medication in various countries. To combat the isolation of ephedrine and *pseudo*ephedrine from medication, manufacturers increasingly add various inhibitors such as aminoalkyl methacrylate copolymer (Eudragit-E), ferrous gluconate, lactose, ethylcellulose, and hydroxypropyl cellulose, which interferes with the conversion of ephedrine or *pseudo*ephedrine to methylamphetamine.

Besides extracting ephedrine and *pseudo*ephedrine from cold medication, clandestine chemists also isolate the compounds from *Ephedra* plant based products. The stem and leaves of the Asiatic and Mediterranean species of *Ephedra* contain ephedrine and *pseudo*ephedrine as well as other alkaloids, such as norephedrine, nor*pseudo*ephedrine, methylephedrine, and methyl*pseudo*ephedrine (Massetti, 1996).

There are eight synthetic methods generally used in the illicit manufacture of methyl-amphetamine hydrochloride; these are presented in Figure 1.11. These methods are easily accessible through the Internet, the scientific literature, patents and books. The method used to synthesize methylamphetamine depends on various factors, such as availability of precursors, essential chemicals, the complexity of the process, equipment and the chemical hazards presented. With increasing Internet usage, many clandestine chemists can purchase the required supplies using Internet auction sites (Massetti, 1996).

MDMA

The clandestine synthesis of 3,4-methylenedioxy-methamphetamine (MDMA) usually begins with the synthesis of its major precursor, phenyl methyl ketone (PMK), which can be achieved in a number of ways (Dal Cason, 1990) A survey of "underground" web sites and recent literature on drug profiling (most of which used *PiHKAL* as a source) indicated that common routes to PMK originate from safrole or isosafrole and proceed via the isosafrole glycol to PMK (Figure 1.12).

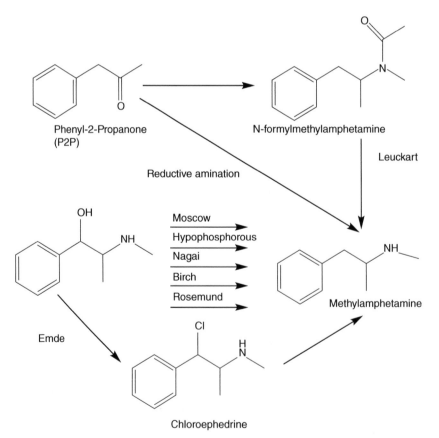

Figure 1.11 Outline of the synthetic routes used in the manufacture of methylamphetamine

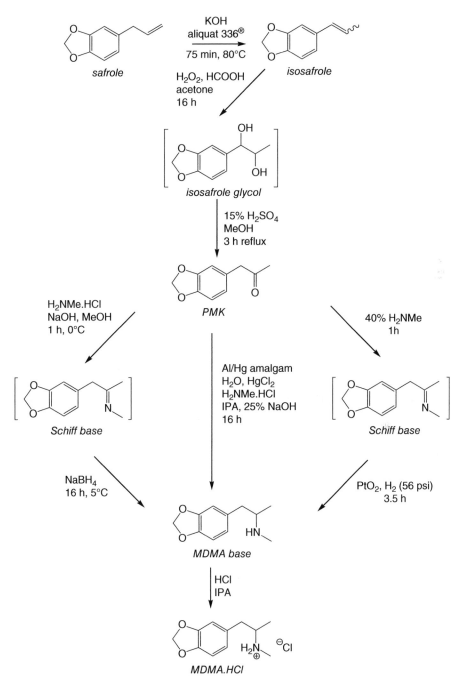

Figure 1.12 Summary of the synthesis pathways to MDMA

The root bark of the sassafras tree contains 5–9% aromatic oil, of which approximately 80% is safrole. Despite the suspected carcinogenic properties (both sassafras oil and safrole are banned as flavors and food additives by the United States Food and Drug Administration), sassafras oil can still be obtained as an "essential oil" in outlets such as health food shops.

PMK can be used to synthesize MDMA by three variations of the popular route of reductive amination. Three different reducing agents can be used: Al/Hg amalgam, $NaBH_4$ ("cold method"), and Pt/H_2 ("high pressure method"). Reportedly, reductive amination with the platinum catalyst has, in recent years, become in the most commonly encountered method of MDMA production in the Netherlands, with $NaBH_4$ sometimes encountered, and Al/Hg rarely encountered (Koper *et al.*, 2007).

1.4.4.3 ATS cutting agents

ATS also contain a range of substances being used as cutting agents. These vary considerably from country to country. Common materials include sugars (mainly lactose of glucose), caffeine, and other anesthetics, such as "Ketamine," and depressant drugs such as barbiturates.

1.4.4.4 Analysis of ATS

Characteristics such as logos, color, size and shape are all recorded as part of the initial examinations carried out. It may be possible to link the logo on tablets to provide intelligence information across drug seizures, although such data should be treated cautiously. The main presumptive tests used for amphetamine compounds are the Marquis and Mandelin tests providing a brown–orange or green response and green response to ATS, respectively. TLC can provide extra information, potentially identifying which of theATS may be present in a particular sample. In this case, is it common to use a developing reagent, such as ninhydrin, acidified iodoplatinate or Fast Black K, to visualize the separated samples.

Both HPLC and GCMS are used when quantification of amphetamines is required. The amphetamines generally chromatograph well on both systems. There can be some co-elution of amphetamine, methylamphetamine, MDA and MDMA occurring and it is wise to always run standards separately to establish the specific retention time of each compound. Mixtures of amphetamines in powders or tablets are not usual but do sometimes occur. Many amphetamine samples also contain caffeine, which has a much greater molar absorptivity coefficient than amphetamine and, as such, will have a much greater peak area than amphetamine, even though the concentration of both within the sample may be similar. This can also cause problems in the resolution of peaks in HPLC analysis.

Amphetamines generally chromatograph well underivatized using GCMS instruments, though many laboratories still chose to derivatize the samples. Derivatization in this case occurs using, for example, trifluoroacetic acid (TFA) where chemical modification occurs at the amine functional group.

Other analytical methods, in particular IRMS and ICP-MS (inductively coupled plasma-mass spectrometry), have been used for the chemical profiling of ATS and, especially, methylamphetamine and MDMA; the reader is directed to the specific literature in this area, as it is beyond the scope of this chapter.

1.4.5 New psychoactive substances
1.4.5.1 Introduction

New psychoactive substances are synthetic substances that have recently become available on the market. As of yet, many are not controlled by the 1961 Single Convention on Narcotic Drugs or the 1971 Convention on Psychotropic substances (UNODC, 2014b). They are often referred to as legal highs, designer drugs or bath salts or spice.

The use of NPS has grown rapidly over the last few years and there are an increasing number of reports on the availability and manufacture of these substances. By 2013, 348 NPS had been reported to UNODC, which is more than the 234 psychoactive substances currently under governance of the 1961 Single Convention on Narcotic Drugs and the 1971 Convention on Psychotropic Substances. New psychoactive substances have been reported in every region, a total of 94 countries worldwide (UNODC, 2014b). Synthetic cathinones make up 25% of the NPS reports to UNODC. Cathinones are analogues of amphetamines and the beta-keto derivatives of phenethylamines; each phenethylamine compound has a parallel cathinone analogue (Schifano *et al.*, 2011) (Figure 1.13).

Synthetic cannabinoids

Synthetic cannabinoids are compounds that have been prepared to have the same functionality as Δ9-tetrahydrocannabinol (THC) in terms of their interactivity with the cannabinoid receptors in the brain. In late 2008, several cannabinoids were detected in herbal smoking mixtures or so-called incense/room odorizers. These products are typically sold via the Internet and in "head shops." Many of the synthetic cannabinoids are structurally unrelated to THC; however, they do often contain a side chain containing between four and nine carbon atoms. The synthetic cannabinoids can be classified into six major structural groups:

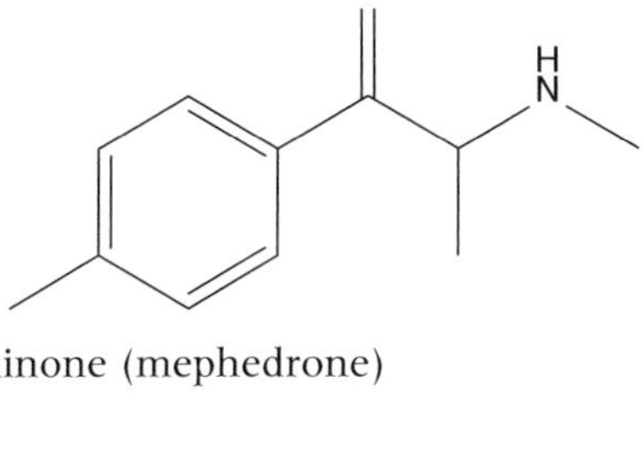

Figure 1.13 4-Methylmethcathinone (mephedrone)

- Naphthoylindoles (e.g., JWH-018, JWH-073 and JWH-398).
- Naphthylmethylindoles.
- Naphthoylpyrroles.
- Naphthylmethylindenes.
- Phenylacetylindoles (i.e., benzoylindoles, e.g., JWH-250).
- Cyclohexylphenols (e.g., CP 47,497 and homologues of CP 47,497).

Substituted cathinones

Substituted cathinones usually take the form of a white/off-white/yellow powder; they are found less frequently as tablets or capsules. The most common route of ingestion is insufflation but they can also be ingested orally, rectally and intravenously. When snorted the effects are felt immediately with a short high and a rapid comedown. Polydrug use is a common habit; it is where multiple drugs are taken in order to increase the high and prolong the effects, thus increasing the potential harm and toxicity effects.

The synthetic cathinone most commonly reported to the United Nations Office on Drugs and Crime in 4-methylmethcathinone, more commonly known as mephedrone. The first online reference to mephedrone was in 2003. However, it became increasingly popular and readily available for purchase in 2007. The rise in popularity has been attributed to a number of factors. Mephedrone was viewed as a legal alternative to illicit substances, meaning it was also viewed as being safer. The decreasing purity in ecstasy and cocaine also led to an increase in mephedrone consumption, as it was seen as a cheaper, more potent alternative. Mephedrone was also readily available to purchase over the Internet and in head shops and was largely driven by web-based marketing, exposing it to a whole new range of audiences (Schifano *et al.*, 2011).

The rise of the Internet as a global marketplace for drug information, distribution and supply presents a unique challenge in disrupting the supply of NPS (EMCDDA, 2014). NPS that are not controlled under international drug laws are being produced outside of Europe and obtained through online retailers. The Internet is set to become the newest method of drug trafficking and law enforcement and policy makers must take this into account in the future.

The speed at which controlled substances are being replaced by new substances provides serious challenges for lawmakers. Once a substance is deemed to present a certain level of harm it is legislated against, but this gives rise to rebranding or development of another alternative with unknown toxicity. When a substance is removed from the drug market it is likely to be replaced almost immediately and the cycle begins again (Winstock and Ramsey, 2010). The time it takes to pass legislation controlling substances cannot keep up with the speed at which new psychoactive substances are appearing on the market, leading policy makers to demand new, faster and more effective ways of drug

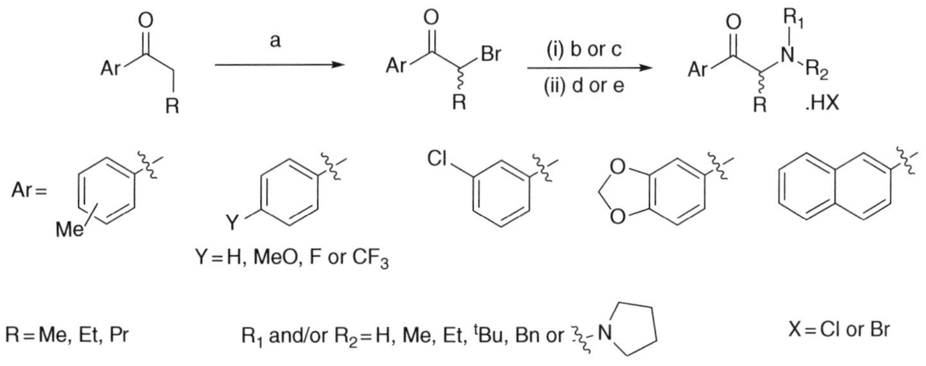

Reagents/Conditions:
(a) $Br_2/HBr/CH_2Cl_2/rt/1$ h; (b) $NHR1R_2 \cdot HCl/NEt_3/CH_2Cl_2/rt/24$ h;
(c) $NHR_1R_2/NEt_3/CHCl_3/reflux/24$ h; (d) HCl-dioxane/propan-2-ol/rt/1 h;
(e) HBr-AcOH/AcOH/rt/1 h.

Figure 1.14 Reaction scheme for the preparation of cathinone derivatives

control. As fast as law makers can identify new substances and assess the risk, another one has already taken its place.

1.4.5.2 Synthesis of cathinones and synthetic cannabinoids
The synthesis of cathinones is chemically straightforward and usually follows a two-step synthesis process. The initial synthesis is of an α-bromoketone followed by synthesis of the relevant cathinone as a hydrochloride or hydrobromide salt (Figure 1.14).

The preparation of synthetic cannabinoids is much more chemically complex and beyond the scope of this chapter.

1.4.5.3 Analysis of cathinones and synthetic cannabinoids
The greatest challenge in the analysis of both cathinones and synthetic cannabinoids is the availability of validated or certified standards. Presumptve tests are listed for the analysis of cathinones. These include, for example, the Zimmerman test, which has been suggested as being consistently effective across a range of cathinone derivatives. There are no presumptive field tests as yet available for the synthetic cannabinoids. TLC analysis is also limited for both sets of compounds although some methods have been published (Nic Daéid *et al.*, 2014).

Both the cathinones and cannabinoids are readily resolved using gas chromatography, but their identification and quantitative analysis is limited by the availability of pure reference samples. In many cases, nuclear magnetic resonance (NMR) spectroscopy is now being used in an attempt to identify the structures of the target molecules. FTIR and Raman are also showing some promise in identifying the compounds once standards are available (Nic Daéid *et al.*, 2014).

1.5 Conclusions

The analysis of drugs of abuse is a constant component of a forensic science laboratory and is one of the most commonly encountered types of analysis preformed. Modern forensic chemistry laboratories are governed by accreditation standards, normally to ISO 17025 or similar. These ensure that the processes carried out are appropriately validated and fit for purpose.

The choice of instrumental technique used in any analysis depends upon the motive behind the analysis and on the nature of the sample. Normal procedure when dealing with bulk samples is to perform presumptive color tests to identify the drug class, possibly followed by TLC to identify the specific member of the drug class. When dealing with trace samples it is often the case that these non-confirmatory tests are circumvented and confirmatory tests only are employed where the choice of technique is based upon the experience of the analyst or laboratory protocol.

Until comparatively recently, our knowledge of the materials encountered has been straightforward and well grounded. However, the now global emergence of new psychoactive substances, where the nature of the compounds is constantly developing, provides new challenges, even in the simple identification of these materials. This requires information sharing across the forensic chemistry community in order to keep up to date on new trends. It also requires a movement towards using technologies such as NMR to confirm the identity of complex isomeric mixtures.

Acknowledgements

Much of the ground work upon which this chapter draws was compiled by the following research students, for which I am extremely grateful: Dr Oliver Sutcliffe, Dr Hilary Buchanan, Dr Vanitha Kunalan, Dr Sara Jayaram, Dr Sarah Baron, Ms Ainslie Smith, Ms Nicola Murphy and Ms Nadia Welch.

References

Anastos, N., Barnett, N.W., and Lewis, S.W. (2005) Capillary electrophoresis for forensic drug analysis: A review. *Talanta*, **67** (2), 269–279.

Bernhauer, D., Fuchs, E.F., and Neumann, H. (1983) Identification of 3-O-acetylmorphine as a decomposition product of diacetylmorphine (heroin) by HPLC and capillary GC – course of reaction of the heroin decomposition and significance of the decomposition products for characterizing illicit heroin samples. *Fresenius Zeitschrift Fur Analytische Chemie*, **316** (5), 501–504.

Besacier, F. and Chaudron-Thozet, H. (1999) Chemical profiling of illicit heroin samples. *Forensic Science Review*, **11** (2), 105–119.

Bora, T., Merdivan, M., and Hamamci, C. (2003) Heroin profiling using alkaloids and metal concentrations. *Forensic Science International*, **136** (Suppl 1), 89.

Brenneisen, R. and Hasler, F. (2002) GC/MS determination of pyrolysis products from diacetylmorphine and adulterants of street heroin samples. *Journal of Forensic Sciences*, **47** (4), 885–888.

Cadola, L., Broseus, J., and Esseiva, P. (2013) Chemical profiling of different hashish seizures by gas chromatography-mass spectrometry and statistical methodology: A case report. *Forensic Science International*, **232** (1–3), E24–E27.

Cai, X.L. and Wu, G.P. (2007) Preliminary study on identification of heroin from different route with clustering analysis by Fourier transform infrared spectroscopy. *Spectroscopy and Spectral Analysis*, **27** (12), 2441–2444.

Casale, J.F. and Waggoner, R.W., Jr. (1991) A chromatographic impurity signature profile analysis for cocaine using capillary gas chromatography. *Journal of Forensic Sciences*, **36** (5), 1312–1330.

Casale, J. F., Ehleringer, J.R., Morello, D.R., and Lott, M.J. (2005) Isotopic fractionation of carbon and nitrogen during the illicit processing of cocaine and heroin in South America. *Journal of Forensic Sciences*, **50** (6), 1–7.

Chiarotti, M. and Fucci, N. (1999) Comparative analysis of heroin and cocaine seizures. *Journal of Chromatography B: Biomedical Sciences and Applications*, **733** (1–2), 127–136.

Collins, M., Casale, E., Hibbert, D.B., *et al.* (2006) Chemical profiling of heroin recovered from the North Korean merchant vessel Pong Su. *Journal of Forensic Sciences*, **51** (3), 597–602.

Dal Cason, T.A. (1990) An evaluation of the potential for clandestine manufacture of 3,4-methylenedioxyamphetamine (MDA) analogs and homologs. *Journal of Forensic Sciences*, **35** (3), 675–697.

Dams, R., Benijts, T., Lambert, W.E., *et al.* (2001) Heroin impurity profiling: trends throughout a decade of experimenting. *Forensic Science International*, **123** (2–3), 81–88.

Dams, R., Benijts, T., Günther, W., *et al.* (2002) Sonic spray ionization technology: performance study and application to a LC/MS analysis on a monolithic silica column for heroin impurity profiling. *Analytical Chemistry*, **74** (13), 3206–3212.

Drozdov, M.A., Kameav, A.V., and T.B. Kimstach (2001), Complex heroin comparison investigation. *Z zagadnien nauk sadowych (Problems of Forensic Sciences)*, **46**, 147–151.

Drummer, O.H (2001) *The Forensic Pharmacology of Drugs of Abuse*. Arnold Publishers.

Ehleringer, J.R., Cooper, D.A., Lott, M.J., and Cook, C.S. (1999) Geo-location of heroin and cocaine by stable isotope ratios. *Forensic Science International*, **106** (1), 27–35.

Ehleringer, J.R., Casale, J.F., Lott, M.J., and Ford, V.L. (2000) Tracing the geographical origin of cocaine. *Nature*, **406**, 311–312.

El Alaoui, M.A., Melloul, M., Amine, S.A., *et al.* (2013) Extraction of high quality DNA from seized Moroccan cannabis resin (hashish). *PLoS One*, **8** (10), e74714.

EMCDDA (2014) *European Drug Report – Trends and Developments*. European Monitoring Centre for Drugs and Drug Addiction (EMCDDA), Lisbon, Portugal.

Freye, E. and Levy, J.V. (2009) *Pharmacology and Abuse of Cocaine, Amphetamines, Ecstasy and Related designer drugs: A Comprehensive Review on Their Mode of Action, Treatment of Abuse and Intoxication*. Springer, New York.

Gilmore, S., Peakall, R., and Robertson, J. (2003) Short tandem repeat (STR) DNA markers are hypervariable and informative in Cannabis sativa: implications for forensic investigations. *Forensic Science International*, **131** (1), 65–74.

Hibbert, D.B., Blackmore, D., Li, J., *et al.* (2010) A probabilistic approach to heroin signatures. *Analytical and Bioanalytical Chemistry*, **396** (2), 765–773.

Huizer, H. (1983) Analytical studies on illicit heroin. 1. The occurrence of O-3-monoacetylmorphine. *Journal of Forensic Sciences*, **28** (1), 32–39.

Idoine, F.A., Carter, J.F., and Sleeman, R. (2005a) Bulk and compound-specific isotopic characterisation of illicit heroin and cling film. *Rapid Communications in Mass Spectrometry*, **19**, 3207–3215.

Idoine, F.A., Carter, J.F., and Sleeman, R. (2005b) Geo-location of heroin and cocaine by stable isotope ratios. *Forensic Science International*, **106** (1), 27–35.

Kaa, E. (1987) Changes in place of origin of heroin seized in Denmark from 1981 to 1986 – chemical fingerprint of 138 samples. *Zeitschrift Fur Rechtsmedizin (Journal of Legal Medicine)*, **99** (2), 87–94.

Kaa, E. (1994) Impurities, adulterants and diluents of illicit heroin. changes during a 12-year period. *Forensic Science International*, **64** (2–3), 171–179.

King, L. (2003) *The Misuse of Drugs Act – A Guide for Forensic Scientists*. Royal Society of Chemistry, Cambridge.

Koper, C., van den Boom, C., Wiarda, W., *et al.* (2007) Elemental analysis of 3,4-methylenedioxy-methamphetamine (MDMA): A tool to determine the synthesis method and trace links. *Forensic Science International*, **171**, 171–179.

LeBelle, M., Callahan, S.A., Latham, D.J., *et al.* (1991). Comparison of illicit cocaine by determination of minor components. *Journal of Forensic Sciences*, **36** (4), 1102–1120.

Le Vu, S., Aquatias, S., Bonnet, N., *et al.* (2006) Chemical content of street cannabis. *Presse Medicale*, **35** (5), 755–758.

Levy, R., Ravbery M., Meirovich L., and Shapira-Heiman O. (1996) A survey and comparison of heroin seizures in Israel during 1992 by Fourier transform infrared spectrometry. *Journal of Forensic Sciences*, **41** (1), 6–11.

Lewis, R., Ward, S., Johnson, R., *et al.* (2005), Distribution of the principal cannabinoids within bars of compressed cannabis resin. *Analytica Chimica Acta*, **538** (1–2), 399–405.

Lindholst, C. (2010) Long term stability of cannabis resin and cannabis extracts. *Australian Journal of Forensic Sciences*, **42** (3), 181–190.

Lurie, I.S. and Toske, S.G. (2008) Applicability of ultra-performance liquid chromatography-tandem mass spectrometry for heroin profiling. *Journal of Chromatography A*, **1188** (2), 322–326.

Lurie, I., Hays, P., and Valentino, A. (2006) Analysis of carbohydrates in seized heroin using capillary electrophoresis. *Journal of Forensic Sciences*, **51** (1), 39–44.

Massetti, J. (1996) Recent laboratory seizures and activity in California clandestine methamphetamine laboratories. *Journal of the Clandestine Laboratory Investigating Chemists*, **6** (4), 13–14.

Morello, D.R. and Meyers, R.P. (1995) Qualitative and quantitative determination of residual solvents in illicit cocaine HCl and Heroin HCl. *Journal of Forensic Sciences*, **40** (6), 957–963.

Nic Daéid, N. and Waddell, R.J.H. (2005) The analytical and chemometric procedures used to profile illicit drug seizures. *Talanta*, **67** (2), 280–285.

Nic Daéid, N., Fallon, V., Brunt, N., *et al.* (2003) Generating DNA profiles from cannabis resin? A potential new linkage tool. *Forensic Science International*, **136** (Suppl 1), 89–89.

Nic Daéid, N., Savage, K.A., Ramsay, D., *et al.* (2014) Development of gas chromatography-mass spectrometry (GC-MS) and other rapid screening methods for the analysis of 16 'legal high' cathinone derivatives. *Science and Justice*, **54**, 22–31.

Nordegren, T. (2002) *The A-Z Encyclopedia of Alcohol and Drug Abuse*. Brown Walker Press, FL.

Odell, L.R., Skopec, J., and McCluskey, A. (2006) A 'cold synthesis' of heroin and implications in heroin signature analysis: Utility of trifluoroacetic/acetic anhydride in the acetylation of morphine. *Forensic Science International*, **164** (2–3), 221–229.

Schifano, F., Albanese, A., Fergus, S., *et al.* (2011), Mephedrone (4-methylmethcathinone, 'meow meow'): chemical, pharmacological and clinical issues. *Psychopharmacology*, **214**, 593–602.

Simonov, E.A., Sorokin V.I., Drozdov, M.A., *et al.* (2000) Comparative analysis of heroin. *Sudebno-Meditsinskaya Ekspertiza*, **43** (3), 23–28.

Trofin, I.G., Dabija, G., and Vaireanu, D.-I. *et al* (2012) The influence of long-term storage conditions on the stability of cannabinoids derived from cannabis resin. *Revista De Chimie*, **63** (4), 422–427.

UN (United Nations) (1961) Single Convention on Narcotic Drugs. https://www.unodc.org/pdf/convention_1961_en.pdf (last accessed 29 June 2015).

UN (United Nations) (1971) Convention on Psychotropic Substances, 1971. https://www.unodc.org/pdf/convention_1971_en.pdf (last accessed 29 June 2015.

UN (United Nations) (1986). *Recommended Methods for Testing Cocaine*. United Nations Division of Narcotic Drugs, United Nations Publications, New York.

UNIDCP (1998) *Recommended Methods for Testing Opium, Morphine and Heroin; Manual for Use by National Drug Testing Laboratories*. United Nations International Drug Control Programme (UNIDCP), Vienna, Austria.

UNODC (nd) *UN Crime and Drug Conventions: Treaties*. United Nations Office on Drugs and Crime (UNODC). http://www.unodc.org/unodc/treaties/ (last accessed 29 June 2015).

UNODC (1988) *Convention Against the Illicit Traffic in Narcotic Drugs and Psychotropic Substances*. United Nations Office on Drugs and Crime (UNODC). https://www.unodc.org/pdf/convention_1988_en.pdf (last accessed 29 June 2015).

UNODC (2005) *Methods for Impurity Profiling of Heroin and Cocaine: Manual for Use by National Drug Testing Laboratories*. United Nations Office on Drugs and Crime (UNODC), Vienna, Austria.

UNODC (2009a) *Guidelines on Representative Drug Sampling*. United Nations Office on Drugs and Crime (UNODC), Vienna, Austria.

UNODC (2009b) *Recommended methods for the identification and analysis of cannabis and cannabis products*. United Nations Office on Drugs and Crime (UNODC), http://www.unodc.org/documents/scientific/ST-NAR-40-Ebook.pdf (last accessed 1 June 2015).

UNODC (2014a) *World Drug Report 2014*. United Nations Office on Drugs and Crime (UNODC). http://www.unodc.org/documents/wdr2014/World_Drug_Report_2014_web.pdf (last accessed 29 June 2015).

UNODC (2014b) *Global Synthetic Drugs Assessment Amphetamine-type Stimulants and New Psychoactive Substances*. United Nations Office on Drugs and Crime (UNODC), Vienna, Austria.

Winstock, A. and Ramsey, J. (2010) Legal highs and the challenges for policy makers. *Addiction*, **105** (10), 1685–1687.

Zhang, D., Sun, W., Yuan, Z., *et al.* (2005), Origin differentiation of a heroin sample and its acetylating agent with 13C isotope ratio mass spectrometry. *European Journal of Mass Spectrometry*, **11** (3), 277–285.

CHAPTER 2

Textiles

Max Houck

Forensic & Intelligence Services, LLC, USA

2.1 Introduction

Textile fibers are "common" in that textiles permeate the modern world in clothing, buildings, and vehicles. Each person has a personal textile environment of clothing, cars, upholstery, things contacted, and people encountered. Textiles constantly shed fibers and these transfer to other people, places, and things. Some textiles are better at shedding fibers – typically loosely woven or knitted textiles – than others, like a tightly woven dress shirt. Textiles also differentially retain fibers depending on their construction, use, and other factors, such as how often they are cleaned.

The analysis of textiles is based on their physical, optical, and chemical properties, using a variety of instruments, including microscopes, spectrometry, and spectroscopy (Robertson *et al.*, 2002). Fibers are a canonical form of forensic evidence despite budget cuts and service reductions (Green, 2014); it only remains for laboratories to recognize the value forensic fiber examinations can bring to casework.

2.2 A science of reconstruction

Perhaps the original scientist-as-detective, Georges Cuvier (1769–Cuvier, 1832), the founder of comparative anatomy and paleontology, used scattered, fractured bits of information to reconstruct the prehistory of the Earth and its animals. In a 1798 paper, Cuvier wrote on his realization of the form and function of bones as it relates to the

Forensic Chemistry: Fundamentals and Applications, First Edition. Edited by Jay A. Siegel.
© 2016 John Wiley & Sons, Ltd. Published 2016 by John Wiley & Sons, Ltd.

overall identifiable anatomy of an animal, leading to the recognition of the creature from which the bone originated:

> "This assertion will not seem at all astonishing if one recalls that in the living state all the bones are assembled in a kind of framework; that the place occupied by each is easy to recognize; and that by the number and position of their articulating facets one can judge the number and direction of the bones that were attached to them. This is because the number, direction, and shape of the bones that compose each part of an animal's body are always in a necessary relation to all the other parts, in such a way that–up to a point–one can infer the whole from any one of them, and vice versa."
>
> (Cuvier, 1832, p. 36)

This has been called Cuvier's Principle of correlation of parts and is a central tenant in biology and paleontology. Rudwick (1998) notes that Cuvier claimed to be able to *identify* an animal taxonomically from a single bone, not completely *reconstruct* it, as the above quote might imply. The reconstruction would only be possible with a sufficient number of bones representing the animal in question.

Cuvier's Principle is used daily in forensic science agencies around the world, although few would recognize it. Forensic scientists take fragmentary bits of things – traces – and interpret larger scale entities and activities from them. Items of evidence are examined and placed into a class or category within a relevant taxonomy; items with the same class characteristics can then be compared on more exacting characteristics, like color. If Locard's Exchange Principle supports how evidence occurs in a meaningful context, then Cuvier's Principle allows for the interpretation of the evidence from the micro to macro level of activity (or "source" to "event", as outlined by Cook *et al.* [1998]). A trilobal fiber is indicative of a carpet, a bright acrylic is indicative of a stuffed toy; a brown modacrylic suggests a wig, and so on. The macro level is not seen but rather interpreted from those traces found at the scene, the symbols of the larger truth revealed by the forensic expert. Thus, forensic science joins its siblings as a predominately historical science, recreating a narrative based on testable facts (Ginzberg, 1979; Buchli and Lucas, 2001). Numerous published examples of fiber casework detail the reconstruction of crimes and criminal activity through forensic analysis of textiles (for a canonical example, see Deadman [1984]).

The forensic process is essentially that of identifying or classifying what an item of evidence is and then comparison of a questioned item, typically from the crime scene, to a presumptively similar sample of known provenance. Both of these steps are discussed in moderate detail here, as they pertain to the analysis of fibers in criminal cases and how that process differs from the analysis of fibers by other industries, particularly the textile industry.

2.2.1 Classification

Classification is the examination of the chemical and physical properties of an object and using them to categorize it as a member of a group. As the process of identification of evidence becomes more specific it permits the analyst to classify the evidence into successively smaller classes of objects. "Class" is a moveable definition – it may not be necessary to classify the evidence beyond nylon fibers if one is looking for animal hairs or polyester fibers. Stating that two objects share a class identity may indicate they come from a common source. What is meant by a "common source" depends on the material in question, the supply chain of production, and the precision of the examinations used to classify the object. The potential complexity of what constitutes a common source can be extensive. The enormous variety of mass produced goods, consumer choices, economic factors, biological and natural diversity, and other traits creates a nearly infinite combination of comparable characteristics for the items involved in any one situation (Houck, 2013a).

The degree to which traits are useful for comparison, discussed next, is their diagnosticity, that is, how well they classify an object. If a trait can reassign an object from one class to another class with fewer members, it is more salient than one that does not. Salience is ranked in order of how many members of a class share that feature. The goal in classification is to place an object, by successively more diagnostic features, into classes with successively fewer members. The salience of a feature should increase inversely with the number of members of a class into which that feature can place an object (Houck, 2013a, 2013b).

2.2.2 Comparison

Classification is the first step in a forensic examination; the next, assuming a shared class identity, is comparison. By itself, comparison seems a meager method; its strength is predicated on the classification that comes before it, placing the questioned and known samples into a class with few enough members that it warrants comparing the two on more specific traits. It is not enough to know that a fiber is nylon but why it has the traits associated with that polymer type, such as interference colors, birefringence, and spectral response. By classifying a fiber as nylon not only allows its comparison to other similar nylon fiber samples, but also a priori excludes it from other categories, such as acrylic, polyester, and so on. The relevance and context of the traits compared is paramount:

> Salience of the elements or traits is of prime importance: there are an innumerable number of arbitrary differences in either elements or relations that could be considered but are not useful given the question at hand ("Are both items smaller than the Empire State Building? Are they redder than a fire truck?"). Ultimately, analogy is a process to communicate that the two comparators (the source and the target) have some relationship in common despite any arbitrary differences. Some notion of possible or hypothetical connection must exist for the comparison to be made. As a forensic example, consider trace debris removed from the clothing of a suspect and the body of a victim: Although there may be no physical evidence (hairs, fibers, glass, soil, etc.) in common, the suspect's clothing and the victim's body have, at least prima facie, a common relationship (the victim is the victim and the suspect is a person of interest in the crime) until proven otherwise. Thus, common relations, not common objects, are essential to analogy and comparison.
>
> (Houck, 2013a, p. 316)

Characteristics of manufactured goods vary based on design, function, form, and costs, among other traits. Design takes intended end use, aesthetic concerns, and cost limitations, as well as price point; an objectonnection must exist for the comparison to d design variances group around necessary and sufficient criteria. Large trilobal fibers make good carpeting but poor garments, in a simple example. Most sets of things in the world to be compared are dissimilar, like "truck" and "duck." This is why forensic comparisons tend to be much stronger in exclusion than inclusion.

2.2.3 Transfer and persistence

Transfer and persistence are fundamental principles in forensic science and provide the keystone for interpreting every kind of evidence (Locard, 1928, 1930). Textile fibers are classic examples of the concepts. Transfer occurs when whole or partial materials leave a source or location and end up in another location, staying in the new location for some period of time. The loss of material from an original source is a type of entropy or taphonomic process; for example, cotton fibers breaking off from a yarn in a textile. A wide variety of factors influence what transfers, how much transfers, where it goes and how long it stays. Fiber type, textile composition, fabric construction, wear, fiber diameter and length, and many others all play a role in transfer and persistence (Roux, and Robertson, 2013). Whether the fibers originate from the fabric in question or are merely "transient" on that fabric can also affect how many fibers are transferred. Material type also influences transfer and persistence but hairs and fibers work under essentially similar environments for these purposes.

Transfer begets persistence. Once a fiber has transferred, it must be detected where it came to rest; if the persistence is so transient that the transfer cannot be detected, the interpretation may be that no transfer occurred. The recovery efforts are important and significant: Poor collection can lead one to believe no evidence was present when, in fact, it was. Loss of transferred materials is almost inevitable and is proportional to activity or motion of the substrate and environment. Loss can be as high as 80% of the transferred material in the first few hours to just a few percent left after 24 hours. On the other hand, fibers or hairs transferred to a stationary substrate, even one outdoors, can persist for some time. Shorter fibers tend to persist longer, especially if they are smaller than 2.5 mm in length. Smooth fibers do not persist as long as those with a roughed surface or crimp. Fibers on an item can redistribute themselves if the substrate is moved; individually packaging items of evidence is critical to successful fiber or hair casework and interpretation.

2.3 Textiles

Normally, any text on fiber examination starts with fibers, the basic units of a textile, and works its way up to fabrics and garments. These macro-level objects, however, are what are encountered in our daily lives as evidence and only through specific methods of collection and detection are fibers ever discovered and deciphered.

More attention is paid to the examination of fibers than to fabrics or, more broadly, textiles, to the detriment of both the examiner and the examination. Some fabrics shed more fibers than others and this sheddabilityo the detrime the detriment of both the examiner and the examination. Some fabrics shed more fibers ng of how fibers become textiles leads to a better understanding of what the fibers themselves actually mean.

2.3.1 Information

The analysis of textiles and fibers, like all forensic sciences, is one of history: What happened to this textile and where did these fibers come from? Classification provides brackets within which to understand possible sources of origin; comparison provides a correlation to potential sources. With textiles, the classification is relatively straightforward. Sourcing, however, is another matter. The production of fibers more than doubled between 1990 and 2010 (to over 102 billion pounds), with the majority of that being synthetic and manufactured (chemical) fibers (Figure 2.1).

Traditionally, companies owned much if not all of their supply chain. Called "vertical integration," it allowed companies to maximize control and efficiency of resources in the predictable economy of the early 1900s. As markets grew and customers had more choices, people became more selective about the products they purchased. Vertical integration was slow and not able to adapt to customer demand. Companies sought more flexibility with lower costs, which meant they needed to distribute their supply chains among other specialized companies. The economy reached a global scale as shipping costs plummeted (Levinson, 2008). This international network of interdependence, with each company doing what it did best, defined the new "virtual integration." The breadth of

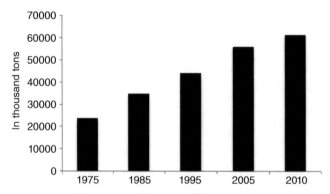

Figure 2.1 The production of textile fibers has grown significantly. This helps and hurts the forensic textile examiner: while innovation and faster product development times with shorter life spans means more diversity and, thus, specificity of sourcing, it also means that the world of textiles is complex and ephemeral. (Source: *Textile World.*)

virtual integration and the globalization of manufacturing lead to a dizzying complexity for even simplest of products:

> Say we get an order from a European retailer to produce 10,000 garments. It's not a simple matter of our Korean office sourcing Korean products or our Indonesian office sourcing Indonesian products. This customer we might decide to buy yarn from a Korean producer but have it woven and dyed in Taiwan. So we pick the yarn and ship it to Taiwan. The Japanese have the best zippers and buttons, but they manufacture them mostly in China. Okay, so we go to YKK, a big Japanese zipper manufacturer, and we order the right zippers from their Chinese plants. Then we determine that, because of quotas and labor conditions, the best place to make the garments is Thailand. So we ship everything there. And because the customer needs quick delivery, we may divide the order across five factories in Thailand... Five weeks after we have received the order, 10,000 garments arrive on the shelves in Europe...
>
> (Magretta, 2000, p. 36)

Six countries work to produce 10,000 garments for European stores in five weeks! Virtual integration vastly complicates product tracking and sourcing for the forensic scientist because of the increased number of separate companies involved in each product and intermediary step (dyeing and installing zippers, for example) rather than one company doing all of the work to complete a final product.

As more new textile products come to market, the complexity of the range of known fibers becomes more difficult to keep current with industry and to find any single source. Fashion companies can now take products from design to store shelves in a matter of weeks and fill orders for goods globally within 48 hours (The Economist, 2006; Chu, 2014). This presents the forensic fiber examiner with a dilemma. Increased innovation and diversity of materials and products means potentially greater specificity to sources; on the other hand, the "world of fibers" grows increasingly and confoundedly complex, making it more difficult to be an expert on the topic. Moreover, it begs the question: How can a system of this complexity be adequately modeled to allow for statistical interpretation?

The morphology and chemistry of textile fibers is discussed briefly in the next two sections. The microscopy and analysis of fibers is covered extensively elsewhere (for example, Robertson *et al.*, 2002) and is not covered in depth here. Advances in fiber interpretations and applications are discussed later in this section.

2.3.2 Morphology

A textile fiber is a unit of matter, either natural or manufactured, that forms the basic element of textiles and has a length at least 100 times its diameter (Hatch, 1993). Fibers differ from each other in their morphology, optical properties, chemistry, and color. Fibers are classified as either natural or manufactured. Natural fibers exist as fibrous materials in their natural state; cotton and wool are good examples. Manufactured fibers are made from materials that, at some point in the manufacturing process, are not fibers; rayon is a good example, being made from dissolved and regenerated cellulosic materials.

All fibers, natural and manufactured, are polymers, formed of hundreds or thousands of repeating chemical units (monomers) that are linked together.

The cross-section is the shape of an individual fiber when it is cut at a right angle to its long axis. Cross-section is an important diagnostic characteristic for fiber identification and sourcing. Natural fibers vary by cross-section and their shape and internal characteristics can distinguish otherwise similar fibers. Using morphology and cross-section to identify plant fibers is particularly useful because they are made entirely of cellulose and appear similar spectroscopically; this is also true of manufactured fibers made of regenerated cellulose, such as rayon or acetate. The microanatomy of plant fibers has been well described and studied (Smole *et al.*, 2013). All plant cells have primary walls that are heterogeneous in structure, being comprised of a composite of cellulose fibrils in a matrix of lignin and polysaccharides; they may be laminate in structure. Plant fibers generally exhibit differential crystallinity that varies by plant species and its fiber structure. The structure of a plant fiber varies with its source, as a seed fiber, a bast (stem) fiber, or a leaf fiber. Cotton is by far the most common plant fiber, if not the most common of all types of fiber.

Shapes for manufactured fibers vary by design; there are hundreds of different cross-sections used for manufactured fibers (Sawyer *et al.*, 2008) (Figure 2.2). The shape of fibers relates to their end use. Natural fibers tend to be used for certain products, such as cordage and rugs, more than others, such as upholstery. Manufactured fibers are designed and produced with their end use in mind (Shih *et al.*, 2014). The shape of

Figure 2.2 The cross-section of a manufactured fiber relates to its end use. About 500 cross-sections are used in the textile industry for various fibers types (*See insert for color representation of the figure.*)

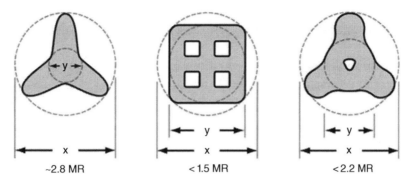

Figure 2.3 The modification ratio measures the outer circumference of a circle drawn around a fiber to the inner circumference. The ratio of the two describes the fibers shape (Source: Antron Carpet Fiber Technical Bulletin, 2011. Invista.)

a fiber produces emergent properties on the textiles they inhabit; surface characteristics, the "hand" (how the fabric feels), and fabric performance are all dependent on fiber cross-section (Bueno *et al.*, 2004; Behera and Singh, 2014). More information on this aspect of fibers can be found in the Fiber Reference Image Library at Ohio State University (https://fril.osu.edu/).

Textiles fibers are typically small in diameter, 10–50 μm, although their length can range from less than a centimeter to, literally, miles. Because of the variation in fiber cross-sections, measuring a fiber's diameter depends on its shape; there is more than one way to measure the diameter of a non-round fiber. A measurement called the modification ratio (MR) is used to describe manufactured fibers' diameter and shape, typically for those with highly non-round cross-sections, such as carpet fibers. The MR is calculated by comparing the circumference of a circle drawn around the outer limits of the fiber to a circle drawn around the inner of the fiber (Figure 2.3). Manufactured fibers can be made in diameters from about 6 μm (microfibers) up to a size limited only by the mechanical properties of the fiber's polymer. Natural fibers vary in diameter from cultivated silk (10–13 μm) to US sheep's wool (up to 40 μm or more); human scalp hairs range from 50 to 100 μm, by comparison.

Fibers are described either as filaments, which have an indefinite or extreme length, or as staple fibers, which are cut to specific lengths. All natural fibers are staple fibers except silk; manufactured fibers begin as filaments but may be cut to staple form (Figure 2.4).

Other fiber morphologies may also be diagnostic. Crimp is the waviness of a fiber. It may be two dimensional (like a wave) or three dimensional (like a corkscrew). Wool, for example, has a natural crimp to it because of the *para-* and *ortho-*molecular structures inside the fibers (Giansetti, 2014). Crimp must be imparted to manufactured fibers, either by production processes that mimic wool (as with bicomponent fibers) or through chemical or thermal changes, such as a "hot box" (Hatch, 1993).

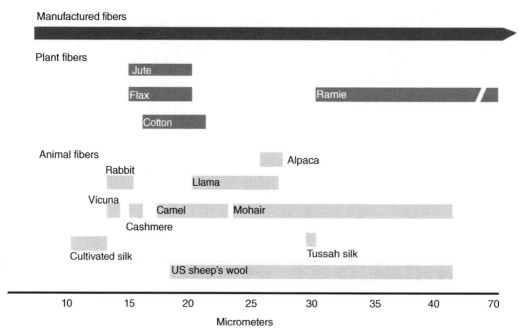

Figure 2.4 Although manufactured fibers can be of any length, natural fibers appear in a variety of lengths depending on their source and genetic variation. All natural fibers except silk are considered to be staple fibers

2.4 Natural fibers

Despite advances in chemistry and manufacturing, natural fibers predominate in textiles. Over half of the fibers produced each year are natural fibers, the majority of which are cotton. Natural fibers come from animals, plants, or minerals (Figure 2.5). The examination of natural fibers relies heavily on microscopy and observation, rather than instrumental analysis. This makes the job of recognizing and identifying the range of animal hairs (which "become" fibers when incorporated into textiles) and plant fibers found in textiles laborious but necessary.

2.4.1 Animal fibers

Animal fibers come either from mammals (hairs) or from certain invertebrates, such as the silkworm. Wool – the hair from sheep, goats, and their kin – and fur – from animals like rabbits, mink, and fox – are encountered most often in textiles. Animal hairs undergo extensive treatment prior to being used as fibers, including washing, chemical treatments, trimming, and dyeing. A reference collection is critical to animal hair identifications and comparisons. The microscopic anatomical structures of animal hairs are important

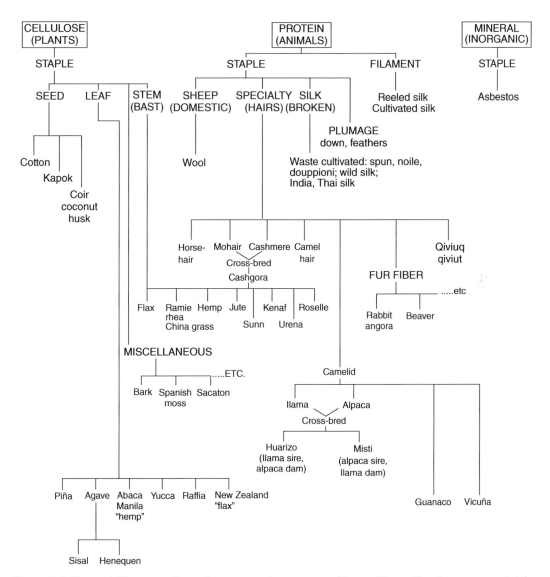

Figure 2.5 Natural fibers are those that appear in nature as fibers. Charts like these are useful for categorizing and making sense of the diversity of fibers. (Source: Humphries, M. (2009) *Fabric Reference*, 4th edn. Prentice Hall, Englewood Cliffs, NJ.)

to their identification (Brunner and Coman, 1974), although DNA methods are being developed in to identify textile animal fiber content (Kerkoff *et al.*, 2009).

Hairs are complicated, composite materials with intricately organized structures; Buffoli and coworkers (2014) have provided a good literature summary on hair, some of which is used here, as have Houck and Siegel (2010). Not all of a hair's microstructures are visible under a microscope and the range of forensically useful characteristics is

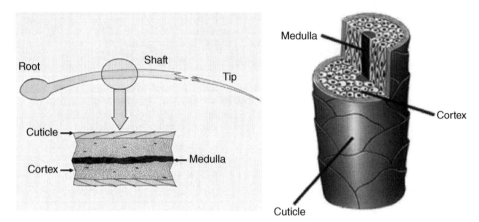

Figure 2.6 Macroscopically, hairs have a root, shaft, and tip. Microscopically, the three main structures are the cuticle, the cortex, and the medulla

somewhat limited. A single hair has a root, a shaft, and a tip. The root is that portion that resides or did reside in the skin, in a structure called the follicle. The shaft is the main portion of the hair and is described in more detail below. The tip is the distal most portion of the hair (Figure 2.6).

The three main structural elements in a hair are the cuticle, the cortex, and the medulla. The cuticle of a hair is a series of layers of flat, overlapping scales that cover the outer hair. Each cuticle cell (one scale) is about 0.3–0.5 μm thick and 50 μm long. The cuticle is important in protecting the hair from physical and chemical damage. Animal cuticle scales and patterns vary by species and may be a useful diagnostic for identifying animal hairs. The cortex makes up the majority of the hair and consists of spindle-shaped macrofibrils bundled within the hair matrix. Pigment granules are dispersed throughout the cortex and vary in size, shape, and distribution. Finally, the medulla is a structure that runs through the center of the hair and varies by species (Figure 2.7).

It is generally easy to distinguish between human and animal hairs by a simple microscopic examination. Determining the animal taxon, however, may not be as easy. Animal hairs are different from human hairs in a number of ways. Animals have three types of hair. Guard hairs are large, stiff hairs that make up the outer part of the animal's coat; these are the hairs that should be used for microscopic identification. Guard hairs may have a widening in the upper half of the shaft, called a shield. Thinner, softer fur hairs fill in the rest of the animal's coat, providing warmth and bulk. The morphology of fur hairs lacks specificity for determining species because it is too generic. Finally, animals have vibrissa (whiskers), the stiff, often white hairs around the eyes and muzzle. Guard hairs will be seen in fur coats and collars, as will fur hairs, but vibrissa are not used in textile manufacture. Wool-bearing animals have been bred for uniformity of hair size, making it easier to construct textiles from them; technically, they have no guard hairs. Certain animals' hair has color bands with abrupt pigmentation transitions along the shaft of the hair.

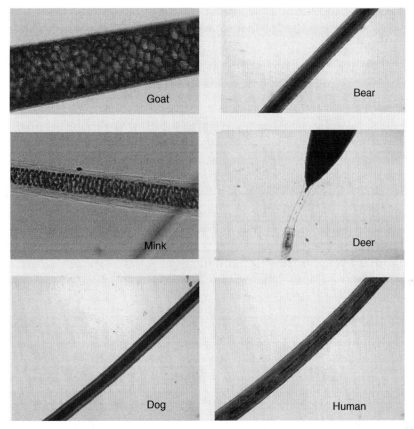

Figure 2.7 Animal hairs and humans are readily distinguished by their morphology, coloration, and medulla size (*See insert for color representation of the figure.*)

While the chemistry of animal hairs is complex, it is not useful information for forensic examinations, except in the most unusual circumstances. The interested reader should refer to Montagna and Ellis (2013).

2.4.2 Plant fibers

The three major sources for fibers derived from plants are the seed, stem (or bast), and leaf (refer again to Figure 2.5). Plant fibers are found in two principal forms: the technical fiber, used in cordage, sacks, mats, and so on, or individual cells, as in fabrics or paper. The examination of technical fibers should include a search for internal structures, such as the lumen (a central channel running through the middle of the fiber), spiral vessels (plant cells with helical walls), or crystals, and the preparation of a cross-section. Technical fibers should be mashed, fabrics teased apart, and paper repulped for the examination of individual cells. The following should be noted: the relative thickness of the cell walls and the size, shape, and thickness of the lumen; cell length; and the presence, type, and distribution of dislocations. The most common plant fibers encountered in casework are cotton, flax, jute, hemp, ramie, sisal, abaca, coir, and kapok.

Natural fibers are also used in wider applications, such as reinforcement in composites (Holbery and Houston, 2006), replacing manufactured fibers or other materials. Forensic scientists need to be aware of these uses and when they may encounter plant fibers apart from textiles.

2.5 Manufactured fibers

Manufactured fibers are the various families of fibers produced from fiber-forming substances, which may be synthesized polymers, modified or transformed natural polymers, or glass (Figure 2.8). Synthetic fibers are those manufactured fibers that are synthesized from chemical compounds (e.g., nylon, polyester). Therefore, all synthetic fibers are manufactured, but not all manufactured fibers are synthetic. Manufactured fibers differ physically in their shape, size, internal properties, and appearance, as well as their obvious chemical differences.

All fibers, natural or manufactured, are polymers, with hundreds or thousands of repeating chemical units, called monomers, linked together. Three polymer types occur in textile fibers. Homopolymers are made of a single monomer repeating along the polymer chain. Copolymers have two or more monomers that make up the polymer chain. Finally, block polymers have groups of homopolymers repeated along the chain. Homopolymers, the most common type, include acetate, aramid, cotton, nylon, olefin, PBI, polyester, rayon, silk, sulfar [poly(phenylene sulfide)], triacetate, and wool. Acrylic, modacrylic, saran [poly(vinylidene chloride) or PVDC)], and vinyon fibers [poly(vinyl chloride)] will have copolymers; the number and type of comonomers are part of the fiber the fiberr and type of comonomers are part of the fiberolyester, rayon, silk, sulfar-silk, Generic names for manufactured and synthetic fibers are given in Table 2.1.

Fibers are made from solid polymers that are rendered into a thick, viscous liquid, called spinning dope. The polymers are either melted, if they are thermoplastic synthetics, such as polyester, or dissolved chemically, such as the wood pulp that becomes rayon. The process for cellulosic fibers like rayon is extensive, involving immersion in caustic soda, mixing with carbon disulfide under controlled temperatures to form cellulose xanthate. A range of methods is used to make a variety of cellulosic rayons, such as high wet modulus, cupramonimum, and high tenacity rayon, among others.

The spinning dope is forced through a shower head-like device, called a spinneret. The spinneret has several to hundreds of tiny holes in it that create the fiber's cross-section. The extruded fibers form continuous filaments of semi-solid polymer. As the filaments emerge from the holes in the spinneret, the liquid polymer is not fully set and is soft; the newly made fibers need to be solidified before they can be used. The fibers may be processed at the same manufacturing plant or may be stored as tow (a kind of unstructured

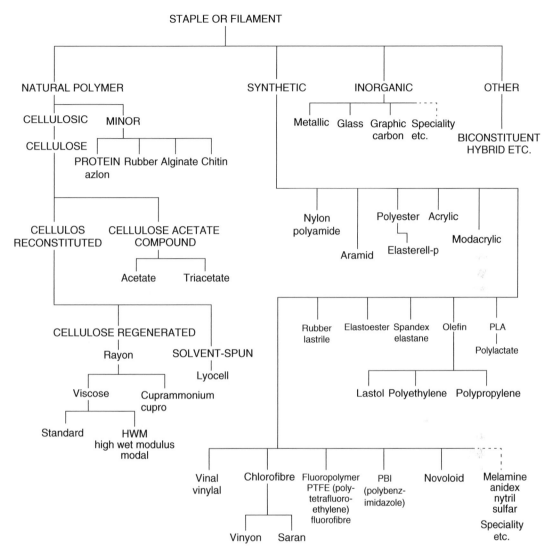

Figure 2.8 Manufactured fibers are rendered from other materials and then remade into a fibrous form. Some manufactured fibers are synthetic, like nylon and polyester. (Source: Humphries, M. (2009) *Fabric Reference*, 4th edn. Prentice Hall, Englewood Cliffs, NJ.)

rope) in large cans for shipping to another facility for further processing. The polymer at this stage is only lightly oriented and typically has to be drawn to increase its strength and durability. There are four main spinning methods: wet, dry, melt, and gel spinning.

Wet spinning is used for polymers that have been dissolved in a solvent. The spinneret is submerged in a chemical bath and the fibers precipitate from solution and solidify as they exit the spinneret head; this is where the method's name comes from. Wet spinning can be used to produce acrylic, rayon, aramid, modacrylic and spandex. Lyocell, a type

Table 2.1 Names and definitions of some common fiber types. (Source: Federal Trade Commission.)

Fiber name	Definition
Acetate	A manufactured fiber in which the fiber-forming substance is cellulose acetate. Where not less than 92% of the hydroxyl groups are acetylated the term triacetate may be used as a generic description of the fiber.
Acrylic	A manufactured fiber in which the fiber-forming substance is any long-chain synthetic polymer composed of at least 85% by weight of acrylonitrile units.
Anidex	A manufactured fiber in which the fiber-forming substance is any long-chain synthetic polymer composed of at least 50% by weight of one or more esters of a monohydric alcohol and acrylic acid.
Aramid	A manufactured fiber in which the fiber-forming substance is any long-chain synthetic polyamide in which at least 85% of the amide linkages are attached directly to two aromatic rings.
Glass	A manufactured fiber in which the fiber-forming substance is glass.
Nylon	A manufactured fiber in which the fiber-forming substance is any long-chain synthetic polyamide in which less than 85% of the amide linkages are attached directly to two aromatic rings.
Metallic	A manufactured fiber composed of metal, plastic-coated metal, metal-coated plastic, or a core completely covered by metal.
Modacrylic	A manufactured fiber in which the fiber-forming substance is any long-chain synthetic polymer composed of less than 85% but at least 35% by weight of acrylonitrile units.
Novoloid	A manufactured fiber in which the fiber-forming substance is any long-chain synthetic polymer composed of at least 85% of a long-chain polymer of vinylidene dinitrile where the vinylidene dinitrile content is no less than every other unit in the polymer chain.
Olefin	A manufactured fiber in which the fiber-forming substance is any long-chain synthetic polymer composed of at least 85% by weight of ethylene, propylene, or other olefin units.
Polyester	A manufactured fiber in which the fiber-forming substance is any long-chain synthetic polymer composed of at least 85% by weight of an ester or a substituted aromatic carboxylic acid, including but not restricted to substituted terephthalate units and para-substituted hydroxybenzoate units.
Rayon	A manufactured fiber composed of regenerated cellulose, as well as manufactured fibers composed of regenerated cellulose in which substituents have replaced not more than 15% of the hydrogens of the hydroxyl groups.
Lyocel:	A manufactured fiber composed of precipitated cellulose and produced by a solvent extrusion process where no chemical intermediates are formed.
Saran	A manufactured fiber in which the fiber-forming substance is any long-chain synthetic polymer composed of at least 80% by weight of vinylidene chloride units.
Spandex	A manufactured fiber in which the fiber-forming substance is any long-chain synthetic polymer composed of at least 85% of a segmented polyurethane.
Vinal	A manufactured fiber in which the fiber-forming substance is any long-chain synthetic polymer composed of at least 50% by weight of vinyl alcohol units and in which the total of the vinyl alcohol units and any one or more of the various acetal units is at least 85% by weight of the fiber.
Vinyon	A manufactured fiber in which the fiber-forming substance is any long-chain synthetic polymer composed of at least 85% by weight of vinyl chloride units.

of cellulosic fiber, is made by dry jet-wet spinning, where the fiber is extruded into air, drawn, and then submerged into the chemical bath. Wet spinning is the oldest method of manufacturing fibers.

Dry spinning is similarly to wet spinning except that the fiber is solidified by evaporating the solvent in a stream of air or inert gas blown over the fibers as they exit the spinneret. Acetate, triacetate, acrylic, modacrylic, PBI, spandex, and vinyon can be made by dry spinning.

For thermoplastic polymers, the fiber-forming substance, typically chips of a specific size, is melted in a hopper and then extruded through the spinneret, after which it cools, becoming more solid. This is called melt spinning, for obvious reasons. Nylon, olefin, polyester, saran and sulfar are produced by melt spinning. Being plastic, melt spun fibers can assume nearly any shape desired by changing the shape of the spinneret holes. Round, trilobal, pentagonal, octagonal, and many other shapes are possible. The shapes relate to the intended end uses. For example, trilobal fibers reflect light and make textiles appear lustrous, pentagonal and hollow carpet fibers show less soil and dirt, and other hollow fibers trap air and provide exceptional insulation in comforters and winter wear.

Gel spinning is used to create high strength fibers. The polymer is not in a true liquid state during extrusion and, thus, the polymer chains are still bound at various points, producing significantly increased tensile strength in the fiber. Liquid crystals in the fiber are more strongly aligned along the fiber's long axis by the shear forces at the spinneret hole during extrusion. Ultra-high-molecular weight polyethylenes, such as Spectra®, which has a strength-to-weight ratio 15 times higher than that of steel, and aramid fibers are produced by gel spinning. Gel spun fibers are used in body armor, cut-resistant gloves, and climbing equipment, among other specialty applications.

While newly made fibers are solidifying, the filaments may be drawn to impart strength. Tension on the fibers orients the crystalline portions of the fibers through viscous flow, increasing crystallinity and giving a considerably stronger fiber or yarn. Some fibers may be drawn many times their original length.

2.6 Yarns and fabrics

Yarn is a term for continuous strands of textile fibers, filaments, or material in a form suitable for weaving, knitting, or otherwise entangling to form a textile fabric. Staple fibers are too short to use without some form of entanglement and yarns allow for further construction. Staple fibers are entwined as they are twisted along an axis, binding together to form a strand. Yarns may be constructed to have an S-twist or Z-twist or zero twist. A yarn may be a single ply (like sewing thread) or constructed with a number of single yarns twisted together to form a plied yarn. Each ply will have its own twist as well as the overall twist of the plied yarn; these twists will be opposite of each other so the yarn does not fray. Originally, yarns were manufactured or spun by hand but this process is now highly mechanized and automated (Figure 2.9).

Figure 2.9 Yarns were originally spun by hand but the process is highly industrialized today. (Source: Wikimedia Commons.)

2.6.1 Fabric construction

Fabric is a textile structure produced by interlacing fibers with a substantial surface area in relation to its thickness. Fabrics are defined by their method of assembly. The three major types of fabrics are woven, knitted, and non-woven.

2.6.1.1 Woven fabrics

Fabrics have been woven for over 8000 years. Woven fabrics are made from two sets of yarns, a longer set called warp yarns and a shorter set called weft yarns, that are interlaced to produce a planar fabric. The patterning of the yarns determines the type of weave. Warp yarns run lengthwise to the fabric and weft yarns run crosswise. If one set of yarns is larger than the other, the weave is described as "unbalanced." An almost unlimited variety of constructions can be fashioned by weaving but nearly all fall into one of three basic types: plain, twill, or satin. In a plain weave, one warp yarn floats over one weft yarn and then passes under the next. In a twill weave, the warp yarn will float over 2–3 weft yarns and then pass under the weft. And, finally, satin weave has the warp yarn float over multiple weft yarns before passing under (Figure 2.10).

Complicated or figurative weaves are created on a jacquard loom, originally designed with punched hole cards to allow the needles to pass or to block them

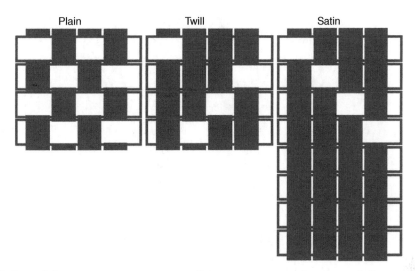

Figure 2.10 In a plain weave, one warp yarn floats over one weft yarn and then passes under the next. In a twill weave, the warp yarn will float over 2–3 weft yarns and then pass under the weft. And, finally, satin weave has the warp yarn float over multiple weft yarns before passing under

creating a pattern. Interestingly, these punched hole cards gave Charles Babbage the idea for programming his difference engine, an early computer; jacquard looms are now run by computers.

2.6.1.2 Knitted fabrics

Knitted fabrics are made of interlocking series of loops of one or more yarns and each series is called a stitch. Unlike woven fabrics, the yarns in knitted fabrics run along a meandering path, called a course and fall into two major categories: warp knitting and weft knitting. A sequence of stitches suspended from a previous course is called a wale (Figure 2.11). In warp knits the yarns generally run lengthwise in the fabric, whereas in weft knits the yarns generally run crosswise to the fabric. Unlike woven fabrics, in which warp and weft are made up of different sets of yarns, courses and wales are formed by a single yarn.

2.6.1.3 Non-woven fabrics

Non-woven fabrics are an assembly of textile fibers held together by mechanical interlocking in a random web or mat, by heat fusing of the fibers, or by bonding with a cementing medium. Felt is a common example but a wide variety of non-woven construction methods are used and other examples are bandage pads, automotive textiles, tote bags, medical fabrics, tea bags, and disposable wipes (Figure 2.12). Non-wovens tend to lack strength unless they are reinforced.

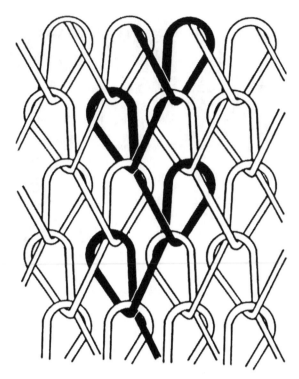

Figure 2.11 The basic pattern of warp knitting has yarns zigzag lengthwise along the fabric, with each loop securing a loop of an adjacent strand from the previous course. (Source: Wikimedia Commons.)

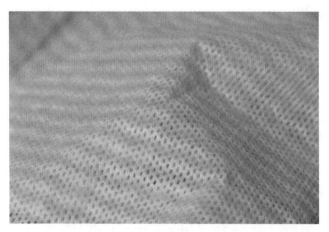

Figure 2.12 Non-woven fabrics are made by any method that can entangle the fibers and have them stay together. In this image of a disposable wipe, the individual fibers have been entangled by water and then pressed together by passing the fibers through rollers. The rollers have small studs which press the holes into the fabric, giving it more surface area and making the fabric more absorbent

2.6.2 Finishes

Finishes are treatments given to fibers, yarns, or fabrics to change or improve their characteristics. Aesthetic finishes change the way the fabric appears. Mechanical treatments can include brushing or calendaring (pressing the fabric between large rotating cylinders). Heat treatments include setting thermoplastic materials and singeing fabrics to remove loose or fuzzy fibers. Finally, chemicals, such as acids, bases, bleaches, polymers, and reactive resins, can be used to change the basic characteristics of the fabric. Functional finishes provide new or enhance existing fabric characteristics. Functional finishes include abrasion resistance, antibacterial properties, stain prevention, creasing, waterproof or repellent properties, flame resistance, and wrinkle resistance. Most fabrics have one or more of these treatments.

2.7 Fiber types

The generic names for manufactured and synthetic fibers were established as part of the Textile Fiber Products Identification Act enacted by Congress in 1954 (Table 2.1). In 1996, lyocel was named as a new, subgeneric class of rayon.

2.7.1 Acetate

The Celanese Corporation first produced acetate in the United States in 1924. To produce acetate, purified cellulose is dissolved from wood pulp by reacting it with acetic acid and acetic anhydride in the presence of sulfuric acid. The cellulose is then partially hydrolyzed to remove the sulfate as well as the appropriate number of acetate groups for the desired properties of the end product. The most common form of acetate, cellulose diacetate, has an acetate group on approximately two of every three hydroxyls. After the fiber is extruded and formed, the solvent is evaporated in warm air (dry spinning), leaving filaments of cellulose acetate.

Acetate does not readily accept dyes normally used for other cellulosic fibers, such as cotton and rayon. Because of this, specific dyes have been formulated for acetate. Rather than being a hindrance, acetate's dye selectivity can be cross-dyed, where yarns of one fiber, such as acetate, and those of another fiber, such as rayon, for example, are woven into a single material in a desired pattern. The fabric is dyed through one process, which reacts with some of the fibers but not the others, revealing the pattern. Acetate fibers can be solution-dyed, where the desired color is imparted as the fiber is extruded; this yields excellent color fastness even with prolonged exposure to sunlight or perspiration and repeated washings.

Major uses of acetate include blouses, dresses, linings, wedding and party attire, home furnishings, draperies, upholstery, and – the largest volume use – cigarette filters.

2.7.2 Acrylic

First commercially produced in 1950 by the DuPont Company, acrylic fibers are produced from acrylonitrile, a petrochemical. Comonomers are usually added to the polymer mix to improve dyeability; pure acrylonitrile is very difficult to dye. Acrylic fibers can be either dry spun or wet spun. Acrylic fibers are unique among synthetic fibers because when they emerge from the spinneret they have an uneven surface. Many bicomponent fibers are made from acrylics with various comonomers.

Acrylic was once regarded primarily as a synthetic replacement for wool and was used mostly in cold weather items, blankets, and sweaters. Acrylic works well in blended fabrics, expanding the look and feel of sweaters and extending their wear life. Acrylic fibers appear in sweaters, socks, fleece wear, circular knit apparel, sportswear and children's wear, blankets, area rugs, upholstery, luggage, awnings, and outdoor furniture. Acrylics can also be used as an asbestos replacement and reinforcement in concrete and stucco construction.

2.7.3 Aramids

Relative latecomers to the fiber industry, aramids were commercially produced in the United States in 1961 by the DuPont Company, which holds the production as a proprietary process. The word "aramid" is a portmanteau of "aromatic polyamide." Aramids are unique in that they handle like other textile fibers but neither ignite or melt in normal levels of oxygen. It has numerous military and public safety applications; Nomex and Kevlar are trade names for popular aramid products.

Aramids are polyamides derived from aromatic acids and amines; in this way, they are similar to nylons, which are aliphatic polyamides. Aramids have exceptional tensile strength and heat resistance because their aromatic rings are more stable and the amide linkages are far stronger because of conjugation with the aromatic structures. The polymer is spun to a solid fiber in a solvent, for example, anhydrous sulfuric acid.

The strength and thermal stability of aramids leads to their application in extreme environments. Aramids may be blended with other fibers for less-demanding uses. Aramids are used in bulletproof and other protective items for firefighters, police, and the military. Layering of aramid fabrics provides an outstanding level of protection against projectiles: a seven-layer aramid fabric can stop a .38-caliber bullet fired from 10 feet (3 meters). Firefighters, race car drivers, and other professionals who need heat and flame protection wear aramid protective clothing. Blended with other fibers, aramid fibers are used as boating sails. Skis with aramid fiber cores have increased flex life and give greater strength and better performance. Industrial uses include composites, asbestos replacement, hot air filtration fabrics, tire and mechanical rubber goods reinforcement, and ropes and cables.

2.7.4 Modacrylic

Union Carbide Corporation first commercially produced modacrylic fibers in 1949. Much like acrylic fibers, modacrylic fibers are made from copolymers of acrylonitrile and other polymers, such as vinyl chloride, vinylidene chloride, or vinyl bromide. Also like acrylic fibers, modacrylics can be either dry spun or wet spun.

Modacrylic fibers are inherently flame resistant and resist acids and alkalis. Modacrylics have low softening temperatures and can be molded or stretched into particular shapes;

if treated, their heat shrinkage can be controlled. If combinations of modacrylics with varying heat shrinkage control treatments are mixed in a fabric, heat treatment results in differential response (shrinkage) and a naturally appearing fur can be created.

Modacrylics are used in trims and linings, fake furs, wigs and hair pieces, children's sleepwear, specialty work apparel, fleeces, awnings, blankets, stuffed toys, carpets, flame-resistant draperies and curtains, and scatter rugs. Its flame and chemical resistant properties promote its industrial use in filters, industrial fabrics, and paint rollers.

2.7.5 Nylon

The DuPont Company first commercially produced nylon in 1939. Nylon is a generic term for a family of linear polyamides, each member being named for the number of carbon atoms separating the two acid groups and the two amines. One of the most common ways of producing nylon is by reacting molecules with an acid (COOH) group on each end with molecules containing amine (NH_2) groups on each end. This reaction results in a salt, known as nylon salt, with an exact 1:1 ratio of acid to base at room temperature. The salt is dried and heated under vacuum to dehydrate it and form the polymer. Nylon 6,6, widely used for fibers, is made from adipic acid and hexamethylene diamine, hence the name; DuPont patented nylon 6,6 in 1938. Another method of producing a nylon chain with repeating units of $(-NH-[CH_2]_n-CO-)_x$ is formed by polymerizing a compound containing an amine at one end and an acid at the other. If n = 5, the nylon is referred to as nylon 6, the second most common form. Commercially, nylon 6 starts with caprolactam using a ring-opening polymerization. Either way, the resulting polyamide melt is spun and drawn after cooling to give the desired properties for each intended use. Industrial and carpet fiber production is continuous, from polymerization, to spinning, and finally drawing. Nylon is relatively easy to dye.

Nylon is a dominant fiber type with a broad range of uses, including blouses, dresses, foundation garments, hosiery, lingerie, underwear, raincoats, ski apparel, windbreakers, swimwear, and cycle wear. High-filament nylon yarns are often blended with spandex and used in athletic apparel, swimwear, and hosiery. Carpets, curtains, and upholstery are common applications for nylon; nylon carpet yarn resists stains, hides soil, and resists mildew and bacteria. Its durability and strength promote its use in tire cords, hoses, conveyer and seat belts, parachutes, racket strings, ropes and nets, sleeping bags, tarpaulins, tents, thread, monofilament fishing line, and dental floss. In the Arctic, the Army uses three-dimensional nylon fabrics for insulated shelters that keep inside temperatures at 50°F when outside temperatures drop to as low as 65° below zero.

2.7.6 Olefins (polypropylene and polyethylene)

Hercules Incorporated first commercially produced olefin fibers in 1958; other specialized types followed in 1961. Olefin fibers, a broad term for both polypropylene and polyethylene, are produced by polymerizing propylene and ethylene gases. Polyethylene has a simple linear structure with repeating units. Polypropylene has a three-dimensional structure founded on a carbon backbone with methyl groups attached to it. For the olefin product to be used as fibers, controlled conditions are required for the polymerization with special catalysts that produce polymer chains with few branches. Olefins resist moisture and chemicals. They are very lightweight, having the lowest specific gravity of all fibers, allowing them to float, a property exploited in several of the fibers' applications. Olefins generally have a low melt point but polypropylene is the more popular for general textile applications, because it has a higher melting point than polyethylene. Olefins are easily thermally bonded, making them useful in non-woven fabrics. Olefins are difficult to dye and are generally dyed during the melt spinning, making them exceedingly colorfast. Olefins can be gel spun and result in highly specialized fibers with outstanding properties.

Olefins are used in active wear and sportswear, socks, thermal underwear, and lining fabrics. They find extensive use in automotive applications, including interior fabrics used in or on kick panels, seat construction, truck liners, and others. In the home, olefins can be found in indoor and outdoor carpets, carpet backing (olefin has largely replaced jute as backing for carpets), upholstery and wall coverings, furniture and bedding. Industrial applications include carpets, non-woven fabrics, ropes, filter fabrics, bagging, and geotextiles (erosion control and shoreline reinforcements, for example). Olefin and acrylic sandbags are used on highways as crash barriers and on levees to prevent flooding. Olefin fiber is used in artificial turf for athletic fields, disposable diapers, housing insulation, protective garments, road paving fabrics, and tear-resistant shipping envelopes.

2.7.7 Polyester

If "cotton is king," then polyester surely is queen: Polyester is the most widely sold manufactured fiber. First commercially produced in 1953 by DuPont Company, polyester is a category of polymers with an ester functional group as its main chain (Table 2.2). Although there are many kinds of polyester, the word typically refers to poly(ethylene terephthalate) (PET), the most common polyester for fiber applications. PET is made by reacting ethylene glycol with either terephthalic acid or its methyl ester in the presence of an antimony catalyst at high temperature and under vacuum. PET is melt spun.

Polyester is used in every form of clothing and household textile. Often spun with natural fibers, polyester can create fabrics with superior environmental resistance; the "classic blend" of 55% polyester/45% virgin wool is common. Other uses for polyester include hoses, power belting, ropes and nets, thread, tire cord, automotive upholstery, sails, and fiberfill for pillows, comforters, and winter wear.

2.7.8 Rayon

The first manufactured fiber, rayon was first commercially produced in the United States in 1910 by FMC Corporation and American Viscose (now Avtex Fibers, Inc.). Discovered by Georges Audemars in 1855, it took over 50 years to develop a viable commercial production scheme. Designed to replace silk, rayon is made from purified cellulose that is

Table 2.2 Types of polyesters. All polyesters have an ester functional group as their main chain. The polyesters that are highlighted are most commonly used for fibers

Composition of the main chain	Number of repeating units	Examples of polyesters	Examples of manufacturing methods
Aliphatic	Homopolymer	Polyglycolide or poly(glycolic acid) (PGA)	Polycondensation of glycolic acid
		Poly(lactic acid) (PLA)	Ring opening polymerization of lactide
		Polycaprolactone (PCL)	Ring opening polymerization of caprolactone
		Polyhydroxyalkanoate (PHA)	
		Polyhydroxybutyrate (PHB)	
		Poly(ethylene adipate) (PEA)	
	Copolymer	Poly(butylene succinate) (PBS)	Polycondensation of succinic acid with 1,4-butanediol
		Poly[(3-hydroxybutyrate)-co-(3-hydroxyvalerate)] (PHBV)	Copolymerization of 3-hydroxybutanoic acid and 3-hydroxypentanoic acid, butyrolactone and valerolactone (oligomericaluminoxane as a catalyst)
Semi-aromatic	Copolymer	Poly(ethylene terephthalate) (PET)	Polycondensation of terephthalic acid with ethylene glycol
		Poly(butylene terephthalate) (PBT)	Polycondensation of terephthalic acid with 1,4-butanediol
		Poly(trimethylene terephthalate) (PTT)	Polycondensation of terephthalic acid with 1,3-propanediol
		Poly(ethylene naphthalate) (PEN)	Polycondensation of at least one naphthalenedicarboxylic acid with ethylene glycol
Aromatic	Copolymer	Vectran	Polycondensation of 4-hydroxybenzoic acid and 6-hydroxynaphthalene-2-carboxylic acid

dissolved to a solution and pumped through the spinneret to form soft filaments. Wood, cotton sweepings, broken waste rayon, and many other forms of cellulose can be used to create staple fiber rayon. Postextrusion, the fibers are regenerated into almost pure cellulose; for this reason, rayon is called a regenerated cellulosic fiber. Rayon is found in a variety of forms that are named according to how the cellulose is solubilized and then regenerated. Rayon fibers are wet spun.

2.7.8.1 The viscose process

Most commercial rayon manufacturing today utilizes the viscose process. This process dates to the early 1900s, with most of the growth in production occurring between 1925 and 1955. In the early period, production was mainly textile filament, although the first staple was produced in 1916. High performance rayons, such as tire cord, did not appear until the late 1930s, with the advent of hot stretching and addition of larger amounts of zinc to the spin bath. Invention of modifiers in 1947 brought on super tire cords and marked the beginning of the high-performance rayon fibers.

The viscose process is the most common method for making rayon and is complicated. For purposes of this description, it will be considered a batch process, although advances in technology have allowed for semi-continuous operation.

Specially processed wood pulp, termed "dissolving cellulose" or "dissolving pulp", is purified for rayon production. Lower grade pulps are used for papermaking and other products; while they could technically be used for fibers, the fineness of the spinneret holes and their tendency to clog prevents lower grade pulp from being used in fiber production. Dissolving cellulose is made of long-chain molecules, generally free from lignin, hemicelluloses, or other short-chain carbohydrates that could foul the process. The cellulose sheets are soaked in caustic soda or sodium hydroxide long enough for soda cellulose, the sodium salt of cellulose, to form. The salt helps control oxidation of the cellulose chains, making sure that cellulose xanthate is formed. The soda cellulose is pressed out of the sheets to remove excess and then the sheets are shredded; at this stage, the cellulose is called "white crumb".

The white crumb is allowed to partially oxidize and degrade to lower molecular weights. To end up with the proper physical properties for a useful fiber, degradation must be carefully controlled, otherwise, the chain lengths will be too short and the viscosities of the spinning dope will not be manageable. After aging for 2–3 days, the white crumb is mixed with gaseous carbon disulfide, forming xanthate ester groups. The crumb is still alkaline and in reacting with the carbon disulfide picks up a yellow color; the material is, therefore, now called "yellow crumb". At this stage, the yellow crumb is basically a block copolymer of cellulose and cellulose xanthate. The yellow crumb is dissolved in aqueous caustic solution. The chains are separated and water pushes them further apart, making the otherwise insoluble cellulose a partially solubilized suspension. Because it is not completely solubilized, the solution has a very high viscosity, hence the name of the process and the original product, viscose.

The viscose is let stand to "ripen," allowing carbon disulfide to escape. This reduces the solubility of the cellulose just enough to permit adequate regeneration and formation of fibers after emerging from the spinneret. After ripening, the viscose is filtered, degassed (to reduce voids, which would cause weak spots in the fiber), and then wet spun into a solution of sulfuric acid, sodium sulfate and, typically, Zn^{2+} ions. The cellulose is regenerated and precipitated from solution as the water diffuses out, the cellulose chains are drawn together, and the rest of the carbon disulfide is lost. The rayon fibers are drawn immediately while the cellulose chains are still flexible, forming interchain hydrogen bonds, and the fiber develops the properties needed as a fiber.

2.7.8.2 Other forms of rayon

The Federal Trade Commission (FTC) has classified other forms of regenerated cellulose fibers as rayon but not as separate subclasses as it did with lyocel. High wet modulus rayon is a highly modified viscose rayon that performs more like cotton than regular rayon does. It is easier to care for and is stronger than regular rayon; it can provide wrinkle resistance without additional finishes. Cuprammonium rayon is made through a process that combines the cellulose with copper and ammonia to make it soluble in caustic soda. The material is wet spun into a bath that removes the copper and ammonia and neutralizes the caustic soda. Cuprammonium rayon makes very fine filaments for lightweight clothing, such as summer dresses and blouses; it may be blended with cotton to make textured fabrics with nubby, uneven surfaces. Fabric made from cuprammonium rayon may be called Bemberg, although that is a trade name owned by the J.P. Bemberg Company. High tenacity rayon is produced by further drawing the rayon after spinning, which yields rayon with twice the strength and over 60% of the stretch of regular rayon.

Rayon is used in a wide variety of apparel, such as blouses, dresses, jackets, lingerie, linings, hats, slacks, sport shirts, sportswear, suit linings, and ties. Many home furnishings employ rayon, for example, bedding, window treatments, slipcovers, tablecloths, and upholstery. It can also be found in an array of disparate applications, such as medical surgical products, non-woven products (disposable wipes, for example), tire cord, and even feminine hygiene products. Flame-resistant rayon/wool blends are used in commercial airline seats.

2.7.9 Spandex

Spandex (also called elastane elsewhere) was first produced commercially in the United States in 1959 by the DuPont Company. When it was introduced to the public in 1962, it revolutionized large segments of the clothing industry. In 2010, an estimated 80% of clothing sold in the United States contained spandex (Penaloza, 2010). Spandex's unique characteristic is that it can be stretched to almost 500% of its length and then regain its original shape. The polymer chain is a segmented block copolymer with long, amorphous segments and short, rigid segments. The rigid segments act as "virtual cross-links," networking all the polymer chains. The interconnectedness of the network stops the polymer chains from slipping past each other and taking on a new shape. When stretched, the bonds between the rigid sections are broken and the amorphous sections straighten out. When the fiber is relaxed, the amorphous segments return to their at-rest state. Spandex is formed in a multistep proprietary process that involves extensive production of a prepolymer (Lewin, 2006); spandex may be spun in a number of methods, including melt, reaction, and solution dry spinning, the latter being by far the most common. It is spun into a monofilament or, more commonly, into multiple fine filaments that are coalesced into a single thread line.

In clothing, spandex provides a combination of comfort and fit, preventing bagging. Spandex can be heat-set, allowing it to be formed into permanent shapes. It is readily dyeable and can be spun into sizes ranging from 10 to 2500 denier. Spandex is found in hosiery, swimsuits, athletic wear, ski pants, golf jackets, disposable diaper, waistbands, and components in undergarments.

2.7.10 Triacetate

The Celanese Corporation first commercially produced triacetate fibers in 1954; triacetate is no longer produced in the United States. Triacetate is derived from cellulose by acetylating cellulose with acetic acid and acetate anhydride. Methylene chloride and methanol are used to dissolve the cellulose acetate into spinning dope. Triacetate is dry spun and results in a fiber with a higher ratio of acetate-to-cellulose than acetate. Triacetate fibers tend to be weaker than acetate but are less vulnerable to melting (during ironing, for example); it holds a permanent pleat. Triacetate is used in dresses, skirts, and sportswear.

2.7.11 Bicomponent fibers

While not an FTC category, bicomponent fibers are commonly enough encountered to warrant at least a brief discussion. A bicomponent fiber is made of two polymers of different chemical or physical properties spun from the same spinneret within the same filament. The pairing of polymers is designed to provide suitability for the end product, such as a heat-setting polymer surrounding one with higher strength to make a thermally bonded non-woven fabric. Off-set thermally labile sheaths can also be made to curl and take on the physical conformation of wool, mimicking the *para-* and *ortho*-structures in the animal's hair; for example:

- polyester core (250°C melting point) with copolyester sheath (melting points of 110–220°C);
- polyester core (250°C melting point) with polyethylene sheath (130°C melting point);
- polypropylene core (175°C melting point) with polyethylene sheath (130°C melting point).

Most commercially available bicomponent fibers are made in a sheath/core, side-by-side, or eccentric sheath/core arrangement (Figure 2.13).

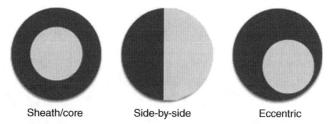

Sheath/core   Side-by-side   Eccentric

Figure 2.13 A bicomponent fiber is made of two polymers of different chemical or physical properties spun from the same spinneret within the same filament. Most commercially available bicomponent fibers are made in a sheath/core, side-by-side, or eccentric sheath/core arrangement

2.8 Chemistry

2.8.1 General analysis

Fibers are traditionally analyzed by a variety of methods, including bright-field microscopy, polarized microscopy, fluorescent microscopy, Fourier transform infrared spectroscopy (FTIR), microspectrophotometry in the UV and visible ranges, among other methods (Carroll, 1992; Robertson *et al.*, 2002). Other methods include solubility and melting point but these have limited relevance for forensic work, as they are destructive and forensic science typically receives very limited samples. The analytical methods available for fiber examination and analysis yield different kinds of information. To maximize resources and to minimize time, the analyst should apply the methods in an order that provides the most exclusionary information first. Fibers should be identified to their FTC generic class or species (nylon 6 verus nylon 6,6, for example), examined for their physical characteristics, and then analyzed for color or dyes (Figure 2.14). In general, at least two methods from each listing should be used to develop concurring results.

At a minimum, a stereomicroscope, a comparison microscope, and a compound light microscope equipped with polarized light capability should be used for fiber examination. Once identified to generic class or better, a comparison of samples from known and questioned sources is conducted to determine whether they exhibit the same characteristics and properties in all tested respects. This comparison involves the recognition and evaluation of class characteristics, which associate materials to a group but never to a single source (Section 2.8.5). Microscopic examination is the quickest, most accurate, and least destructive means of determining the characteristics and polymer class of textile fibers (SWGMAT, 1999).

Optical properties, such as refractive index, birefringence, and color are those traits that relate to a fiber's structure or treatment revealed through observation (Houck and Siegel, 2010). Some of these characteristics aid in the identification of the generic polymer class of manufactured fibers. Others, such as color, are critical discriminators of fibers that have been dyed or chemically finished. A visual and analytical assessment of

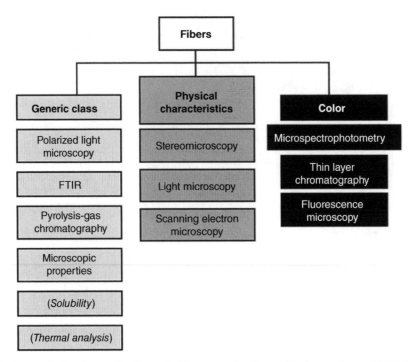

Figure 2.14 An analysis scheme for forensic fiber examinations. (Redrawn from SWGMAT, 1999.)

fiber color must be part of every fiber comparison. The fluorescence of fibers and their dyes is another useful point of comparison.

As-spun fibers tend to be very shiny and lustrous, called "bright" in textile parlance. Delusterants are finely ground particles that are introduced into the spinning dope to reduce the fiber's brightness. Materials such as titanium dioxide are ground to 0.1–1.0 µm in size and are added up to about 2% of the fiber content (Lewin, 2006). These particles help to diffract light passing through the fibers and reduce their luster. The size, shape, distribution, and concentration of delusterants should be noted in a forensic examination.

2.8.2 Instrumental analysis

Based on a fiber's polymer composition, it will react differently to various instrumental methods, such as FTIR or pyrolysis-gas chromatography (P-GC). The application of these methods has been dealt with in some detail in the literature for many years (Tungol *et al.*, 1995; SWGMAT, 1999; Robertson *et al.*, 2002). FTIR microscopy is a powerful method to characterize the chemical composition of natural and synthetic fibers, whether organic or inorganic. Modern instruments have a high lateral resolution and even a single fiber (about 100 µm in length) is a sufficient sample to analyze (SWGMAT, 1999). Importantly, FTIR is non-destructive and samples can be reanalyzed or subjected to further methods. Minimal sample preparation is required with the attenuated total reflectance (ATR) technique.

2.8.3 Color

Color is one of the most diagnostic characteristics in a fiber examination. Almost all manufacturing industries are concerned with product appearance and everything that is manufactured has a color, either naturally or imparted. Certain colors are chosen for some products rather than others for aesthetic reasons – "safety cone orange" is not popular as a carpet color, for example – and these colors may suggest the source product. A nearly infinite range of colors can be produced in textile fibers and the visual and optical analysis of fiber color is easy and repeatable (Blackledge, 2007). A great deal of work has been done on textile color analysis in recent years and should lead to more useful interpretations of fiber evidence (Morgan, 2014).

Colored fibers are either dyed or are pigmented. A dye is an organic chemical that absorbs and reflects certain wavelengths of visible light. Pigments are water insoluble particles incorporated into the fiber during production or bonded to the surface of the fiber; some fibre types, like olefins, are not easily dyed and are often pigmented. Dozens of dyers world-wide are registered with the American Association of Textile Chemists and Colorists (AATCC) and almost hundreds of trademarked dyes are registered with the Association; some dyes have as many as 40 variants. More than 7000 dyes and pigments are currently produced worldwide (Apsell, 1981). Most textiles are dyed with multiple dyes; even a simple dye may be put through 8–10 processing steps to achieve a final dye form, shade, and strength. Upon reflection, it is apparent that continuously dyeing fibers to a specific standard is virtually impossible and batch dyeing is the rule rather than the exception. This variability works in the forensic scientists favor, as the number of producible colors is nearly infinite, difficult to get exactly the same, and color is an easy discriminator.

The three main methods of evaluating color in fibers are visual examination, chemical analysis, and instrumental analysis. While the first and easiest should be a visual examination using a variety of microscopes, this is a qualitative technique to sort the likely samples from the exclusions. Instrumental analysis of possible positive associations must follow a visual examination. The benefits are that it is quick and a simple comparison is an excellent screening technique. However, human color perception is variable and subject to day-to-day and observer-to-observer variations. Color vision acuity varies and as many as 6% of males have some form of color vision deficiency (Wong, 2011). Moreover, color vision changes with age. Along with annual proficiency testing, color testing, even as simple as the Ishihara Color Test (Figure 2.15), should be mandatory for all forensic scientists annually.

After a visual examination, color must be evaluated instrumentally or the dyes analyzed chemically. Spectrophotometric analysis in the ultraviolet and visible ranges is the standard method in forensic science. The microspectrophotometer (MSP) is an instrument that measures color of minute samples, including individual fibers. The MSP is a standard spectrophotometer with a microscope attached to focus on the sample. A spectrophotometer compares the amount of light passing through air with the amount of light transmitted through or reflected off a sample. The percentage of light reflected or transmitted is expressed as a ratio, which is calculated at each

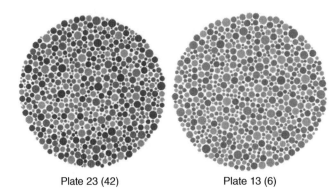

Plate 23 (42)   Plate 13 (6)

Figure 2.15 Two plates from the Ishihara color vision test. The correct answers are shown in parentheses. (Source: Wikimedia Commons.) (*See insert for color representation of the figure.*)

wavelength of the spectrum analyzed. Most materials are heterogenous at the microscopic scale and require spectral scanning at multiple locations on a fiber or fibers to generate representative values for the whole sample in question. Single fibers may not have taken up the dye(s) uniformly and natural fibers exhibit even greater variance due to non-uniform cross-sections along their length and their inherent heterogeneity. Real and apparent variations in dyeing shade at different places along a fiber are possible. Sampling, even at the microscopic level, is critical to a valid analysis.

While not technically color analysis, the chemical analysis of the dyes in fibers involves extracting the dye and characterizing or identifying its chemistry. Thin layer chromatography (TLC) is a common method although others may be successfully employed. Because the entire process of dyeing a fiber is geared to making the fiber–dye combination colorfast, it can be difficult to extract a dye from a fiber. Very light or very small fibers have little dye in them and poor results are common. Dye analysis is also a destructive method.

2.8.4 Raman spectroscopy

Raman spectroscopy is coming into its own for fiber and fiber-dye analysis. Although not common in most forensic laboratories in the United States, it is used quite common in Europe and elsewhere. It has the capability of detecting not only polymer type but also dye concentrations as low as 0.005% for some dyes (Massonnet *et al.*, 2012) with ease of process and less sample handling (Yu and Sandercock, 2012). After polymer type, color is probably the best distinguishing characteristic of a fiber. Raman spectroscopic studies have reinforced the results of traditional UV-Vis microspectrophotometry color analysis and can aid that process by detecting minor dye components missed by MSP:

> For these reasons Raman spectroscopy is a very convenient technique to confirm or perhaps clarify MSP results, especially for fiber types with common MSP spectra. Furthermore, MSP-Vis also showed some limitations with very light or very dark colored fibers whereas Raman spectroscopy could still discriminate between fiber types.

> (Palmer, 2014, p. 193)

In situ analysis of single fibers is accurate, non-destructive, and can be highly discriminatory. For example, Appalaneni and coworkers (2014) analyzed single fibers dyed to closely color match with acid blue (AB) 25 and 41 and direct blue (DB) 1 and 53 and correctly identified fibers with no false positives. As Raman spectroscopy increases in utility and as instrumentation prices drop, it is anticipated that this instrument will be found more often in forensic laboratories.

2.8.5 Interpretation

DNA is a relatively simple molecule that behaves in a predictable pattern and is easily interpreted by population frequencies based on a limited sampling of that population (Butler, 2005). Fibers, on the other hand, comprise dozens of polymer types, dyed with thousands of dyes to produce millions of colors, and are finished with numerous treatments for fashion and use (Burkinshaw, 1995). Fiber production is a global process with minimal useful tracking or documentation at the forensic level; for example, how many nylon 6,6 fibers with a trilobal cross section 16 μm in diameter were dyed with Direct Orange 23 dye to a specific shade (which can be affected by many factors, such as pH, soaps, dye uptake, pretreatment, and others)? This is essentially an unknowable number. The range and potential combinations of fibers, dyes, and other factors precludes an accurate accounting of fibers with traditional data basing, including popularity of specific retail items, distribution systems, short productions runs, and counterfeits. Even if a database could be created (as has been done by various groups), the event horizon of textile products and their distribution preclude any database being comprehensive over time.

If a comprehensive or even representative database is not forthcoming, what does a positive fiber association mean? Sampling studies to develop an understanding of fiber frequencies include looking for specific fibers on a wide variety of objects (target fiber studies), cross-checking fibers in particular locations (car seats, for example; Roux and Margot, 1997), and frequency studies. Burd and Kirk performed the first target fiber study in 1941. In this study, 193 bolts of cloth used for making men's suits were used to provide 26 distinct types of blue fibers to act as target fibers. Then, the bolts of cloth were examined to see how many colors of the blue fibers were present. They found that less than 6% of any one color matched the cloths, basing their comparisons only on the results of comparison microscopy. Many such studies have been conducted over the years, with increasing complexity. A study of unrelated textiles from random criminal cases intercompared 2043 fibers; none had the same optical properties and microscopic characteristics in over two million comparisons (Houck, 2003). Another study calculated the frequency of finding at least one red woolen fiber on a car seat at 5.1%; if more than five are found, however, the relative frequency plummets to 1.4% (Roux and Margot, 1997). The authors of that study determined that, except for white, indigo (denim), or grey/black cotton, no fiber should be considered as "common." A recent study was able to discriminate 31 different samples of single polyester fibers from automobile trunk liners (Hiroma, 2010). Fabric finishing can also aid in the distinctiveness of fiber evidence.

For example, "nubs," the small bits of polymer left on garments that have been "singed" (a thermal process that smooths the fabric), lend additional evidential value because not all fabrics are treated by this method (Deedrick, 2001). Beyond as-manufactured traits, use can also help to distinguish fibers and their environments (source level). Comparison of residual surfactants from different laundering soaps helped to characterize white cotton fibers, which are otherwise indistinguishable, by detergent type (Kato *et al.*, 2011).

All of these studies demonstrate that it is exceedingly rare for two fibers at random to have the same microscopic and chemical traits. Although fibers are still class-level evidence, the class of potential sources can be exceedingly small.

2.9 The future

For textiles, the future is diverse, fast, and complex. Staying abreast of innovative fiber and fashion technology is barely possible. Technologies change faster than one can keep up and, of course, consumer demands for the "next big thing" push the textile industry further and faster:

> Every market into which the consumer's fashion sense has insinuated itself is, by that very token, subject to this common, compelling need for unceasing change in the styling of its goods...No single style of design, no matter how brilliantly it is conceived, can claim any independent fashion significance at all, nor can it possess more than a fugitive lease on life.
>
> (Robinson, 1965, p. 52)

Fibers' "lease on life" can make or break a forensic case. The chemistry of fibers puts them squarely in the sciences, with both qualitative and quantitative methods readily available and, in fact, expanding. Interpretation is still complex but there are signs of hope on the horizon with statistics derived from multimodal instrumental analysis (Morgan, 2014).

References

Appalaneni, K., Heider, E.C., Moore, A.F., and Campiglia, A.D. (2014). Single fiber identification with nondestructive excitation–emission spectral cluster analysis. *Analytical Chemistry*, **86** (14), 6774–6780.

Apsell, P. (1981) What are dyes? What is dyeing? 1981. In: *Dyeing Primer* (ed. J.R. Aspland), American Association of Textile Chemists and Colorists, Research Triangle Park, NC.

Behera, B. K., and Singh, M. K. (2014) Role of filament cross-section in properties of PET multi-filament yarn and fabric. Part II: effect of fibre cross-sectional shapes on fabric hand. *The Journal of the Textile Institute*, **105** (4), 365–376.

Blackledge, R. (2007) *Forensic Analysis on the Cutting Edge: New Methods for Trace Evidence Analysis*. John Wiley & Sons, Inc., Hoboken, NJ.

Brunner, H. and Coman, B. J. (1974) *The Identification of Mammalian Hair*. Inkata Press, Melbourne, Australia.

Buchli, V. and Lucas, G. (2001) The absent present, in *Archaeologies of the Contemporary Past* (eds V. Buchli and G. Lucas), Routledge, New York, pp. 3–18.

Bueno, M.A., Aneja, A.P., and Renner, M. (2004) Influence of the shape of fiber cross-section on fabric surface characteristics. *Journal of Materials Science*, **39** (2), 557–564.

Buffoli, B., Rinaldi, F., Labanca, M., *et al.* (2014) The human hair: from anatomy to physiology. *International Journal of Dermatology*, **53** (3), 331–341.

Burd, D.Q. and Kirk, P.L. (1941). Clothing fibers as evidence –A study of the frequency of occurrence of blue wool fibers. *Journal of Criminal Law and Criminology*, **32**, 353–357.

Burkinshaw, S. (1995) *Chemical Principles of Synthetic Fibre Dyeing*. Springer, New York.

Butler, J.M. (2005) *Forensic DNA Typing: Biology, Technology, and Genetics of STR Markers*. Academic Press.

Carroll, G.R. (1992) Forensic fibre microscopy, in *Forensic Examination of Fibres* (ed. J. Robertson), Ellis Horwood, New York, NY, pp. 99–126.

Chu, K. (2014) Why Zara is a 'fast-fashion' pioneer. *Wall Street Journal*; http://blogs.wsj.com/corporate-intelligence/2014/06/24/why-zara-is-a-fast-fashion-pioneer/ (last accessed 6 June 2015).

Cook, R., Evett, I.W., Jackson, G., *et al.* (1998) A hierarchy of propositions: deciding which level to address in casework. *Science and Justice*, **38** (4), 231–239.

Cuvier, G. (1832) *The Animal Kingdom Arranged in Conformity with its Organization*. Treater and Co., London.

Deadman, H. (1984) Fiber evidence and the Wayne Williams trial. *FBI Law Enforcement Bulletin*, March and May, 1–17.

Deedrick, D. (2001) Fabric processing and 'nubs', in *Mute Witnesses: Trace Evidence Analysis*, (ed. M. Houck), Academic Press, New York, pp. 1–20.

Giansetti, M. (2014) *Study and Scientific Rationalization of the Last Finishing Stages for High Quality Wool Fabrics*. PhD dissertation, Politecnico di Torino, Italy.

Ginzberg, C. (1979) Clues: Roots of a Scientific Paradigm. *Theory and Society*, **7** (3), 273–288.

Green, F. (2014) State forensic lab cuts back on some services to police, *Richmond Times-Dispatch*; http://www.richmond.com/news/state-forensic-lab-cuts-back-on-some-services-to-police/article_03876d75-6c29-5559-99de-8f514b9ac083.html (last accessed 6 June 2015).

Hatch, K. (1993) *Textile Science*. West Publishing, St. Paul, MN.

Hiroma, Y., Hokura, A., and Nakai, I. (2010) The forensic identification of trunk mat fibres by trace element analysis of single fibres with Laser Ablation ICP-MS. *Bunseki Kagaku*, **59** (9), 759–769.

Holbery, J. and Houston, D. (2006) Natural-fiber-reinforced polymer composites in automotive applications. *JOM (The Journal of The Minerals, Metals & Materials Society)*, **58** (11), 80–86.

Houck, M.M. (2003) Inter-comparison of unrelated textiles. Forensic Science International, **135** (2): 146–149.

Houck, M.M. (2013a) Interpretation/The Comparative Method, in *Encyclopedia of Forensic Sciences*, 2nd edn (eds J.A. Siegel and P.J. Saukko), Elsevier, Amsterdam, pp. 314–317.

Houck, M.M. (2013b) Forensic Classification of Evidence, in *Encyclopedia of Forensic Sciences*, 2nd edn (eds J.A. Siegel and P.J. Saukko), Elsevier, Amsterdam, pp. 318–321.

Houck, M.M. and Siegel, J.A. (2010) *Fundamentals of Forensic Science*. Elsevier, Amsterdam.

Kato, T., Hasegawa, M., and Kagawa, M. (2011) The discrimination of white cotton fibres by the detection of residual surfactants. *Japanese Journal of Forensic Science and Technology*, **16** (1), 29–42.

Kerkhoff, K., Cescutti, G., Kruse, L., and Müssig, J. (2009) Development of a DNA-analytical method for the identification of animal hair fibers in textiles. *Textile Research Journal*, **79** (1), 69–75.

Levinson, M. (2008) *The Box: How the Shipping Container Made the World Smaller and the Economy Bigger*. Princeton University Press, Princeton, NJ.

Lewin, M. (2006) *Handbook of Fiber Chemistry*, 3rd edn. CRC Press, Boca Raton, FL.

Locard, E. (1928) Dust and its analysis: An aid to criminal investigation. *Police Journal*, **1**, 177.

Locard, E. (1930) The analysis of dust traces. Part I. *American Journal of Police Science*, **1** (3), 276–298.

Magretta, J. (2000) Fast, global, and entrepreneurial: Supply chain management, Hong Kong style, an interview with Victor Fung, in *Harvard Business Review on Managing the Value Chain*, Harvard Business Press, Cambridge, MA, pp. 29–59.

Massonnet, G., Hemmings, J., Leijenhorst, H., *et al.* (2012) Raman spectroscopy and microspectrophotometry of reactive dyes on cotton fibres: Analysis and detection limits. *Forensic Science International*, **222**, 200–207.

Montagna, W. and Ellis, R.A. (eds) (2013) *The Biology of Hair Growth*. Elsevier, Amsterdam.

Morgan, S. (2014) *Evaluation of Statistical Measures for Fiber Comparisons: Interlaboratory Studies and Forensic Databases*. Report 248386. National Institute of Justice: Washington, DC.

Palmer, R. (2014) The forensic examination of fibres and textiles review: 2010–2013, in *17th Interpol International Forensic Science Managers Symposium Review Papers* (ed. N. Nic Daiéd); http://www.interpol.int/INTERPOL-expertise/Forensics/International-Forensic-Science-Symposium2 (last accessed 6 June 2015).

Penaloza, M. (2010) Spandex stretches to meet U.S. waistlines. National Public Radio, December 11. http://www.npr.org/templates/transcript/transcript.php?storyId=143003539 (last accessed 6 June 2015).

Robertson, J., Roux, C., and Wiggins, K. (2002) *Forensic Examination of Fibres*, 2nd edn. CRC Press, Boca Raton, FL.

Robinson, D. (1965) The meaning of fashion, in *Inside the Fashion Business*, (eds J. Jarnow, B. Judelle, and M. Guerreiro), John Wiley & Sons, Inc., New York, p. 52.

Roux, C. and Margot, P. (1997) An attempt to assess the relevance of textile fibres recovered from car seats. *Science and Justice*, **37**, 225–230.

Roux, C. and Robertson, J. (2013) Transfer, in *Encyclopedia of Forensic Sciences*, 2nd edn (eds J. Siegel and P. Saukko), Elsevier, Amsterdam, pp. 113–116.

Rudwick, M. (1998) *Georges Cuvier, Fossil Bones, and Geological Catastrophes: New Translations and Interpretation of the Primary Texts*. University of Chicago Press, Chicago, IL.

Sawyer, L., Grubb, D., and Meyers, G. (2008) *Polymer Microscopy*. Springer, New York.

Shih, W.Y.C., Agrafiotes, K., and Sinha, P. (2014) New product development by a textile and apparel manufacturer: a case study from Taiwan. *The Journal of the Textile Institute*, **105** (9), 905–919.

Smole, M., Hribernik, S., Kleinschek, K., and Kreže, T. (2013) Plant fibres for textile and technical applications, in *Advances in Agrophysical Research* (ed. S. Grundas), InTech, doi: 10.5772/52372; http://www.intechopen.com/books/advances-in-agrophysical-research/plant-fibres-for-textile-and-technical-applications (last accessed 6 June 2015).

SWGMAT (1999) Forensic fiber examination guidelines. *Forensic Science Communications*, **1** (1); http://www.fbi.gov/about-us/lab/forensic-science-communications/fsc/april1999/houcktoc.htm (last accessed 6 June 2015).

The Economist (2006) Shining examples: How three large and successful companies are using their supply chains to compete. *The Economist*, **June 15**, S1–20.

Tungol, M.W., Bartick, E.G., and Montaser, A. (1995) Forensic examination of synthetic textile fibers by microscopic infrared spectrometry. *Practical Spectroscopy Series*, **19**, 245–245.

Yu, M. and Sandercock, M. (2012) Principal component analysis and analysis of variance on the effects of 'Entellan New' on the Raman spectra of fibres. *Journal of Forensic Sciences*, **57** (1), 70–74.

Wong, B. (2011) Points of view: Color blindness. *Nature Methods*, **8** (6), 441.

CHAPTER 3
Paint and coatings examination

Paul Kirkbride

School of Chemical and Physical Sciences, Flinders University, Australia

3.1 Introduction

Many manufactured articles that are involved in crime, for example tools used in breaking and entering, the objects themselves that are broken and entered into, and vehicles that are involved in accidents or ram-raids, have some form of coating that acts to protect, aesthetically improve, or add some special functionality to the article. Usually these coatings are paint.

Although current paint formulations are extremely sophisticated and include many advanced modern materials, the basic concept of paint – a liquid mixture that hardens into a solid coating after its application to a surface – has been known for many millennia. Neolithic cave paintings have been found in Europe and there are records of Australian aboriginal usage of paint from well before that time.

In common with modern paints, ancient paints contained a binder, which is the material that forms the film and causes it to adhere to an object. Binders, ancient or modern, are either transparent or have an unattractive hue; therefore, paints also include pigments, which are materials that confer color or whiteness to the paint as well as other desirable properties, such as opacity, or to bulk-out or "extend" the film. In this chapter the definition of a pigment will follow that referred to in the Paint and Coating Testing Manual (Koleske, 2012), that is, a pigment is any material that is colored, black or white, organic or inorganic that retains a crystalline form in paint. The fact that pigments maintain a crystalline form in the liquid paint and the solidified film is what distinguishes them from dyes, which are soluble in the liquid paint. By definition, extenders are also pigments. While many ancient pigments are still in use today (such as iron oxide), many new pigments are now in use and traditional binders (such as egg products, animal fats, blood, drying oils, etc.) have been supplanted by modern polymeric resins and chemically

Forensic Chemistry: Fundamentals and Applications, First Edition. Edited by Jay A. Siegel.

modified oils. In addition to binder and pigment, modern paints usually also contain a variety of additives that are added in small quantities to protect the film from attack by UV light or mildew, to modify its viscosity, to improve its antirust performance or to improve the rate at which the film hardens.

Paint is, therefore, a complex mixture of organic and inorganic compounds, some of which are present in many tens of per cent by weight, some of which are present at a level of less than 1%, and some of which are present adventitiously as impurities in the main ingredients. The paint industry has its own techniques for paint analysis but many of these techniques are related to industrial processes, such as controlling the production process or examining the performance of dried films under exposure to water, UV radiation, and so on. The aim of industrial analysis is not the same as the typical forensic aim, which is to compare one paint against another in order to ascertain whether there is an association between them, nor does industrial analysis involve the examination of microscopic fragments that have attached to a surface as a result of contact. These forensic paint comparison tasks usually require specialist techniques, which are described in this chapter. Whilst the techniques of paint analysis are fundamental to forensic paint comparison, knowledge of the limitations of technique and conveying the relevance and probative value of analytical results to law enforcement investigators and triers of fact is arguably more important. Both of these are also discussed in this chapter.

A paint specimen for which the origin is known will be referred to as a "known specimen" in this chapter. This usage is synonymous with the terms "reference," "control," or "exemplar". Conversely, a paint specimen whose original source is unknown will be referred to as an "unknown specimen" in this chapter rather than the other common usage "questioned sample". Just two types of paints – automotive paints and architectural paints – are discussed.

3.2 Paint chemistry

3.2.1 Binders

When liquid paint is applied to a surface one of two things happens to the film-forming polymers present. In "enamel," "stoving" and "two-pack" paints, the polymer in the film that forms on the surface has a molecular weight much higher than that in the liquid film. This arises through cross-linking reactions that form bonds between polymer molecules that are already quite high in molecular weight (these are referred to as "pre-polymers" in this chapter). The term "non-convertible" is used to refer to these paints that cross-link and form a film. Films such as these can be thought of as being composed of a single molecule and are, therefore, very hard wearing and resistant to chemicals.

In lacquers and domestic latex (or emulsion) paints, the polymer retains its original nature and "simply" precipitates out on the surface while the solvents that make the liquid paint a liquid evaporate. In latex paints, where the polymer is dispersed in the liquid paint as tiny droplets (i.e., the paint is an emulsion) simple evaporation results in a collection of discrete "beads" on the surface. A continuous film is produced when an additive in the paint, a coalescing agent, causes the "beads" to merge and polymer chains to become entangled with one another, thus forming a robust film. Paints such as these are called "convertible." In some modern domestic latex paints a small amount of cross-linking might also take place as a result of oxygen absorption from air or as a result of the effects of UV light in exterior finishes. It is important to draw a distinction between domestic latex paints and the water-based automotive paints emulsions that are also called latexes. As indicated later, extensive cross-linking takes place in all modern factory automotive paints, even those that use latex formulations. It is also important to realize that the word latex in any paint application does not indicate that the paint contains natural rubber latex.

A very important group of non-convertible paints, and one that has had a long history, is the alkyd paint group. In an alkyd the prepolymer is a resin formulated from a polyester and a monoglyceride, which, in turn, is prepared by the transesterification reaction between an oil and a polyhydric alcohol such as glycerol, penterythritol, 1,6-hexanediol, 2-methyl-1,3-propanediol, neopentylglycol, 2,2-dimethyl-4-methylpentan-1,3-diol, cyclohexanedi-methanol or trimethylolpropane (see Figure 3.1, in which glycerol and linseed oil are used as the example in the formation of the triglyceride, and ethylene glycol and phthalic anhydride are used to prepare the polyester).

Figure 3.1 shows a reaction between preformed monoglycerides and polyesters. This is for the sake of simplicity; in the industrial process it is usual for the transesterification to be commenced and for the reagents that form the polyester to be added at some time later. As indicated above, the range of alcohols, dibasic carboxylic acids and acid anhy-drides from which alkyd resins are derived is very broad. Furthermore, when alcohols such as glycerol and pentethyritol are used, further variation is possible because the proportion of free to esterified hydroxy groups can be controlled, which effectively controls the extent of cross-linking in (or the average molecular weight of) the alkyd prepolymer. In regards to the monoglyceride, a great deal of variation on the theme outlined in Figure 3.1 is possible. First, the oils that act as feedstock for the monoglyceride are so-called "drying" oils, which have multiple sites of unsaturation as illustrated in Figure 3.1 for linseed oil. The reason why these are called "drying" oils and why they are important is explained later. It is sufficient to indicate here that these oils are natural triglyceride esters formed by the esterification of glycerol by fatty acids and they usually are derived from plant sources, such as linseed and soya, but can also originate from sources such as fish. These oils naturally are mixtures of a range of triglycerides formed from a range of fatty acids. As indicated above, a number of polyhydric alcohols can be used to formulate monoglyc-erides and although the reaction is depicted as going to completion in Figure 3.1 in reality in the paint factory that is not the case. What is produced is a mixture, which in the case of glycerol is a mixture of monoglyceride, diglyceride and some unchanged oil. When all the chemical permutations and combinations are considered it can be seen that there

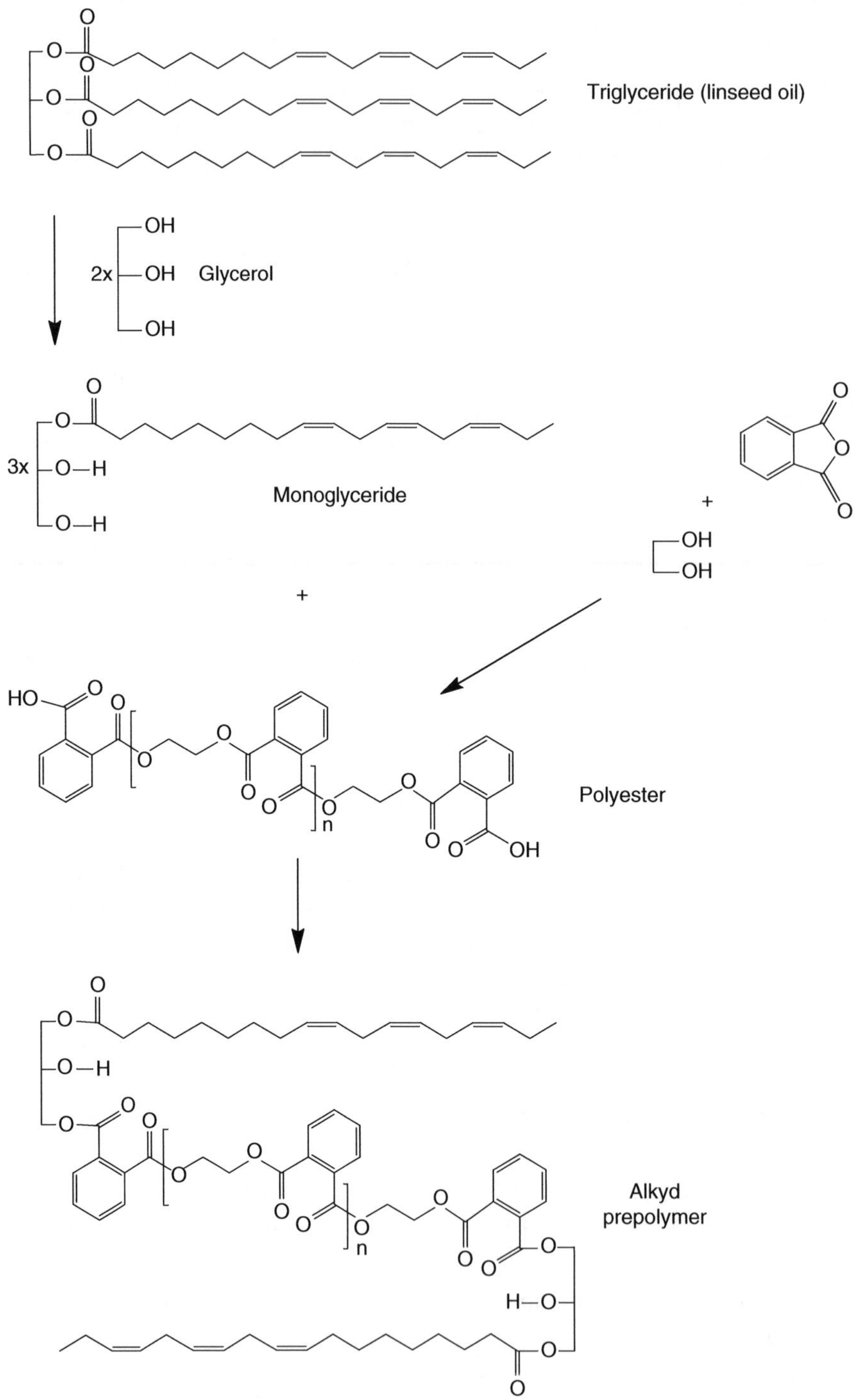

Figure 3.1 Reaction between the major drying oil present in linseed oil (a triglyceride comprising glycerol and α-linolenic acid residues) and glycerol to form a monoglyceride. Note that one molecule of linseed oil is treated with two molecules of glycerol to form three molecules of monoglyceride. Subsequent reaction between the monoglyceride and a polyester formed by the reaction between phthalic anhydride and ethylene glycol is also shown, together with the final alkyd prepolymer

is great scope for variation within the broad family of alkyd prepolymers, and even within a particular formulation there is a very complex blend of macromolecules. In any event, all alkyd prepolymers will have termini that include hydroxyl, carboxyl and fatty acid moieties; this is the binder prepolymer ready to form a paint.

Alkyds are important automotive finishes. In that application, some chemistry that takes place at high temperature is employed to crosslink the prepolymer and make a durable coat. This makes a "stoving" or "baked enamel" finish (called this because a stove or oven is used to bake the paint) and the identification of such a paint in an investigation indicates that an original equipment manufacturer (or OEM) paint is involved. The usual "stoving" additive is called "melamine" but is actually an etherified melamine of the type indicated in Figure 3.2, where R is usually butyl. When liquid melamine-alkyd paint is placed into a stove, any free hydroxyl or carboxyl groups in the paint act as nucleophiles towards the melamine and displace its alkoxy group to effect cross-linking (top half of Figure 3.2). An alcohol (butanol if R = butyl in Figure 3.2) is formed, which boils off from the film at the elevated temperature. What is produced is a glossy, hard finish. What takes place

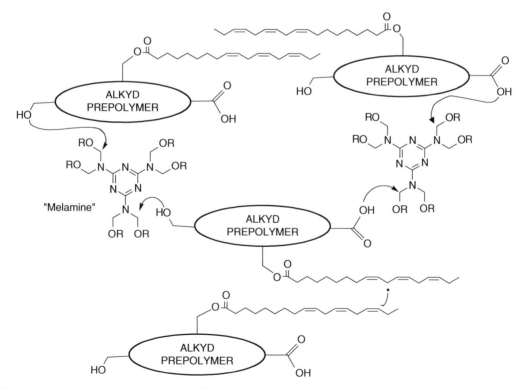

Figure 3.2 Stylized diagrams of the alkyd prepolymer showing its reactive hydroxyl, carboxyl and fatty acid moieties. The top half of the figure depicts the cross-linking reaction between "melamine" and the nucleophilic moieties attached to the prepolymer. The bottom half shows the oxygen catalyzed free radical cross-linking of the prepolymer

next is a further, slow cross-linking of the paint film mediated by the "drying" oil residues present. The critical feature of these oils is that they have sites of unsaturation (double bonds) that have the dual properties of (i) being able to react with atmospheric oxygen to form hydroperoxides that decompose to form free radicals and (ii) taking part in cross-linking reactions with those free radicals (Figure 3.2, bottom half). Certain metallic salts of carboxylic acids (such as zinc, manganese, lead or cobalt naphthenate) catalyze cross-linking. This additional cross-linking step or "air-drying" capability is why the oils used in alkyd paints are called "drying" oils. The process continues for many months. Quite functional alkyd paints can be made without the addition of melamine and the need for stoving. These paints, which dry purely by the cross-linking of the drying oil residues present, are used in domestic architectural applications (where they are called oil-based paints or enamel paints) or in automotive resprays.

High temperatures and extended reaction times are required to prepare the alkyd pre-polymer, which can be detrimental to the drying properties of the paint. Instead of treating monoglycerides with relatively unreactive carboxylic acids and anhydrides it is possible to react them with more reactive compounds – isocyanates – that are very reactive towards oxygen and nitrogen nucleophiles. What is formed are polyurethane (PU)-modified monoglycerides, also called polyurethane oils (Figure 3.3). Polyurethane oils can be included in the general reaction scheme outlined in Figure 3.1. These PU-modified

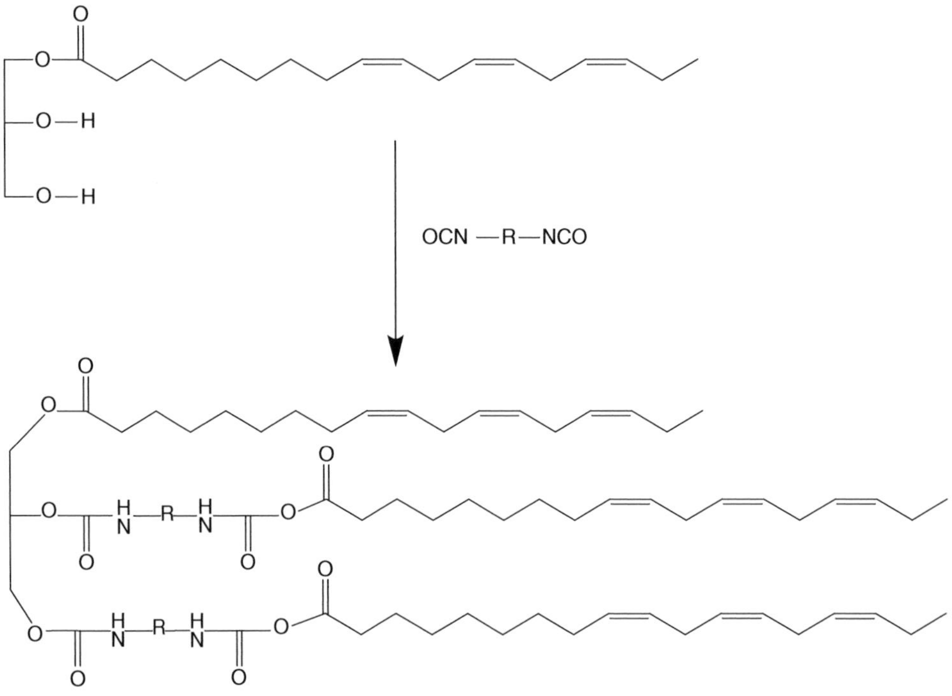

Figure 3.3 The formation of a polyurethane oil

alkyds produce very durable, air-drying finishes that are useful in applications where hard-wearing properties are required, such as finishes for flooring.

The reactive properties of isocyanates and the resilience of their PU products have been exploited further in the production of a number of other types of paint that make use of isocyanates. The American Society for Testing Materials classifies polyurethane coatings into six types, four of which are described here (polyurethane lacquers are omitted and urethane oils have been described above).

Figure 3.4 depicts the production of a simple moisture-curing or "one-pack" (1K) polyurethane paint. The first step in the process is the reaction between a polyhydric alcohol (which can be many and various, including castor oil or polyethers) and a multifunctional di-isocyanate, usually toluene di-isocyanate (TDI). The result is a film-forming tri-isocyanate of relatively high molecular weight. When it is applied to a surface the liquid paint reacts with atmospheric moisture as depicted in the bottom half of Figure 3.4. The first step of cross-linking is attack on a free isocyanate by water to form an unstable carbamic acid intermediate. This intermediate in turn undergoes decarboxylation reaction to form a nucleophilic species that, in turn, attacks a free isocyanate, accomplishing the second step of cross-linking.

One-pack PU paints are often used in floor varnishes, where their toughness is valuable. A limitation of these paints is their curing mechanism – it is very hard to exclude moisture from the can to stop the paint from hardening before use and in conditions of low humidity film cure time could be excessive. A very important type of 1K PU paint makes use of "blocked" isocyanate formulations. In the blocked formulations the free isocyanate groups have been treated with a compound (such as butanone oxime) that forms a more stable adduct that is not sensitive to moisture and, therefore, has a longer shelf life and is much easier to handle. Paints that employ blocked isocyanates have a polyalcohol blended in that "unmasks" or "unblocks" the isocyanate, but only at elevated temperatures. These types of paints are, therefore, an example of "stoving" paints. The masking agents confer such stability to the prepolymer that they can be produced as water-based paints, which are attractive to automotive manufacturers, as factory organic emissions can be greatly reduced.

As well as one-pack PU paints there are two-pack (or 2K) versions, where two reactive components are mixed together immediately prior to application. One type of 2K finish involves one component formulated with the chemistry outlined in Figure 3.4 and the other component containing a tertiary amine catalyst. The amine causes cross-linking analogous to that depicted in Figure 3.4 but which occurs more reliably in dry atmospheres and quite quickly. The other type of 2k finish makes use of the reactivity of isocyanates towards oxygen nucleophiles, in particular hydroxyl groups. In these finishes, one pack is a polyisocyanate as described previously. The other pack is a hydroxyl-rich resin formulated from polyol-polybasic acid polyesters, or 2-hydroxyethyl methacrylate (see the following discussion relating to acrylic resins) or, less frequently, polyethers. When the two packs are mixed together, hydroxyl and isocyanate moieties react to form a

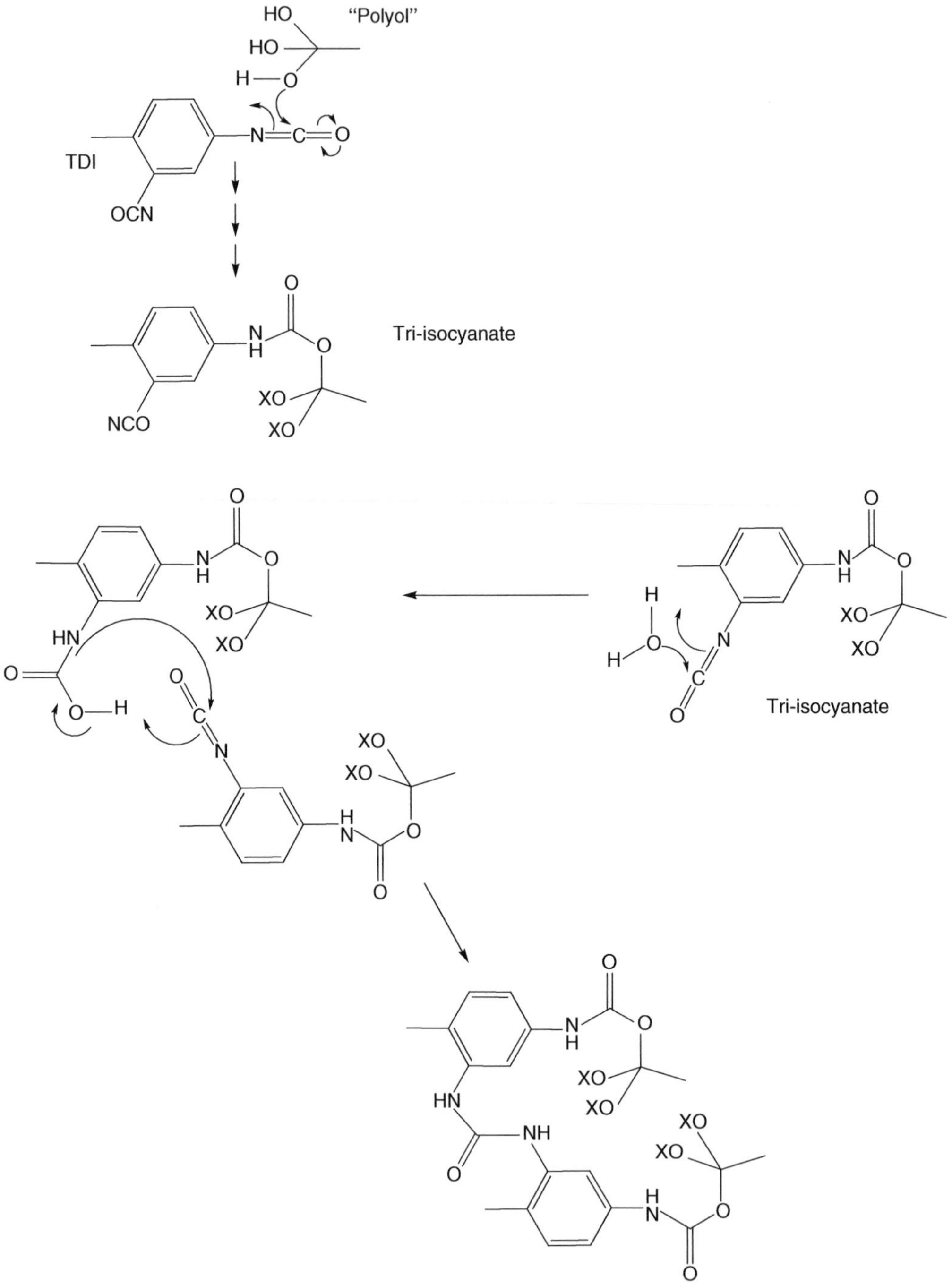

Figure 3.4 A polyol reacts with three equivalents of toluene di-isocyanate (TDI) to form a tri-isocyanate (top) where X represents arylcarbamoyl residues from TDI. When exposed to moisture (center) the tri-isocyanate forms an unstable carbamic acid that releases carbon dioxide and produces a nucleophile that attacks another tri-isocyanate. (For clarity not all electron movements are illustrated in bottom part of the figure.)

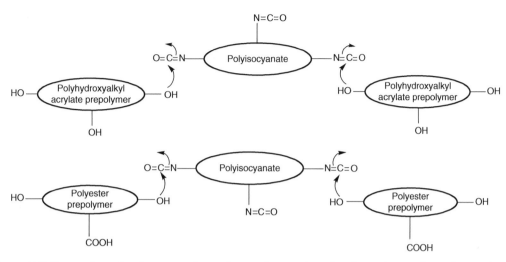

Figure 3.5 Formation of polyester polyurethanes (bottom) and polyacrylate polyurethanes (top). (For clarity not all electron movements are illustrated.)

cross-linked polyurethane film (Figure 3.5). The acrylic polyurethanes are often used in high quality automotive resprays or OEM finishes on items that cannot be stoved, such as bumpers, wing mirrors, and so on.

Acrylic resins are extremely important in both automotive and architectural finishes. These resins are based upon monomers that have the acrylate moiety in common and one of a number of functionalities at the α-position (e.g., methacrylate resins) or within the alkoxy group of acrylate esters (e.g., methyl, ethyl, 2-ethylhexyl, butyl, hydroxy-ethyl esters) (Figure 3.6). When treated with a free radical initiator the acrylate mono-mers readily polymerize to form acrylate resins. These resins can be formulated into latexes for architectural applications, dissolved in solvents to form acrylic lacquers for automotive application (used as OEM finishes in years gone by and in some modern refinishes), and suspended in vehicles for modern artists' paints. As already indicated, the polymer in a lacquer paint film has the same molecular weight distribution as the polymer in the liquid paint and, therefore, the same chemical properties, in particular solubility. As a consequence, these types of acrylic paints are not particularly resistant to solvents. The durability of acrylic paints can be improved by the inclusion of "melamine" and by "stoving." In order for an acrylic resin to react with "melamine" it must have a nucleophilic moiety – acrylate resins destined for "stoving" usually have hydroxyethyl-methacrylate (HEMA) blended into the polymer (Figure 3.6) or epoxy resins (see later). Acrylate and methacrylate resins find application in water-based emulsion paints for domestic architectural applications (so-called "acrylic emulsions" or "acrylic latexes"). Methacrylate resins are quite glassy and confer hardness to the paint while acrylates are softer and confer flexibility; often a blend of the two is present in paint.

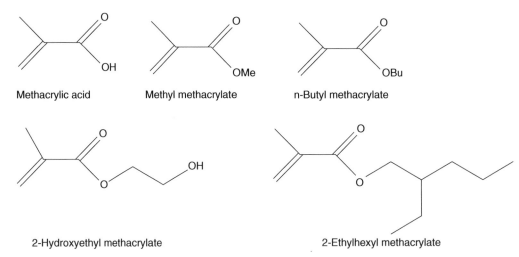

Methacrylic acid   Methyl methacrylate   n-Butyl methacrylate

2-Hydroxyethyl methacrylate   2-Ethylhexyl methacrylate

Figure 3.6 Common acrylic resin monomers

Another simple type of resin, which also finds extremely wide application in water-based emulsion paints for domestic interior architectural applications, is polyvinylacetate or PVAc. Although polyvinylacetate is quite soft, it is also quite glassy and requires the addition of other monomers (such as ethyl acrylate, 2-ethylhexylacrylate, dibutylma-leate or a substance going by the trade name vinyl versatate™) or high boiling point solvents (such as phthalate esters) to act as plasticizers (Figure 3.7). As previously indicated, polyhydric alcohols (usually Texanol™) are included in the latex paint to encourage coalescence of the emulsion after its application and after evaporation of some of the water base. Vinyl acetate can also be blended with vinyl chloride and ethylene chloride to form a water-dispersable copolymer of extremely high viscosity or jelly-like consistency that finds application as a spill-proof, solid emulsion timber undercoat. The acrylic resins are much more resistant to weathering than those based on polyvinylace-tate and are they are used in exterior water-based emulsion paints.

A wide variety of paints used in automotive and architectural applications make use of epoxy resins. Figure 3.8 illustrates a basic reaction that is used in the preparation of many epoxy resins used in paint – the reaction between the bifunctional electrophile epichlorohydrin and nucleophilic oxygen in substrates such as phenols but possibly also carboxylic acids and alcohols. Both the hydroxyl and epoxy moieties in epoxy resins can react with unsaturated fatty acids, which produces air-curing paints resem-bling alkyds. This can be taken one step further through the introduction of "melamine" for automotive stoving applications and through the introduction of acrylic acid copolymer into acrylate resins, which allows cross-linking between carboxylic acid and epoxy termini to form the so-called "acid-epoxy" formulations. An alternative chemistry involves copolymerization of glycidyl methacrylate (Figure 3.8) with other acrylate esters in order to produce resins with diverse functionality for cross-linking.

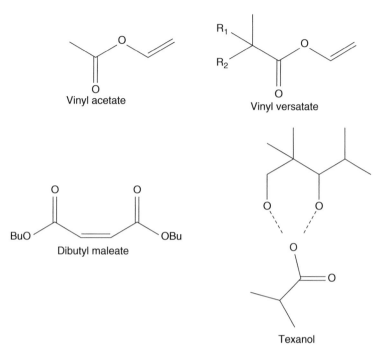

Figure 3.7 Compounds related to PVA emulsion paints. The dotted lines in the Texanol™ structure indicate that it is a mixture comprising the 1- and 3-isobutyrate esters. (Vinyl versatate™ is a trade name and the alkyl substituents are not specified.)

Similar to the alkyds and polyurethanes, the versatility of epoxide chemistry and the production of hydroxyl moieties when epoxides react allows for many permutations and combinations in paint formulation. In automotive applications epoxy paints have gained wide usage in primers but acrylic epoxies are offered as clearcoats as well (Ryland *et al.*, 2001; Suzuki, 2014a).

The natural polymer cellulose, modified in two ways, can be encountered in paint. When treated with a mixture of nitric acid and sulfuric acids, nitrocellulose is produced. A high level of nitration (>12% nitrogen content) produces an energetic material that is used as ammunition propellant but a lower level of nitration (<12%) produces a polymer that makes an effective lacquer for furniture or automotive refinishing (de la Ossa *et al.*, 2011). Cellulose can be esterified with a number of anhydrides to produce a number of important polymers, such as cellulose acetate and cellulose acetate butyrate; the latter finds application in paints. The proportion of free hydroxyl, acetate and butyrate groups determines the properties of the polymer, with higher proportions of butyration leading to greater insolubility in organic solvents and water resistance and higher levels of free hydroxyl conferring higher solubility in organic solvents, especially alcohols (Koleske, 2012). These properties can be modified further by the addition of carboxymethyl groups (i.e., to form carboxymethylcellulose actetate butyrate, CMCAB) that enable the resin to be dispersed

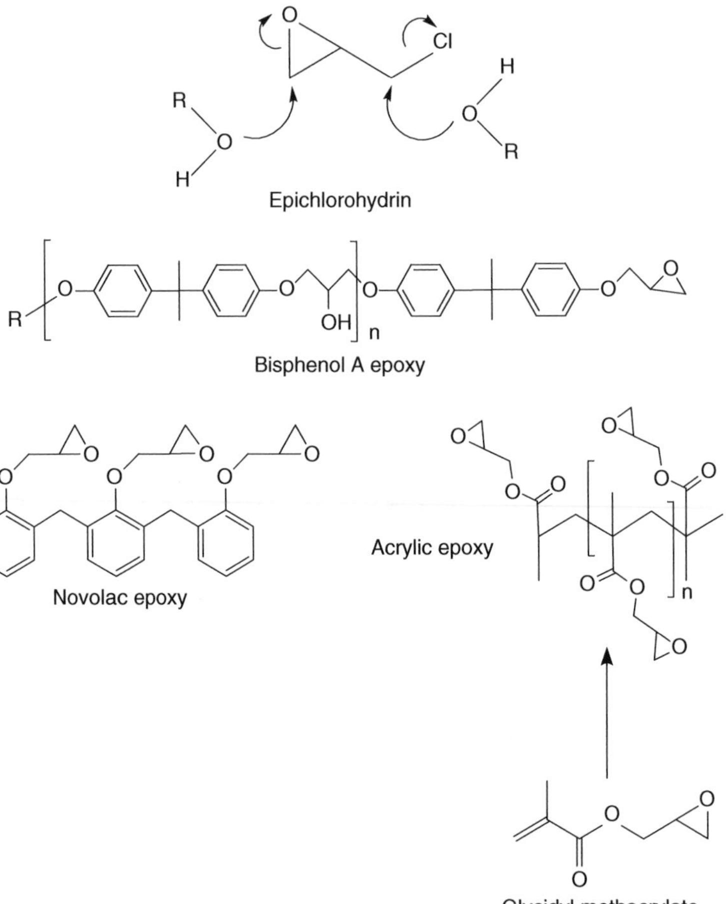

Figure 3.8 Top is the general reaction between epichlorohydrin and oxygen nucleophiles. Below are three important epoxy resin products derived from 1) epichlorohydrin and bisphenol A (which is a diphenol), 2) epichlorohydrin and novolac (which is a mixture of tri- and tetra-phenols), and 3) by polymerization of glycidyl methacrylate. (Note that R represents either an epoxy or phenolic terminus.)

in water. Cellulose acetate butyrate can be used as the primary film-forming polymer in lacquers or, when it has sufficient levels of free hydroxyl present, as a cross-linking polymer.

3.2.2 Dyes and pigments

As indicated above, in this chapter a pigment is any material that is colored, black or white, organic or inorganic that retains a crystalline form in paint. Dyes, on the other hand, dissolve in the paint. Many pigments are inorganic compounds. These include the silicates mica (hydrated potassium aluminum silicate), talc (hydrated magnesium silicate), kaolin (hydrated aluminum silicate) and montmorillonite (also called bentonite, hydrated sodium calcium aluminum magnesium silicate), the oxides anatase and rutile (titanium dioxide, either raw or encapsulated with silica), silica (silicon dioxide, which might be synthetic or the natural crystalline materials diatomaceous earth and quartz) and iron oxide (many forms), the sulfates gypsum (calcium sulfate) and baryte (barium sulfate), and carbonates calcite and aragonite (calcium carbonate) and dolomite (calcium magnesium carbonate). With the exception of iron oxide (which comes in various shades of red and brown), the above pigments are white and are present in paint to add opacity, bulk (in the case of undercoats), reduce gloss (if matt or satin paints are required), whiteness (in the case of white paints) or reduce sagging when liquid paint is applied to vertical surfaces (porous synthetic silica takes on this role in automotive clearcoats). In the case of colored films, typical inorganic colored pigments that might be encountered are iron oxide, chromates (including lead chromate in older red and orange paints, lead sulfate/lead chromate in yellows and strontium chromate in yellows), ferrocyanides (such as Prussian blue), bismuth vanadate/molybdate (in yellow paints), chromium oxides (in older green paints), and cadmium sulfides (reds to yellow). Lead silicate (in older vehicles), zinc phosphate, mixtures of zinc, aluminum, calcium and strontium phosphates, molybdenum, bismuth salts, zinc and zinc aluminum chromates can be found in the base layers applied to sheet metal in automotive applications.

There are organometallic pigments (such as the very popular phthalocyanine blues and greens) and hybrid inorganic–organic substances called "lakes," where a colored organic compound is adsorbed onto an inorganic particle such as alumina. There are many organic dyes and pigments in use based upon quinacridone, perylene, and pyrrolopyrrole substrates for automotive use and benzimidazolone, naphthol, monoarylide and azo substrates for architectural uses.

In addition to the "solid color" pigments mentioned above, there are a number of "effect" pigments. The simplest of these are aluminum flakes. These take the form of either tiny (5–50 µm) silver "cornflakes," which feature ragged edges and convoluted surfaces with multiple reflective patches, or "silver dollars," which are thicker and have a flat reflective surface and smooth edges (Figure 3.9). The latter are formed by vapor

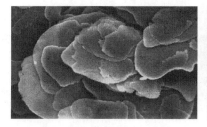 

Figure 3.9 Images of aluminum silver dollars (left) and cornflakes (right). (Buxbaum and Pfaff, 2005. Copyright Wiley-VCH Verlag GmbH & Co. KGaA. Reproduced with permission.)

metal deposition while the former are machine-ground using a fatty lubricant. If this lubricant is an unsaturated fatty acid then the flakes disperse in paint and can be blended with colored pigments in order to achieve a special effect. On the other hand, flakes made with a saturated fatty acid lubricant float to the surface of their paint layer and, as a consequence, they cover up what is beneath and are used to produce silver-colored paints only. Flakes might also be anodized to yield a greenish hue that produces an overall "champagne" effect, and when they are applied in aqueous finishes flakes must have a passivation surface coating to stop the aluminum reducing water to produce hydrogen.

A sparkling effect similar to that achieved using aluminum flake can be achieved using tiny particles of mica, which are small, bronze-like, smooth plates. Transparent iron oxide is a transparent flake that has red to yellow colors. It strongly absorbs UV light and, therefore, protects exterior finishes in automotive and architectural applications.

Modified mica finds extensive use in modern automotive paints. Suzuki (2014a), although his article mainly relates to the use of bismuth oxychloride as a pigment, provides a wealth of information in regards to the very popular pigments that are produced when mica is coated with a thin, highly-controlled transparent film of titanium dioxide (a photomicrograph of such a particle is shown in Figure 3.10). The resultant particle is highly lustrous and sometimes vividly colored. Paint films incorporating them can take on an iridescent color or a pearl- or silky-effect as a result of light reflection and scattering. Unlike conventional pigments, the effect of these pigments is produced when a particular wavelength of incident white light suffers destructive interference at the surface of the particle, meaning, therefore, that light reflected from the particle will be a color complementary to that which has been "absorbed" (or "destroyed"). The color of the particle can be controlled by varying the thickness of the coating. As color is derived from interference, these types of pigments are called interference pigments or, because they produce a pearl-like sheen, they are called pearlescent pigments. The most common pearlescent pigments are silver or white; they have the same thin coating of titanium dioxide and the difference between them is simply the size of the particle. Other particles

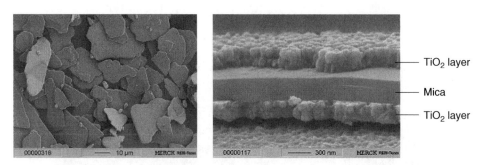

Figure 3.10 Scanning electron microscope photomicrograph of a titanium dioxide coated mica pigment particle (right) and a collection of many (left). (Buxbaum and Pfaff, 2005. Copyright Wiley-VCH Verlag GmbH & Co. KGaA. Reproduced with permission.)

with thicker films take on a particular hue such as magenta, green, and so on. As these particles are quite transparent the color of the substrate beneath is also very important, as it will form a contrasting "ground" or "masstone" color to the light reflected off the particles. Additional effect can be achieved by coating particles with colored pigments. One particular type of finish produces a "color shift", where the entire film color changes from, for example, violet to green, as the angle of viewing changes. Other effect pigments can be produced by coating alumina or mica flakes with iron oxide, titanium dioxide or silica or a combination of these.

Although it is used as an effect pigment in cosmetics, bismuth oxychloride has only been employed rarely in automotive applications over a short period of time (Suzuki, 2014a). Therefore, although paint containing this pigment is unlikely to be encountered in case-work, when it is there will be a high probative value or strong potential for investigative support. This situation could change if a "new generation" of cerium hydroxide-stabilized bismuth oxychloride pigments (for example Mearlite SSQÔ produced by BASF) become attractive to the automotive industry for the production of so-called "liquid metal" finishes.

A paint effect that was popular for a number of years involved the usage of micronized titanium dioxide and aluminum flake, which produced milky-looking light metallic silvery-gray shades with a touch of blue.

There are far too many molecular structures for organic pigments and pigment variants to cover in this chapter and the interested reader is referred to the works by Suzuki and colleagues over many years (Suzuki *et al.*, 1996a, 1996b, 1999a, 1999b, 2014a, 2014b; Suzuki and McDermot, 2006; Suzuki and Marshall, 1997, 1998) for more information, or to Buxbaum and Pfaff (2005), Faulkner *et al.* (2009) or Herbst and Schwartz (2005).

3.2.3 Additives

The main additives in automotive clearcoats are substances that counteract the effects of sun damage. These are of two forms: light stabilizers (usually referred to as HALS, an acronym derived from hindered amine light stabilizers) and UV absorbers. The latter are organic substances that have a high extinction coefficient in the UV range and, therefore, act as a "sun block." Many UV absorbers are based upon the hydroxyphenylbenzotriazole structure (the Tinuvins™), benzotriazoles, hydroxybenzophenones and oxanilides (Hostavin™ and Sanduvor™ products); structures of some of these compounds are shown in Table 3.1. HALS (Table 3.2 shows the molecular structures of a selection of Tinuvin™ HALS) function by quenching free radicals that are produced by UV photolysis of the paint. Although HALS regenerate after interacting with free radicals they do eventually break down and, similar to UV absorbers, migrate through the film if they are not covalently bonded to the binder (Nichols and Kaberline, 2013).

Clearcoats from some manufacturers feature added silicones. However, rather than being an additive that is simply a spectator that is free to migrate like the light stabilizers, the silicones are present as a functional group attached to acrylate monomers and take part in cross-linking reactions.

Table 3.1 Common UV absorbers used in paint. The basic hydroxyphenylbenzotriazole structure (top) is modified at R_1 and R_2 with the moieties listed in the table. Tinuvin 1130 contains approximately 50% of R_2 with the top structure and approximately 38% with the bottom structure, where BZT refers to a hydroxyphenylbenzotriazolopropionyl moiety

Name	R_1	R_2
Tinuvin 328		
Tinuvin 900		
Tinuvin 384		
Tinuvin 1130		

In acrylic automotive lacquers (resprays and older OEM finishes) one or more plasticizers (usually phthalate esters) and cellulose acetate butyrate (to reduce detrimental effects of gasoline) might be present.

In automotive paints there are also many other additives to control rheology and catalyze cross-linking. There are also organic solvents or additives to assist the usage of water as a solvent. However, the amounts of these substances present, or their volatility in the case of solvents, mean that they are not likely to be detected in forensic examinations.

A number of texts describing paint chemistry are available in the literature, including Florio and Miller (2004), Buxbaum and Pfaff (2005), Streitberger and Doessel (2008), Talbert (2008), Wicks *et al.* (2007), Koleske (2012) and Bentley and Turner (1997).

Table 3.2 Common hindered amine light stabilizers used in paint

Name	Structure
Tinuvin 123	
Tinuvin 440	
Tinuvin 292	
Tinuvin 144	

3.3 Automotive paint application

The requirements for application of domestic architectural paint are not particularly stringent and it is often the case that architectural paint encountered in forensic cases has been applied by amateurs. On the other hand, application of automotive paint is a highly technical operation typically carried out by professionals.

The modern industry norm for original equipment manufacturers (i.e., the finish that is applied in the vehicle manufacturing plant) is to use a clearcoat of between 30 and 60 µm thickness to protect and beautify the decorative coats and a conversion coating of about 5 µm underneath to protect the metal and assist with adhesion of the film to it. Some clearcoats in modern cars have a tint that acts to enhance the color of the basecoat. Between the conversion coat and the clearcoat a number of layers and painting techniques are employed. Many older vehicles (even if they had a metallic finish) did not have clearcoat finishes.

The simplest approach is in vehicles that are decorated with a simple solid color. In this case an electrocoat is applied to the conversion coating and, in turn, an antichip (if present, usually on lower body panels) or primer–surfacer coat is applied followed by a basecoat (which is the decorative, colored layer) and, finally, a clearcoat is applied. In modern factories the clearcoat is applied to the basecoat "wet on wet."

Effect pigments are usually contained in the basecoat. In order to achieve a higher degree of sparkle, metallic basecoats are often formulated to reduce to a thin film (about 15μm thick) when dry, so that the aluminum plates are forced to orientate themselves parallel to each other and the film surface. Pearl-effect pigments can be included with aluminum plates in a single basecoat but it is also the case that pearl coatings can have three layers above the primer: a foundation coat containing opaque pigments that is underneath a clearcoat that contains pearl-effect pigments (and allows visibility of the foundation coat) that is underneath the clear topcoat.

Automotive paints for OEM application are now manufactured by relatively few, transnational suppliers. This would tend to suggest that the variation between paints from different vehicles is now quite small. Fortunately, there is still variation that can be detected using forensic analytical techniques. This is because major suppliers in a number of cases still have regional subsidiaries that must customize formulations to suit the particular application equipment used by their regional clients. For example, paint company A might have a subsidiary (paint company B) in country B, which might supply paint to three regional automotive factories and a number of accessory manufacturers each with different types of paint spray equipment. Company A might release a new color and each of the regional factories might choose to manufacture a vehicle using that color. Company B will provide different formulations of the paint to suit the various factories and manufacturers and tint the paint to match exactly exemplars of the color from company A.

3.4 Forensic examination of paint

3.4.1 General considerations

Forensic examination of paint has the potential for a very wide scope of analysis and comparison. First, paints are a complex mixture of organic compounds (such as one or more polymers, pigments dyes, and traces of other substances such as light stabilizers) and inorganic compounds (such as pigments). Second, paint is rarely present as a single layer on items of forensic interest and, furthermore, the make-up of one layer is not a determinant of the exact composition of an adjacent layer, especially in domestic paint films. Finally, in operation, paints suffer degradation, small molecules migrate, active functional groups continue to cross-link, solvents evaporate and substances are applied in order to arrest degradation and restore or preserve the film's appearance; the film might even have acquired an aftermarket respray. To a greater or lesser degree, these chemical features that distinguish one paint coating (i.e., the entirety of the layers on the object) from another can be detected by analytical and/or observational techniques.

The aim of the previous section was to describe the breadth of chemical and physical attributes of paint that can be exploited – or at least *potentially* exploited – by forensic scientists for the purpose of differentiating between known and unknown paint samples. This section covers the commonly applied forensic techniques for that purpose and their limitations.

The reports published on-line by Collaborative Testing Services Incorporated (CTS, 2014) provide an interesting view of the "popularity" of various methods of analysis that have currency for forensic paint examination. CTS provides proficiency tests to many forensic laboratories around the world. Participants in a given test are all provided identical sets of paint samples, usually three or four small painted substrates, and are asked to submit the samples to their usual paint comparison procedure. Participants are then asked to provide answers to the typical forensic question, "Could paint sample A share the same source as paint sample B?" Table 3.3 shows the range of tests that participating laboratories used in their completion of proficiency tests in 2014 and 2013.

Other techniques that can be applied to forensic paint analysis, such as laser ablation–inductively coupled plasma mass spectrometry, laser induced breakdown spectrometry and X-ray diffraction (XRD) techniques have been reported very rarely in CTS reports and, as a consequence, their treatment in this text will be restricted to a brief discussion of their potential.

A universally correct or "world's best" practice for forensic paint examination has not been agreed upon. However, readily-accessible guidelines have been published in regards to forensic paint examination processes by the US-based paint subgroup of the Scientific Working Group on Materials Analysis (SWGMAT, 2000) and by the American Society for Testing and Materials (ASTM, 2014).

Table 3.3 Tests used by participating laboratories in CTS proficiency tests in 2014 and 2013

Year (trial, participant numbers)	SM (%)	IR (%)	EDS (%)	PLM (%)	FM (%)	S (%)	MSP (%)	PGC (%)	XRF (%)	R (%)
2014 (14-545, 119)	100	99	59	40	35	31	24	29	10	8
2013 (13-546, 85)	99	99	53	29	24	24	25	20	13	7

Abbreviations: SM, stereomicroscopy; IR, infrared spectrometry; PLM, polarized light microscopy; PGC, pyrolysis gas chromatography; FM, Fluorescence microscopy; EDS, energy-dispersive X-ray spectrometry; MSP, microspectrophotometry; R, Raman spectrometry; S, solubility tests; XRF, X-ray fluorescence spectrometry.

Compared to other materials such as glass, even single layer paint is a highly heterogeneous material. As a consequence, it is important that the analyst assesses the intrasample variance by carrying out repetitive analyses of the known and unknown sample. It is also important when using analytical techniques that are capable of "zooming in" on very small regions within the whole specimen (referred to as a "region of interest" in this chapter) that the analyst also "zooms out" to carry out an analysis of a large enough region of interest in order to arrive at representative overall composition of the paint.

A term that will be used throughout this chapter in regards to the analytical methods described is discrimination power (Dp). Discrimination power is estimated by submitting a large number of known specimens from a population to the particular method (Smalldon and Moffat, 1973). The analytical data produced are then screened to identify how many specimens are not distinguishable by the method, and a numerical index, the Dp, is derived by dividing the number of distinguishable pairs by the total number of pairs in the original sample set. For example, a population of 10 paints collected at random is subjected to examination using a stereomicroscope. There are 45 unique and relevant comparison pairings in this population (i.e., pairings that exclude the comparison of any sample with itself, which is trivial, and count the pairing of paint X and paint Y as identical to the comparison of paint Y with paint X. The number of relevant and unique pairings is simply $(n-1)n/2$, where n is the number of samples in the population). If four pairs of paints are indistinguishable, then the Dp in this case is $(45 - 4)/45$, or 0.91. Many studies in the literature report Dp values that have been calculated after a barrage of analytical tests (such as microscopy followed by infrared microspectrometry followed by scanning electron microscopy-energy dispersive X-ray spectrometry). While this is of great value when the paint examiner comes to assessing the probative value of their findings, it does not provide a comparison between techniques as to their relative powers. Relative power can be important in regards to process efficiency; the most efficient process is to carry out the most discriminating test first, followed by the second most discriminating test and so on. Where possible, the Dp for individual analytical techniques as well as barrages of tests will be provided in this chapter.

Depending upon the operational idiosyncrasies of the jurisdiction, the person who conducts the laboratory analysis might not be the person who collects the known and unknown paint samples from the crime scene or object. Irrespective of who carries this out, sample collection can profoundly influence laboratory testing and, as such, it is the critical task in forensic paint examination. It is possible that there is a physical fit between the known and unknown paint, for example the shape of a foreign paint chip on the victim's vehicle might match the damage to the paint on the suspect's vehicle. This is extremely strong evidence and, therefore, it is important to record photographically this physical fit prior to attempting to recover the foreign paint, just in case it gets damaged during recovery. Foreign paint is often present as smears and these are difficult to recover. Where possible, it is desirable to submit the entire object bearing the foreign paint to the laboratory, where it can be collected under controlled conditions. If this is not possible, a scalpel can be used to cut away the substrate paint and its smear for submission to the laboratory. Tape-lifting sounds like a good idea but it can be very difficult to remove microscopic samples of paint from adhesive tape and this should be considered a last resort. The circumstances of the transfer control the quantity, quality, and size of the unknown fragments but the person collecting paint from the crime scene can control the quality and quantity of known paint samples taken for comparison. The key task is to ensure that the known sample truly represents the paint that might have been transferred. In this regard, then, the known paint should be collected close to, but not within, the damaged area because contamination might have taken place during contact. Clearly, it is desirable that all paint layers present on the putative source are collected. If an auto-mobile collision is involved, paint should be taken from the panel believed to be involved in the collision, as it is not unknown for adjacent panels to differ in their OEM finish (in a factory individual panels will be refinished, if required, rather than the entire vehicle) or it could be that the panel is a replacement of the OEM one or has had an aftermarket respray. Furthermore, different panels suffer different exposure to sun, automotive fluids, or engine heat, which can alter the chemical characteristics of the sample. All these precautions guard against a false exclusion of the putative source.

3.4.2 Microscopy

Examination of known and unknown paints using a stereomicroscope with magnifica-tion from about 10 to 80 times is the first analytical step that should be conducted in a forensic paint investigation. It is at this stage that many cases can be "written off" and suspects eliminated from further consideration when known and unknown paints differ in their layer structure. Studies by Gothard et al. (Gothard, 1976; Gothard and Maynard, 1996), Ryland and Kopek (1979), and Edmondstone et al. (2004) demonstrate a high Dp for simple stereomicroscopy. After examination of two sets of 500 vehicles Gothard found that paint flakes of more than six layers were unique. Others studies confirm Gothard's findings, albeit with smaller data sets. The work of Tippett et al. (1968), in regards to examination of paint from 2000 houses in Wales, found that three matching pairs of paint were in the data set but these were found to be repeat samples from the same three buildings.

If known and unknown paints are found to be the same with regards to layer number, color and sequence then stereomicroscopy can be used to look for differences in the texture, gloss level and other visible features of the specimens.

Using transmitted light and more sophisticated equipment, brightfield, darkfield, fluorescence and polarized light microscopy can be used to seek differences between known and unknown specimens. However, except for fluorescence microscopy, this does require the paint sample to be treated to yield a specimen that transmits light and to be mounted on a glass slide. Using fluorescence and polarized light it can be possible to differentiate specimens on the basis of their mineral compositions. Many research publications refer to the examination of cross-sections, either produced by microtomy or polishing of specimens mounted in resin. It can be unwise, difficult or impossible to produce cross-sections of evidentiary unknown paint, however, and the best that might be achievable is to examine the specimen using techniques that can be applied to the specimen as found.

The reader is referred to further reading in regards to microscopy in the works of Hamer (1980, 1982), De Forest (2002) and the chapter in this book.

3.4.3 Vibrational spectrometry

As shown by the data associated with CTS proficiency tests (Section 3.4.1), infrared spectrometry is a very popular technique for the examination of paint, and justifiably so. The differentiation of paints originating from different polymer classes (even those comprising mainly ester functionality, such as alkyds and acrylics) is a trivial exercise and many extenders and inorganic pigments are instantly recognizable in the spectral data. Infrared microspectrometers are becoming less costly and much easier to use; therefore, infrared spectrometry can be and is applied routinely to "forensic size" specimens.

As already indicated, many common paints, with the exception of those based on epoxy resins and nitrocellulose, are based upon esters or urethanes. As a consequence, most paints show a strong absorbance between 1650 and 1750 cm^{-1} (carbon–oxygen double bond stretch, sometimes a doublet for polyurethanes) and multiple absorbances between 1000 and 1300 cm^{-1} (carbon–oxygen single bond stretches). Although paints are complex, it is quite a simple matter to distinguish between acrylic, PVAc, polyurethane and alkyd paints by eye. Peaks at 815 and 1550 cm^{-1} indicate the presence of melamine cross-linking. Table 3.4 shows the characteristic group frequencies that can be used to classify common binders.

Table 3.4 Characteristic group frequencies for common binders

Binder type	Mid-IR absorbance (cm^{-1})
Acrylic	1730, 1450, 1380, 1260, 1170, 1150
Orthophthalic alkyd	1725, 1450, 1380, 1270, 1130, 1070, 740, 700
Isophthalic alkyd	1725, 1475, 1375, 1305, 1237, 1135, 1074, 730
Terephthalic alkyd	1725, 1270, 1250, 1120, 1105, 1020, 730
Epoxy	1510, 1240, 1180, 830
Nitrocellulose	1650, 1280, 840
Poly(vinyl acetate)	1735, 1235, 1370, 1025, 1110, 1435, 945
Polyurethane modification	1690, 1460
Styrene modification	1490, 1450, 760, 700
Urea modification	1655
Melamine modification	1550, 815
Benzoguanamine modification	1590, 1540, 825, 780, 710
Silicone modification	1020, 1100, 1260, 800

Many paints are formulated by blending polymers from different groups, for example by blending polyurethanes with acrylics, alkyds with polyurethane; these, too, can be readily distinguished, as indicated by Maric *et al.* (2014) in regards to automotive primer–surfacers. Ryland illustrated the power of infrared spectrometry in 2001 when he showed that it was possible to differentiate between various clearcoat formulations and in 1995 when he differentiated red OEM acrylic-melamines produced by two different paint manufacturers for use in the same make and model of vehicle (Ryland, 1995; Ryland *et al.*, 2001).

Inorganic pigments, with the exception of titanium dioxide and the iron oxides, produce strong, characteristic absorbances in the mid-infrared range, as shown in Table 3.5 (absorbance in the far-infrared region is also presented in the table but is discussed later). The fact that infrared spectrometry is sensitive to both pigment and binder both helps to assist in the discrimination of one paint from another but, as the peaks can be quite broad as well as intense in the carbon–oxygen single bond stretching region, some critical diagnostic spectral features can mask each other. Suzuki in his many articles (Suzuki, 1996a, 1999a, 1999b; Suzuki and Marshall, 1997, 1998) and Massonnet and Stoecklein (1999a) describe the scope for identification of organic pigments in paints.

There have been several Dp studies regarding IR analysis of paints that serve to quantify the value of infrared spectrometry to forensic paint investigations. In a study of 54 red automotive basecoats Massonnet and Stoecklein (1999a) reported a Dp of 0.95 for light red paints, 1.0 for metallic red paints and 0.96 for dark red paints. A Dp of 0.95 was measured by Buzzini and Massonnet (2004) for a set of 40 green non-professional spray paints. In a study of 51 red spray paints (domestic and non-professional automotive) Govaert and Bernard (2004) measured a Dp of 0.77. Seccombe (2001) in her PhD dissertation analyzed a wide range of automotive paint colors using a number of techniques. She found values for infrared microspectrometry of 0.97 for red paint, 0.89 for white, 0.95 for black, 0.93 for gray, 0.92 for blue, 0.95 for green, 0.90 for yellow, and 1.00 for brown, although only small numbers of yellow and brown paints were examined. In a study of 34 red paints and 70 white ones, Eyring *et al.* (2007) produced data that showed a Dp of greater than 0.99 for both colors using attenuated total reflection (ATR) microspectrometry, while in their study of single-layer white architectural paints Wright *et al.* (2013) achieved a Dp of 0.94.

In general, then, infrared microspectrometry is capable of very high discrimination between paints. The work of Maric *et al.* (2012) and Lavine *et al.* (2014) has shown that the differences between paint from different automobile factories, even within a single manufacturer, can be resolved using infrared spectrometry, even when only the clearcoat is examined. Even though discrimination was achieved using chemometrics, the spectral features responsible for the variance are visible to the naked eye. For example, Maric classified clearcoats from 130 different vehicles into nine groups using chemometrics but close inspection of representatives of the groups (the so-called centroid object of the

Table 3.5 Characteristic absorbance frequencies for inorganic pigments

Substance	Mid-IR absorbance* (4000–700 cm^{-1})	Far-IR absorbance* (<700 cm^{-1})
Mica (muscovite)	1010, 1030 and 920 (shoulders), 3610, 740	470, 530, 400
Mica effect pigments	965, 1010 and 1030 (shoulders), 700	470, 530, 400
Talc	1010, 3675	670
Calcite	1440, 1795, 880, 710	
Feldspar	995, 1130, 780, 720	580, 410, 530
Titanium dioxide (Rutile)	Rising absorbance	
Anatase	towards 700	
Silica (synthetic, can be present as an encapsulating layer)	1100 (asymmetric), 795 (broad and weak)	465 480, 620, 385
Silica (crystalline – diatomaceous earth)	1100 (asymmetric), 800, 780, 700	480
	1100 (asymmetric), 800	
Kaolin	1035, 1010, 1115, 915, 3695, 3620, 3650, 3670	542, 470, 695
Bentonite	1055, 915, 833, 795	522, 468
Aluminum oxide	1058, 1157, rising absorbance towards 700	
Chrome yellow	855 (shoulder spike at 830)	390
(Silica encapsulation[1])	1100[1]	465[1]
Chromium trioxide	635, 585	415, 450, 305
Copper phthalocyanine		
Zinc phosphate	1110, 1025, 945	635
Bismuth oxychloride	–	276
Barium sulfate	1080, 1120, 1180, 980	610, 640
Iron hexacyanoferrate (Prussian blue)	209	605, 495
Nickel titanate and chromium titanate (yellow/yellow-orange)	Rising baseline at low frequency end of range	550/470 (broad, overlapped), 400, 345
Bismuth vanadate (bright yellow, often mixed with bismuth molybdate[2], or encapsulated with silicates, zinc phosphate, or alumina)	810/750 (broad overlapped),	365[2], 330[2]
Red iron oxide	900, 795	550, 475, 320
Yellow iron oxide		485, 280, 605
Molybdate orange	860	465[3]
(silica encapsulation[3])	1090[3]	
Calcium sulfate	1160, 1120	675, 600
Calcium silicate	1020, 1090, 920,	470, 570, 650, 680

*In decreasing order of intensity

groups) by eye allows the observation of about seven groups within the data, mainly on the relative contributions to the spectral data made by melamine, styrene and the acrylic resin, especially the relative melamine contribution. This is quite remarkable, given that OEM clearcoats effectively come in only two broad formulations: melamine cross-linked styrene–acrylic, or melamine cross-linked styrene–acrylic–polyurethane (Lavine *et al.*, 2014). In later work, Maric *et al.* (2013) found using synchrotron infrared microspectrometry that melamine appears to migrate from basecoat to clearcoat. As the relative contribution of melamine to the overall spectrum of a paint sample is one of the key features that discriminates one clearcoat from another, the implications of the findings of Maric *et al.* are clear – whether as a result of defining a region of interest within an intact specimen or as a result of excising a small piece of clearcoat for analysis – it is best to avoid collecting a spectrum from the clearcoat that is close to the basecoat. Given that it is the clearcoat that suffers the most from sun damage and from the application (deliberate or otherwise) of waxes, automotive fluids, dust, and so on, it is also wise to not analyze the clearcoat too close to its surface (this warning is of especial relevance to the use of ATR spectrometry). In his PhD dissertation, Maric (2014) indicates that the effects of weathering upon the HALS or UV absorbers in clearcoats are not discernible by infrared microspectrometry.

As can be appreciated from the paint formulation chemistry outlined in the previous section, however, there is wide scope for subtle variation in binder composition within a paint class that would not be detected using infrared spectrometry. For example, it is conceivable that two alkyd paints could be formulated using exactly the same feedstock save for monoglycerides formulated from quite different oils – these would only differ in regards to the chain length of the fatty acid residue and any unsaturation still present after curing. These differences are too subtle for infrared spectrometry to detect. Other subtle binder variations, such as those that arise from chain length or cross-linking variations in prepolymers, differences in polyhydric alcohols and small variances in the proportions of the prepolymers used would also be very hard to detect using infrared spectrometry. Ryland (1995) demonstrated this by showing the spectra for paint from a 1977 Ford (green acrylic-melamine enamel manufactured by PPG) and from a 1978 Ford (green acrylic-melamine enamel manufactured by Cook Paint and Varnish Co.); they could not be distinguished by infrared spectrometry (the same paints are readily differentiated, however, using pyrolytic techniques). Clearly, the pigments responsible for light color shades would not be detectable by infrared spectrometry.

Suzuki in several articles (Suzuki, 1996b, 2014a, 2014b; Suzuki and McDermot, 2006) discusses the application of far-infrared spectrometry (or "extended range" infrared spectrometry) to many paint pigments. This spectral range contains a wealth of information in regards to many compounds (Table 3.5) but it is not a range routinely available to forensic paint examiners carrying out infrared microspectrometry, because the usual detector employed for this (a narrow-range mercury-cadmium telluride detector) is effective only to about 700 cm^{-1}. Raman spectrometry (or access to a synchrotron beamline) is the only effective option for exploiting this valuable region for the examination of typical "forensic size" specimens.

One of the challenges faced in the forensic analysis of paint evidence is the management of small specimens for analysis. The desired outcome is to make use of effective techniques to handle the sample and to achieve good quality spectral data that are representative of the paint sample. A typical piece of evidence will be a small chip comprising several layers. For transmission microspectrometry the thickness for the specimen must be no more than 10–20 μm, otherwise the beam will be totally absorbed; ironically, most trace evidence is much too thick to analyze "as is" by transmission microscopy. Exposed surfaces can be analyzed "as is" using ATR microspectrometry simply by pressing the sensing crystal onto the exposed surface. This is usually possible for the top layer (avoiding the top surface of the top layer if at all possible) and bottom layers of a piece of paint evidence and other layers if they are also exposed. It is often the case, however, that all the paint layers are not conveniently exposed. Under these circumstances there is no option but to treat the sample in some way in order to facilitate analysis. Microtomy is an excellent way to expose each layer. If the sections are thin enough for transmission infrared microspectrometry, or if the section is flattened between diamonds in order to make them thin enough, or if the section is well enough supported to allow ATR infrared microspectrometry to be carried out, then excellent data will be collected. However, microtomy is very difficult to carry out on typical unknown forensic paint samples because they are just too small to handle effectively and the risk of destroying the specimen is too great. Embedding a sample in resin and then polishing it back to reveal each layer is also an excellent way of preparing a sample but it limits analysis to ATR spectrometry and there is the risk that the mounting resin will infiltrate the sample and contaminate spectral data. Although it sounds tedious (and is), the common forensic practice is to cut small pieces from each layer and either apply them to a metal slide and carry out ATR spectrometry or crush each of them in turn in a microdiamond cell until they are thin enough to pass infrared light for transmission measurement. In the case of very small samples this takes some practice, patience, a delicate needle-like probe to hold the specimen down and an Exacto™ knife or scalpel for the cutting operation. It can be quite convenient to either use the same specimen for subsequent SEM-EDX analysis by transferring it to a stub with double-sided adhesive tape after it has been subjected to IR microspectrometry or to cut an extra specimen while the sample is being prepared for microspectrometry.

For some samples it can be quite difficult to cleanly cut one layer away from another and a specimen with a paint boundary will be obtained. Clearly with microtomed sections all the paint boundaries will be present. It is very tempting to believe that if one sets up a region of interest within the specimen defined by the microscope's apertures, then a spectrum is obtained solely arising from the region of interest and from none of the surroundings. This "what you see is what you get" supposition is not absolutely correct for infrared microspectrometry, nor is it the case for other techniques described later.

In infrared microspectrometry the radiation used to collect the spectral data is of much longer wavelength than that used by humans to view the specimen and define the region of interest. What is seen by humans to be a clearly delineated region of interest with the remainder masked-off is far more ill-defined in the infrared region due to the effects of diffraction. The impacts of diffraction can be appreciated by consideration of the point-spread function (PSF) created by the optical elements of the microscope. If a point source of infrared radiation at the specimen plane is imaged by a typical infrared microscope objective, then at the focal plane behind the objective (i.e., where the aperture curtains are located) the point is spread into a more diffuse function as depicted by the curve in top part of Figure 3.11, which shows the intensity of the infrared light at the objective's rear focal plane along the X and Y axes (i.e., the PSF). In practical terms, as a result of being imaged by the objective, the point is diffracted into a patch of light surrounded by rings of light. Therefore, if the specimen is thought of as an array of point sources in the X,Y plane, then it follows that the light coming from any one point (which of course carries spectral information) is mixed with the light from its neighboring points, even if those neighbors are apparently masked by the microscope's apertures. This is sometimes called "spectral leakage" and it is most relevant at boundaries within a

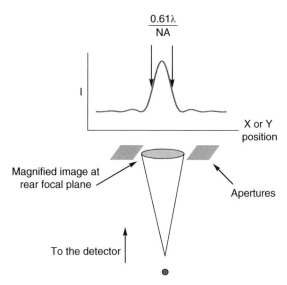

Figure 3.11 The point spread function (PSF) arising at the rear focal plane of an objective when a point source of light is magnified through that objective. The curve shows the intensity of light in the X,Y focal plane and the pattern is symmetric about the Z axis at X = Y = 0. The equation and arrows depict the width at half height of the central maximum of the PSF, where NA refers to the numerical aperture of the reflecting objective and λ is the wavelength of the point source of light (note: the objective is omitted for simplicity). Using this equation with $\lambda = 10\ \mu m$ (1000 cm^{-1}), a point source imaged through a typical infrared microspectrometer objective diffracts into a spot with a dimension of about 9 μm between the arrows and with a diameter of the first "ring" of light greater than 20 μm. About 80% of the source energy is present in the central spot

specimen, which for a chip of paint means the boundary between the chip and air, or between the layers within the chip. The most important practical consideration is the range over which spectral leakage operates and what influences it.

The equation presented in Figure 3.11 defines how wide the pattern is, but as it is calculated in relation to visual resolution (i.e., it defines the distance between two points of light that be just discerned as two points) and is applicable for conventional refracting objectives, the constant (0.61) is not exactly applicable to the use of reflecting objectives nor for spectral spatial resolution (i.e., defining when spectral leakage becomes unacceptable). The equation also does not accurately define the case for many infrared microspectrometers that use apertures both between the specimen and the source and between the specimen and the detector. The use of such apertures reduces the spread of the pattern but also spreads more energy into the rings around the central maximum, so that only about 50% of the source energy is in the central maximum rather than about 80%. Although an exact equation is not available for the PSF applicable to infrared microspectrometry, Carr (2001) and Sommer and Katon (1991) have empirically demonstrated that spectral leakage can be expected to become apparent when apertures are placed closer than 5 to 20 μm from a boundary in a paint specimen (a range is given because the PSF is wavelength dependent, therefore more spectral leakage will be evident at the long wavelength end of the spectrum).

In the case when a specimen comprises a piece of a single layer, the relevant boundary is the edge of the specimen. If the region of interest is drawn tight on the specimen edge, then light from outside the region of interest (i.e., the unattenuated IR beam) will reach the detector. This is "stray light", which is manifested as a loss of "contrast" in the spectrum (i.e., large peaks in the spectrum are forced to lower absorbance values and small peaks seem to be accentuated; Sommer and Katon, 1991; Kirkbride and Tungol, 1999).

Another issue of importance arises from the use of very closely-spaced apertures (i.e., when very small regions of interest are defined). Apertures defining a region of about 1λ (i.e., less than about 10 μm for the mid-IR range) only transmit 25% of the incident source radiation at that wavelength, which of course means that the resulting spectrum acquired for that region of interest exhibits poor signal to noise. A region of interest about 3λ in size still only transmits 70% of the source energy.

In transmission microscopy, therefore, it is a wise move to produce a specimen as wide as possible, so that a large region of interest can be drawn well away from any boundaries and maximum energy throughput can be attained. A good approach, if excising a sample from a layer, is to not try and cut a specimen thin enough for direct measurement, as it will most likely also have a quite small surface dimension and a small region of interest must be set (i.e., closely-spaced apertures). It is better to cut out a thicker piece and flatten it to the correct thickness, thereby producing a larger surface dimension that will allow a large region of interest to be set. Even microtomed sections might require spreading out if one or more layers is quite thin. As can be seen from Figure 3.11, the PSF is inversely proportional to the NA of the objective. This becomes relevant in ATR spectrometry because the high refractive index of the crystal in the objective greatly inflates the NA and, thereby, greatly reduces the PSF and, therefore, spectral leakage.

The principles of diffraction apply in microspectrophotometry (MSP) and Raman spectrometry but the impacts are greatly reduced because light of much shorter wavelength is used.

Of the two techniques, transmission microspectrometry and ATR microspectrometry, the latter offers a lot to the examiner. It is usually the quickest method, is influenced least by diffraction and does not suffer from fringes due to optical interference (a phenomenon especially apparent when clearcoats are pressed between smooth diamonds).

An innovation in the field of infrared microspectrometry has been the introduction of focal plane array (FPA) detectors. These detectors allow the collection of infrared spectra from many adjacent pixels over a large field of view under the microscope without aperturing (so-called "chemical imaging"). For heterogeneous specimens such as paint, that means that many spectra can be acquired for each of the layers present in one experiment. Compared to the traditional approach (i.e., using apertures to select an array of individual contiguous regions of interest and acquiring spectra at each) a time reduction of about 100 times can be achieved (Miller and Dumas, 2006). As indicated by Flynn *et al.* (2005), FPA infrared microspectrometry offers great convenience for the paint analyst – assuming that an appropriate specimen, ideally a cross-section, can be prepared.

For additional operational advice relating to infrared microspectrometry readers are referred to the publications of the American Society for Testing and Materials (ASTM, 2013b) and the Scientific Working Group on Materials Analysis (SWGMAT, 2009).

Compared to infrared microspectrometry, Raman micospectrometry is much less frequently used in forensic paint examination, as can be seen in the CTS proficiency test responses referred to previously. Raman spectrometry is described as the vibrational spectrometry complement of infrared spectrometry. The latter is sensitive to bonds within a substance that have a dipole moment and, therefore, is very sensitive towards the C–O, C=O, and similar functionalities in paint while the former is less sensitive to these moieties but more sensitive towards aromatic rings, –N=N–, C–H and similar non-polar but polarizable functionality. A paint usually has a mixture of polar and non-polar substances and functionalities; therefore, both infrared and Raman spectrometry have something to offer in regards to the analysis of paint. Furthermore, Raman spectral peaks are usually narrower than those from infrared spectrometry (thus one can expect fewer overlaps between spectral features), Raman peak intensities are more precise, and the intrasample variance with infrared spectrometry is relatively high; consequently, Raman spectrometry is said to offer the potential for better discrimination (Bell, 2005a). As an illustration of this, Bell *et al.* found that a group of 39 architectural paint resins (31 of which were alkyd resins) could be discriminated into six groups with each technique. However,

Raman spectrometry was able to discriminate more members within each group than infrared spectrometry. This reference provides a good example of how difficult it can be with infrared spectrometry to discriminate between alkyds, in this case those that contain differing amounts of monoglyceride incorporation in otherwise identical resins. Raman spectrometry, on the other hand, was shown to discriminate between them on the basis of relative signal intensities arising from alkyl C–H and aryl C–H stretching bands (the difference is subtle though).

Paints contain more than just resins, however, and it can be the case that the Raman signals due to the resin are small compared to signals arising from pigments, especially organic pigments. This has been demonstrated by De Gelder *et al.* (2005), Bell *et al.* (2005a, 2005b), and Massonnet and Stoecklein (1999b), all of whom have presented spectra of real paint specimens that are dominated by signals due to the pigments present; discrimination between paints was still effective, however. Surprisingly, given the ability of infrared microspectrometry to discriminate between automotive clearcoats (Maric *et al.*, 2012) and the sensitivity of Raman spectrometry to the non-polar bonds in styrene and melamine, the work of De Gelder *et al.* (2005) has shown that Raman spectrometry is not particularly effective for the discrimination of clearcoats or even primers, although the same authors did show that it was effective for the discrimination of basecoats. Titanium dioxide is a strong Raman scatterer and is a widely used pigment. It exists in two forms, anatase and rutile, which are easily discriminated using Raman spectrometry but are identical under infrared spectrometry, where their presence is usually only visible as a rising baseline at the long wavelength end of the MCT (mercury cadmium telluride) detector range. Another pigment that can be detected readily by Raman spectrometry is carbon black. It is present in many paints and is "invisible" to many other analytical techniques, not just infrared spectrometry. Iron oxides, which are also ubiquitous and absorb in the far infrared region beyond the capability of MCT infrared detectors, are also readily detected using Raman spectrometry.

Unlike infrared microspectrometry, where high spectral quality requires careful specimen preparation, Raman spectrometry has fewer requirements. The lasers used for Raman spectrometry operate in the visible region (e.g., green or blue) to the near IR (1024 nm) and, as a consequence, the point spread function for Raman microspectrometry (and therefore the spatial resolution) is much smaller than that for infrared microspectrometry. Some Raman microscopes can operate in confocal mode, where pinholes (analogous to the apertures in an infrared microspectrometer) control not only the point spread function of the scanning laser beam across the X,Y plane of the specimen (i.e., across its surface) but also only allow the detector to receive light from a small range in the Z axis (i.e., within the specimen). Effectively, the confocal Raman microspectrometer only receives signal from an oblate spheroid (i.e., a football shape) fixed at a particular depth in the specimen and to a lesser extent from shells around it (i.e., as though the point spread function shown in Figure 3.11 is centered on each of the X, Y and Z axes of the specimen and elongated along the Z axis, as in Figure 3.12). As a consequence,

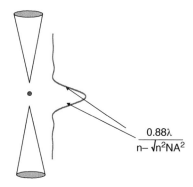

$$\frac{0.88\lambda}{n-\sqrt{n^2NA^2}}$$

Figure 3.12 Depiction of the PSF along the Z axis for a point source in confocal microscopy (symbols as for Figure 3.11, with n being the refractive index of the specimen, in this case, paint). The PSF along the Z axis is more spread out than that in the X,Y plane, which is as shown in Figure 3.11. The resulting PSF in three dimensions is a convolution of the patterns for the Z axis and for the X,Y plane and can be thought of as an oblate spheroid (with its long axis along the Z axis) surrounded by concentric "shells" of diminishing intensity

Raman microspectrometry is capable (in theory at least) of providing a signal from a volume of paint of a few or a few tens of μm^3.

This "depth profiling" capability is valuable in forensic paint examination because it offers the possibility of recording spectral data from the top coat and layers below it simply by shining the laser beam perpendicularly onto the surface of the paint and stepping the point of focus through the layers while recording spectral data. De Gelder *et al.* (2005) report a simpler approach whereby the spectra of both clearcoat and basecoat in intact paint chips are acquired in a single acquisition. Buzzini *et al.* (2006) describe the use of Raman microspectrometry to record selectively the spectra of foreign paint smears *in situ* on their substrate, such as on a painted crowbar or on large paint samples collected from a victim's vehicle.

Similar to the case with regards to infrared microspectrometry, however, in Raman microspectrometry some of the total signal emerges from parts of the specimen some distance from the user-defined region of interest or depth of interest. This is because the laser penetrates much further into the specimen if the beam is not confocally controlled (especially in transparent media, such as clearcoats) and, as a result of multiple scattering events, the beam interacts with (and therefore causes emission of signal from) regions of the specimen away from where the beam strikes (Everall, 2008, 2009, 2013). This is so effective that an entire branch of Raman spectrometry (called spatially-offset Raman spectrometry) has developed where the Raman signal is collected from regions away from where the laser beam impacts. This phenomenon is independent of and additional to the spread of the beam as a result of its point spread function when confocal Raman microspectrometry is used. Paint has a high refractive index compared to the air surrounding the specimen and has a strong influence upon the focus of the beam once it penetrates into the specimen. Corrected objectives can compensate for this to a certain extent but many Raman microspectrometers are fitted with uncorrected objectives. The result is that with many spectrometers the region from which spectral data are collected is at a significantly shallower depth than the operator believes (Everall, 2013).

Perhaps the biggest shortcomings in regards to Raman spectrometry are that the signal strength can depend upon the wavelength of the laser used for excitation and a significant number of specimens fluoresce to such an extent that a Raman signal cannot be collected at all. This limitation can be overcome to some extent by having access to a number of different lasers operating at different frequencies in the hope that one will allow collection of good spectral data. It has also been reported (Buzzini *et al.*, 2006, Zieba-Palus and Michalska, 2014) that fluorescent compounds in paint can be photo-bleached (i.e., the laser is left to irradiate the specimen for some time, causing irreversible photolysis of the compound responsible for the fluorescence). It seems that Raman microspectrometry is not a replacement for infrared microspectrometry, nor is the converse true, and there is not enough evidence yet to indicate that pigment analysis using Raman microspectrometry can replace UV-visible microspectrometry for the color analysis of paint. Instead, Raman microspectrometry (employing a number of lasers) and infrared microspectrometry together provide as comprehensive a view of a paint's make up as can be achieved using vibrational spectrometry.

Raman spectrometry has been assessed as to its discrimination power with regards to paints. Bell *et al.* (2005b) found Raman spectrometry to be about equivalent to infrared spectrometry in regards to discriminating alkyd paints. In a study of 54 automotive red basecoats, Massonnet and Stoecklein (1999b) measured a Dp of 0.89 for light red paints, 0.9 for red metallic paints and 0.97 for dark red paints. A Dp of 0.91 was measured by Buzzini and Massonnet (2004) in a survey of 40 green spray paints.

3.4.4 SEM-EDX and XRF

Examination of CTS proficiency trials indicates that the usage of scanning electron microscopy-energy dispersive X-ray spectrometry (SEM-EDX) is popular for the forensic analysis of paint. Although it is an expensive technique, its popularity is warranted as it is a practical, versatile, and probative technique for a number of reasons.

First, the electron microscope is useful for imaging the specimen. Even though the image does not carry any color information, it can often reveal layers in a specimen that are very difficult to discern using visible light microscopy. This can be enhanced even further if one has access to a backscattered electron detector, elemental mapping capability or if polished cross-sections can be prepared. Under the SEM, individual pigment granules can be located and examined under magnification much higher than can be achieved using light microscopy.

Second, the technique offers unrivalled practicality for the elemental analysis of paint specimens of "forensic" size. While it is the case that inductively-coupled plasma and X-ray fluorescence (XRF) techniques offer much better elemental limits of detection (ppm or below) than SEM-EDX (which offers limits in the thousands of ppm range), this is not as big a limitation as it sounds. The high limits of detection are really only a factor if large regions of interest are analyzed or if traces of relevant elements are present homogeneously distributed though the specimen. With SEM-EDX it is possible to analyze a large region of interest in order to identify the elements present in high abundance and then zoom in to much smaller regions of interest, such as a pigment granule or an aluminum or mica flake, and obtain an elemental analysis of the particular particle. Even though this approach will only detect elements within the particle to about the level of thousands of ppm, the effective limit of detection with regards to the entire specimen can be in the ppm range. A good example of this capability can be found in the work of Suzuki (2014a), which relates to the detection of bismuth oxychloride. While the detection of that particular pigment in and of itself has some relevance to forensic paint investigation (described later), the article is noteworthy because it illustrates the strengths of SEM-EDX. In order for bismuth oxychloride to be effective as a pigment for automotive applications it is given a thin, cerium-rich coating. When SEM-EDX is used to analyze a large region of interest the level of cerium is way below the limits of detection and there is no signal for cerium, but bismuth and chlorine are detected, which provides guidance that bismuth oxychloride is present. However, it is possible to focus the electron beam onto the individual granules of pigment (backscatter electron imaging is of assistance in this regard) and acquire an EDX spectrum of just the particle. Under these circumstances bismuth, chlorine and cerium are all detected with ease. This is in contrast to the results achieved using XRF, which has limits of detection in the ppm range but is restricted to the analysis of a large regions of interest, where cerium can be detected but at the limits of performance.

While the ability to detect elements at trace level is of primary utility, the capability of SEM-EDX to demonstrate colocation of elements within individual particles is an additional benefit. In this particular example, SEM-EDX provides evidence that cerium-coated bismuth oxychloride particles are present in the paint; XRF cannot distinguish that situation from a trace of cerium evenly distributed throughout the paint. A more common example would be the detection of calcium, barium, and sulfur in paint. This indicates that barium sulfate is present but it might not be obvious whether calcium sulfate is present as well or calcium carbonate or both. With SEM-EDX it would be possible to zoom in on individual particles and identify whether sulfur is colocated with calcium or not, thereby resolving a dilemma that XRF cannot. If the analyst has access to elemental mapping capability on their instrument then it is possible to depict the colocation of elements pictorially. Many inorganic pigments are coated, such as lead chromate and molybdate orange (with silica), and bismuth vanadate (silicates, alumina, etc.), while others, such chromium and nickel titanates, are "doped" with other elements such as antimony, niobium and manganese (Suzuki and McDermot, 2006). EDX analysis with a spot beam impacting on individual grains will help to fully characterize pigments such as these and confirm or enhance pigment identification carried out using infrared microspectrometry or Raman microspectrometry. The combination of elements associated with common paint pigments is provided in Table 3.6.

Table 3.6 Elements associated with common paint pigments

Pigment	Major elements
Mica (muscovite)	Al, Si, K
Bentonite	Na, Al, Mg, Si
Talc	Mg, Si
Kaolin	Al, Si
Dolomite	Ca, Mg
Feldspar	Si, K, Na, Al,
Nickel titanates	Ti, Ni (possibly including a trace of Sb or Nb)
Chrome titanates	Ti, Cr (possibly including a trace of Sb or Nb)
Bismuth vanadate	Bi, V (likely to also contain Mo, Al and Zn from the stabilizer coating, and possibly Si, or Zn and P, or Al if the pigment is encapsulated)
Molybdate orange	Mo, Pb, Cr, S
Chrome yellow	Pb, Cr, S (and possibly Si for encapsulated pigment)

SEM-EDX is a relatively straightforward technique, but similar to infrared microspectrometry and Raman microspectrometry the analyst needs to be aware when analyzing spots or regions of interest within a specimen that the region defined by the software is not necessarily exactly the region in the specimen whence the X-rays emerge. The electron beam penetrates some distance into the specimen – EDX is not exactly a near-surface analytical technique. The depth of penetration depends upon the energy of the beam and the atomic weight of the elements present in the specimen; of course, where the electrons penetrate is where the X-rays are generated. X-rays generated close to the surface of the specimen will suffer little attenuation by the specimen but X-rays generated deeper within the specimen will suffer greater attenuation by it. The exact shape and size of the interaction volume for X-ray emission is quite complex and depends upon the energy of the electron beam (which is usually high in paint examination in order excite emission from heavy elements) and the atomic weight of the specimen (which can vary from low in the case of a clearcoat to high in the case of a paint highly filled with a substance such as barium sulfate). It is, therefore, wise to bear in mind that a spot analysis will most likely yield a spectrum derived from both the particle of interest and its surroundings (therefore an analysis of the spot's surroundings is always warranted) and that analyses close to layer interfaces might suffer from "spectral leakage." It might also be informative to re-analyze the region of interest using as low a beam energy as practicable in order to reduce the size of the excitation volume.

SEM-EDX does not require much in the way of sample preparation (other than coating with a thin layer of carbon to dissipate the charge acquired from the electron beam) nor is it a destructive test. Electron beams cannot be controlled in the way that laser beams can be in confocal Raman microspectrometry; for practical purposes if the electron beam cannot interact with an exposed surface of the specimen in the region of interest then an X-ray spectrum cannot be acquired. Usually a paint specimen is multilayered and the objective is to acquire an X-ray spectrum from each layer. On occasions a chip of paint for analysis will fortuitously have a shape that allows it to be placed on a piece of double-sided adhesive tape attached to a pin stub and there is enough of each layer exposed to allow the beam to interact with each layer. With some electron microscopes the specimen can be tilted to facilitate the interaction of a large beam with the specimen layers. If luck is not on the analyst's side, then some sample preparation will be required. A simple option is to recover specimens that have been dissected from each of the layers and sub-jected to infrared transmission microspectrometry, or to dissect new specimens from each layer for SEM if the size of the sample permits. Another option is to dissect the specimen in a way that facilitates EDS analysis and yields "offcuts" of each layer suitable for infrared transmission microscopy. Such a dissection involves slicing through the specimen to produce a sloping cut at a shallow angle that exposes each layer. Alternatively, "stair steps" can be cut into the specimen by the following process. Cut through the top coat to its bottom interface close to one edge of the specimen and scrape or chip out the top coat to the edge of the specimen. Then do the same with the next layer, making the cut through it closer to the edge and so on. Polished or cut cross-sections make an excel-lent specimen for EDX but, as mentioned already, this can be difficult to achieve with "forensic size" specimens. As EDX does not require the beam to pass through the specimen and does not require forceful contact between an objective and the specimen (both of which are required in infrared microspectrometry), another sample preparation technique is available. This involves the use of a microvice that can clamp the paint chip vertically so that a cross-section is presented to the beam. An improvised method for this is to make a vertical cut through the flat surface of an SEM stub to produce a vertical surface to which can be applied a small piece of double-side adhesive tape that, in turn, can be attached the paint chip so that it presents its cross-section to the beam.

For chips about 1 mm square (i.e., big in forensic terms) cross-sections can be cut relatively easily using a microtome, even a cheap hand microtome. First, the paint must be captured between two pieces of craft shop cellulose acetate sheet (about 10×10 mm and 0.4 mm thick) and some acetone "glue". This can be accomplished by taking the two pieces of sheet and clamping them together along one edge using a bulldog clip. Prize the sandwich open with a probe at the edge that is not clamped, apply a small drop of acetone into the gap that is produced, withdraw the probe then put another bulldog clip onto the edge of the sandwich that has been treated with the acetone and allow the sandwich to set for a few hours. After that time remove the bulldog clips, prize open that part of the sandwich that has not been treated with acetone, add a drop of acetone to the gap, place the chip of paint close to the edge of the gap and clamp the sandwich with a bulldog clip and let it set for a couple of hours. Once set, place the sandwich in a microtome and slice some cross-sections; the cellulose acetate will only have very weak adhesion (if any) to the paint cross-section and the whole (or just the paint cross-section) can be placed directly onto a stub for analysis (the same technique can be used to prepare cross-sections for transmission infrared microspectrometry).

It is important to recognize that the X-ray detector has a relatively small surface area and it is, therefore, important to ensure that the X-rays emitted by the specimen can actually reach the detector and not be "shadowed" by other, higher regions of the specimen or the sample holder. This is of importance when "stair steps" are cut into the specimen or if a cut stub or vice are used and a metal edge is shadowing the specimen.

As indicated, a valuable advantage offered by SEM-EDX is the ability to carry out spot analysis on individual particles within a paint specimen. It is useful, however, to obtain an overall analysis of the major elements present in each layer of paint within a specimen. SEM-EDX is effective for this type of analysis but it is important to ensure that the heterogeneity (or "granularity") of the unknown paint is compensated for. This can be done either by carrying out analysis on a region of about 50–100 μm square (or an equivalent area for a rectangle) or by carrying out multiple measurements of smaller areas if a large area is not available for analysis. The former will provide an analytical result that represents the bulk composition and this will be relatively simple to compare with the results obtained from a similar region of interest in the known specimen. The latter will provide the analyst with a number of spectral snapshots covering a range of elemental profiles and the task will be to assess whether that range encompasses the elemental profile acquired from the known paint.

Although SEM-EDX is widely used for forensic paint comparison, there would appear to be only one study in which the discrimination power of SEM-EDX has been estimated (Seccombe, 2001). In this study, very low powers of discrimination were found for non-metallic blue paints (0.17) and for metallic green paint (0.20) but much higher values were estimated for non-metallic black paint (0.97) and for metallic blue paint (0.93).

Readers are referred to the guidelines produced by the American Society for Testing and Materials (ASTM, 2013a) and Scientific Working Group on Materials Analysis (SWGMAT, 2002) for additional information in regards to SEM-EDX analysis of paint.

A number of other elemental analysis techniques are used for forensics paint examination although they are much less popular than SEM-EDX. As discussed above, XRF offers the paint examiner very low limits of detection. Moderating this capability is the relatively large size of the X-ray beam (even in microfluorescence instruments) and its penetration into (in reality, through) low atomic weight specimens such as paint. This limits practically the technique to the examination of thick paint layers, the total analysis of chips, or pieces of each layer dissected from the whole sample. Two laser-based techniques, laser ablation-inductively coupled plasma mass spectrometry (LA-ICP-MS) and laser induced breakdown spectrometry (LIBS), offer great potential for paint examination. Although a relatively large analytical spot size is required if sub-ppm limits of detection are to be achieved, the beam does not penetrate the entire specimen as is the case with XRF, nor does it have the diffuse point spread function problems associated with Raman and infrared microspectrometry or the large excitation volume associated with electron beams. The technique is destructive, though, and significant quantities of soft materials such as paint are ablated with each pulse of the laser beam. This does bring benefits though; it is possible to train the laser beam on the top surface of a multilayer paint specimen and record elemental depth profile detail while the laser tunnels its way through the specimen. This obviates the need to dissect the specimen or manipulate it to present a cross-section to the beam. Readers who are interested in further information about XRF are referred to the works of Suzuki (Suzuki and McDermot, 2006; Suzuki 2014a, 2014b) and the articles by Hobbs and Almirall (2003) and Deconinck *et al.* (2006) in regards to LA-ICP-MS.

X-ray diffraction is another technique that is also not widely used and while it does not provide elemental profiles of paint it does provide extremely valuable information in regards to crystalline substances present, which in paint means the pigments. Similar to XRF, in XRD the specimen is interrogated with a relatively large X-ray beam. In the latest instruments the beam is just small enough to allow the interrogation of a single paint layer in a cross-section, but in the case of thin films and older-generation instruments the paint must be dissected and specimens relating to individual layers must be presented to the instrument in order to obtain unambiguous information for each layer. The characteristics of XRD are that the signals can be used semi-quantitatively and, depending upon the mixture of pigments, the presence of a minor component can be detected down to the low percent level. For the elemental analysis techniques mentioned in this section the analyst must infer the identity of pigments from the elemental profiles obtained; for example, the presence of barium and sulfur would allow an inference of the presence of barium sulfate. The unique capability offered by XRD is that it can be used to positively identify and discriminate between many minerals. In theory, therefore, XRD should be capable of a very high discrimination power for paint analysis, but numerical values have not yet been calculated. For additional information on XRD, the reader is directed to Clearfield *et al.* (2008), Dinnebier and Billinge (2008), and Koleske (2012).

3.4.5 Pyrolytic techniques

Pyrolytic techniques are relatively popular for forensic paint examination with up to one in three laboratories that participated in CTS trials making use of it.

The basis of pyrolytic techniques is to heat the specimen to a high temperature in the absence of oxygen and collect and examine the vapors produced (i.e., the pyrolyzate). Infrared spectrometry and direct infusion mass spectrometry have been used to examine pyrolyzates but by far and away the most effective method is to direct pyrolyzate into a gas chromatograph, preferably one equipped with a mass spectrometer (where the technique is referred to as Py-GC-MS) and acquire a chromatogram (referred to as a pyrogram).

There are two ways in which the specimen can be heated. One method employs a flattened ferromagnetic wire folded around the chip of paint to be pyrolyzed. This wire is then placed into a furnace where a high frequency electromagnetic field inductively heats the wire almost instantaneously up to its Curie point (the temperature at which it loses is ferromagnetism, 770°C in the case of iron wire), which causes instant pyrolysis of the paint. That technique is referred to as Curie point pyrolysis. The other technique is to place the paint into a small tube and place it into close contact with a wire coil that is resistively heated to the chosen temperature. Conduction of heat from the coil through the tube into the paint causes pyrolysis.

The benefit of Py-GC-MS is that pyrolysis does not produce a random collection of molecules in the pyrolyzate, the composition of the pyrolyzate is rational and usually reproducible. Therefore, the technique is far more than a pattern-matching exercise where the pyrogram pattern produced by the unknown paint is compared with the pyrogram pattern produced by the known paint.

The simplest pyrograms in regards to paint are produced from the acrylic resins; the effect of heat is to bring about a depolymerization of the resin. Pyrolysis of pure poly(methyl methacrylate), for example, produces a pyrogram containing just one peak, that corresponding to methyl methacrylate. Pyrolysis of copolymers produces a pyrogram with peaks corresponding to each monomer and if plasticizer is present then it will also appear in the pyrogram. It is, therefore, very simple to "deconstruct" acrylic-containing paints into their components and use this as the basis for comparison between known and unknown paints. Styrene is a very common comonomer in paint and under the influence of heat it yields its monomer, styrene, but in the pyrolysis chamber very rapidly some of the monomer dimerizes and trimerizes.

Other polymer systems undergo much more complex chemistry under pyrolysis. For example, poly(vinyl acetate) does not undergo simple depolymerization to yield vinyl acetate, and alkyd resins produce quite complex pyrograms. While polyurethanes do produce the polyol and isocyanate from which they are prepared, other products form as well. In order to simplify the chemistry involved, reduce the polarity of products, and increase their molecular weight it is common practice to carry out thermally-assisted hydrolysis-alkylation (THA) pyrolysis for alkyd, vinyl, epoxy and other highly function-alized or highly polar paints. The basis of the THA chemistry is depicted in Figure 3.13.

Although it is possible to compare complex paints without resorting to THA, pyrograms are much cleaner if this technique is used and much more informative. As pyrolysis is destructive, and a relatively large piece of paint is required to carry it out, it might only be possible to subject unknown paints to a single pyrolysis experiment. Therefore, it is recommended that pyrolysis of an unknown paint only be carried out after infrared microspectrometry has informed the analyst as to the formulation of the paint. If it is found that the paint is of the acrylic type then it is not necessary to resort to THA but if the paint is of the alkyd, urethane or epoxy type then THA is recommended. When there is enough known paint then a number of pyrolyses can be carried out using it to

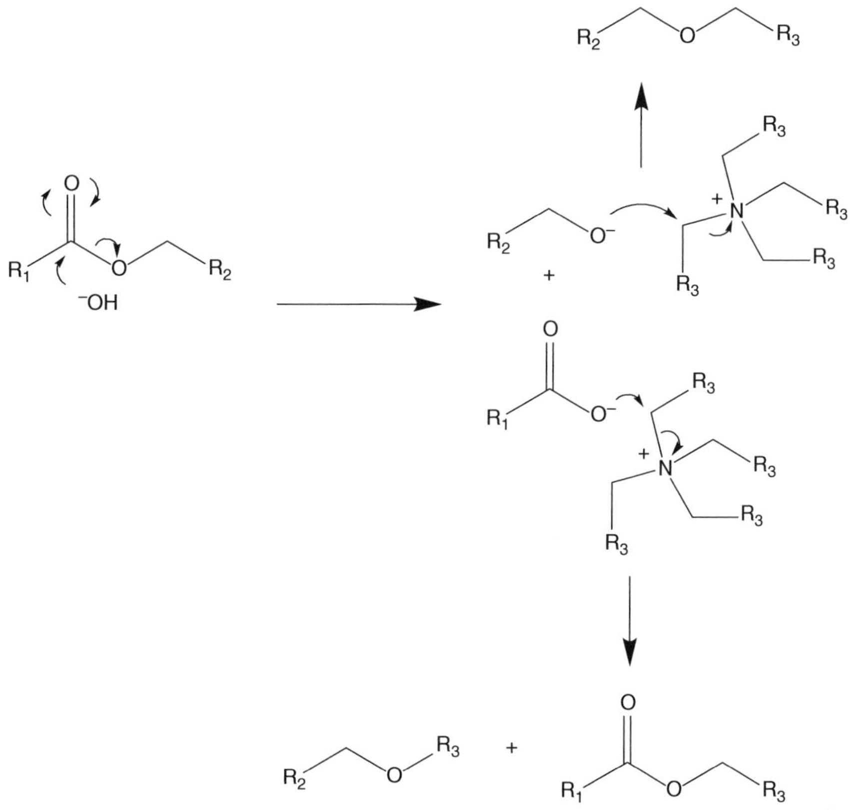

Figure 3.13 Thermally-assisted hydrolysis-alkylation reaction scheme using the alkylation of an ester as an example. R_3 usually is H, methyl, ethyl or butyl

establish whether THA is required and to characterize the intrasample pyrogram variability (usually manifested as variability in peak areas and presence/absence of trace-level compounds).

Other than the temperature at which pyrolysis is conducted and the type of pyrolyzer used, the analyst has relatively little control over the pyrolysis process. This is not the case with the associated GC or MS equipment, which can be varied widely to influence selectivity and resolution. As most pyrolyzates contain oxygenated and, therefore, polar compounds (especially if THA is not used), it is effective to use a stationary phase with some polarity, rather than the more common polydimethylsiloxane or 5% phenyl–95% dimethyl-siloxane phases in which many pyrolyzates have low solubility and poor retention. A stationary phase that performs very well in Py-GC-MS is that based on the "1701" formulation of 14% cyanopropylphenyl–86% dimethyl-siloxane (OV1701, DB1701, HP1701 etc.).

As is usually the case with chromatography-based techniques, certain quality control measures are important for Py-GC-MS. First, a positive control is vital. This should be a specimen that yields results that are familiar to the analyst; for example, a known quantity of a particular resin that yields a pyrogram comprising a known number of peaks at known retention time of known intensity. Such a specimen will indicate whether there are leaks in the system (reduction in peak intensity and perhaps lengthened retention times), whether pyrolysis is not taking place or the correct temperature is not being achieved (no pyrogram or unusual relative peak intensities) or whether there are chromatographic problems (such as incorrect settings of GC or MS parameters, or active sites in the pyrolyzer, GC liner or column, which manifest as tailing peaks or peak absences). Second, system blanks are very important in order to identify signals in the pyrogram that arise from anything associated with the process (contaminants in the pyrolysis head, injector liner, THA reagents or the GC column, or fingermarks or other impurities on pyrolysis wires or tubes) rather than the paint itself. Ideally a system blank should be run after the positive control, as this will allow carryover of analytes to be identified. It is also good practice to run a system blank between pyrolysis of the known and unknown specimens in order to demonstrate that pyrolyzates from the known paint did not carryover into the pyrogram of the unknown paint.

Py-GC-MS is extremely powerful for forensic paint comparison and offers some unique capability. It has excellent capability for high resolution qualitative analysis and, as peaks rarely overlap and their areas can be measured accurately, Py-GC-MS can be used to compare known and unknown paints on the basis of the relative abundances of compounds in their pyrolyzates. For example, it is a trivial exercise to separate and identify the phthalate ester plasticizers in paint by GC-MS. As the infrared absorption bands for these plasticizers overlap with those arising from paint resins it is very difficult to even detect the presence of plasticizer in some paints, let alone detect whether a mixture of them is present (which there can be, either as a result of deliberate mixing or because the plasticizer used is not 100% pure). In the case of alkyd resins, Py-GC-MS can resolve the presence of multiple polyhydric alcohols or dibasic acids and indicate their relative abundances in the mixtures. Also in regards to alkyd paints, Py-GC-MS can resolve the fatty acid residues from the monoglycerides used, which in turn will provide evidence as to the triglyceride oil(s) used in the paint and the "length" of the oil (i.e., the ratio of unsaturated to saturated fatty acid residues, with a "long" oil alkyd having a high level of unsaturation). The GC column, if it is temperature programmed correctly and if the correct stationary phase is used, will be capable of resolving saturated fatty acids from unsaturated ones and the mass spectrometer will readily allow their identification. As unsaturated fatty acids will gradually cross-link over time, Py-GC-MS offers the capability to compare the "age" of the known paint with that of the unknown paint on the basis of the ratio of saturated to unsaturated fatty acids present in each. Additives such as UV absorbers are easily detected by pyrolysis. Finally, Py-GC-MS has limits of detection low enough and resolution high enough to allow the detection of traces of compounds present in the paint film; this includes residual solvents in fresh paint such as turpentine or coalescing agent.

Clearly in the case of a multilayer paint sample it is desirable to dissect representatives from each layer and pyrolyze them separately. As a few micrograms are required for pyrolysis this means that the unknown paint sample must be quite large. Given the recommendations above in regards to quality control, the requirement to dissect the specimen, and the long run time taken to acquire a single pyrogram (about 15–20 minutes) it can be seen that from the beginning of analysis to its end Py-GC-MS is slower than infrared microspectrometry and much slower Raman microspectrometry. The paint examiner must take great care to not contaminate pyrolysis tubes or wires with fingermarks and, when dissecting a sample, ensuring that fragments from adjacent layers do not contaminate the analytical specimen because, unlike microspectrometry (where contaminated sections of the specimen can be seen and avoided), what goes into the pyrolyzer will produce some peaks in the pyrogram.

There have been several qualitative descriptions of the discrimination power of pyrolysis techniques. Ryland has shown that Py-GC-MS can easily discriminate clearcoats that are difficult to differentiate using IR spectrometry (Ryland *et al.*, 2001) and in 1995, as described previously, Ryland found that Py-GC was easily capable of discriminating two green Ford acrylic-melamine automotive paints that were indistinguishable using infrared spectrometry (Ryland, 1995). Zieba-Palus *et al.* (2008) examined 150 clearcoats and reported that in most cases the paints could be discriminated; a Dp value was not reported however. Muehlethaler *et al.* (2013) described a European round-robin assessment of batch-to-batch variation for four spray paints (four colors, three manufacturers, 23 specimens with orthophthalic alkyd or nitrocellulose formulations). Simple pyrolysis rather than THA was used (arguably not the best choice for the formulations examined) and another limitation was that only a small sample set was used in the research. It was reported that results varied markedly across participating laboratories in regards to the presence and relative abundance of pyrolysis products. Furthermore, different discrimination powers were achieved for each of the different colors; pyrolysis could not be used to distinguish any of the batches of papaya colored paint but distinguished all of the white paints. The final conclusion in regards to detecting batch-to-batch variation in paints was that tests that distinguish pigments (Raman, MSP, and microscopy) gave better discrimination than those probing either the binder (infrared microspectrometry and pyrolysis) or elemental compositions. In a somewhat dated publication, Fukuda (1985) describes the analysis of 31 white paints, 31 metallic silver paints and 16 clearcoats. Although Dp values were not reported, the publication is detailed enough to allow the reader to calculate Dp values, which were 0.98 for the white paints, 0.96 for silver paints and 0.97 for clearcoats. In a more conventional power of discrimination study (involving analysis of paints gathered at random rather than analysis of batches of the same paint), Seccombe (2001) found Dp values ranging from 0.67 (for 27 samples of grey paint) to 0.77 (for 13 samples of white paint) using Curie point GC with flame ionization detection. Seccombe highlighted two white alkyd paints that were indistinguishable using infrared microspectrometry but easily discriminated by

Py-GC, even with a non-polar column (5% phenyl–95% methyl-silicone) and direct pyrolysis rather than THA.

Leading references in regards to pyrolysis as applied to the analysis of paints are Wampler *et al.* (1997), Wampler (2004), Challinor (1991, 2001a, 2001b, 2007), and the Scientific Working Group on Materials Analysis (SWGMAT, 2013).

3.4.6 Color analysis

The fact that many painted articles can readily be distinguished from one another purely by their color is self-evident – especially to jurors. It is also self-evident that automobiles come in a variety of shades and, moreover, some shades wax and wane in popularity. The range of architectural paint colors is less broad and light shades are more popular; therefore, color examination is likely to be a less probative for discrimination of architectural paints.

Human eyes, even though they are capable of discriminating approximately 100,000 colors side-by-side (Boynton, 1979), can be fooled. The human perception of color can be considered as a convolution of the spectral properties of the source of light (e.g., sunlight or tungsten light), the light-reflecting or light-absorbing properties of the object being viewed, and the sensitivity of the eyes to light (which is not constant across the visible spectrum). As a result, it is the case that different combinations of different wavelengths can result in the same color sensation – this phenomenon is referred to as metamerism (Cousins [1989] provides an excellent example of two paint samples that are metameric). Finally, in forensic paint comparison the pieces of unknown paint are often quite small in size and, therefore, the only way that the examiner can view the evidence in any detail is through a microscope. Consequently, unless the pieces of unknown paint are large, the only way that an examiner can compare the color of known and unknown paint by eye is to use a comparison microscope. Even if the light sources in a comparison microscope are well balanced (and for valid color comparison they must be), the comparison process is subjective and the only objective record that the examiner can provide for the jury or counsel is a photograph. For all of the preceding reasons, an objective means for comparison of one color with another for forensic purposes is important.

Microspectrophotometry (MSP) of the known and unknown paints, carried out in the visible or ultraviolet–visible region, is the most popular, direct, objective way that color comparison can be accomplished. An indirect method is to use Raman spectrometry and compare the pigment compositions of the known and unknown paint. While the statistics relating to CTS proficiency tests indicate that Raman spectrometry is not widely used, surprisingly, it appears that more laboratories use visual means to compare the color of paint samples than MSP, perhaps because of the expense of the equipment required to carry out MSP. Whatever the reason for the popularity of MSP, it is nevertheless a valuable and powerful technique.

MSP probes the specimen with short wavelength radiation, similar to the situation in Raman microspectrometry. As a consequence, the point spread function applicable to the analysis is small and spectral "leakage" between adjacent layers is much reduced (however, it is still not recommended to measure spectra with a region of interest that terminates at a layer boundary). Unlike infrared microspectrometry, where diffuse reflectance spectrometry is usually not carried out for paint analysis, MSP can and is used quite effectively in either transmission or diffuse reflectance configuration, but transmission measurement produces lower noise and narrower peaks, and it is more reproducible (Eyring *et al.*, 2007). Samples microtomed to about 1 µm for dark-colored paints or thicker for pale-colored ones can be effective for transmission measurements – if the sample allows light to pass when viewed with transmitted light under the microscope then it should be suitable for MSP. In addition to the method described in the infrared microspectrometry section (Section 3.4.3), the methods of Wilkinson *et al.* (1988, who used gelatine mounting medium) and Eyring *et al.* (2007, who used Technovit 2000LC resin, which can be quickly set using intense blue light) have been used to achieve appropriate cross-sections for MSP. Pale-colored specimens flattened on diamond windows for infrared transmission microspectrometry can be used without further modification for transmission or reflectance MSP. The ideal situation for comparison of paint using transmission MSP is to measure cross-sections of known and unknown samples of the same thickness; that way if the paints do have a common source then the absorbance values in the pair of spectra should be close, as predicted by the Beer–Lambert law.

It is important in reflectance MSP to ensure that the contributions of stray light and specular reflectance are equal (ideally zero) for the known and unknown paint samples in order to avoid falsely concluding that the two samples are not associated. Instrument design helps in this regard; microscope objectives for reflectance MSP are of the reflected darkfield type (Figure 3.14). These objectives make use of an annulus of light (rather than a circular spot) to illuminate the specimen and light that is reflected specularly from the specimen is rejected and that which diffusely reflects at close to 45° is collected. The examiner does their part to ensure that specular reflectance is avoided by measuring polished cross-sections for the two paint samples but, as mentioned previously, this might not always be an appropriate tactic for small samples of paint. If this is not possible then the only option for reflectance MSP is to search for a flat region of interest in the unknown paint specimen that appears to be aligned perpendicular to the instrument's beam.

Similar to the other techniques described above, the heterogeneity of paint and the variation in the surface finish across the specimen must be taken into account when carrying out MSP (and especially in reflectance MSP), as the analytical region of interest is usually quite small (approximately 10 µm or less). The best approach is to record a number of measurements from different locations within the specimen under analysis in order to define the intrasample variance. However, unlike any of the

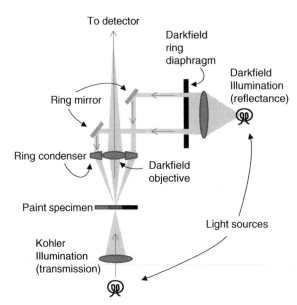

Figure 3.14 Schematic diagram of a typical microspectrophotometer (*See insert for color representation of the figure.*)

spectrometric techniques discussed above, the features in MSP data are extremely broad and overlapped[1]. This can make it difficult to compare spectra from known and unknown paint directly by eye. The examiner first must look for obvious features, such as a different number of resolved or partly resolved peaks in the known and unknown samples or additional shoulders, and then look for more subtle features, such as different slopes in the spectral cutoff region and cross-overs (metameric inversions) that are usually only apparent when spectra are overlaid. As demonstrated by Eyring *et al.* (2007), because absorbance is related to transmittance logarithmically, absorbance overlays are much more effective for revealing variations in strong spectral features. However, distinguishing features in visible spectra can be weak; in this case, transmittance overlays help to reveal variance much more readily than absorbance overlays.

The task can be assisted by mathematical treatment of spectral data. The simplest approach is to convert the spectral data into its first or fourth derivative form and compare those data for the known and unknown samples. Features such as baseline offsets or a rising baseline due to scattering will be ameliorated by plotting derivative data. Any even-numbered derivative will reduce the width of spectral peaks, thus allowing deconvolution of overlapping peaks, but the process strongly accentuates any narrow peaks that are present with the broader features being discriminated against. The main factor counting against the widespread usage of derivative spectrometry is that it degrades

---

[1] An exception applies to interference pigments, which give rise to relatively narrow peaks.

the signal-to-noise ratio (noise, being narrow in peak width, is accentuated by derivatization), which can be naturally quite low in MSP. Another mathematical approach is to convert MSP data into a numerical representation (a process referred to as colorimetry) – the most popular is to produce CIE L*, a*, and b* chromaticity coordinates. An in-depth discussion of colorimetry is beyond the scope of this chapter; interested readers are directed to the review by Thornton (1997) as a starting point. The variable L* covers the lightness or darkness of a color (ranging from 0 for pure black to 100 for pure white), a* covers the gamut of colors along a cyan-magenta continuum with gray at the center of the range, and b* covers a gamut of color along a yellow-blue continuum with gray at the center of the range. The CIELAB color space has these axes arranged orthogonally and when the space is illustrated two dimensionally it shows a* plotted against b* for some value of L* such as 50. Personal computers attached to microspectrophotometers can readily convert a transmission or reflection spectrum into its L*, a* and b* coordinates. Therefore, the range, average, and standard deviations of values for known and unknown specimens can be compared objectively. The American Society for Testing and Materials Guide E2808 (ASTM, 2011) specifies that the range for the known and unknown paints should coincide if the two are associated and it stresses that colorimetry should be used in conjunction with a visual analysis of spectral data rather than as substitute for it, as metameric colors can have the same L*, a*, and b* values. A number of articles have been published that deal with the use of colorimetry to identify a short-list of vehicles that could be the source of an unknown paint for intelligence or investigative support (Nowicki and Patten, 1986; Locke *et al.*, 1987, 1988; Taylor *et al.*, 1989).

The forgoing discussion has dealt with paints that are obviously colored. True black objects, true white ones and those colored with a mixture of the two (i.e., gray objects) do not exhibit any transmission or reflectance variation across the spectral range. However, many black, white and gray paints are not necessarily "true" and exhibit some peaks in the visible spectral region. This is because whites can contain tints such as red and carbon black to alter their appearance. As titanium dioxide is slightly blue its influence on the overall paint shade can be diminished by the introduction of some yellow pigment. Likewise, black paints can contain mixtures of colored pigment granules, which can be observed using polarized light microscopy. It is not, therefore, as seems to be commonly believed, pointless to subject white, black or gray paints to MSP (Kopchick and Bommarito, 2006).

Very dark paints can be difficult to analyze using MSP. To rectify this by lightening the shade, Cousins (1989) recommended rubbing the paint sample on a white tile and recording a reflectance measurement off the tile, or grinding the paint and mixing it with a pure white extender such as magnesium carbonate. Another technique for soluble pigments that has been revisited in the literature over a number of years (Massonnet and Stoecklein, 1999c; Fuller, 1985) is to extract the pigment(s), separate them using thin layer chromatography (TLC), and then carry out reflectance MSP. This approach would seem to offer the possibility of augmenting MSP with more sophisticated, molecular analysis techniques for the analysis of pigments such as laser desorption ionization mass spectrometry (Weyermann *et al.*, 2012) or time-of-flight secondary ion mass spectrometry (Coumbaros *et al.*, 2009), both of which have been shown to be effective for the analysis of dyes on TLC plates.

UV absorbers and HALS are not colored but clearly they function by interacting with UV light. This means that if MSP is carried out using a microscope and accessories (in particular microscope slides and covers slips) that are capable of transmitting UV light, then it is possible to differentiate clearcoats on the basis of their UV absorbers and HALS (Table 3.1 and 3.2) and it is possible to differentiate clearcoats that are stabilized from those that are not. Stoecklein and Fujiwara (1999) and Liszewski *et al.* (2010) provided good examples of spectra and an excellent discussion on the subject.

Laing *et al.* (1982) in a study of gloss domestic architectural paints measured a Dp of 0.97 using reflectance MSP and 0.89 for transmission. Rather than assess discrimination by eye, the authors used a mathematical algorithm that measured the "distance" between the spectral data of the known and unknown paint samples. Seccombe (2001) calculated the Dp for MSP for red, green, and yellow paints (only four samples of the latter) as 0.8, 0.64 and 1, respectively. In a study involving macroscopic spectrophotometry (11 mm spot size), Lee and Sandercock (2011) took the approach of converting reflectance data from 156 automotive paint samples into their CIE L*, a*, and b* chromaticity coordinates and used a criterion of $\Delta E>2$ as representative of paints that can be discriminated by eye. Their work showed that the chance of a random color match between two vehicles was 0 for gold colored vehicles, 0.28 for greens, 0.04 for blues, 0.06 for reds, 0.07 for silver/greys, 0.22 for whites, and 0.46 blacks. MSP can be expected to show much lower precision than macroscopic spectrophotometry; therefore, the work of Lee and Sandercock probably represents an upper limit to what might be achieved using MSP.

Where it is possible, known samples should be collected from areas that have suffered weathering and fading equivalent to that suffered by paint in the area from which the unknown paint is thought to originate, otherwise MSP might lead to a falsely negative comparison.

Good general references to MSP are provided by the American Society for Testing Materials (ASTM, 2011) and the Scientific Working Group on Materials Analysis (SWGMAT, 2007).

3.5 Paint evidence evaluation and expert opinion

If a report is being compiled for court it is presumably the case that the analyst has reached a conclusion that the questioned paint could have originated from the object involved in the crime, for example, the victim's car or house. There are many components that make up a court report that are not discussed here, such as chain of custody records, description of items, the purpose of the examination and so on. Instead, this section concentrates solely upon that part of the report that deals with the statement of the type "the questioned paint could have come from" the particular crime object.

An examination of proficiency test responses (CTS, 2014) shows that the vast majority of respondents reported simply that the questioned paint could have come from the putative source, or that the questioned paint is consistent with its origin being the putative source, or that it cannot be excluded as the putative source. In many of those responses the examiner included in their report words to the effect that sources of paint other than the putative source cannot be excluded, in deference to the fact that paints are produced in large batches and will be applied to a number of motor vehicles or a number of architectural fittings. One respondent did not include that caveat, and simply said that the unknown paint did originate from the same source as the known. Approximately 10% of respondents offered nothing more in their report than that the known and unknown paints shared similar characteristics, were indistinguishable, showed no significant differences, or were consistent.

It might be the case that the respondents would provide additional information in court, if asked, as to what weight and significance should be attached to the word "could" or what interpretation should be attached to the words "indistinguishable" or "no significant difference," assuming of course that the examiner is called to court. One approach to help the court is to provide advice as to how many sources can be eliminated, given that a finite number of tests have been performed, each of which cannot individualize a paint. In the example of an automotive paint transfer to a victim's car it is clear that not just any other vehicle could have been the source; the true source vehicle comes from a cohort of vehicles with paint color indistinguishable from that of the unknown paint and with composition indistinguishable based upon a barrage of tests (for example, infrared microspectrometry and SEM-EDX). The court should be informed if the above factors indicate that the transferred paint is rare in the community (and therefore that the significance of the match is high) or whether it is common. As well as the rarity of the transferred paint, the court also might need some assistance in grappling with whether the characteristics of the transferred paint are commensurate with the alleged activity associated with the offence. This, too, is within the capability of the jury. They should be able to comprehend, for example, that a large quantity of transferred paint is consistent with a high-energy collision, and if the amount of paint transferred in the case that they are hearing is consistent with the proposed scenario then the jury should be able to understand that the evidence is sound and significant. A relevant consideration, though, is whether the jury is capable of integrating in a logical way all of the information that will be presented by the paint examiner.

The 2014 CTS trial responses showed that a small minority of examiners (less than 5%) drew their opinion from a hierarchical list of levels of association (Bommarito, 2006), ranging from Type 1 (conclusive association, such as a match in fracture pattern between the unknown paint and its source) through Type 3 (where the unknown paint could have come from the putative source but the possibility of another origin cannot be ruled out) to Elimination (where it can be conclusively stated that the unknown paint did not originate from the putative source). In the absence of additional information from the examiner, statements such as these still leave it to the jury to decide just how likely it is to encounter another source of paint indistinguishable from the unknown one.

Less than 3% of respondents took it upon themselves to provide a "ready-made" interpretation of their analytical results for the users of their reports. This was done by indicating the extent to which their examinations supported the proposition that the unknown paint came from the putative source. This approach has value because at least the jury will be presented with a logically constructed opinion that might be tested under cross-examination rather than having to formulate their own, which will not be tested. None of the respondents indicated that numerical methods were used to arrive at their opinion, which might explain why there was such a diverse range of opinions expressed under the highly controlled circumstances of a proficiency test; one respondent indicated support for the proposition, one indicated moderately strong support, and one indicated strong support. McDermott and Willis (1997) in a seminal study found that paint examiners who mostly relied upon qualitative methods to arrive at their opinion consistently underestimated, or expressed conservatively, the strength of paint association in a number of scenarios. While this is laudable inasmuch as it favors the defendant, such biases are not in the interests of justice (or in the interests of the victim).

A significant body of knowledge is building with respect to the usage of "logically correct" methods based upon empirical data for evaluation of many evidence types, including paint. Nevertheless, unlike biological evidence and glass for example, the wealth of relevant empirical data that can be applied to numerically-based frameworks for paint evidence is comparatively small. The following discussion is based upon the works of McDermott et al. (McDermott and Willis, 1997; McDermott et al., 1999) and Buzzini et al. (2005) and the reader is strongly urged to read those articles and the references cited therein.

An approach recommended in the literature for numerical evaluation of paint (and other types of) evidence is for the examiner to calculate a mathematical operator called the likelihood ratio (LR) applicable to their findings. The significance of the LR is that it is a simple multiplier that can be used to "update" the opinion that a trier of fact already holds (as a result of the evidence that they have heard previously) in light of the new evidence that they are receiving (in this case, the paint analysis findings). This is referred to as the trier's "prior odds" being updated by the LR to arrive at the trier's "posterior odds." This is written mathematically as:

$$Posterior\ odds = LR \times Prior\ odds$$

which is known as the LR form of Bayes' Theorem.

Clearly, if a LR is very large then multiplication of the trier's prior odds by a large number results in posterior odds much higher than the prior odds – in this case as a result of the new evidence the trier should arrive at the conclusion that the evidence against the accused is much stronger than it was before the new evidence was heard.

As implied by the name, the LR is a ratio; it is the ratio of the probabilities of two hypotheses relating to the particular evidence in question, in this case, paint. These two hypotheses are often referred to as the prosecution hypothesis and the defense hypothesis. As the prosecution hypothesis is usually "the defendant committed the crime" and the defense hypothesis for practical purposes is "my client is innocent" – both of which are ultimate questions that are the province of the triers of fact and not the forensic scientist – it is better to consider the LR to involve probabilities of source (or not) or the probabilities of contact (or not). Under these conditions, the LR is expressed as:

$$LR = \frac{P\left(Evidence|source\ is\ victim's\ and/or\ suspect's\ car\right)}{P\left(Evidence|source\ is\ not\ victim's\ and/or\ suspect's\ car\right)}$$

The two probability statements should be read in the following way. For the numerator, it is usual that this refers to the probability that the evidence is of the form observed (e.g., layers of paint of a certain size, color layer sequence and composition) given that the vehicles owned by the suspect and/or the victim are the source(s) of paint. The qualification "and/or" is required because paint transferal is not simple – it can involve one-way transfer from the victim's car to the suspect's, or vice versa, or two-way transfer from each vehicle. For the denominator, it is usual that this is the probability that evidence of the same form is observed given that a vehicle at random rather than the suspect's or victim's vehicle is involved.

In a hit and run collision the simplest scenario involves one way transfer of paint: foreign paint is found on the victim's car and the allegation is that it arose from the suspect's car; or foreign paint is found on the suspect's car and the allegation is that it came from the victim's car. Although intuitively it feels like the LR of these two scenarios should be equivalent, they are not. In the scenario where foreign paint is found on the victim's vehicle we have additional information on-hand; it is usually the case that the victim knows for sure that the foreign paint was not present on their vehicle before the collision. In this instance, the numerator of the LR can be stated as "what is the probability of observing the paint evidence given that it was transferred from the suspect's vehicle?" Clearly this has to be 1; if there is no possibility that the foreign paint arose from another vehicle prior to the accident and the suspect's vehicle is indeed the source of the foreign paint on the victim's vehicle, then that paint evidence would certainly be of the form found. In regards to the denominator, if the suspect's car is not the source, then another car at random with paint of that form must be the source. The LR then becomes simply:

$$LR = 1\,/\,fsp$$

where fsp is the frequency of occurrence at random of paint of the type found adhering to victim's vehicle. The word "type" here relates to paint with the characteristics (IR spectrum, color, etc.) revealed during analysis.

As alluded to above, the LR for the scenario where foreign paint found on the suspect's car matches the paint on the victim's car is different to the LR for the scenario where paint found on the victim's car matches that on the suspect's car. This is because there are two ways in which the paint evidence would be as it is observed if the victim's car is the source. The first way is the trivial one; that is, that the paint evidence is as observed because the paint was transferred to the suspect's car as a result of its contact with the victim's car (as before, the probability of observing this evidence would be 1). The second way that the paint evidence could be observed is if the foreign paint was present on the suspect's vehicle before contact with the victim's car and paint was not transferred during the collision with the victim's car. This second situation is not applicable in the scenario above where foreign paint was known to have transferred to the victim's car as a result of a collision. The numerator of the LR in the second scenario must, therefore, have three components: one that deals with the probability of observing the particular paint evidence if the paint was deposited before the accident; one that deals with the probability of observing the particular paint evidence if the victim's car is the source (i.e., 1); and one that deals with the probability that a collision of the type that the victim's car was subjected to *did not* result in paint being transferred. The probability of observing the evidence if the paint was in place before the accident can be estimated from surveys that count the number of vehicles "on the road" that have foreign paint adhering to them (let this frequency be fp). Surveys from Ireland (McDermott *et al.*, 1999), New Zealand (Seccombe, 2001), and Australia (Monahan, 2010) have provided these data and the values were 9.4%, 36.7%, and 13.5%, respectively. Currently, there are no data relating to the probability of paint being transferred during a particular type of accident; McDermott *et al.* (1999) have estimated this probability in the case of just one layer of paint being transferred (ft) as 0.8, and 0.4 as the probability of more than one layer being transferred. Therefore, the probability that an accident would *not* result in a single layer of paint or multiple layers of paint being transferred are 1 − ft (which are 0.2 or 0.6, respectively). As before, the denominator is quite simple; it is the probability of observing the particular paint evidence if a car at random rather than the victim's car is the source of the paint on the suspect's car. Therefore, the expression for the LR for this scenario becomes:

$$LR = \frac{(1-\text{ft}) \times 1 \times \text{fp}}{\text{fvp}}$$

As both 1 − ft and fp are less than 1, the numerator in the above expression must also be less than 1. As a consequence, if all other factors are equal, the LR for one-way transfer from the suspect's vehicle to the victim's vehicle will always be greater that the LR for one-way transfer in the other direction. If the scenario involves two-way transfer then the applicable LR is simply the product of the two expressions given above or:

$$LR = \frac{(1-\text{ft})\times 1 \times \text{fp}}{\text{fvp}} \times \frac{1}{\text{fsp}}$$

Critical to the estimation of LRs for any forensic scenario involving automotive paint is estimation of the rarity of relevant paint in the community (i.e., fsp and fvp). McDermott *et al.* (1999) calculated LRs simply on the basis of the color of the transferred paint and the frequency of that particular color in the motor vehicle population of their jurisdiction (Ireland). Their work includes a table of frequencies for many automotive colors in Ireland in the late 1990s. Using the data published for the scenario where a single layer of foreign paint is transferred only to the victim's car, values for the LR of between 8 and 1000 were calculated for common colored paints (e.g., white) and rare colored paints (e.g., light orange), respectively, so it can be seen that the LR is quite sensitive to fsp. Other surveys relating to the popularity of automotive paint colors at different points in time in different countries are available (Ryland *et al.*, 1981; Seccombe, 2001; Edmonstone *et al.*, 2004; Monahan, 2010; Lee and Sandercock, 2011). Monahan (2010) carried out a comparison between a color survey conducted in the Australian Capital Territory and records collected by the road transport authority in that jurisdiction. There was a close agreement between the survey and official figures, and the choice between the two data sets had very little practical impact upon the final LRs calculated. Surveys are time consuming to conduct, especially on an annual basis. Although it depends upon the idiosyncrasies in data collection in other jurisdictions, government figures might be a valuable and relatively simply-accessed source of fc data for forensic paint evidence evaluation.

A number of authors (Ryland and Kopec, 1979; Gothard and Maynard, 1996; Seccombe, 2001; Edmonstone *et al.*, 2004) applied a full suite of techniques to sets of paints and examined the pairs of paints left undiscriminated after all the tests have been applied – their finding was that the undiscriminated pairs were vehicles of the same color, make, and model. In more recent studies (Maric *et al.*, 2012; Lavine *et al.*, 2014) infared spectrometry and chemometrics were found to be sufficient to discriminate clearcoats to the make, model, and manufacturing plant. This level of discrimination arises as a result of the necessity for paint manufacturers to "fine tune" their formulations to suit regional differences in the spray paint equipment (Section 3.3). If a barrage of tests can differentiate paints to this level then fps or fpv is "simply" the product of the coincidence frequency of the particular color of the paint multiplied by the relevant coincidence frequency of the make and model and plant of origin of the vehicle in question. McDermott *et al.* (1999) showed how to use make use of manufacturer frequencies to arrive at LR values. The LR applicable for the one-way transfer of white paint (the most common color) to the victims vehicle from a Toyota (the most common make at 12.5% of the population) is about 64 (compared to about 8 when the make is not considered) while one-way transfer of light orange paint (rare) from a Saab (a rare vehicle at about 0.7% of the population) gives a LR of about 140,000 (compared to about 1000 when make is not considered). Currently there are no published figures regarding the relevant coincidence frequencies for vehicle make, model, and manufacturing plant nor any publications dealing with the elaboration of these statistics into LR values – and it is not a trivial exercise. It will have to be established whether differentiation is dependent on temporal factors (i.e., is it possible to distinguish paint on vehicles from make A, model B, plant C in 2013 from paint on vehicles from the same make, model and plant in 2014?) and then it will be necessary to estimate the size of each of distinguishable make, model and plant cohort vehicles "on the road" (which will vary over time as the vehicle population evolves, and of course, will vary from country to country). Although the ability of paint examination to predict the make, model, and plant from an unknown paint cannot yet be used numerically to evaluate evidence for court purposes, it is applicable to assisting police investigations or contributing to intelligence.

Another, arguably more practical, way to evaluate paint evidence is to make use of published empirical discrimination power (Dp) values. Gothard *et al.* (Gothard, 1976; Gothard and Maynard, 1996) examined collections of 500 automotive paint specimens as to their layer structure and counted the frequency of occurrence of the many layer sequences found. They found that specimens comprising more than six layers of paint were very rare in a random population (i.e., most vehicles only had their OEM paint layers present) and, furthermore, each of the >6 layer sequences occurred only once in the data set. Similar conclusions have been drawn by Ryland and Kopek (Ryland and Kopec, 1979) with a smaller data set. On the basis of this work then, in the situation where a paint fragment of six or more layers is encountered in a case, there is justification to make fpv or fps quite small. Although Gothard concluded that specimens comprising more than six layers can be considered unique, the database he used necessarily was finite (500 specimens, of which 95 had 6 or more layers). There is the possibility that had Gothard examined 501 paints then he could have encountered a pair of paints with >6 layers that matched, so the number 501 can be used to calculate a minimum Dp value. A set of 501 paints represents 125,250 pairwise comparisons; thus, in the case where a multilayer paint chip (>6 layers) is involved and no other tests are carried out other than examining the layer structure, then fsp is no greater than 1/125,250 and, therefore, the LR is at least 125,250.

It is usually the case that paint examiners consider results derived from a barrage of analytical tests rather than simply the layer sequence or color of the paint in question. One would expect intuitively that as more and more tests are conducted upon the questioned paint, and assuming that in each of these tests the paint resembles that of the putative source, the analyst would become more confident in asserting that the questioned paint originated from the putative source. In regards to the equations derived above, the usage of a barrage of tests would have the effect of reducing the number of paints at random that resemble the questioned paint, that is, fpv and fps are being made smaller and the LRs are being made larger. Many publications listed here give specific Dp values applicable to particular barrages for a particular context (e.g., automotive paint of a particular color), which gives the paint examiner the option of a directly-applicable Dp if they happen to use exactly the same barrage of tests and related case context. Let us assume that we have used a particular barrage of tests to examine foreign paint on a victim's vehicle, a published Dp is available for this barrage of tests (0.99) and the frequency of the color of the foreign paint in the relevant jurisdiction is 0.1, then the LR is 1000:

$$LR = 1 / 0.1 \times (1 - 0.99)$$

The other expressions for the LR given above for other paint transfer scenarios can be elaborated in an analogous way. Unfortunately, not enough Dp values calculated using large data sets are available to allow operational adoption of this practice at the moment.

Buzzini *et al.* (2005) present several expressions for likelihood ratios applicable to single- and two-way transfer of paint in burglary scenarios involving painted crowbars and painted architectural fittings. The expressions derived are analogous to the ones shown above based on the framework of McDermott, but they also include three extra refinements. One is a factor relating to the expected size of the transferred evidence. Clearly if the size of the paint chip recovered is larger or smaller than can be expected from the circumstances surrounding the burglary then that provides less support for the prosecution hypothesis (i.e., a lower LR) than if the transferred paint is the "right" size. Another refinement is that consideration is given to the possibility that a crowbar might have a number of unrelated groups of foreign paint present. This is relevant because burglars might well use the one tool for a sequence of break-and-enter offences. The LR values derived in this article by Buzzini also take into consideration the Dp of infrared spectrometry (but no other techniques were performed). The survey results by Tippett *et al.* (1968) of architectural paints from 2000 houses in Wales (where only three matching pairs of paint were in the data set and these were repeats from the same three buildings) or Wright *et al.* (2013) (where 94% of white architectural paint could be discriminated by a barrage of tests) could be applied in LR calculations. If the crime scenario involves the detection of paint on the clothes of the suspect rather than an implement then the work of Moore *et al.* (2012) is a fundamental importance.

As can be seen from the CTS responses, though, there is an aversion in forensic science to express opinions in numerical or LR form. The Association of Forensic Science Providers (2009), based upon an idea proposed by Evett *et al.* (Evett, 1998; Evett *et al.*, 2000), has suggested an alternative where the examiner calculates a LR applicable to the case under examination and then converts that number into a verbal equivalent for communication to the jury. For example, if a LR of between 1000 and 10,000 is calculated, then that evidence lends strong support for the proposition of an association. This approach does seem to offer simplicity of communication (as much research has shown that jurors have difficulty coping with numbers and probability) combined with the robustness, transparency, and rigor of a logically correct evidence evaluation framework. The use of such an approach is not without its critics, however. The verbal approach incorporates artificial steps where, for example, an examiner logically should express a LR of 9999 quite differently to one of 10,001. In the United Kingdom a panel of Appeal Justices criticized an examiner for not making it clear that they made use of a verbal scale that was based upon a mathematical derivation of a LR (see Thompson [2012] and the references cited therein; England and Wales Court of Appeal [2010]). Whilst the Justices did make several mistakes of logic in their judicial opinion, their comments in regards to a lack of transparency in regards to the methodology used to evaluate evidence are salutary.

Whereas the pace of development in regards to the analytical technology used in forensic paint comparison appears to be slowing, the opposite is the case in regards to the development of paint evidence evaluation and presentation practices. The focus is now being turned to the refinement of models for treatment of evidence, and to the identification and resolution of gaps in empirical data that support the models – the field looks set for continued development.

References

Association of Forensic Science Providers. (2009) Standards for the formulation of evaluative forensic science expert opinion. *Science and Justice*, **49** (3), 161–164.

ASTM (American Society for Testing and Materials) (2011) *ASTM E2808-11, Standard Guide for Microspectrophotometry and Color Measurement in Forensic Paint Analysis*. ASTM International, West Conshohocken, PA, USA.

ASTM (American Society for Testing and Materials) (2013a) *ASTM E2809-13, Standard Guide for Using Scanning Electron Microscopy/X-Ray Spectrometry in Forensic Paint Examinations*. ASTM International, West Conshohocken, PA, USA.

ASTM (American Society for Testing and Materials) (2013b) *ASTM E2937-13, Standard Guide for Using Infrared Spectroscopy in Forensic Paint Examinations*. ASTM International, West Conshohocken, PA, USA.

ASTM (American Society for Testing and Materials) (2014) *ASTM E1610-14, Standard Guide for Forensic Paint Analysis and Comparison*. ASTM International, West Conshohocken, PA, USA.

Bell, S.E.J., Fido, L.A., Speers, S.J., *et al.* (2005a) Forensic analysis of architectural finishes using Fourier transform infrared and Raman spectroscopy, Part I: The resin bases. *Applied Spectroscopy*, **59** (11), 1333–1339.

Bell, S.E.J, Fido, L.A., Speers, S.J., *et al.* (2005b) Forensic analysis of architectural finishes using Fourier transform infrared and Raman spectroscopy, Part II: White paint. *Applied Spectroscopy*, **59** (11), 1340–1346.

Bentley, J. and Turner, G.P.A. (1997) *Introduction to Paint Chemistry and Principles of Paint Technology*, 4th edn. Chapman and Hall, London, UK.

Bommarito, C. (2006) *A new approach to report writing in comparative trace evidence examinations.* Advanced Trace Evidence Symposium, Midwest Association of Forensic Scientists and Midwest Forensic Resource Centre, Ames, IA, USA.

Boynton, R.M. (1979) *Human Color Vision.* Holt, Reinhart and Winston, New York, USA.

Buxbaum, G. and Pfaff, G. (eds) (2005) *Industrial Inorganic Pigments.* Wiley-VCH Verlag GmbH & Co. KGaA, Wenheim, Germany.

Buzzini, P., Massonnet, G., and Monard Sermier, F. (2006) The micro Raman analysis of paint evidence in criminalistics: case studies. *Journal of Raman Spectroscopy*, **37** (9), 922–931.

Buzzini, P., Massonnet, G., Birrer, S., *et al.* (2005) A survey of crowbar and household paints in burglary cases – population studies, transfer and interpretation. *Forensic Science International*, **152** (2–3), 221–234.

Buzzini, P. and Massonnet, G. (2004) A market study of green spray paints by Fourier transform infrared (FTIR) and Raman spectroscopy. *Science and Justice*, **44** (3), 123–131.

Carr, G.L. (2001) Resolution limits for infrared microspectroscopy explored with synchrotron radiation. *Review of Scientific Instruments*, **72** (3), 1613–1619.

Challinor, J.M. (1991) Structure determination of alkyd resins by simultaneous pyrolysis methylation. *Journal of Analytical and Applied Pyrolysis*, **18** (3–4), 233–244.

Challinor, J.M. (2001a) Pyrolysis techniques for the characterisation and discrimination of paint, in B. Caddy (ed.) *Forensic Examination of Glass and Paint*, Taylor and Francis, London, UK.

Challinor, J.M. (2001b) Review: The development and applications of thermally assisted hydrolysis and methylation reactions. *Journal of Analytical and Applied Pyrolysis*, **61**, 3–34.

Challinor, J.M. (2007) Examination of forensic evidence, in T.P. Wampler (ed.) *Applied Pyrolysis Handbook*, 2nd edn, CRC Press, Taylor and Francis, FL, USA.

Clearfield, A. Reibenspies, J.H., and Bhuvanesh, N. (eds) (2008) *Principles and Applications of Powder Diffraction.* John Wiley & Sons Ltd, Chichester, UK.

Coumbaros, J., Kirkbride, K.P., Klass, G., *et al.* (2009) Application of time of flight secondary ion mass spectrometry to the in situ analysis of ballpoint pen inks on paper. *Forensic Science International*, **193** (1–3), 42–46.

Cousins, D.R. (1989) The use of microspectrophotometry in the examination of paints. *Forensic Science Review*, **1** (2), 142–162.

CTS (Collaborative Testing Service Incorporated) (2014) *Forensic Summary Reports*, http://www.ctsforensics.com/reports/main.aspx (last accessed 9 June 2015).

Deconinck, I, Latkoczy, C., Guenther, D., *et al.* (2006) Capabilities of laser ablation-inductively coupled plasma mass spectrometry for (trace) element analysis of car paints for forensic purposes. *Journal of Analytical and Atomic Spectrometry*, **21** (3), 279–287.

De Forest, P.R. (2002) Foundations of forensic microscopy, in R. Saferstein (ed.) *Forensic Science Handbook*, Vol. **1**, 2nd edn, Prentice Hall, NJ, USA.

De Gelder, J., Vandenabeele, P., Govaert, F., and Moens, L. (2005) Forensic analysis of automotive paints by Raman spectroscopy. *Journal of Raman Spectroscopy*, **36** (11), 1059–1067.

Dinnebier, R.E. and Billinge, S.J.L. (eds) (2008) *Powder Diffraction: Theory and Practice.* The Royal Society of Chemistry, Cambridge, UK.

Edmondstone, G., Hellman, J., Legate, K., *et al.* (2004) An assessment of the evidential value of automotive paint comparisons. *Journal of the Canadian Forensic Science Society*, **37** (3), 147–154.

England and Wales Court of Appeal. (2010) *R v T. Crim 2439, Redacted Judgement Approved by the Court for Handing Down.* http://www.bailii.org/ew/cases/EWCA/Crim/2010/2439.pdf (last accessed 9 June 2015).

Everall, N.J. (2008) The influence of out-of-focus sample regions on the surface specificity of confocal Raman microscopy. *Applied Spectroscopy*, **62** (6), 591–598,

Everall, N.J. (2009) Confocal Raman microscopy; performance, pitfalls, and best practice. *Applied Spectroscopy*, **63** (9), 245a–262a.

Everall, N.J. (2013) Optimising image quality in 2D and 3D confocal Raman mapping. Journal of Raman Spectroscopy, **42** (6), 1330–1334.

Evett, I.W. (1998) Towards a uniform framework for reporting opinions in forensic science casework. *Science and Justice*, **38** (3), 198–202.

Evett, I.W., Jackson, G., Lambert, J.A., *et al.* (2000) The impact of the principles of evidence interpretation on the structure and content of statements. *Science and Justice*, **40** (4), 233–239.

Eyring, M., Lovelace, M., and Sy, D. (2007) A study of the discrimination of some automotive paint films having identical color codes. NIJ Trace Evidence Symposium, Clearwater Beach, FL, USA. http://projects.nfstc.org/trace/docs/Trace%20Presentations%20CD-2/eyring.pdf (last accessed 9 June 2015).

Faulkner, E.B. and Schwartz, R.J. (eds) (2009) *High Performance Pigments*. Wiley-VCH Verlag GmbH & Co. KGaA, Weinheim, Germany.

Flynn, K., Oleary, R., Lennard, C., *et al.* (2005) Forensic applications of infrared chemical imaging: multi-layered paint chips. *Journal of Forensic Science*, **50** (4), 832–841.

Florio, J.J. and Miller, D.J. (eds) (2004) *Handbook Of Coating Additives*, 2nd edn. Marcel Dekker Inc., New York.

Fukuda, K. (1985) The pyrolysis gas chromatographic examination of Japanese car paint flakes. *Forensic Science International*, **29**, 227–236.

Fuller, N.A. (1985) Analysis of thin-layer chromatograms of paint pigments and dyes by direct microspectrophotometry. *Forensic Science International*, **27** (3), 189–204.

Gothard, J.A. (1976) Evaluation of automobile paint flakes as evidence. *Journal of Forensic Sciences*, **21** (3), 636–641.

Gothard, J.A. and Maynard, P. (1996) *Evidential Value of the Examination of Motor Vehicle Paints*. National Institute of Forensic Sciences, Melbourne, Australia.

Govaert, F. and Bernard, M. (2004) Discriminating red spray paints by optical microscopy, Fourier transform infrared spectroscopy and X-ray fluorescence. *Forensic Science International*, **140** (1), 61–70.

Hamer, P.S. (1980). *A Guide to Vehicle Paint Examination by Transmitted Light Microscopy*. Home Office Research and Development Report 8, The Metropolitan Police Forensic Science Laboratory, London, UK.

Hamer, P.S. (1982). Pigment analysis in the forensic examination of paints. III. A guide to motor vehicle paint examination by transmitted light microscopy. *Journal of the Forensic Science Society*, **22**, 187–192.

Herbst, W., Hunger, K., Wilker, G., *et al.* (2005) *Industrial Organic Pigments: Production, Properties, Applications*, 3rd edn. Wiley-VCH Verlag GmbH & Co. KGaA, Weinheim, Germany.

Hobbs, A. and Almirall, J.R. (2003). Trace elemental analysis of automotive paints by laser ablation-inductively coupled plasma-mass spectrometry (LA-ICP-MS). *Analytical and Bioanalytical Chemistry*, **376**, 1265–1271.

Kirkbride, K.P. and Tungol, M.W. (1999) Infrared microscopy of fibers, in Robertson, J.R. Grieve, M. (eds), *Forensic Examination of Fibers*, Taylor and Francis, Philadelphia, USA.

Koleske, J.V. (ed.) (2012) *Paint and Coating Testing Manual: Fifteenth Edition of the Gardner-Sward Handbook*. ASTM Manual Series, MNL17, American Society for Testing Materials, West Conshohocken, PA, USA.

Kopchick, K.A. and Bommarito, C.R. (2006) Color analysis of apparently achromatic automotive paints by visible microspectrophotometry. *Journal of Forensic Science*, **51** (2), 340–343.

Laing, D.K., Dudley, R.J., Home, J.M., and Isaacs, M.D.J. (1982) The discrimination of small fragments of household gloss paints by microspectrophotometry. *Forensic Science International*, **20**, 191–200.

Lavine, B.K., Fasasi, A., Mirjankar, N., *et al.* (2014). Search prefilters for mid-infrared absorbance spectra of clear coat automotive paint smears using stacked and linear classifiers. *Journal of Chemometrics*, **28** (5), 385–394.

Lee, C. and Sandercock, P.M.L. (2011) A survey of automotive topcoat colors in Edmonton, Alberta. *Journal of the Canadian Forensic Science Society*, **44** (4), 130–143.

Liszewski, E. A., Lewis, S.W., Siegel, J.A., *et al.* (2010) Characterization of automotive paint clear coats by ultraviolet absorption microspectrophotometry with subsequent chemometric analysis. *Applied Spectroscopy*, **64** (10), 1122–1125.

Locke, J., Cousins, D.R., Russell, L.W., *et al.* (1987) A data collection of vehicle topcoat colors. 1. Instrumentation for color measurements. *Forensic Science International*, **34,** 131–142.

Locke, J., Wilkinson, J.M., and Hanford, T.J. (1988) A data collection of vehicle topcoat colors. 2. The measurement of color samples used in the vehicle refinishing industry. *Forensic Science International*, **37**, 177–187.

Maric, M. (2014) *Chemical Characterisation and Classification of Forensic Trace Evidence*. PhD Dissertation, Faculty of Science, Curtin University, Perth, Australia.

Maric, M., van Bronswijk, W., Lewis, S.W., *et al.* (2012) Rapid characterisation and classification of automotive clear coats by attenuated total reflectance infrared spectroscopy. *Analytical Methods*, **4** (9), 2687–2693.

Maric, M., van Bronswijk, W. Lewis, S.W., *et al.* (2013) Characterisation of chemical component migration in automotive paint by synchrotron infrared imaging. *Forensic Science International*, **228**, 165–169.

Maric, M., van Bronswijk, W., Lewis, S.W., *et al.* (2014) Synchrotron FTIR characterisation of automotive primer surfacer paint coatings for forensic purposes. *Talanta*, **118**, 156–161.

Massonnet, G. and Stoecklein, W. (1999a). Identification of organic pigments in coatings: applications to red automotive topcoats. Part II: Infrared spectroscopy. *Science and Justice*, **39** (2), 135–140.

Massonnet, G. and Stoecklein, W. (1999b) Identification of organic pigments in coatings: applications to red automotive topcoats Part III: Raman spectroscopy (NIR FT-Raman). *Science and Justice*, **39** (3), 181–187.

Massonnet, G. and Stoecklein, W. (1999c) Identification of organic pigments in coatings: Application to red automotive topcoats. Part I: Thin layer chromatography with direct visible microspectrophotometric detection. *Science and Justice*, **39** (2), 128–134.

McDermott, S.D. and Willis, S.M. (1997) A survey of the evidential value of paint transfer evidence. *Journal of Forensic Science*, **42** (6), 1012–1018.

Mc Dermott, S.D., Willis, S.M., and McCullough, J.P. (1999) The evidential value of paint. Part II: A Bayesian approach. *Journal of Forensic Science*, **44** (2), 263–269.

Miller, L.M. and Dumas, P. (2006) Chemical imaging of biological tissue with synchrotron infrared light. *Biochimica et Biophysica Acta*, **1758**, 846–857.

Monahan, S. (2010) *A Survey of the ACT Automotive Paint Situation: a Bayesian Approach*. Canberra Institute of Technology, Bachelor of Applied Science (Forensic Investigation) Specialist Project Report, Canberra, Australia.

Moore, R., Kingsbury, D., Bunford, J., *et al.* (2012). A survey of paint flakes on the clothing of persons suspected of involvement in crime. *Science and Justice*, **52** (2), 96–101.

Muehlethaler, C., Massonnet, G., Deviterne, M., *et al.* (2013) Survey on batch-to-batch variation in spray paints: A collaborative study. *Forensic Science International*, **229** (1–3), 80–91.

Nichols, M.E. and Kaberline, S.L. (2013) Migration of hindered amine light stabilizers in automotive clearcoats. *Journal of Coatings Technology and Research*, **10** (3), 427–432.

Nowicki, J. and Patten, R. (1986) Examination of US automotive paints: 1. Make and model determination of hit-and-run vehicles by reflectance microspectrophotometry. *Journal of Forensic Sciences*, **31** (2), 464–470.

de la Ossa, M.A.F., Lopez-Lopez, M., Torre, M., *et al.* (2011) Analytical techniques in the study of highly-nitrated nitrocellulose. *Trends in Analytical Chemistry*, **30** (11), 1740–1755.

Ryland, S., Bishea, G., Brun-Conti, L., *et al.* (2001) Discrimination of 1990s original automotive paint systems: A collaborative study of black nonmetallic base coat/clear coat finishes using infrared spectroscopy. *Journal of Forensic Sciences*, **46** (1), 31–45.

Ryland, S. (1995) Infrared microspectroscopy of forensic paint evidence, in H.J. Humecki (ed.) *Practical Guide to Infrared Microspectroscopy*, Practical Spectroscopy, Vol. **19**, Marcel Dekker Inc., New York, USA.

Ryland, S. and Kopec, R. (1979) The evidential value of automobile paint chips. *Journal of Forensic Sciences*, **24** (1), 140–147.

Ryland, S., Kopec, R.J., and Somerville, P.N. (1981) The evidential value of automobile paint. Part II: Frequency of occurrence of topcoat colors. *Journal of Forensic Sciences*, **26** (1), 64–74.

Seccombe, A.L. (2001). *Discrimination and Evidential Value of Vehicle Paint in Forensic Casework.* PhD Dissertation, Chemistry Department, University of Auckland, Auckland, New Zealand.

Smalldon, K.W. and Moffat, A.C. (1973) The calculation of discrimination power for a series of correlated attributes. *Journal of Forensic Science*, **13** (4), 291–295.

Sommer, A. and Katon, J. (1991) Diffraction-induced stray light in infrared microspectroscopy and its effect on spatial resolution. *Applied Spectroscopy*, **45** (10), 1633–1640.

Stoecklein, W. and Fujiwara, H. (1999) The examination of UV-absorbers in 2-coat metallic and non-metallic automotive paints. *Science and Justice*, **39** (3), 188–195.

Streitberger, H.J. and Doessel K.F. (eds) (2008) *Automotive Painting and Coatings, 2nd Completely Revised and Extended Edition*, Wiley-VCH Verlag GmbH & Co. KGaA, Weinheim, Germany.

Suzuki, E.M. (1996a) Infrared spectra of US automobile original topcoats (1974–1989).1. Differentiation and identification based on acrylonitrile and ferrocyanide C-N stretching absorptions. *Journal of Forensic Sciences*, **41** (3), 376–392.

Suzuki, E.M. (1996b) Infrared spectra of US automobile original topcoats (1974–1989). 2. Identification of some topcoat inorganic pigments using an extended range (4000–220cm^{-1}) Fourier transform spectrometer. *Journal of Forensic Sciences*, **41** (3), 393–406.

Suzuki, E.M. (1999a) Infrared spectra of US automobile original topcoats (1974–1989): VI. Identification and analysis of yellow organic automotive paint pigments-isoindolinone yellow 3R, isoindoline yellowk, anthrapyrimidine yellow, and miscellaneous yellows. *Journal of Forensic Sciences*, **44** (6), 1151–1175.

Suzuki, E.M. (1999b) Infrared spectra of US automobile original topcoats (1974–1989): V. Identification of organic pigments used in red nonmetallic and brown nonmetallic and metallic monocoats – DPP Red BO and Thioindigo Bordeaux. *Journal of Forensic Sciences*, **44** (2), 297–313.

Suzuki, E.M. (2014a) Infrared spectra of US automobile original finishes (1998–2000). IX. Identification of bismuth oxychloride and silver/white mica pearlescent pigments using extended range FT-IR spectroscopy, XRF spectrometry, and SEM/EDS analysis. *Journal of Forensic Sciences*, **59** (5), 1205–1225.

Suzuki, E.M. (2014b) Infrared spectra of US automobile original finishes (Post-1989). VIII: In situ identification of bismuth vanadate using extended range FT-IR spectroscopy, Raman spectroscopy, and X-Ray fluorescence spectrometry. *Journal of Forensic Sciences*, **59** (2), 344–363.

Suzuki, E.M. and McDermot, M.X. (2006) Infrared spectra of US automobile original finishes. VII. Extended range FT-IR and XRF analyses of inorganic pigments in situ – nickel titanate and chrome titanate. *Journal of Forensic Sciences*, **51** (3), 532–547.

Suzuki, E.M. and Marshall, W.P. (1997) Infrared spectra of US automobile original topcoats (1974–1989). 3. In situ identification of some organic pigments used in yellow, orange, red, and brown nonmetallic and brown metallic finishes – benzimidazolones. *Journal of Forensic Sciences*, **42** (4), 619–648.

Suzuki, E.M. and Marshall, W.P. (1998) Infrared spectra of US automobile original topcoats (1974–1989): IV. Identification of some organic pigments used in red and brown nonmetallic and metallic monocoats-quinacridones. *Journal of Forensic Sciences*, **43** (3), 514–542.

SWGMAT (Scientific Working Group on Materials Analysis) (2000) *Forensic Paint Analysis and Comparison Guidelines*. http://www.swgmat.org/Forensic%20Paint%20Analysis%20and%20 Comparison%20Guidelines.pdf (last accessed 9 June 2015).

SWGMAT (Scientific Working Group on Materials Analysis) (2002) *Standard Guide for Using Scanning Electron Microscopy/X-ray Spectrometry in Forensic Paint Examinations*. http://www.fbi.gov/about-us/ lab/forensic-science-communications/fsc/oct2002/bottrell.htm (last accessed 9 June 2015).

SWGMAT (Scientific Working Group on Materials Analysis) (2007) *Standard Guide for Microspectrophotometry and Color Measurement in Forensic Paint Analysis*. http://www.swgmat.org/ Standard%20Guide%20for%20Microspectrophotometry%20and%20Color%20 Measurement%20in%20Forensic%20Paint%20Analysis.pdf (last accessed 9 June 2015).

SWGMAT (Scientific Working Group on Materials Analysis) (2009) *Standard Guide for Using Infrared Spectroscopy in Forensic Paint Examinations*. http://www.swgmat.org/SWGMAT%20 infrared%20spectroscopy09-09-2004.pdf (last accessed 9 June 2015).

SWGMAT (Scientific Working Group on Materials Analysis) (2013) *Standard Guide for Using Pyrolysis Gas Chromatography and Pyrolysis Gas Chromatography-Mass Spectrometry in Forensic Paint Examinations*. http://www.swgmat.org/Paint%20PGC%20Doc%2011-25-13%20final.pdf (last accessed 9 June 2015).

Talbert, R. (2008) *Paint Technology Handbook*. CRC Press, Taylor & Francis Group, Boca Raton, FL, USA.

Taylor, M.C., Cousins, D.R., Holding, R.H., *et al.* (1989) A data collection of vehicle topcoat colors. 3. Practical considerations for using a national database. *Forensic Science International*, **40** (2), 131–141.

Thompson, W.C. (2012) Discussion paper: Hard cases make bad law – reactions to R v T. *Law, Probability and Risk*, **11**, 347–359.

Thornton, J.I. (1997) Visual color comparisons in forensic science. *Forensic Science Review*, **9** (1), 38–57.

Tippett, C.F., Emerson, V.J., Fereday, M.J., *et al.* (1968) The evidential value of the comparison of paint flakes from sources other than vehicles. *Journal of the Forensic Science Society*, **8** (2), 61–65.

Wampler, T.P. (2004) Practical applications of analytical pyrolysis. *Journal of Analytical and Applied Pyrolysis*, **71** (1), 1–12.

Wampler, T.P., Bishea, G.A., and Simonsick, W.J. (1997) Recent changes in automotive paint formulation using pyrolysis gas chromatography mass spectrometry for identification. *Journal of Analytical and Applied Pyrolysis*, **40** (1), 79–89.

Weyermann, C.L., Bucher, L., Majcherczyk, P., *et al.* (2012) Statistical discrimination of black gel pen inks analysed by laser desorption/ionization mass spectrometry. *Forensic Science International*, **217** (1–3), 127–33.

Wicks, Z.W., Jones, F.N., Pappas, S.P., and Wicks, D.A. (2007) *Organic Coatings: Science and Technology*, 3rd edn. John Wiley & Sons, Inc., New Jersey, USA.

Wilkinson, J.M., Locke, J., and Laing, D.K. (1988) The examination of paints as thin sections using visible microspectrophotometry and Fourier transform infrared microscopy. *Forensic Science International*, **38**, 43–52.

Wright, D.M., Bradley, M.J., and Hobbs Mehltretta, A. (2013) Analysis and discrimination of single-layer white architectural paint samples. *Journal of Forensic Sciences*, **58** (2), 358–364.

Zieba-Palus, J. and Michalska, A. (2014) Photobleaching as a useful technique in reducing of fluorescence in Raman spectra of blue automobile paint samples. *Vibrational Spectroscopy*, **74**, 6–12.

Zieba-Palus, J., Zadora, G., Milczarek, J.M., and Koscieniak, P. (2008) Pyrolysis-gas chromatography/mass spectrometry analysis as a useful tool in forensic examination of automotive paint traces. *Journal of Chromatography A*, **1179** (1), 41–46.

CHAPTER 4

Forensic fire debris analysis

Reta Newman

Pinellas County Forensic Laboratory, USA

4.1 Introduction

There were an estimated 1,240,000 fires reported in the United States in 2013 (NFPA, 2014a). Of those, 44,800 were eventually classified as arson (FBI, 2014). The process of determining the incendiary nature of a fire has many critical elements, of which the analysis provided by the forensic laboratory plays a relatively small, but important role.

Fire investigators combine an analysis of the dynamics of fire progression and damage at the fire scene with investigate analysis of potential motive and circumstance to determine if a fire is suspicious in nature. In a suspicious fire, the role of the laboratory is to analyze physical evidence collected as deemed relevant to the investigation. While in the course of any investigation, multiple types of forensic analysis may be necessary or appropriate (DNA, latent prints, glass analysis, etc.), in a fire investigation the primary role of the forensic laboratory is to determine the presence and nature of potential ignitable liquid residues (ILRs) in samples collected from the scene or in the course of the investigation.

4.2 Process overview

Fire debris analysis can be broken into four primary and sequential steps: collection, extraction, analysis, and interpretation (Figure 4.1). Each is critical to obtaining accurate and meaningful results. Each has specific considerations that, if not addressed, can have serious impact on the laboratory findings and the implications to the total investigation.

Forensic Chemistry: Fundamentals and Applications, First Edition. Edited by Jay A. Siegel.

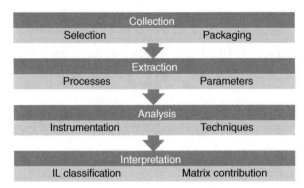

Figure 4.1 The analytical process associated with forensic fire debris analysis

At this point it is important to note the difference between an accelerant and an ignitable liquid. Accelerants are incendiary materials, typically ignitable liquids, that are used to intentionally and deliberately start or propagate a fire. An ignitable liquid is simply a liquid that burns readily. Our modern environment is inundated with incidental ignitable liquids; a few examples include wood stains, charcoal starters, oil-based paints, fuels, insecticide solvents, cleaning solvents, mastics, stainproofers, and waterproofers. While an accelerant is most commonly an ignitable liquid, ignitable liquids found in the course of a fire investigation are not *de facto* accelerants. Laboratory analysis provides for the recognition and classification of ignitable liquids only. That information, put in the context of the total investigation, will determine whether the ignitable liquid is, in fact, an accelerant.

4.3 Sample collection

The recognition and classification of ignitable liquids begins well outside of the laboratory. Sample selection and sample collection are critical to the success of ignitable liquid detection. *NFPA 921 Guide for Fire and Explosion Investigations* provides the scientific framework for investigating the "origin, cause, responsibility, and prevention" of fire and explosions (NFPA, 2014b). Using the principles defined therein, a fire investigator will attempt to determine the likely origin and progression of a fire. If any aspect of the investigation leads to a suspicious fire in which the use of an accelerant is suspected, samples are collected for laboratory analysis.

In most cases, visual inspection is neither sufficient nor practical for identifying samples where potential accelerants may have survived the heat of the fire and suppression efforts. Sample selection is based upon logical assumptions, along the fire path, where an accelerant may have been used and where it would have the most likely probability of surviving. Many jurisdictions supplement that with the use of accelerant detection canines (ADCs). Well-trained canines are the most valuable tools for identifying samples with the highest potential of ILR retention. That said, due to the inherent lack of specificity, a "hit" by an ADC is not sufficient for conclusively determining the presence of an ILR or an accelerant (IAAI, 1995), which can only be confirmed by laboratory analysis (NFPA, 2008).

Samples of interest identified at the scene are packaged and submitted to the laboratory in airtight containers that, in themselves, must be free of incidental ILRs. Sampling devices, such cotton swabs or absorbent pads, may be necessary in some situations. Various and abundant petroleum-based ignitable liquids are used in manufacturing, thus individual lots of all materials should be tested for contaminants.

Metal paint cans, glass jars, and specially developed polymer evidence bags are used almost exclusively for packaging evidence for ILR analysis. All have advantages and disadvantages. Rigid cans and jars are better suited for common debris where sharp items are often present, but are limited by their capacity. Large items are better suited for bags but can be compromised by sharp debris.

4.4 Ignitable liquid classification

The next step in fire debris analysis is extraction or sample preparation for subsequent instrumental analysis, but it would be impossible to discuss these techniques and their implications without an understanding of the chemical properties and analytical processes used in identification and classification of ignitable liquids.

Ignitable liquids are broadly divided into to two major classes: petroleum based and non-petroleum based. The vast majority of those encountered in fire debris and fire investigations are petroleum based. They are widely available and inexpensive, which makes them ideal accelerants. Furthermore, petroleum-based ignitable liquids are widely used in manufacturing, construction, and production of common household products (Lentini *et al.*, 2000), which makes incidental identification a factor. One study has shown that petroleum-based solvents used in stains can be extracted and identified more than two years from the date of application (Lentini, 2001). The significance of this is discussed later; for now, it is simply important to note that petroleum-based ignitable liquids are prevalent in structure and vehicle fires, not only for their appeal as an accelerant but also as an inherent component of sample substrates.

Non-petroleum-based ignitable liquids are less commonly encountered, but must also be considered both as a potential accelerant and inherent ignitable liquid. Non-petroleum products are more difficult to recognize and often require additional analytical techniques for clear identification.

ASTM E1618-14 Standard Test Method for Ignitable Liquid Residues in Extracts from Fire Debris Samples by Gas Chromatography-Mass Spectrometry (ASTM, 2014) provides a classification scheme that separates petroleum-based ignitable liquids by boiling range and general chemical composition; these include petroleum distillates, isoparaffinic products,

Table 4.1 ASTM E1618 ignitable liquid classes and hydrocarbon ranges

Ignitable liquid class
Gasoline
Petroleum distillates
Isoparaffinic products
Aromatic products
Naphthenic paraffinic products
Normal alkane products
Oxygenated solvents
Other – Miscellaneous products

Hydrocarbon ranges	
Light:	C4–C9
Medium:	C8–C13
Heavy:	C9–C20+ or C11+

aromatic products, and naphthenic–paraffinic products. Non-petroleum-based ignitable liquids are included in the classification; however, they are grouped in either an "oxygenates" class or placed in a general "miscellaneous" class that is separated only by boiling range (Table 4.1).

Petroleum products are those that are directly produced through the refinement of crude oil. The most common is gasoline. It is followed closely by petroleum distillates, which are used in a variety of products, including vehicle and home heating fuels, and are basis of all other refinery products. Further processing of distillates results in refinery streams that primarily comprise aromatic compounds, normal alkane compounds, isoalkanes compounds, or combinations of isoalkane and cycloalkanes compounds (Speight, 2014). These are often further blended to meet the specifications of the final products. Gasoline is simply a blend of various refinery streams.

Refined petroleum products typically contain relatively simple hydrocarbons, specifically normal alkanes, isoalkanes, cycloalkanes, simple aromatics (alkyl substituted benzene rings), indanes, and polynuclear aromatics (Table 4.2). The petroleum industry commonly uses archaic terms of chemical nomenclature that have carried over into the forensic platform. "Alkanes" are often referred to as "paraffins"; thus the terms "isoalkane" and "isoparaffin" are synonymous in describing a saturated branched chain alkane. Similarly, "cylcolakane" and "cylcoparrafin" describe cyclic saturated hydrocarbons. Additionally, the term "naphthene," not to be confused with "naphthalene," is also used in lieu of "cycloalkane." In referring to normal-alkane hydrocarbon ranges, chemical

Table 4.2 Hydrocarbons found in petroleum-based ignitable liquids

Hydrocarbon classes	Examples	
Alkanes		
normal alkanes	undecane	
isoalkanes (isoparaffins)	isopentane	
Cycloalkanes (cylcoparaffins)	methylcyclohexane	
Mononuclear aromatics (alkyl benzenes)	p-xylene	
Indans		
indanes	indane	
indenes	2-methyl indene	
Polynuclear aromatics (PNA)	2-methylnaphthalane	

http://www.emolecules.com

name and/or formula are generally replaced with an acronym comprised of "C" and the number of carbons. For example, *n*-hexane (C_6H_{14}) is referred to as C6, undecane ($C_{11}H_{24}$) is referred to as C11 and eicosane ($C_{20}H_{42}$) is C20, and so on.

The nature of ignitable liquids fit ideally to analysis by gas chromatography (GC). Ignitable liquids are not generally single compounds; the number of compounds in a relatively small boiling point range can number in the hundreds. Gas chromatography, using non-polar columns, allows for reasonable separation of the compounds. However, given the number of compounds with similar boiling points, baseline resolution is not typically possible and, fortunately, unnecessary.

Most ignitable liquid identification and classification is achieved by using a technique called pattern recognition. The presence of any one, or two, or twenty, compounds is not as significant as their presence in relation to other compounds in a given mixture. Figure 4.2 depicts a portion of the GC data for a common ignitable liquid. The highest peak is 1,2,4-trimethylbenzene; and while a product of this ignitable class will always

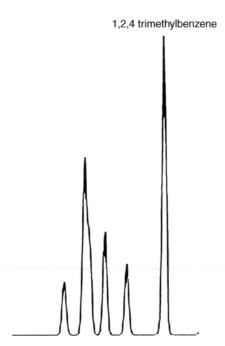

1,2,4 trimethylbenzene

Figure 4.2 Typical C3 alkylbenzene peak pattern of petroleum-based ignitable liquids

have this compound present, it is the presence in a specific pattern that is relevant to the identification of a petroleum product. 1,2,4-trimethylbenzene is also commonly extracted from debris samples, unrelated to an ignitable liquid, as a pyrolysis product from the degradation of many burned materials (Stauffer, 2001). Most petroleum-based ignitable liquid identifications are based on specific patterns unique to specific boiling ranges of ignitable liquids.

Accurate recognition, identification, and classification require the ability to detect chromatographic patterns based upon hydrocarbon class. The lack of baseline resolution and the disproportionate representation of one hydrocarbon class over others in a given product can limit the viability of gas chromatographic data alone. Coupling the GC with a mass spectrometer (MS) and extracting ions indicative of chemical class provides for pattern recognition analysis at enhanced levels.

The ion fragmentation of each class of hydrocarbon is predictable and reproducible. The fragmentation of an alkylbenzene by electron impact (70 ev) mass spectrometry will always include the ions 91, 105 and 119, assuming that the molecular weight of the compound exceeds that of the ion. By extracting ions 91, 105 and 119 from the total ion chromatogram, a chromatographic pattern that is limited to the subset of compounds that contain those ions can be created and evaluated. It is important to note that while this is typically referred to the "aromatic or alkylbenzene" profile, the compounds repre-sented in the pattern are not unique to alkylbenzenes. All compounds from ignitable liquids or debris contributors with any of the ions specified will be included in the pattern. Ion chromatograms can be evaluated by each single ion or, more commonly, class indicative ions are summed to produce a select chromatographic data set that encompasses the entire range of compounds in sample.

Figure 4.3a represents the total ion chromatogram (TIC) for a diesel fuel, a heavy petroleum distillate; Figure 4.3b is the chromatographic pattern resulting from the extraction and summation of ions 91, 105 and 119, primarily representing the alkylbenzenes present in this product. In the analysis process, extracted ion profiles (EIP), also called extracted ion chromatograms (EIC), can be created and analyzed for each hydrocarbon class. The ions indicative of the other hydrocarbon classes found in petroleum-based ignitable liquids are provided in Table 4.3. Figures 4.3c–4.3f represent the peak patterns generated from the ions extracted from diesel fuel and associated with alkanes, cycloalkanes, polynuclear aromatics, and indanes, respectively.

The use of extracted ion chromatography allows the analyst to break complex peak patterns into relevant subsets that aid recognition and classification of reference ignitable liquids. More importantly, it allows for clarification and recognition of specific patterns when the data set is complicated by matrix contributions.

The diagnostic patterns associated with the various ignitable liquid classes are specific to boiling point range. For example, Figure 4.4 represents the alkylbenzene compounds that are found in petroleum-based products between C6 and C14. Any petroleum products that encompass a corresponding range, and that contain aromatic compounds, will have these specific patterns. The abundance of grouping will be dependent on the boiling range of the ignitable liquid; however, the pattern and relative abundances of compounds in specific groupings, i.e., C3 alkylbenzenes, C4 alkylbenzenes, and so on, will be consistent.

Each ASTM E1618 class is characterized not only by composition but is also further subdivided into designated boiling point ranges: light, medium, and heavy (ASTM, 2014). Petroleum products that exist in the boiling range of interest to ILR analysis fall in to the range of C5–C25 and most typically are found in the C6–C20 range. ASTM E1618 defines light ignitable liquids as compounds and mixtures that fall primarily into the boiling range of C4–C9. Medium products are C8–C13 range products that are limited to a span of 3–4 normal alkanes. Heavy products are either broad, encompassing greater than five normal alkanes in the C9–C20+ range, or are heavier products that start above C11 (eicosane) (ASTM, 2014). The defined ranges refer to most but not all of the compounds included in the product. For example, while light petroleum products predominately fall into the C4–C9 range, there are often minor compounds present into the C10–C11 range. It is noteworthy that the forensic community, not the petroleum industry, defines these ranges and classifications. While the classification scheme includes most of the petroleum products that a forensic analyst will encounter, there are some that do not cleanly fit into either a specific classification or boiling range.

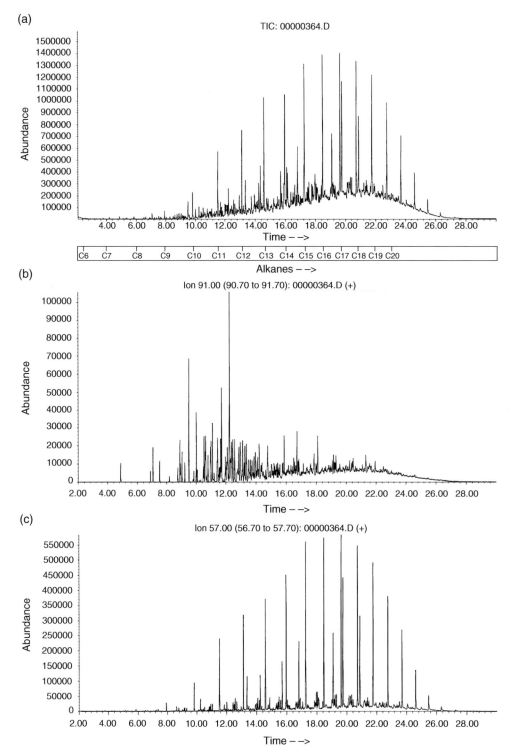

Figure 4.3 GC-MS extracted ion data associated with a heavy petroleum distillate (diesel fuel): (a) total ion chromatogram; (b) extracted aromatic (alkylbenzene) profile; (c) extracted alkane profile;

(d)

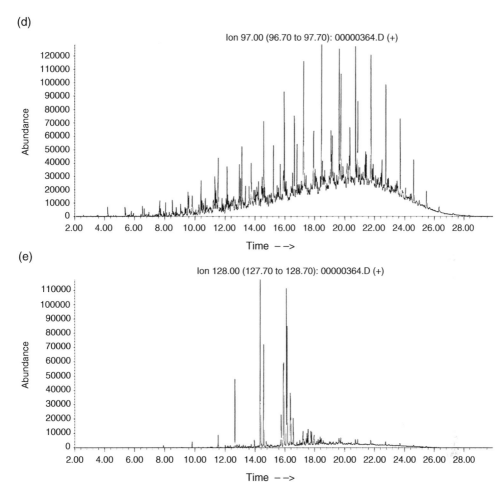

(e)

Figure 4.3 (*Continued*) (d) extracted cycloalkane profile; (e) extracted polynuclear aromatic profile. © 2005 University of Central Florida. All Rights Reserved. Image provided by the National Center for Forensic Science (http://ncfs.ucf.edu/)

Table 4.3 Ions indicative of hydrocarbon class

Ions indicative of hydrocarbon classes	
Alkanes (*n*-alkanes and isoalkanes)	57, 71, 85, 99
Cycloalkanes and alkenes	55, 69, 83, 97
Aromatics (alkylbenzenes)	91, 105, 119, 133*
Polynuclear aromatics	128, 142, 156
Indanes	117, 118, 131, 132, 145*, 159*

*class indicative, but not commonly used in analysis

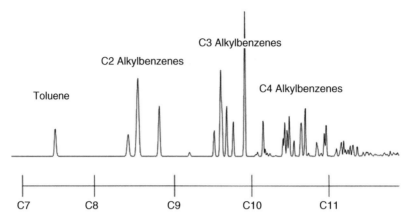

Figure 4.4 Aromatic peak patterns associated with petroleum-based ignitable liquids

4.5 Petroleum-based ignitable liquids

The ASTM E1618 petroleum-based ignitable liquid classes are: distillates, gasoline, aromatic products, isoparaffinic products, naphthenic–paraffinic products and *n*-alkane products. Distillates are products of the most basic refinery process, distillations, and are the starting point of all other petroleum products of interest to ILR analysis. Distillates contain all the chemical components of crude oil within a given boiling range. All other petroleum products are modified distillates; modified either by the separation of specific hydrocarbon classes, chemical processes to convert to different classes, and/or blending of various refinery streams, all of which originate from distillates (Speight, 2014).

Distillates (Figure 4.5) are characterized by disproportionately abundant homologous series of *n*-alkanes in a Guassian distribution interdispersed with less abundant isoalkanes and cylocalkanes in specific patterns. Aromatic compound groupings (alkylbenzenes, indanes, polynuclear aromatics) will be present based upon boiling range unless specific refinery processes were used to remove them resulting in the subclass of de-aromatized distillates.

Various refinery processes are used for manipulating the distillate fractions of crude oil to create streams that include hydrocarbon classes of commercial, hence profitable, significance. This mainly relates to those needed for the production of gasoline and other fuels. Streams created from these processes are also marketed as a variety of commercial products (i.e., charcoal starters, automotive parts cleaners, candle oils, paint thinners, etc.) that are of great interest in ignitable liquid analysis.

Gasoline is the most refined and the only ignitable class that is uniquely identifiable as a specific product, although, not to a brand or source. There is published methodology for comparing neat liquids to determine the potential for common histories. However,

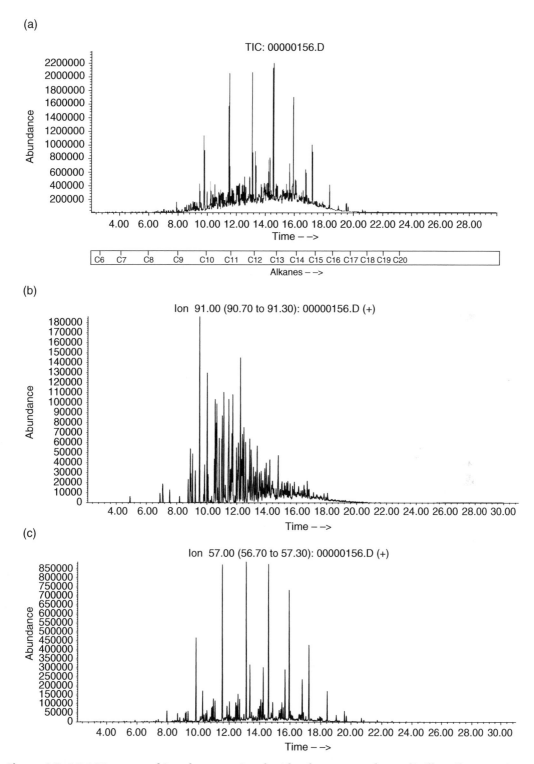

Figure 4.5 GC-MS extracted ion data associated with a heavy petroleum distillate (kerosene). Note the homologous series of spiking normal alkanes. (a) Total ion chromatogram; (b) extracted aromatic (alkylbenzene) profile; (c) extracted alkane profile;

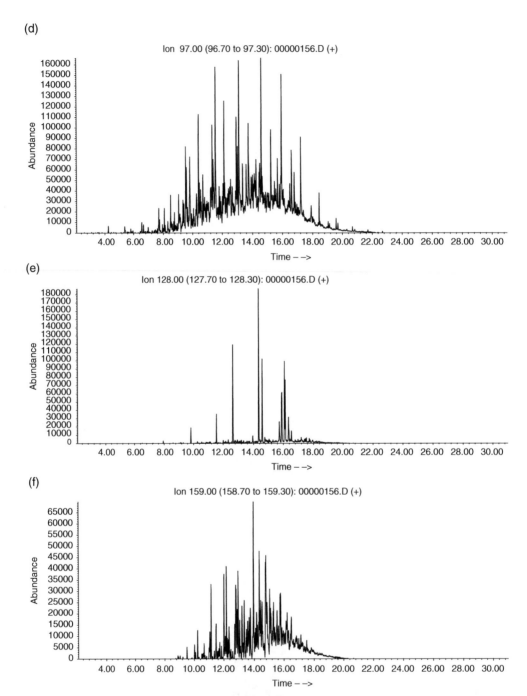

Figure 4.5 (*Continued*) (d) extracted cycloalkane profile; (e) extracted polynuclear aromatic profile; (f) extracted indane profile. © 2002 University of Central Florida. All Rights Reserved. Image provided by the National Center for Forensic Science (http://ncfs.ucf.edu/)

this type of analysis is not considered reliable for ILR comparisons (Mann, 1987; Barnes *et al.*, 2004). Gasoline is created by blending various refinery fractions to meet specifications as a fuel and various regulatory requirements. The most prominent compounds in gasoline are aromatics, which are necessary to increase the octane rating, thus the efficiency, of the fuel. Various aliphatic compounds are present, although the amount and composition varies by manufacturer and location. Modern gasoline is also commonly enhanced with ethanol, detection of which is depended on degree of evaporation, suppression efforts, extraction parameters, and chromatographic conditions.

The chromatographic data associated with gasoline (Figure 4.6) includes abundant aromatics, visualized by prescribed patterns in both the TIC and alkylbenzene profiles; the presence of indanes and, depending on boiling range, polynuclear aromatics in prescribed patterns; and the presence of alkanes and cycloalkane compounds/patterns. Note that the alkanes and cycloalkanes vary in abundance and by pattern in gasoline samples. Isoalkanes are added in various amounts to further improve octane rating, but the amount and process varies. Alkane and isoalkane profiles are necessary for gasoline identification; however, the patterns vary significantly from sample to sample. A lack of alkane compounds would support identification of an aromatic solvent.

Aromatic solvents are comprised of aromatic compounds. These include alkylbenzenes, indanes, and polynuclear aromatics based upon the boiling range of the product. The marked absence of aliphatic compounds is an important diagnostic feature (Figure 4.7). Similarly, isoparaffinic products are comprised of the isolkanes separated or manufactured from distillates (Figure 4.8). Instrumental data are typified by peaks in a distinct boiling range, with the TIC, alkane and cylcoalkane profiles differing only in abundance. There are no cycloalkanes in an isoparaffinic product; however, a "cyloalkane" profile mirroring the alkane profile is always present. The ions indicative of a cylcoalkane (55, 69, 83) are also present, though significantly less abundant, in all alkanes (Figure 4.9). The "cycloalkane" profile in the case of an isoparaffinic product is the alkane profile.

Naphthenic paraffinic products are comprised of abundant cycloalkanes (naphthenes) and isoalkanes; *n*-alkanes may also be present at significantly diminished abundances compared to the associated distillate (Figure 4.10). Commercial products are most commonly found in the medium or heavy range. Because the number of isoalkane and cycloalkane isomers increases dramatically based on carbon number, the number of compounds in these products can be immense. The result is a TIC that is generally represented as an unresolved, but Gaussian, envelope of peaks. The extracted ion profiles for alkanes and cycloalkanes are abundant and markedly different (Figure 4.11) from each other due to the representation of both isoalkanes and cycloalkanes. The aromatic profiles are notably absent.

N-alkane products are comprised solely of normal alkanes (Figure 4.12). They are typically limited to 3–5 compounds and, as a result, pattern recognition is of limited value. *n*-alkane products are confirmed by the specific identification of each compound through a combination of mass spectral data and chromatographic retention time.

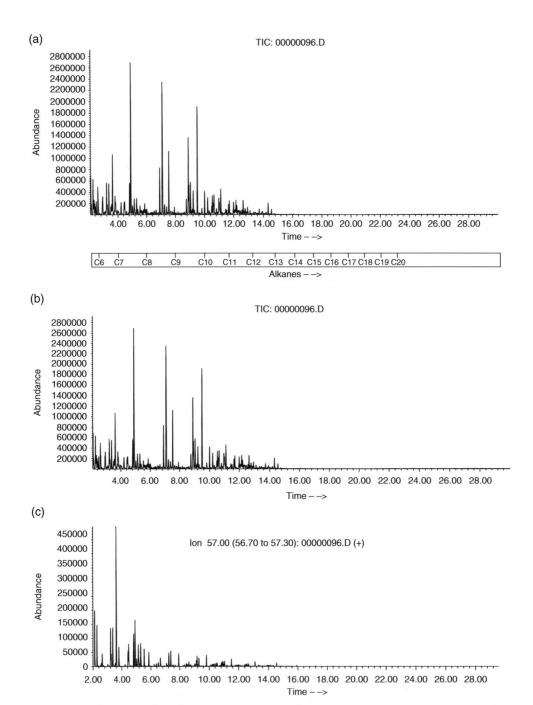

Figure 4.6 GC-MS extracted ion data associated with gasoline. Note the abundance of alkylbenzenes in distinctive peak patterns. (a) Total ion chromatogram; (b) extracted aromatic (alkylbenzene) profile; (c) extracted alkane profile;

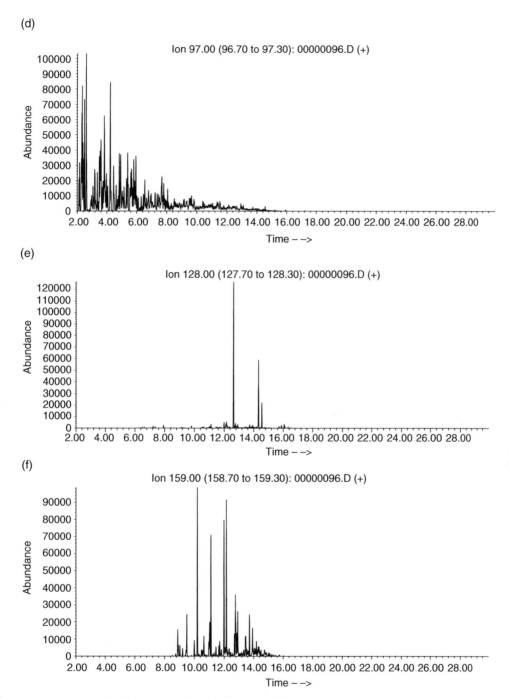

Figure 4.6 (*Continued*) (d) extracted cycloalkane profile; (e) extracted polynuclear aromatic profile; (f) extracted indane profile. © 2001 University of Central Florida. All Rights Reserved. Image provided by the National Center for Forensic Science (http://ncfs.ucf.edu/)

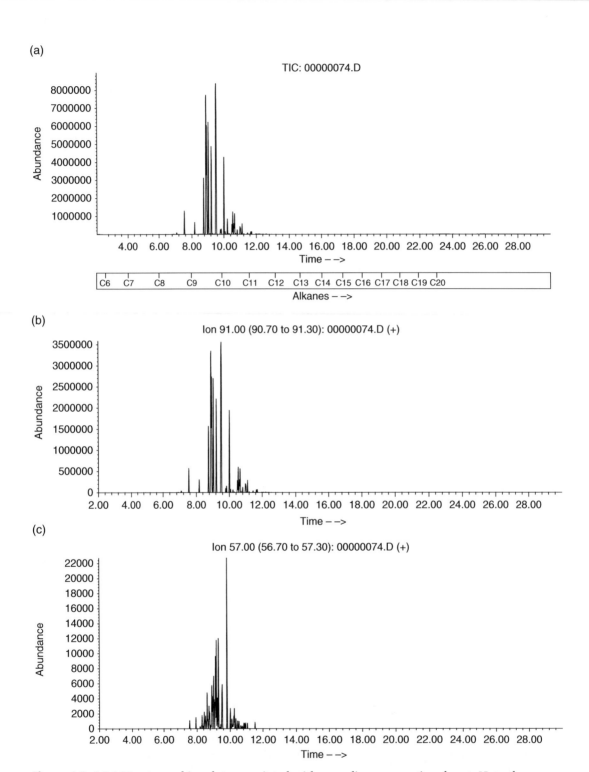

Figure 4.7 GC-MS extracted ion data associated with a medium aromatic solvent. Note the abundance of alkylbenzenes in distinctive peak patterns and the notable absence of aliphatic compounds. (a) Total ion chromatogram; (b) extracted aromatic (alkylbenzene) profile; (c) extracted alkane profile;

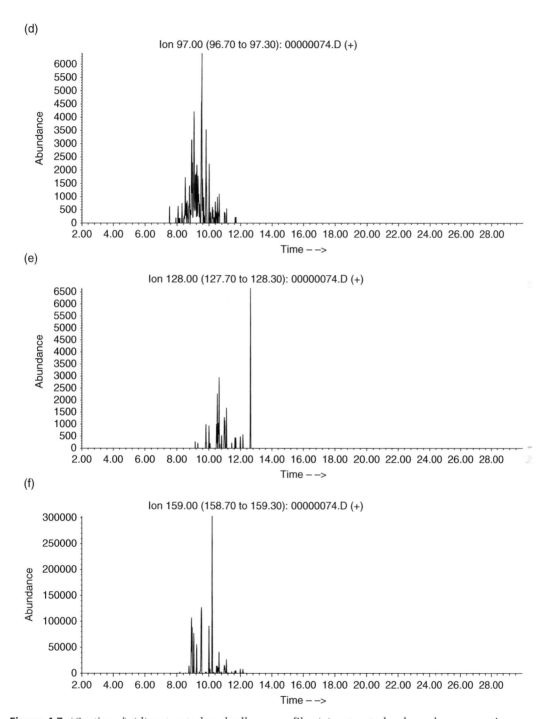

Figure 4.7 (*Continued*) (d) extracted cycloalkane profile; (e) extracted polynuclear aromatic profile; (f) extracted indane profile. © 2001 University of Central Florida. All Rights Reserved. Image provided by the National Center for Forensic Science (http://ncfs.ucf.edu/)

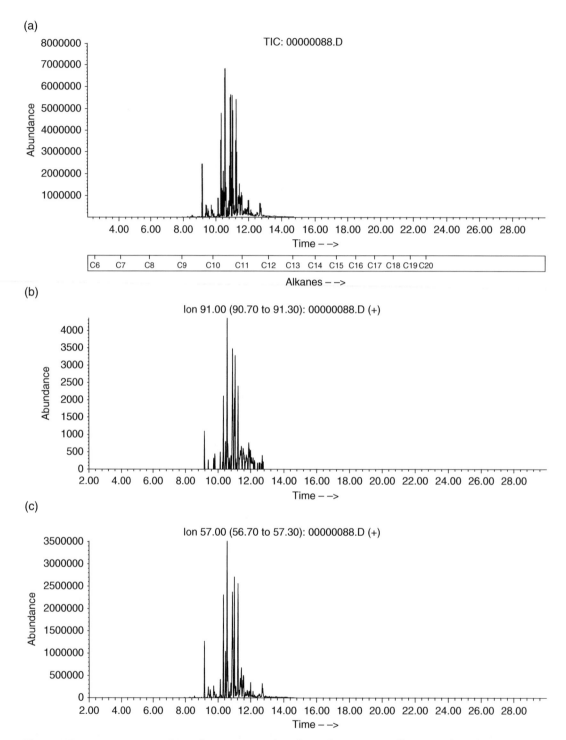

Figure 4.8 GC-MS extracted ion data associated with medium isoparaffinic product. Note the abundance of alkanes and the similarity among the TIC, alkane and cycloalkane profiles. (a) Total ion chromatogram; (b) extracted aromatic (alkylbenzene) profile; (c) extracted alkane profile;

(d)

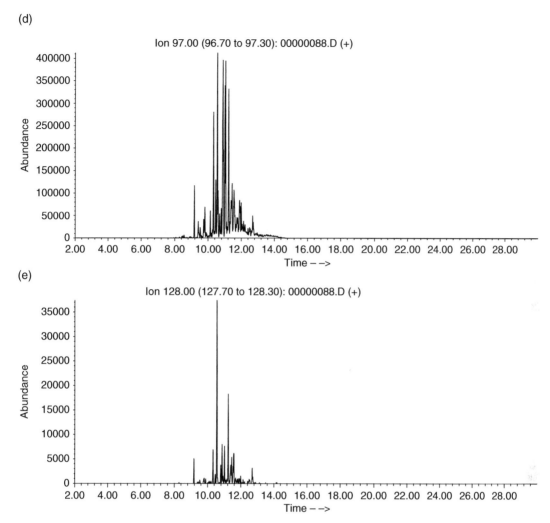

(e)

Figure 4.8 (*Continued*) (d) extracted cycloalkane profile; (e) extracted polynuclear aromatic profile. There are indane indicative ions in this product. © 2001 University of Central Florida. All Rights Reserved. Image provided by the National Center for Forensic Science (http://ncfs.ucf.edu/)

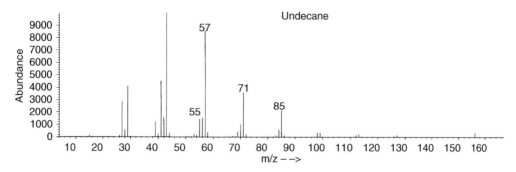

Figure 4.9 Mass spectrum of a normal alkane (undecane). Note the presence of ions 55, 69, and 83, which are highly indicative of alkenes but are also present, although less prominently, in alkanes

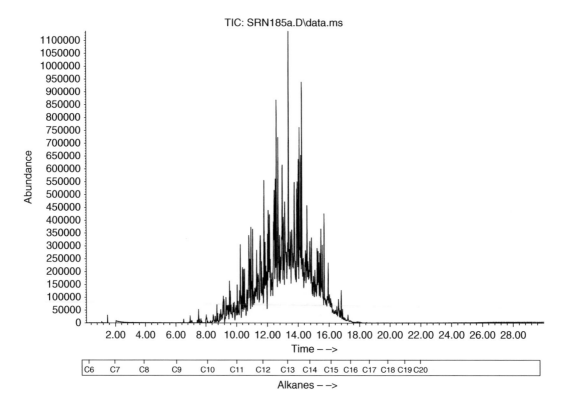

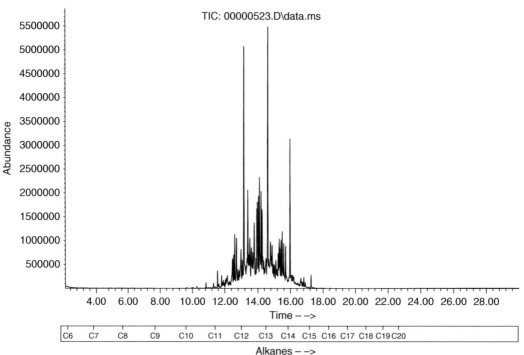

Figure 4.10 Total ion chromatogram of a heavy naphthenic paraffinic product (top) compared to a heavy petroleum distillate. Note that the only significant difference is the absence or diminution of the normal alkanes. © 2009 University of Central Florida. All Rights Reserved. Image provided by the National Center for Forensic Science (http://ncfs.ucf.edu/)

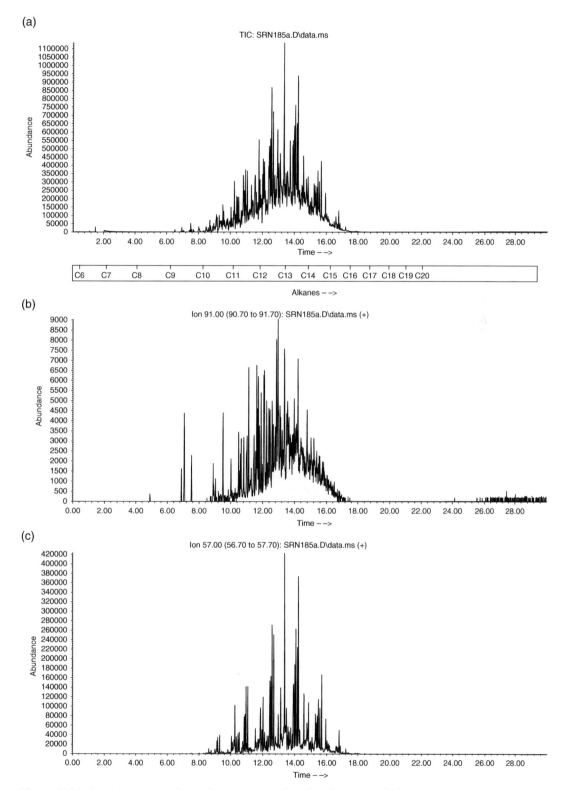

Figure 4.11 GC-MS extracted ion data associated with a heavy naphthenic paraffinic product. Note the abundance of alkane and cycloalkanes compound. (a) Total ion chromatogram; (b) extracted aromatic (alkylbenzene) profile; (c) extracted alkane profile

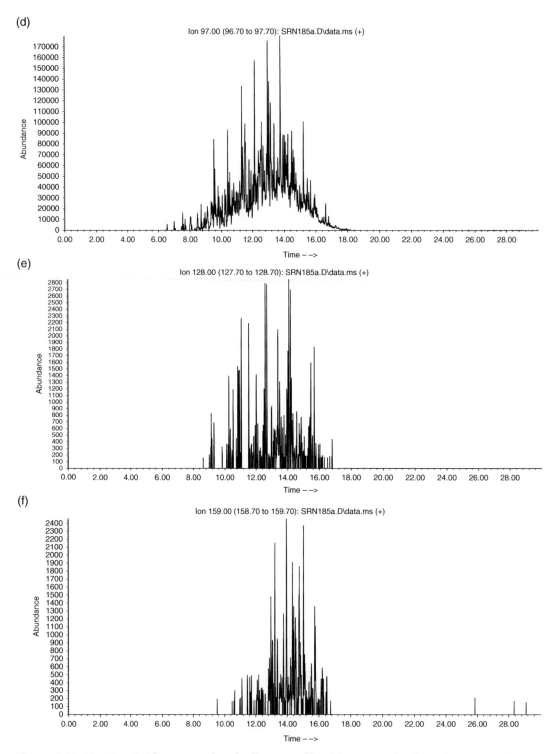

Figure 4.11 (*Continued*) (d) extracted cycloalkane profile; (e) extracted polynuclear aromatic profile; (f) extracted indane profile. © 2012 University of Central Florida. All Rights Reserved. Image provided by the National Center for Forensic Science (http://ncfs.ucf.edu/)

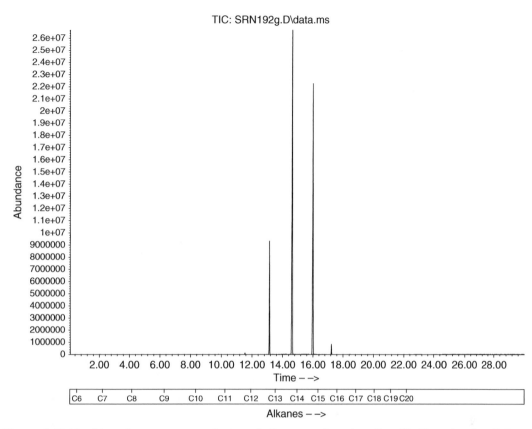

Figure 4.12 Total ion chromatogram of normal alkane product (candle oil). Note the simplicity of the peaks. © 2012 University of Central Florida. All Rights Reserved. Image provided by the National Center for Forensic Science (http://ncfs.ucf.edu/)

A comprehensive collection of representatives of all ignitable liquid classes/boiling range subclasses is not possible within the space constraints of a single chapter. Every laboratory should maintain a library of reference ignitable liquids that includes multiple exemplars from all classes and subclasses at different states of evaporation. There are two published references that include extensive libraries of ignitable liquid TIC/EIC data: the *GC-MS Guide to Ignitable Liquids* (Newman *et al*,, 1998) and *Ignitable Liquid Reference Collection (IRLC)* (NCFS, 2014). The ILRC collection is an on-line and interactive database that provides total ion chromatographs and extracted ion profiles representing the most current ASTM E1618 classifications. A summary of the chemical and chromatographic characteristics of the petroleum-based ignitable liquids classes is provided in Table 4.4.

The concept that forensic scientists created this classification system is critical. The petroleum industry provides products that meet a given set of specifications. For example, the specifications of a charcoal starter would include boiling point and flash point. Many different refinery products can be, and are, used as charcoal starters (Figure 4.13) and given that combinations of waste streams may be economical; the level of purity of any

Table 4.4 Summary of chemical and chromatographic characteristics of petroleum-based ignitable liquids

Characteristics of ignitable liquid classes	Chemical composition	Data characteristics
Gasoline	Blend of various refinery streams Highly aromatic Includes aliphatic compounds	Distinctive aromatic peak grouping Abundant alkylbenzenes Indanes present Polynuclear aromatics may present Alkanes/cycloalkanes present – not abundant
Distillates	Highly aliphatic May contain aromatic compounds	Homologous series of spiking n-alkanes Significant iso and cycloalkane patterns Alkylbenzenes may be present Indanes may be present Polynuclear aromatics may present
Isoparaffinic products	Exclusively aliphatic 100% isoalkanes	Distinct clean TIC pattern Alkane pattern matches "cylcoalkane" pattern Aromatics absent
Aromatic products	Exclusively aromatic	Distinct clean TIC pattern Characteristic alkylbenzene and/or polynuclear aromatic pattern – based on range. Notable absence of aliphatic compounds
Naphthenic paraffinic products	Exclusively aliphatic Contains cyclo and isoalkanes	Unresolved envelope of peaks Generally Gaussian in appearance Predominant alkane and cycloalkane patterns
n-alkane products	Exclusively aliphatic 100% n-alkanes	Clean TIC with few peaks Identification by MS and GC retention time

given product varies. As a result, many products do not fit exactly or cleanly into a specific ASTM E1618 class or may be mixtures of multiple classes. For example, the presence of a significantly low level of aromatic compounds in an otherwise isoparaffinic product is not unusual. Classification is then by the predominate fractions present. When

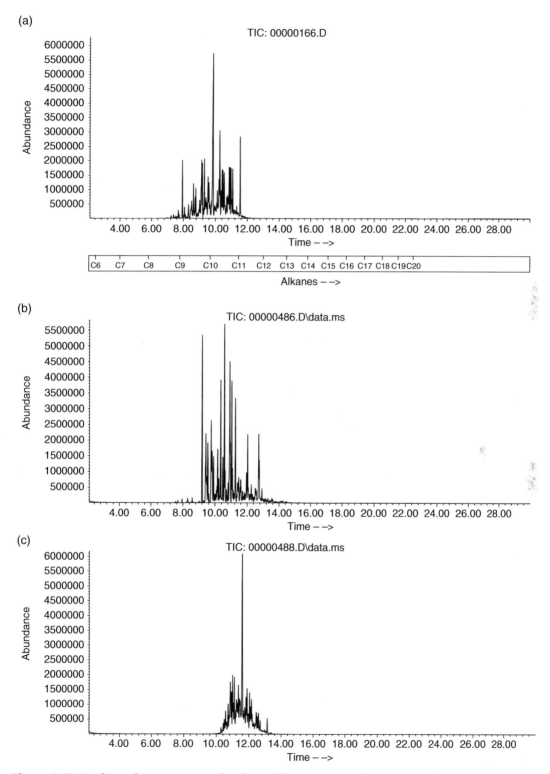

Figure 4.13 Total ion chromatograms for three different charcoal starters. © 2002, 2008, University of Central Florida. All Rights Reserved. Image provided by the National Center for Forensic Science (http://ncfs.ucf.edu/)

Table 4.5 Examples of common products associated with petroleum-based ignitable liquids

	Light	Medium	Heavy
Distillates	Camping fuels	Charcoal starters	Kerosene
	Lighter fluids	Paint thinners	Diesel fuel
	Engine starter fluids	Mineral spirits	Jet fuel
	Spot removers	Insecticides	Charcoal starters
	VM and P Naphthas	Wood stain solvents	Waterproofers
		Gas treatments	Stainguards
			Ink solvents
Isoparaffinic products	Aviation fuels	Charcoal starters	Commercial solvents
		Paint thinners	Manufacturing
		Mineral spirits	lubricants
		Fuel additives	
Aromatics products	Degreasing solvents	Degreasers	Insecticides
	Automotive parts	Commercial solvents	Industrial solvents
	cleaners	Automotive parts cleaners	
	Industrial solvents	Fuel additives	
	Paint/varnish removers	Insecticides	
Naphthenic-paraffinic products		Charcoal starters	Insecticides
		Insecticides	Lamp oils
		Lamp oils	Ink solvents
Normal alkane products	Industrial solvents	Copier toners	Candle oils
			NCR paper
			Copier toners

an unexpected chemical class of compounds is present at more than a trace level, mixtures/blends must be assumed. However, currently there are not clear delineations within the classification system.

Ignitable liquid identification can be specific only to class. In any given ignitable liquid class, with the exception of gasoline, there can be numerous products, brands of products, and product uses (Table 4.5). Product identification or ignitable liquid source is not possible. Analytical findings are typically reported to include the ignitable liquid class and representative samples of potential products.

4.6 Non-petroleum-based ignitable liquids

Non-petroleum products fall into the "miscellaneous class" of ASTM E1618, but for discussion purposes can be subdivided into four predominate types: oxygenates, terpenes, alternative fuels, and vegetable oils.

Oxygenates are oxygenated compounds or mixtures that include oxygenated compounds. Examples include common household products and industrial solvents such as acetone, isopropanol, and methyl ethyl ketone. Due to their higher volatility, detection and identification may require alternate GC parameters. Identification is based upon GC retention time and mass spectra. When oxygenated solvents are detected, caution must be given to the significance of the findings, as oxygenated compounds can be generated as incomplete combustion products. ASTM E1618 recommends that the detection of an oxygenated solvent should not be considered significant unless the abundance is at least one order of magnitude above other pyrolysis and combustion contributions visualized in the data (ASTM, 2014).

Terpenes are extracts from woods, typically soft woods. While turpentine is the most commonly associated retail product, limonene, a terpene derived from citrus, is commonly found in both ignitable and non-ignitable retail products. The use of softwood in construction commonly results in the detection of terpenes in structural fires. The prevalence of incidental terpenes in a structure fire can make the identification of turpentine suspect.

In the last decade, alternative fuels, including biofuels and ethanol-enhanced gasolines, have been introduced into the market in efforts to reduce global reliance upon crude oil. Ethanol enhanced gasoline (E85) is a blend of 85% ethanol with 15% gasoline that runs in specially modified engines. Most commercially available gasolines contain up to 10% ethanol. Thus, the presence of ethanol in gasoline-related fires is not uncommon or unexpected. Detection of the ethanol is dependent on the analytical parameters and the amount of evaporation due to heat and exposure (Kuk and Spagnola, 2008).

Biofuels, specifically biodiesel, represent a new challenge and interpretation strategy to fire debris analysts. Biodiesel are produced from vegetable oils or animal fats or oils. B100 is a 100% biodiesel product; B20 is a 20% biodiesel product blended with petroleum-based diesel fuel. Biodiesel is produced by the transesterification of triglycerides and fatty acids to create fatty acid methyl esters (FAMEs). Limited detection may be possible in neat liquids with the chromatographic conditions used for traditional ignitable liquids (Stauffer and Byron, 2007); however, in samples recovered from debris, alternative sample preparation and chromatographic conditions are necessary (ASTM, 2013).

On a similar note, some natural fats and oils (most notably linseed oil) are associated with self-ignition fires. The exothermic reaction associated with the oxidation of these oils can produce sufficient heat to spontaneously ignite oil-soaked natural fabrics. Analysis requires the conversion free fatty acids and or triglycerides to FAMEs for analysis by GC-MS (ASTM, 2103).

4.7 Sample preparation

The term "fire debris" implies charred remnants of fire. In reality, fire debris samples include any items associated with a fire investigation that are submitted for analysis. Aside from charred debris, evidence items can include items like clothing (often from potential suspects), furnishings and construction materials, soil, biological tissues, and various liquids. The analytical process for creating a sample appropriate for GC-MS analysis depends on two primary factors: the physical condition of the evidence and the ignitable liquid target.

The physical state of the evidence is easy to discern, neat organic liquids are processed by different sample preparation methods than aqueous liquids or evidence that contains solid materials. The ignitable liquid target can be more elusive. If the investigation includes vegetable oils, different sample preparation processes are necessary (ASTM, 2013). Typically, the initial target is the isolation and identification of petroleum-based ignitable liquids or low boiling oxygenates.

There are three modern methods for the extraction of common ignitable liquid residues from debris samples: headspace concentration, solvent extraction, and simple headspace sampling. Passive headspace concentration with activated charcoal is the mostly widely used method in the United States for most applications. The use of solvent extraction and simple headspace is typically limited to a specific subset of sample types.

Simple headspace analysis (ASTM, 2012a) is the direct extraction of volatilized compounds from the headspace of a closed container. Its use is typically limited to targeted analysis of low molecular weight oxygenated and organic ignitable liquids. The container can be heated (60–80°C) to increase the volatile composition in the headspace. A headspace sample is drawn into a syringe and directly injected into the GC injection port for analysis. The technique works well for highly volatile compounds in high concentrations. However, it is not suitable for products with higher molecular weight components (reduced volatility) or where the target amount of ignitable liquid is significantly diminished.

Solvent extraction (ASTM, 2010) is the dissolution of compounds in an appropriate solvent. Pentane, carbon disulfide, diethyl ether and methylene chloride are considered suitable solvents. However, given the volume of solvent required and the negative health and environmental impacts of carbon disulfide and methylene chloride, pentane is the most appropriate solvent for fire debris extraction by this method. Solvent extraction is ideal for simple solid matrices (i.e., glass fragments from a Molotov cocktail) or smaller debris samples with obvious high concentrations of ignitable liquids (i.e., strong incidental odor of fuel). It can also be useful for samples where preliminary analysis by a concentrated headspace technique indicates a heavy petroleum product and the investigation necessitates differentiation of products based on total boiling range. For example, kerosene products are typically heavy petroleum distillates that encompass the C9–C17 range; diesel fuel is a broader product that encompasses the C9–C25 range. Passive headspace techniques with complex (charred) debris can be limited on the higher end to extraction of C17–C18 (Fultz, 1994). In the event that differentiation is necessary, solvent extraction would be required.

There are multiple problems with the use of solvent extraction. First, a relatively large amount of potential toxic or environmentally unfriendly solvent is required. Second, the

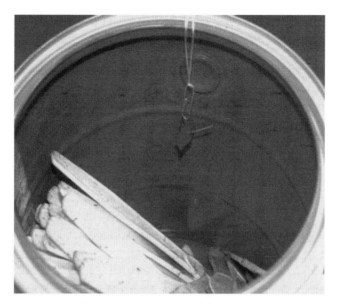

Figure 4.14 Activate charcoal strip suspension in fire debris container

extraction is not limited to volatile compounds; any chemical in the sample matrix that is soluble in the solvent will be represented in the final extract, thus adding complexity to the data and potentially masking ignitable liquid components and jeopardizing identification.

The most common and effective techniques for ignitable liquid residue extraction are passive headspace concentration with activated charcoal (common in the United States) and dynamic headspace concentration with Tenax, a polymer based adsorbent (more common in Canada and Europe). Both involve the use of adsorbents to collect and concentrate volatilized compounds in the container headspace.

The extraction technique for passive headspace (ASTM, 2012b) concentration with activated charcoal is simple, relatively inexpensive, and non-labor intensive. In most laboratories, an activated charcoal strip (ACS), comprised of active cocoanut charcoal imbedded on a polymer base, is suspended into the headspace of a sealed evidence container from a metal hook (paperclip, fishhook, safety pin) tied to a string (Figure 4.14). Other charcoal packages can be used; however, activated charcoal strips are most the most commonly employed.

The container is then heated for a period of time to allow volatilization of compounds in the debris and for the adsorption of a representation of those compounds on the adsorbent. There are five critical parameters that will determine the composition of the adsorbed species: adsorbent capacity, adsorption temperature, adsorption duration, desorption solvent, sample composition, and ignitable liquid concentration (Newman and Dietz, 1996; Dolan and Newman, 2001). To achieve the best representation of ignitable liquid residues in the debris, these parameters must be optimized. Obviously, the analyst can neither control, nor typically predict, the type and concentration of any ILR present in the sample. Thus, optimization must come from the adsorbent capacity, temperature, duration, and desorption solvent.

Adsorption in a closed system is a dynamic process. Volatilized compounds that come in contact with the adsorbent will attach and detach based on the adsorbent's affinity for the adsorbed compound and the temperature of the system. Physical adsorption is based upon attachment derived from intermolecular forces; in general, the strength of an adsorption bond is only 2–3 times that of the heat of condensation. Just as a compound can be evaporated from and condense onto the sample matrix, it will also attach (adsorb) and detach (desorb) from the adsorbent. Highly volatile compounds will have significantly less "residence time" on the adsorbent than a less volatile compound. When a heavier compound displaces a lighter compound at an adsorption site, the nature of the adsorbed mixture will shift towards the higher molecular weight (less volatile compounds). Figure 4.15a is neat mixture of ignitable liquids comprised of 1:1:1 50% evaporated gasoline, kerosene and diesel fuel mixture. Figure 4.15b is the resultant data from the extraction of 5 uL of the mixture from a container by passive headspace concentration with ACS with an adsorption duration of 16 hours at 60°C. Figure 4.15c is a similar extraction but an initial concentration of 25 uL. The nature of the adsorbed species has shifted to the higher molecular weight compounds. This process of displacement is accelerated by increased temperature, longer duration and/or sample concentration. Displacement can be a significant factor in data presentation and interpretation anytime the adsorption capacity of the adsorbent is exceeded.

Displacement is not limited to volatility. Activated carbon has a higher affinity for aromatic compounds compared to aliphatic compounds. At higher concentrations, duration or temperatures, the aromatic compounds will be disproportionately represented in the sample extract, further complicating the final interpretation. Maximizing the amount of adsorbent, moderating adsorption parameters, and adjusting for highly concentrated samples are all necessary to obtaining the best representation of any ILR present in debris samples.

Fortunately, for most debris samples, extraction parameters at moderate temperatures (60–80°C) for a longer duration (16 hours/overnight) using the maximum strip size (140 mm^2) produce reliable results. In samples in which either the initial assessment or the initial extraction indicate a high concentration of ILR, adaptation of the parameters (lower temperature, shorter duration) may be necessary to obtain realistic or representative data.

The desorption solvent used to elute the adsorbent also has implications for the final results. Pentane, carbon disulfide (Dolan and Newman, 2001), diethyl ether (Lentini and Armstrong, 1997) and methylene chloride (Hicks *et al.*, 2003) are all used in the field. Desorption efficiency is determined both the solubility of the analyte (adsorbed species) and the adsorptivity of the solvent to the adsorbent. Ignitable liquids are soluble in all of the solvents listed; however, the affinity of the solvent for active charcoal varies.

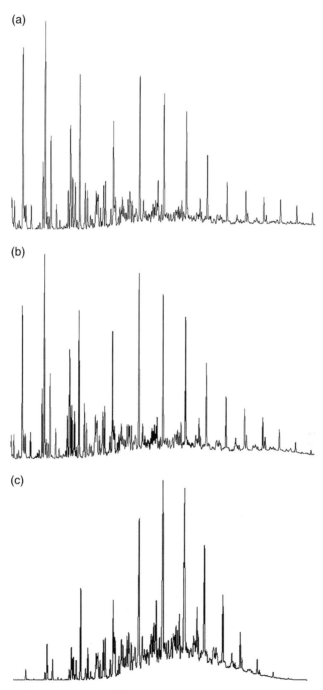

Figure 4.15 Total ion chromatographs for SAM mixture (50% evaporated gasoline, kerosene and diesel fuel 1:1:1) extracted under different conditions: (a) TIC of the neat liquid diluted in pentane; (b) TIC of ACS extraction of 5 uL at 60°C for 16 h; (c) TIC of ACS extraction of 25 uL at 60°C for 16 hs. Note the displacement of lower boiling compounds, which is the result of sample concentration that exceeds the capacity of adsorbent. Displacement can be accelerated with higher adsorption temperatures and/or extended adsorption durations

In order to effectively remove all of the absorbed species, the solvent must bind strongly with the adsorbent to prevent readsorbtion of sample compounds. Aromatic compounds bind strongly to activated charcoal, much more so than alkanes. Pentane, being both an alkane and a low molecular weight compound, will elute a disproportionate representation of the alkane compounds in the extract (Dolan and Newman, 2001). While pentane would be preferred based upon safety, cost, and environmental impact, consideration must be given to the accuracy of the analyzed sample. Diethyl ether and methylene chloride perform better than pentane; however, for a true representation of the extract, carbon disulfide is the best choice. Due to the significant health risk associated with carbon disulfide, minimal volumes (generally 250 ul or less are sufficient) and strict engineering controls are a necessity.

An alternative, but equally advantageous, technique involves the use of a porous polymer adsorbent Tenax TA (2,6-diphen-p-phenylene oxide). Tenax is used in a dynamic headspace concentration method where a headspace sample is drawn out of the heated sample container to the adsorbent using a syringe-like device. The sample is heated to the desired temperatures and a syringe is introduced through a hole in the container. An adsorbent tube containing Tenax is housed in the barrel of the syringe, exterior to the sample container. The adsorbed species is thermally desorbed and introduced to the GC column. The same concerns of displacement, in the form of breakthrough, and selective adsorption exist and must be considered when optimizing the process and interpreting the results.

Another passive headspace concentration technique that can be employed for screening debris for the presence, range, or relative concentration of ignitable liquids is solid phase microextraction (SPME). SPME uses an adsorbent film, typically polydimethylsiloxane, coated upon on polymer filament. A SPME device is employed to expose the filament to the headspace of a (typically) heated evidence container. Volatile compounds are collected and the sample is analyzed by thermal desorption. Due to the very limit amount of adsorbent, displacement is difficult to control, even at lower concentrations. However, the results provide fast and diagnostic information that can be used to optimize other adsorption or extraction technique parameters (ASTM, 2008).

4.8 Sample analysis and data interpretation

The resulting GC-MS data from an extracted sample represent contributions from the matrix and any ignitable liquid products present. Matrix contribution can come from multiple sources, including the raw materials, pyrolysis and combustion products associated with the raw materials and any ignitable liquids that have contacted the raw materials. In some cases, degradation of the sample due to matrix type and condition can also contribute to the extracted sample data and their subsequent interpretation.

Most raw materials have volatile compounds inherent to the material or acquired during the process of manufacturing the material. For example, wood, notably soft-wood, contain terpenes, which are volatile compounds extracted and detected in the analytical process. Carpets will contain volatile compounds used in its manufacture, as well as various postmanufacture processes such as stain guards or UV protectants.

Matrix contribution is further enhanced by thermal degradation and incomplete combustion of these materials. The thermal degradation of synthetic materials is especially noteworthy, as many common polymers used to manufacture common commercial and household items degrade in to compounds common to petroleum-based ignitable liquids (Stauffer, 2001).

Fire debris is not homogenous. A given fire scene sample will typically contain multiple matrix contributors (e.g., carpet, carpet pad, wood subflooring, upholsteries, etc.), and thus multiple sources of interfering compounds. Fortunately, with some exceptions, the matrix-specific compounds do not show up in specific or reproducible patterns. Since many of the compounds detected can be found in the matrix as well as ignitable liquids, analysts look to identify by the known peak patterns associated with ignitable liquids.

Other matrix contributors will produce compounds commonly found in ignitable liquids, especially aromatic compounds. However, the peak grouping will be skewed compared to an ignitable liquid. Significant deviation from those patterns and adjacent peak ratios, even in the presence of compounds found in ignitable liquids, must lead to the conclusion that an ignitable liquid (IL) is not present or identifiable. An additional factor is the potential for microbial hydrocarbon degradation in samples in which soil is a primary substrate (Mann and Gresham, 1990; Turner and Goodpaster, 2011).

The identification of an IL in a fire debris sample requires the direct comparison of the sample data to that of a reference ignitable liquid (RIL) analyzed with the same chromatographic conditions. The use of reference libraries to identify potential RILs is valuable. However, identification based solely on comparison to published data is generally unacceptable.

Examples of the process of evaluating and interpreting chromatographic data for an ignitable liquid in the presence of fire debris are provided in the following case studies. In all instances the samples were extracted by passive headspace concentration with activated charcoal strips at 60°C for 16 hours and eluted with about 200 μL of carbon disulfide with a 50 ppm PCE internal standard. For brevity, each case is compared to the patterns associated with a specific ignitable liquid class. In actual casework, the data would be evaluated based upon the characteristics of all IL classes.

Case 1

(Figure 4.16a represents the data collected from a sample comprising charred debris that includes carpet and carpet padding.) The total ion chromatograph is characterized by an abundance of peaks. A library search indicates that the majority of compounds are simple and complex aromatic and aliphatic compounds. The peak grouping associated with the

C3 alkyl benzenes indicate the potential for a petroleum-based ignitable liquid. However, the complexity of the data does not provide sufficient information for identification. The summed extracted ion profiles indicative of key hydrocarbon classes result in more definitive information. Whereas the C4 alkyl benzene pattern (Figure 4.16b) was not

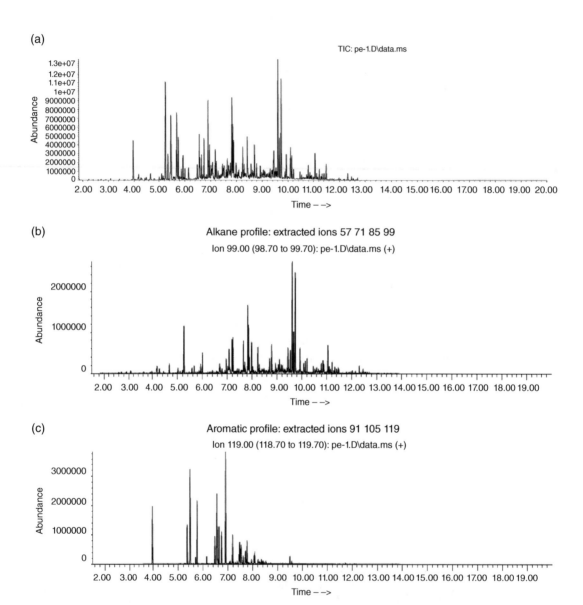

Figure 4.16 GC-MS extracted ion data associated with Case 1. Note the abundance of alkane and cycloalkanes compound. (a) Total Ion chromatogram; (b) extracted aromatic (alkylbenzene) profile; (c) extracted alkane profile;

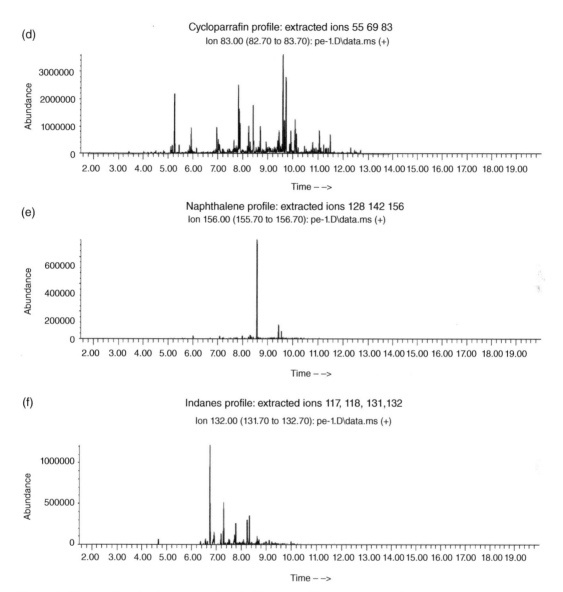

Figure 4.16 (*Continued*) (d) extracted cycloalkane profile; (e) extracted polynuclear aromatic profile; (f) extracted indane profile

discernible on the TIC, it can easily be deduced in the aromatic profile. Similarly, the extraction ion profiles allow for elucidation of key indane and polynuclear aromatic compound peak groupings (Figures 4.16c–4.16d). In comparing the boiling range of the peaks of interests, the peak patterns associated with the EIPs and comparing those to a reference ignitable liquid (Figure 4.6), gasoline is identified

Case 2

Figure 4.17 represents data generated from an extraction of charred debris that included melted plastic. The TIC is characterized by a relatively resolved homologous series of compounds in a Gaussian distribution. A library search reveals the presence, among other compounds, of abundant normal alkenes. The subset of data provided by the

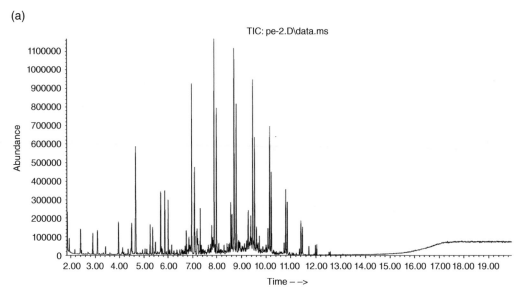

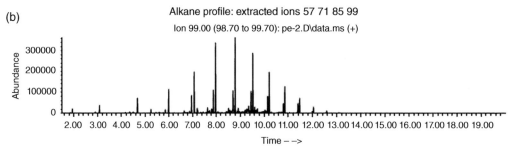

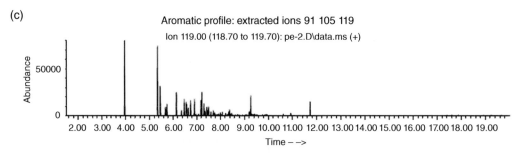

Figure 4.17 GC-MS extracted ion data associated with Case 2: (a) Total ion chromatogram; (b) extracted aromatic (alkylbenzene) profile; (c) extracted alkane profile;

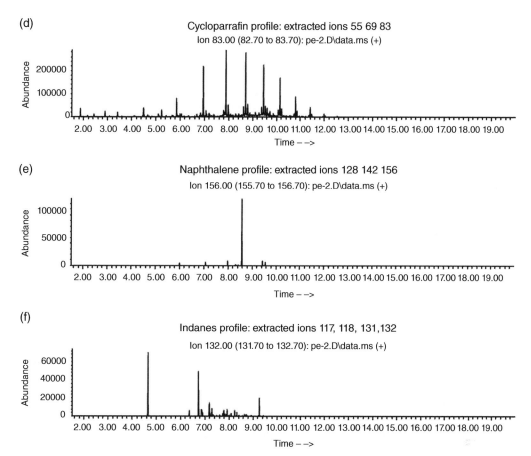

Figure 4.17 (*Continued*) (d) extracted cycloalkane profile; (e) extracted polynuclear aromatic profile; (f) extracted indane profile

extraction ion profiles show clear doublets in the "alkane" profile and triplets in the "cycloparrafin" profile. The ions indicative of cycloalkanes (55, 69, and 83) are also common to alkenes. Additionally, those ions are minor contributors to the fragmentation of alkanes. This provides a clear example where extracted ions are not specific to, but only indicative of, hydrocarbon classes. This data presentation is common to the thermal degradation of high-density polyethylene plastic.

Also noteworthy in case 2 are the aromatic profiles. There is the clear presence of a C2, C3, and C4 alkylbenzene peak grouping in the simple aromatic profile, as well as a notable presence of compounds in the polynuclear (naphthalene) and indane profiles. This could indicate the presence of a petroleum product. Direct comparison of those profiles to that of gasoline, however, clearly shows that while all the same compounds appear to be present, the relative peak ratios within groupings, most notably the C3 and C4 alkylbenzene groupings of the aromatics, are not consistent and, thus, the determination of the presence of an ignitable liquid in this sample is not supported (Figure 4.18).

(a)

Ion 119.00 (118.70 to 119.70). pe-2.D\data.ms (+)

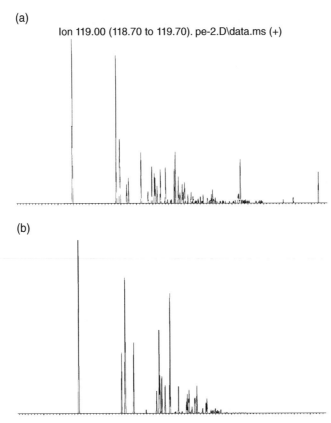

(b)

Figure 4.18 (a) Aromatic (alkylbenzene) profile for Case 2 compared to (b) the aromatic profile of gasoline

4.9 Summary

The identification or lack of identification of an ignitable liquid in a sample collected in conjunction with a fire investigation must be put into proper context. While the fire debris analyst may play a role, the ultimate responsibility for the determination of fire causation lies with the scene investigator and must be assessed in conjunction with other factors in the case.

The lack of identification of an ignitable liquid does not preclude arson or the use of an accelerant. The extent of fire damage, the type of fire suppression incorporated, the selection of samples, the type of accelerant used, and the complexity of matrix contributions to the data can result in negative findings.

Conversely, ignitable liquids can be identified in non-accelerated fires. Positive findings simply reflect the presence and class of the ignitable liquid but not the provenance. Ignitable liquids are inherent to the environment of advanced civilization.

Petroleum-laced backgrounds are common in a number of matrices including wood (stains and sealants), flooring and subflooring (adhesives, insecticides), paper products (ink vehicles), and shoes (various processes) (Lentini *et al.*, 2000). Additionally, most establishments have any number of incidental petroleum products stored on site. The identification of gasoline in a fire sample collected from a garage near a lawn mower would not be unexpected and would not lead, by itself, to a conclusion that arson was involved. That said, gasoline found in multiple samples throughout a structure file of a real estate office complex would be suspicious. Like most forensic evidence, the result of fire debris analysis is circumstantial and laboratory results are only a small but important part of the investigative process.

References

ASTM (2010) *ASTM E1386-10 Standard practice for separation of ignitable liquid residues from fire debris samples by solvent extraction*. ASTM International, West Conshohocken, PA.

ASTM (2008) *ASTM E2154-01 Standard practice for separation and concentration of ignitable liquid residues from fire debris samples by passive headspace concentration with solid phase microextraction (SPME)*. ASTM International, West Conshohocken, PA.

ASTM (2012a) *ASTM E1388-12 Standard practice headspace vapors from fire debris samples*. ASTM International, West Conshohocken, PA.

ASTM (2012b) *ASTM E1412-12 Standard practice for separation of ignitable liquid residues from fire debris samples by passive headspace concentration with activated charcoal*. ASTM International, West Conshohocken, PA.

ASTM (2013) *ASTM E2281-13 Standard practice for extraction and derivitization of vegetable oils and fats from fire debris and liquid samples with analysis by gas chromatography-mass spectrometry*. ASTM International, West Conshohocken, PA.

ASTM (2014) *ASTM E1618-14 Standard Test Method for Ignitable Liquid Residues in Extracts from Fire Debris Samples by Gas Chromatography-Mass Spectrometry*. ASTM International, West Conshohocken, PA.

Barnes, A., Dolan, J., Kuk., R., and Siegel, J. (2004) Comparison of gasolines using gas chromatography-mass spectrometry and target ion response, *Journal of Forensic Sciences*, **49** (5):1018–1023.

Dolan, J. and Newman, R. (2001) Solvent options for the desorption of activated charcoal in fire debris analysis. Paper presented at 55th Annual Meeting of the American Academy of Forensic Sciences, Seattle, WA.

FBI (2014) Crime in the United States 2013. Federal Bureau of Investigation http://www.fbi.gov/about-us/cjis/ucr/crime-in-the-u.s/2013/crime-in-the-u.s.-2013/property-crime/arson-topic-page (last accessed 11 June 2015).

Fultz, M. (1994) The effects of sample preparation on the identification of class 4 and class 5 petroleum products. Paper presented at the 40th Annual Meeting of the American Academy of Forensic Sciences, San Antonio, TX.

Hicks, G., Pontbriand, A., and Adams, J. (2003) Carbon disulfide versus dichloromethane for use of desorbing ignitable liquid residues from activated charcoal strips. Paper presented at the 55th Annual Meeting of the American Academy of Forensic Sciences, Chicago, IL.

IAAI (International Association of Arson Investigators Forensic Science Committee) (1995) IAAI Forensic Science Committee position on the use of accelerant detection canines. *Journal of Forensic Sciences*, **40** (4), 532–534.

Kuk, R. and Spagnola, M. (2008) Extraction of alternative fuels from fire debris samples. *Journal of Forensic Sciences*, **53** (5), 1123–1129.

Lentini, J. (2001) Persistence of floor coating solvents. *Journal of Forensic Sciences*, **46** (6), 1470–1473.

Lentini, J. and Armstrong, A. (1997) Comparison of the eluting efficiency of carbon disulfide with diethyl ether: the case for laboratory safety. *Journal of Forensic Sciences*, **42** (2), 307.

Lentini, J., Dolan, J., and Cherry, C. (2000) The petroleum-laced background. *Journal of Forensic Sciences*, **45** (5), 968–989.

Mann, D. (1987) The comparison of automotive gasolines using capillary gas chromatography I: Comparison methodology. *Journal of Forensic Sciences*, **32** (2), 606–615.

Mann, D. and Gresham, W. (1990) Microbial degradation of gasoline in soil. *Journal of Forensic Sciences*, **35** (4), 913–923.

NCFS (2014) *Ignitable Liquids Reference Collection*. National Center for Forensic Science (NCFS), University of Central Florida. http://ilrc.ucf.edu/ (last accessed 11 June 2015).

Newman, R.T., Dietz, W.R., and Lothridge, K. (1996) The use of activated charcoal strips for fire debris extractions by passive diffusion. Part 1 – The effects of time, temperature, strip size and concentration in the use of activated charcoal strips in fire debris analysis. *Journal of Forensic Sciences*, **41** (3), 361–370.

Newman, R., Gilbert, M., and Lothridge, K. (1998) *GC-MS Guide to Ignitable Liquids*. CRC Press, Boca Raton, FL.

NFPA (National Fire Protection Association) (2008) *NFPA 921: Guide for Fire and Explosion Investigations*. National Fire Protection Agency, Quincy, MA, pp. 137–138.

NFPA (National Fire Protection Association) (2014a) Fires in the US. http://www.nfpa.org/research/reports-and-statistics/fires-in-the-us (last accessed 11 June 2015).

NFPA (National Fire Protection Association) (2014b) *NFPA 921: Guide for Fire and Explosion Investigations*, National Fire Protection Agency, Quincy, MA.

Speight, J. (2014) *The Chemistry and Technology of Petroleum*, 5th edn. CRC Press, Boca Raton, FL.

Stauffer, E. (2001) Identification and characterization of interfering products in fire debris analysis. Masters thesis, International Forensic Research Institute, FIU, Miami, FL.

Stauffer, E. and Byron, D. (2007) Alternative fuels in fire debris analysis: Biodiesel basics. *Journal of Forensic Sciences*, **52** (2), 371–379.

Turner, D. and Goodpaster, J. (2011) Effect of microbial degradation on the chromatographic profiles of Tiki torch fuel, lamp oil, and turpentine. *Journal of Forensic Sciences*, **56** (4), 984–987.

CHAPTER 5

Explosives

John Goodpaster

Department of Chemistry and Chemical Biology, Indiana University–Purdue University Indianapolis (IUPUI), USA

5.1 The nature of an explosion

5.1.1 Types of explosions

Explosions are inherently powerful and dangerous to people and property. Simply stated, an explosion is a sudden and violent escape of gases from a central point accompanied by high temperature, violent shock and loud noise. From the point of view of an explosives investigator, the National Fire Protection Association (NFPA) defines an explosion as follows (NFPA, 2011):

> A sudden conversion of potential energy (chemical or mechanical) into kinetic energy with the production and release of gas(es) under pressure. These gases then do mechanical work, such as defeating their confining vessel or moving, changing or shattering nearby materials.

Note that the sudden failure of a pipe due to hydrostatic pressure is not considered an explosion as a gas is not involved. In turn, even sudden and violent fires that are not confined are not explosions, as they do not generate any overpressure.

Overall, true explosions can result via three main mechanisms: nuclear, mechanical and chemical. Nuclear explosions involve either the fission or fusion of atoms. This splitting or joining of nuclei results in tremendous release of energy. Investigating these types of explosions will generally not involve explosives investigators or forensic chemists. A mechanical explosion results from a gradual buildup of pressure in a rigid container due to the physical change of its contents from a liquid to a vapor (e.g., a hot water heater). An explosion results upon the sudden failure of the container and release of its contents. Finally, a chemical explosion results from the extremely rapid conversion of a solid or liquid energetic material into reaction products that have a much greater volume than the substance from which they are generated.

Forensic Chemistry: Fundamentals and Applications, First Edition. Edited by Jay A. Siegel.
© 2016 John Wiley & Sons, Ltd. Published 2016 by John Wiley & Sons, Ltd.

One subtype of chemical explosion occurs when a mixture of a gaseous fuel with air is somehow confined. If initiated, the mixture will violently combust and generate overpressure until it is suddenly released when the confinement is breached. These explosions are sometimes called "combustion explosions" or "fuel–air explosions". A dust explosion, which results from the sudden ignition of a large volume of combustible particulates, also falls into this category. These types of incidents generally do not benefit from chemical analysis, as the fuel will no longer be present after the explosion. Chemical analysis of potential components from devices used to initiate such an explosion would be valuable, however.

Lastly, all explosives that are commonly used for military and commercial purposes are chemical explosives. These explosive formulations can be pure compounds or mixtures, inorganic or organic in composition, and either of military, commercial or improvised origin. Chemical explosives will also differ in their sensitivity, explosive power and stability. A forensic chemist can play a large and important role in the identification of these types of explosives as well as in the identification of the remains of an explosive device. Hence, these topics are the main focus of this chapter.

5.1.2 Explosive effects

There are a number of processes that occur during a chemical explosion. The first, initiation, starts the chemical reaction that leads to an explosion. The energy barrier that must be overcome during initiation is the energy of activation of the explosive reaction. Therefore, there must be some form of energy that is introduced into the explosive, such as heat/flame, impact, friction, electrical discharge or a shock wave, that is generated by another explosive.

As shown in Figure 5.1, an explosion will dramatically affect its surroundings, via primary blast pressure, secondary blast pressure, a thermal/incendiary event and the formation of fragments/shrapnel. Although blast pressure is the greatest threat from detonating explosives, all explosions will unleash potentially lethal effects.

Primary blast pressure is the initial blast effect where hot, expanding gases create a pressure wave that radiates outward. Overall, blast waves form in a fraction of a second, exert enormous pressures on the atmosphere surrounding the blast and travel at thousands of miles per hour. It is also the case that the initial blast pressure will always project at 90° angles to the surface of the explosive, as shown in Figure 5.2 for two idealized cases.

Primary blast pressure also has distinct positive and negative phases. In the positive pressure phase, the compressed layer of air (also known as the shock front) applies a sudden shattering, hammering blow to any object in its path. This phase is extremely brief but the overpressure can cause catastrophic damage to people and objects. Overpressure can be hazardous to humans even at relatively low levels (e.g., injuries due to flying glass occur at overpressures between 0.6 and 3 psi). Overpressures at or above 14.5 psi are considered potentially lethal (primarily from lung hemorrhage).

Figure 5.1 Explosion of 20,000 pounds of Composition C-4 inside a small truck. Note that the shock wave is visible due to its extreme effect on the refractive index of the surrounding air. The "thermal event" occurs within the volume defined by the shock wave. Photo courtesy ATF laboratory (*See insert for color representation of the figure.*)

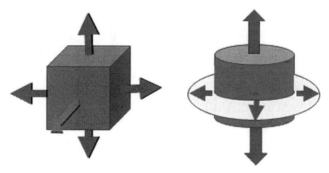

Figure 5.2 Depiction of primary blast pressure resulting from a cubical and cylindrical explosive charge

Overpressures above 25 psi will be lethal for almost all victims. Overpressure will also seriously damage or destroy structures. For example, serious structural damage will begin to occur at overpressures at or above 2.3 psi. Houses will be destroyed at 5–7 psi and overpressures at or above 10 psi can totally destroy buildings (NFPA, 2011).

In an open space, the pressure wave quickly loses power as it travels outwards, hence damage from overpressure decreases with distance from the explosion. At the same time, the outward compression of air in the positive pressure phase will form a vacuum at the point of the explosion. After the outgoing pressure wave dwindles, the compressed and displaced atmosphere reverses its movement and rushes inward to fill this void. This negative pressure phase is less powerful and longer lasting than the positive pressure phase.

Although not an initial effect of an explosion, secondary blast pressure begins within seconds after the primary blast pressure and can significantly increase damage. Secondary blast pressure is a general term for several phenomena that involve blast waves, namely reflection, focusing, shielding and venting.

In reflection, blast pressure waves encounter a hard surface, such as a wall, reflect off of that surface and travel in another direction. The angle of the reflected blast wave will be smaller than the angle of the incident blast wave, and the pressure behind the reflected wave will be greater than the original blast wave.

The phenomenon of shielding occurs when the blast pressure wave encounters a rigid and immovable object. For example, if a square, solid concrete post is placed in the path of the blast pressure wave, the wave will strike the post and the post will in effect "cut a hole" in the pressure wave. The area immediately behind a shielding object is afforded some protection from the pressure of the explosion. At some point beyond the object, however, the split blast pressure wave will reform and continue, but with diminished force. This idea is used in the physical security of many buildings where large, squat concrete pillars are placed in between the building and surrounding streets, where a potential car or truck bomb could be deployed.

Focusing is related to both to reflection and shielding and occurs according to the surrounding of the main charge. For example, an explosive charge surrounded by sand bags on three sides will generate a blast wave that is focused towards the unblocked side. Commercial and military explosives can be engineered to deliver a focused blast wave, such as in shape charges that can be used to cut with remarkable precision through solid objects like doors or walls.

Venting occurs when the blast pressure is able to escape a confined space. This phenomenon is exploited during the destruction of large quantities of explosives such as fireworks. In these cases the material is burned in a strong but perforated container (e.g., a metal dumpster with holes cut into the walls). The venting process allows the material to burn without building up pressure and exploding as the gases escape.

The thermal/incendiary effect of an explosion, sometimes called the "thermal event", is usually seen as the bright flash or fireball at the moment of the explosion. This event is extremely brief, measured in fractions of seconds. There is also a difference between deflagrating explosives, which produce relatively cooler, longer thermal events and detonating explosives, which produce shorter, hotter thermal events. This can have implications in bombings, as detonating explosives are less likely to start fires than are deflagrating explosives due to the shorter exposure of the surroundings to the heat of the explosion. Regardless, if a fire is started, the debris may provide additional fuel and contribute to spreading the fire. When fires start in a structure that has been bombed, they are usually traceable to broken or shorted electrical circuits and ruptured natural gas lines rather than to incendiary thermal effects.

Fragmentation occurs when the material that had been a part of the bomb casing or objects nearby are ruptured in the blast. Fragments can reach supersonic velocities and will travel in a straight line of flight until they fall to earth or strike an object, whereupon they will either ricochet or become embedded. It is also possible for fragments, which are at a high temperature due to the blast, to cause secondary fires. For example, a fragment may puncture an automobile fuel tank and ignite the gasoline, or fragments may start grass fires some distance from the explosion. Importantly, when an explosive device is on or close to the ground, a significant number of fragments and other trace evidence are propelled into the ground and can be found within the blast seat/crater.

Most improvised explosive devices (IEDs) incorporate a container to either hide the contents or provide confinement for a low explosive. Containers that provide confinement include pipes, tubing/conduits, grenade hulls, pressurized containers (gas cylinders), ammunition cans, bottles, and so on.

Given their frequency, the fragmentation of pipe bombs with low explosive fillers has been well studied. Extensive experience and several scientific studies have shown that the fragmentation of steel pipe is affected primarily by the amount and type of explosive filler as well as the means of initiation (e.g., a fuse versus a detonator). Therefore, a chemist can deduce the nature of the filler based upon fragment morphology alone, prior to any chemical analysis.

For example, pipe bombs filled with inorganic propellants that are less energetic (e.g., black powder and black powder substitutes) will yield a relatively low number of large fragments. The pipe may not fully fragment but rather split along its seam. The end cap faceplates will often be blown out. The fragments themselves will have square, 90° edges and exhibit a heavy grey or black residue, sometimes with a "rotten egg" smell from sulfur in the explosive. The solid reaction products that result from the deflagration will often cause corrosion of the fragments.

In contrast, pipe bombs filled with organic propellants that are more energetic (e.g., smokeless powder) will yield a larger number of smaller fragments. Single-base smokeless powder (composed largely of nitrocellulose [NC]) produces fragments with 90° breaks but no apparent residue – the fragments may even appear "shiny" on the surface that was facing the interior of the pipe. Double-base smokeless powder (composed of nitrocellulose and nitroglycerin [NG]) will produce fragments that have edges with 45° reversing slants ("stepping") and the pipe walls will be noticeably thinned due to the extreme pressure buildup inside the pipe. If a detonator is used to initiate double-base smokeless powder inside a pipe, it has been observed that it can "go high order" (i.e., detonate). In this case, the smokeless powder behaves as a secondary high explosive. As a result, there will be extreme fragmentation of the pipe and the damage will be similar to that obtained with a similar mass of NG dynamite.

Several studies involving the mass of pipe bomb fragments have been completed, including measuring the mass distribution of the fragments (Oxley *et al.*, 2001). In particular, this study established the use of a Fragment Weight Distribution Map (FWDM), the slope of which is directly related to the explosive power of the filler. This and subsequent studies (Bors *et al.*, 2014a, 2014b) have quantitated the extent to which explosives such as Pyrodex yield fewer larger fragments whereas smokeless powder yields numerous smaller fragments. Lastly, if sufficient container fragments are recovered, it is possible to attempt reconstruction of the device. Other features of the container, including maker's marks, size designation, and so on, can be noted.

Lastly, shrapnel is defined as additional objects, such as nails, ball bearings, and so on, that are attached to an explosive device and are propelled outward by the energy released during the explosion. Shrapnel may be placed on the exterior or interior of the device. Shrapnel is explicitly designed to have antipersonnel effects by cutting, slicing, or penetrating victims in the vicinity of the device.

5.2 Physical and chemical properties of explosives

Legally, an explosive is broadly defined as any chemical compound, mixture or device, the primary or common purpose of which is to function by explosion (US Federal Code, 18 U.S.C. Chapter 40). Explosive formulations can be categorized in a number of different ways, although most lay persons and investigators differentiate explosives based upon their manufacture and purpose. This results in three main categories of explosive (Table 5.1)

Table 5.1 Categorization of explosives by source and end use

Category	Application(s)	Examples	
Commercial	Ammunition Mining Construction	Black powder Black powder substitutes Smokeless powder Perchlorate explosive mixtures Pyrotechnics	Dynamite Emulsions Water gels Ammonium Nitrate/Fuel Oil (ANFO) Binaries (ammonium nitrate + nitromethane)
Military	Ammunition Demolition Munitions	TNT Composition B (TNT/RDX) Composition C-4 (RDX) Detonating cord (PETN) Semtex (PETN/RDX) Flexible sheet explosive (PETN)	
Improvised	Improvised explosive devices	Home-made propellants	TATP HMTD Urea nitrate Concentrated hydrogen peroxide

As viewed by chemists, explosives are materials that undergo a rapid and energetic transformation into more stable substances, liberating large quantities of heat and gas. Explosives can be organized into six distinct chemical "families", which are based upon chemical structure. Explosives can be further classified based upon their explosive power and explosive mechanism. This results in two main types of explosives: low explosives and high explosives, which are discussed in more detail here.

The explosive families, their common structural elements and examples of low and high explosives that fall into that family are summarized in Table 5.2. This classification scheme is more relevant for determining the analytical procedures and methods to be used on any given sample.

5.2.1 Low explosives

Upon initiation, low explosives will deflagrate, which is defined by the National Fire Protection Association (NFPA) as a reaction in which the velocity of the reaction front through the unreacted medium is less than the speed of sound. This process is relatively slow, as it relies upon the burning of individual particles of explosive and the preheating of adjacent particles until they ignite, eventually propagating a flame front. Note that a deflagrating explosive will only cause an explosion if it is somehow confined (e.g., inside

Table 5.2 Categorization of explosives by chemical composition and explosive power

Family	Structural element	Low explosives	High explosives
Nitroalkane	$R-NO_2$	N/A	Nitromethane Nitroethane
Nitrate ester	$-O-NO_2$	Nitrocellulose (NC)	Nitroglycerine (NG) Pentaerythritol tetranitrate (PETN)
Nitroaromatic	$Ar-NO_2$	2,4-dinitrotoluene (DNT)	2,4,6-trinitrotoluene (TNT)
Nitramine	n	N/A	Cyclotrimethylene trinitramine (RDX) Cyclotetramethylene tetranitramine (HMX)
Acid salts	XNO_3 / $XClO_4$*	Potassium nitrate (KNO_3) Potassium chlorate ($KClO_3$) Potassium perchlorate ($KClO_4$)	Ammonium nitrate (AN) Urea nitrate (UN)
Organic peroxides	$-O-O-$	N/A	Triacetone triperoxide (TATP) Hexamethylene triperoxide triamine (HMTD)

*X = Na, K, Ba, Sr, NH_4

Table 5.3 Common inorganic low explosives with their associated oxidizers and fuels

Explosive	Oxidizer(s)	Fuels
Black powder	KNO_3	sulfur, charcoal
Pyrodex®	KNO_3, $KClO_4$	sulfur, benzoic acid, dicyandiamide
Triple Seven®	KNO_3, $KClO_4$	benzoic acid, nitrobenzoic acid, dicyandiamide
American Pioneer Powder®	KNO_3, $KClO_4$	ascorbic acid
Perchlorate explosive mixtures	$KClO_4$	sulfur, aluminum

a rigid container). In this case, the internal pressure rapidly increases until the container suddenly fails and releases the heat and gas into its surroundings. Although they are less powerful, low explosives are quite sensitive in that they can be easily ignited by spark, shock, or flame.

5.2.1.1 Inorganic low explosives

Low explosives that are primarily inorganic in composition are physical mixtures that contain separate oxidizers and fuels. Some of the most commonly encountered inorganic low explosives are shown in Table 5.3.

Note that these types of explosives are not completely efficient and relatively large amounts of residual explosive and/or solid reaction products are commonly found on postblast debris. This is, of course, readily exploited by a forensic chemist.

During deflagration, the fuel and oxidizer act in a synergistic fashion whereby the oxidizer releases oxygen, the oxygen reacts with the fuel, and the heat of combustion of the fuel further decomposes the oxidizer. Deflagration of an inorganic low explosive can be viewed as a two-step process that creates a positive feedback loop, as shown here for a generic oxidizer and carbonaceous fuel:

(i) Oxidizer + heat = Product + O_2
(ii) Fuel + O_2 = CO_2 + H_2O + heat

As mentioned, oxidizers are compounds (typically solids) that decompose at elevated temperature liberating oxygen gas. By definition, oxidizers act as oxidizing agents and include compounds such as nitrates, chlorates and perchlorates. Some of the oxidizers found in inorganic low explosives are:

Ammonium nitrate (NH_4NO_3)
Ammonium perchlorate (NH_4ClO_4)
Barium chlorate ($Ba(ClO_3)_2$)
Barium chromate ($BaCrO_4$)
Barium nitrate ($Ba(NO_3)_2$)
Lead chromate ($PbCrO_4$)
Potassium bichromate ($K_2Cr_2O_7$)

Potassium chlorate ($KClO_3$)
Potassium dichromate (KCr_2O_7)
Potassium nitrate (KNO_3)
Potassium perchlorate ($KClO_4$)
Strontium nitrate ($Sr(NO_3)_2$)

Note that oxidizers containing barium and strontium are commonly found in pyrotechnic compositions, where their primary purpose is to emit green and red light, respectively. Therefore, another oxidizer such as potassium chlorate or potassium perchlorate will also be present to drive the deflagration reaction.

Important properties of an oxidizer that affect the characteristics of an explosive mixture are its melting point, heat of decomposition and oxygen content. As melting point increases, ignitability decreases, whereas as heat of decomposition increases, ignitability increases. Finally, as oxygen content increases, explosive power increases.

Table 5.4 illustrates the effect of the anion in a series of potassium salts that serve as oxidizers (Conkling and Mocella, 2011). In general, these oxidizers are comparable in terms of their oxygen content (and hence explosive power). However, chlorate-based oxidizers are much more ignitable due to their relatively low melting point. In addition, the decomposition reactions for these oxidizers are actually exothermic rather than endothermic. This leads to specific safety concerns for explosive mixtures containing chlorate (ClO_3), as the potential for accidental ignition is significant. Explosive mixtures containing perchlorate (ClO_4) are generally preferred, as they are much less ignitable due to the higher melting point of the oxidizer. They also provide the largest oxygen content, which translates into greater explosive power.

In contrast to oxidizers, fuels are compounds that react with oxygen to form stable gaseous products and heat. Example fuels include:
Aluminum (Al)
Antimony (Sb)
Antimony sulfide (Sb_2S_3)
Carbon (charcoal) (C)
Iron (Fe)
Magnalium (Mg/Al alloy)
Magnesium (Mg)
Phosphorus, red (P)
Sulfur (S)
Titanium (Ti)

Important properties of a fuel that affect the characteristics of an explosive mixture are the melting point, combustion temperature, gas production, thermal conductivity and particle size. The most ignitable fuels have an optimal combination of a low melting point, high thermal conductivity and small particle size. Fuels that yield the greatest explosive power have high combustion temperatures and large gas production. The data shown in Table 5.5 illustrate these effects (Conkling and Mocella, 2011). Note that

Table 5.4 Physical and chemical properties of common oxidizers (Conkling and Mocella, 2011)

Oxidizer	Formula	Melting point (°C)	Heat of decomposition (kcal/mol)	Oxygen content (%)
Potassium nitrate	KNO_3	334	+75.5	40
Potassium chlorate	$KClO_3$	356	−10.6	39
Potassium perchlorate	$KClO_4$	610	−0.68	46

Table 5.5 Physical and chemical properties of common fuels (Conkling and Mocella, 2011)

Fuel	Formula	Melting point (°C)	Heat of combustion (kcal/g)
Carbon (charcoal)	C	decomposes	7.8
Sulfur	S	119	2.2
Magnesium	Mg	649	5.9
Aluminum	Al	660	7.4
Titanium	Ti	1660	4.7

Table 5.6 Heats of reaction for common inorganic explosives (Conkling and Mocella, 2011)

Composition	Application	Heat of reaction (kcal/gram)
KNO_3 + C + S	Black powder	0.66
$NaNO_3$ + Al	White light composition	2.00
$KClO_4$ + Mg	"photoflash"	2.24

elemental sulfur is used in explosive mixtures such as black powder or pyrotechnics as a "tinder", whereby its low melting point allows the composition to react more rapidly. These data also illustrate that one reason that metallic aluminum is a popular fuel is its relatively low melting point and high heat of combustion as compared to other metals.

Ultimately, the amount of energy released by a low explosive mixture depends on the specific combination of the oxidizer and fuel. As shown in Table 5.6, metallic fuels allow for the production of large amounts of heat as compared to non-metallic fuels such as carbon and sulfur.

5.2.1.2 Organic low explosives

Low explosives that are primarily organic include all forms of smokeless powder, which is based upon nitrocellulose. These explosives serve as the propellant for all modern ammunition. In general, smokeless powders are more energetic than inorganic propellants and much more efficient. As an example, burning black powder only yields

50% gaseous reaction products whereas the yield from nitroglycerin (one of the energetic components of double-base smokeless powder) is effectively 100%. As a result, the volume of gas produced from burning 100 grams of nitroglycerin is 403 liters, whereas the same mass of black powder only generates 27 liters of gas.

The major categories of smokeless powder are determined by what energetic compounds they contain. Single-base smokeless powder contains only nitrocellulose. Double-base powder contains nitrocellulose and nitroglycerin. Triple-base powder contains nitrocellulose, nitroglycerin and nitroguanidine. The latter type is generally not found in explosives investigations as it is only used as a propellant in large caliber military munitions.

Smokeless powders also contain stabilizers, plasticizers, deterrent coatings and flash inhibitors, such as:

Stabilizers: diphenylamine (DPA), ethyl centralite (EC), methyl centralite (MC)
Plasticizers: dibutylphthalate (DBP)
Deterrent coatings: vinsol resin, synthetic polymers
Flash inhibitors/inorganics: K_2SO_4, potassium hydrogen tartrate, KCl, KNO_3

Smokeless powders are quite diverse in their morphology (Figure 5.3). The shape and size of the particles directly influences their burn rate and gas output, which are crucial for good performance in modern pistol, rifle and shotgun ammunition.

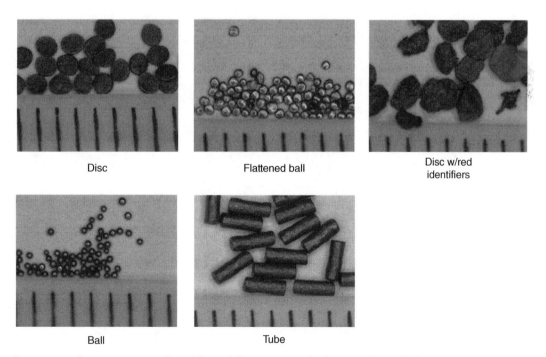

Disc   Flattened ball   Disc w/red identifiers

Ball   Tube

Figure 5.3 Photomicrographs of five different morphologies of smokeless powders (scale is in mm) (*See insert for color representation of the figure.*)

5.2.2 High explosives

High explosives undergo detonation, which is a nearly instantaneous transformation of the explosive substance into gas at very high temperature. Important characteristics of high explosives include detonation velocity, which is the rate at which the detonation wave propagates through a high explosive. Brisance is a term describing the shattering capability of a high explosive. Brisance is typically associated with high explosives with high detonation velocities. It is a desirably property when destroying large volumes of rock (i.e., in mining and construction) or in military applications.

In contrast to low explosives, some high explosives can be thought of as containing a fuel and oxidizer bound in the same chemical structure. This takes the form of highly oxidized and electron withdrawing functional groups such as nitro (NO_2) and highly reduced and electron donating functional groups such as methyl (CH_3). These types of explosives are sometimes called "molecular explosives" and include well-known military explosives such as trinitrotoluene (TNT), pentraerythitol tetranitrate (PETN) and 1,3,5-trinitro-1,3,5-triazine (RDX).

In a detonation, the velocity of the reaction front through the unreacted explosive is greater than the speed of sound (344 m/s or 1129 ft/s). As shown in Table 5.7, the

Table 5.7 Detonation velocities for several high explosives (Cooper, 1996)

Explosive	Application	Detonation velocity (m/s)
Hexamethylene triperoxide diamine (HMTD)	Improvised HE	4500
Urea nitrate (UN)	Improvised HE	4700
Ammonium nitrate (AN)	Emulsions Water gels Blasting agents	5270
2,4,6-trinitrotoluene (TNT)	Cast explosives	6900
Trinitroglycerine (NG)	Dynamite	7700
Pentaerythritol tetranitrate (PETN)	Detonating cord Sheet explosive Cast explosives (Pentolite)	8400
Cyclonite (RDX)	Detonating cord Composition C-4 Cast explosives (Composition B)	8750
Octogen (HMX)	Detonating cord (shock tube) Cast explosives (Octol)	9400

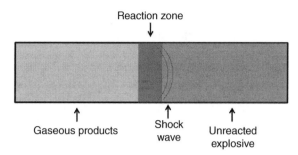

Figure 5.4 Depiction of a detonation wave propagating through a high explosive (*See insert for color representation of the figure.*)

highest detonation velocities are observed in military high explosives, which also exhibit the highest brisance, or shattering capability (Cooper, 1996).

Detonation is a process where a compression wave moves through an unreacted explosive material. The wave creates a distinct reaction zone where the explosive experiences extremely high pressure and heat. This causes the unreacted explosive to immediately transform into gaseous products, further propagating the detonation wave (Figure 5.4).

However, not all detonating explosives function as designed, leading to a differentiation between a high order and a low order explosion (which is not to be confused with low and high explosives). A high order explosion is a complete consumption of a detonating explosive at its optimum velocity. A low order explosion is an incomplete consumption of the detonating explosive, or it is complete at lower than optimum velocity. Low order explosions can be caused by a number of factors, such as deterioration of the explosive, poor contact between the initiator and the explosive, disruption of the shock wave due to air bubbles or impurities in the explosive, insufficient power in the initiator or any combination of these factors.

5.2.2.1 Primary high explosives

Some high explosives are quite sensitive and will detonate if exposed to spark, shock or flame. These materials are called "primary" or "initiating" high explosives. Examples of primary high explosives include the compounds used in ammunition primer compositions (e.g., mercury fulminate, silver fulminate, lead azide, silver azide, and lead styphnate). Other primary high explosives include nitroglycerin (when it is a pure liquid), triacetone triperoxide (TATP) and hexamethylene triperoxide diamine (HMTD).

The latter two explosives are organic peroxides that have been used as main charges or in smaller amounts as improvised detonators. These explosives have received a large amount of attention as they are highly sensitive to spark, shock and friction. In addition, these explosives will detonate with detonation velocities approaching that of TNT. The synthesis of both TATP and HMTD is relatively straightforward and, hence, can be completed by the curious or criminal bomb makers. When pure, these materials are white, crystalline powders and can be handled carefully. However, the quality of the chemicals

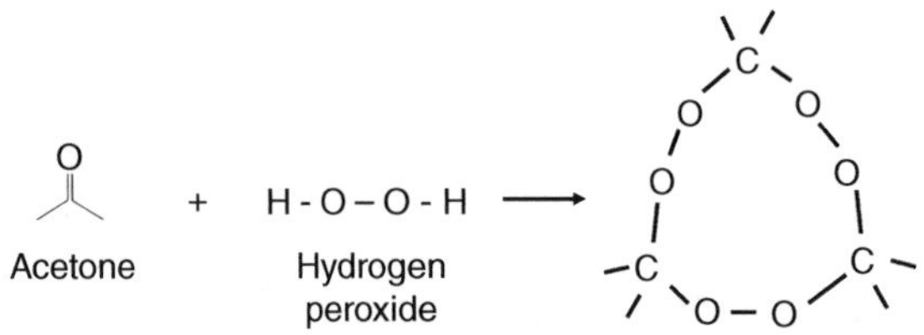

Figure 5.5 Synthesis of TATP from acetone and hydrogen peroxide

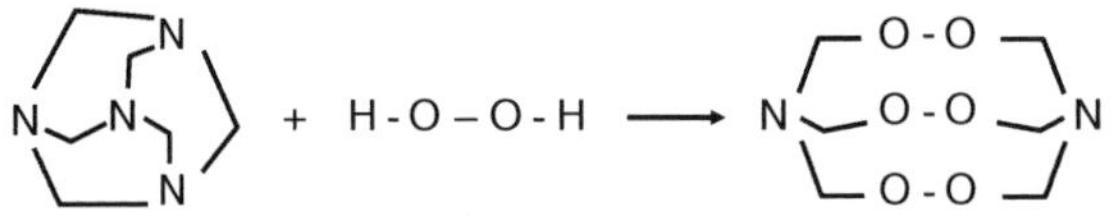

Figure 5.6 Synthesis of HMTD from hexamine and hydrogen peroxide

used in their synthesis and the extent of their purification will determine the stability and sensitivity of the final product. Impure batches will be unstable and are highly dangerous and unpredictable.

The synthesis of TATP can be carried out in a single vessel using acetone, concentrated hydrogen peroxide and an acid catalyst. The reaction then proceeds with the final product precipitating out of the solution (Figure 5.5).

TATP is unusual in that the solid form can sublime at room temperature and pressure. This sublimation is rather rapid (e.g., the majority of the mass of a TATP sample that is left open to the air at room temperature will have sublimed after only two weeks). This can be hazardous, as TATP inside an air tight jar will sublime and deposit TATP crystals on the jar lid and/or its threads. This can lead to accidental explosions when the jar is opened.

The synthesis of HMTD is shown in Figure 5.6. Unlike TATP, HMTD is not volatile and can be safer to handle. HMTD can chemically degrade over time, however, reducing its effectiveness.

5.2.2.2 Secondary high explosives

Most high explosives need a strong impulse in order to detonate, hence they are called "secondary" or "non-initiating" high explosives. Secondary high explosives include various molecular explosives, which are often formulated as plastic-bonded explosives (PBX). PBX can be quite complex, containing binding agents, waterproofing agents and color intensifiers. Binding agents include wax, resins, plastics and oils that make finely divided explosive particles adhere to each other. This also prevents segregation of particles and ensures a uniformly blended composition. In some instances, binding agents can reduce the sensitivity of the composition.

Secondary high explosives can also be mixtures. One of the most well-known high explosive mixtures is ammonium nitrate and fuel oil or "ANFO". ANFO is sold and transported as a preprepared mixture of prilled ammonium nitrate with fuel oil as a minor component coating the prills. Alternatively, some high explosive mixtures are sold and transported as separate components and are only combined immediately prior to use. These are known as "binary high explosives"; a common formulation makes use of powdered ammonium nitrate as an oxidizer and liquid nitromethane as a fuel.

5.2.2.3 Oxygen balance

Although molecular explosives have both the fuel and oxidizer bound into the same molecule, the amount of oxygen released may be either too much or too little to fully convert the molecule to end products. This leads to the concept of oxygen balance, which is defined as:

$$oxygen\,balance = \frac{mass\,of\,oxygen\,released\,or\,consumed}{mass\,of\,explosive}$$

Oxygen balance can be calculated for any molecular explosive as follows:

- **(i)** Write the decomposition reaction of the explosive, with CO_2, H_2O and N_2 as the only final products.
- **(ii)** Balance the reaction for C (as CO_2), H (as H_2O) and N (as N_2).
- **(iii)** Add or subtract molecular oxygen (O_2) from the products to balance the number of moles of oxygen in the overall reaction.
- **(iv)** Calculate the difference between how many moles of oxygen are found in one mole of the explosive and the number of moles of oxygen produced or consumed in the decomposition,
- **(v)** Calculate the mass of O released or consumed divided by the molecular weight of the explosive,

As an example, the oxygen balance of RDX (MW 222 g/mol) would be calculated from this balanced chemical equation:

$$C_3H_6N_6O_6 = 3CO_2 + 3H_2O + 3N_2 - 1.5O_2$$

Therefore, the oxygen balance of RDX is:

$$oxygen\,balance = \frac{-1.5\,moles\,O_2\,consumed \times 32\,grams\,/\,mole}{222\,grams\,/\,mole} = -21.6\%$$

Table 5.8 Calculated oxygen balance for various high explosives

Explosive	MW	C	H	N	O	Oxygen balance (%)
Ammonium nitrate	80	0	4	2	3	20.0
Nitroglycerin	227	3	5	3	9	3.5
PETN	316	5	8	4	12	−10.1
RDX	222	3	6	6	6	−21.6
HMX	296	4	8	8	8	−21.6
Nitromethane	61	1	3	1	2	−39.3
TNT	227	7	5	3	6	−74.0

Oxygen balance can also be calculated directly using the molecular weight (MW) of the explosive and the number of moles of carbon (C), hydrogen (H) and oxygen (O) present in one mole of the explosive:

$$oxygen\,balance\,(\%) = -\frac{1600}{MW}\left(2C - \frac{H}{2} - O\right)$$

Regardless of calculation method, the oxygen balance of explosives can vary greatly as shown in Table 5.8.

In theory, an oxygen balance of zero will yield the greatest explosive power, as the explosive is most efficiently converted into gaseous products. The concept of oxygen balance can be used to advantage by combining different explosives to achieve an overall oxygen balance that is as close to zero as possible.

For example, TNT can be used alone as a secondary high explosive. TNT, however, has a very large negative oxygen balance, meaning that oxygen from the surrounding atmosphere must be consumed in order for the explosive to be fully decomposed into gaseous products. In contrast, Composition B is a secondary high explosive that is approximately 50% TNT (oxygen balance of −74) and 50% RDX (oxygen balance of −21.6). The oxygen balance of Composition B is more favorable and can be calculated as the weighted average of the oxygen balance of the two explosives by themselves:

$$oxygen\,balance\,(\%) = 0.5(-74.0) + 0.5(-21.6) = -47.8\%$$

Similarly, the optimal composition of binary high explosives, such as mixtures of ammonium nitrate (positive oxygen balance of 20) and nitromethane (negative oxygen balance of −39.3), can be calculated by setting the total oxygen balance to zero and solving for the mass fractions of each explosive:

$$oxygen\,balance\,(\%) = 0 = (x)(20) + (1 - x)(-39.3)$$

$$0 = 20x - 39.3 + 39.3x$$

$$39.3 = 59.3x$$

$$x = 0.66$$

Therefore, the optimal composition is $66\% \, (w/w)$ ammonium nitrate
and $34\% \, (w/w)$ nitromethane

It is important to keep in mind, however, that achieving an overall oxygen balance of zero is not the sole determining factor for the mass factions in high explosive mixtures. In practice, explosive formulations may differ from the theoretical ideal. For example, Kinestik® is a commercial binary high explosive whose final composition upon mixing is approximately 80% (w/w) ammonium nitrate and 20% (w/w) nitromethane.

5.2.2.4 Power index

The total amount of energy released by an explosive can be expressed in terms of the Gibb's free energy equation, which separates the enthalpic and entropic components of a spontaneous chemical reaction:

$$\Delta G = \Delta H - T\Delta S$$

In this equation the enthalpic term is directly related to the thermal event of an explosion whereas the entropic term is directly related to the amount of work (in the thermodynamic sense) that the explosion can exert on its surrounds. By definition, explosive chemical reactions are spontaneous as they are highly exothermic ($\Delta H < 0$) and release tremendous amounts of gaseous reaction products ($\Delta S > 0$). The intensity of the "thermal event" of an explosive can vary and, in some cases, this can be extremely dangerous if explosives are used underground where flammable gases may be present.

Explosive power is influenced by the enthalpy and entropy of an explosives reaction. Explosive power is the total amount of energy released by an explosive that is available to do work. This can be described in terms of the number of moles of gas produced and their temperature. This can be described using the Ideal Gas Law:

$$PV = nRT$$

where R is the gas constant and P and V are the pressure and volume of the generated gas. These two terms combine to equal the total amount of "PV work" that an explosion will generate (in Joules). The number of moles of gas generated (n) and the temperature (T) will ultimately determine this quantity. Note that this is a conceptual relationship only as the extreme conditions following an explosion are clearly far from those of an ideal gas.

Explosive power can also be expressed as the product of the heat evolved in an explosion (Q) and the volume of gaseous products (V). When this quantity is calculated

Table 5.9 A comparison of power index and detonation velocity for several military high explosives

Explosive	Power index	Detonation velocity (m/s)
TNT	331	6850
PETN	452	7920
HMX	455	9100
RDX	457	8440

for an explosive and normalized relative to picric acid, the Power Index (PI) for that explosive can be determined:

$$Power\ Index\ (PI) = \frac{(QV)_{explosive}}{(QV)_{picric\ acid}}$$

The power indices and detonation velocities for several military high explosives are shown in Table 5.9. Note that explosives that have quite different detonation velocities such as RDX, HMX and PETN can have similar power indices.

5.3 Protocols for the forensic examination of explosives and explosive devices

5.3.1 Recognition of evidence

Although laboratory examination of explosives can yield important information, it is typically the final step in an explosives investigation. At the crime scene, qualified explosive ordinance disposal (EOD) personnel will render any live explosive devices safe and the remains of these devices will be collected as evidence. In addition, any potential evidence of devices that functioned as designed will be collected. All evidence should be photographed, packaged, labeled, recorded in an evidence log and secured.

Recognition of the remains of an IED can be challenging but it is vital for complete reconstruction and possible tracing of an IED to a bomb maker. For example, it can be difficult to differentiate potential parts of the device from parts of the target (e.g., automobile parts). As device components are found, it is useful to sort them by category (e.g., container, firing train, power source, triggering mechanisms). At that point, investigators can attempt to reconstruct the firing system and fusing mechanism of the IED.

It is generally true that low mass fragments will be found closer to the seat of the blast than high mass fragments, which tend to gather significant momentum during the blast. Hence, extensive sifting of material from the blast seat will often yield small pieces of

Figure 5.7 Conducting a line search during a postblast training exercise (*See insert for color representation of the figure.*)

valuable evidence and a well organized search of the immediate vicinity using a line search or other procedure should also be carried out to locate larger device fragments (Figure 5.7). As a general rule, the scene radius should be established at a distance that is 150% of the distance of the farthest fragment that has been found.

5.3.2 Portable technology and on-scene analysis

In general, the ability to conduct forensic testing in the field can be desirable as little to no sample degradation occurs, results are known immediately and samples can also be retained for further testing. Although on-site testing for explosives residue is not routine in the United States, several techniques exist with that potential.

For example, color tests can be used for on-scene testing of materials suspected to contain explosives. One such commercial product (the Expray Explosives Detection Kit) utilizes three color tests in series. The first contains tetrabutylammonium hydroxide, which reacts with polynitroaromatics and forms Meisenheimer complexes that range in color from dark brown for TNT to orange for tetryl. The second reagent contains sulfanilamide and N-ethylenediamine, which undergoes the Griess reaction and forms a pink dye with the nitrate ions formed during the first reaction. Finally, the third reagent uses zinc dust to reduce nitrates to nitrite ions, which then also react with the Griess reagent. The use of this three-part color test to estimate concentrations of explosives in environmental samples has been evaluated by the US Army Corps of Engineers (Bjella, 2005).

A test kit that uses chemistry that is similar to the Expray has been developed by researchers at Lawrence Livermore National Laboratory (LLNL) (Reynolds *et al.*, 2006). This kit is called the Easy Livermore Inspection Test for Explosives (ELITE) and employs a swipe that is used to collect potential residue. The swipe is then exposed to a set of chemical reagents and any color change is noted. Detection of a number of organic and inorganic explosives has been demonstrated (Johnson, 2006).

LLNL has also developed a portable GC-MS system (Reynolds *et al.*, 2006). The clear advantage of this system is that it offers unambiguous identification of substances in the field. It does require, however, external power and extensive training of the user. Capillary electrophoresis (CE) also lends itself to miniaturization and portability. Recent research has developed and demonstrated the use of a portable CE system in the identification of inorganic ions at postblast scenes (Hutchinson *et al.*, 2008). Most recently, a portable mass spectrometer has been described that can be worn in a backpack and used to directly sample surfaces for explosives residue (Hendricks *et al.*, 2014).

5.3.3 In the laboratory

It is only after scene processing is complete that items are submitted to a laboratory. Any evidence submitted to a laboratory should be accompanied by some sort of evidence control/chain of custody document. This will include information on the submitter, suspect, victim, incident, evidence and what exams are requested. Upon reaching a laboratory, standard protocols for receipt, storage and examination of evidence will be used.

In general, the laboratory analysis of explosives evidence will follow certain "testing pathways". Initial laboratory analysis involves documenting the appearance of the explosives evidence to be analyzed and examining it for trace evidence, whether chemical or biological in nature. This may include all ancillary parts, electronics and/or pieces of the explosive device collected as evidence, or samples from the surrounding explosion scene that may contain residual trace explosives residue or trace evidence left by the perpetrator.

For example, a general approach to a postblast case would take the following steps:

 (i) Provide a general description of the exhibit.
 (ii) Photograph evidence.
 (iii) Remove visible explosive particles (if present).
 (iv) If another discipline is required during examination (such as latent prints or tool marks), contact the appropriate personnel.
 (v) Perform explosive analysis.
 (vi) Examine and describe device components in detail – identify based on physical appearance (e.g., marks, logos) and/or based on chemical characteristics.

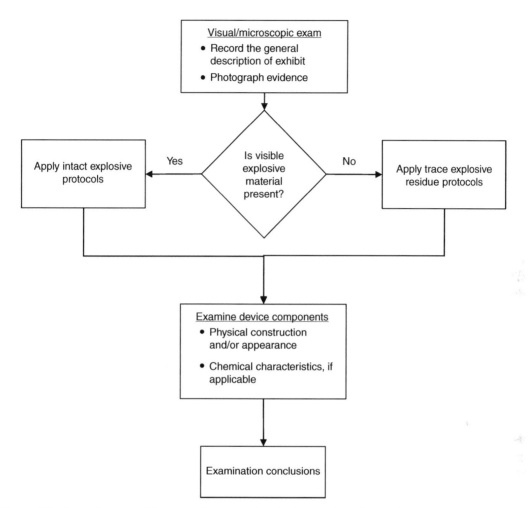

Figure 5.8 General protocol for examination of explosives and explosive devices

A graphical example of a protocol for the laboratory examination of explosives evidence is shown in Figure 5.8. This is followed by a discussion of the analysis of intact explosives, post-blast residues and device components.

Of course, not every case will include every element shown above. However, after completing this general approach, a forensic chemist should be able to reliably address questions such as:

- What explosive was used in the device?
- What device components were used (e.g., manufacturer and brand)?
- Is there any association between the physical and chemical characteristics of the IED and materials in the possession of a suspect?

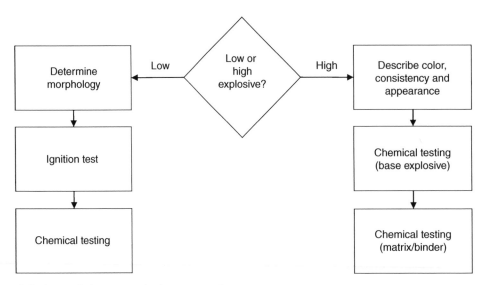

Figure 5.9 Example intact explosive protocol

5.3.3.1 Intact explosives protocols

Low explosives are generally propellant powders whose shape, color and size can reveal much about their origin and manufacturer. High explosives are typically dense solids whose base explosive is mixed with a binder or other matrix. Information about the possible manufacturer of a high explosive is generally obtained through chemical analysis. As a result, the ultimate methodology that is used for an intact explosive sample will differ significantly between low and high explosives (Figure 5.9).

The process by which a low explosive is identified will tend to differ based upon its overall chemical composition (e.g., inorganic versus organic). For example, a typical protocol for an inorganic low explosive powder such as black powder will differ from the protocol for an organic low explosive such as smokeless powder:

Black powder
  (i) Visual/microscopic examination
  (ii) Ignition test
  (iii) Chemical testing
  (iv) SEM-EDS to identify K, N, O, S, C
  (v) XRPD to identify KNO_3, S

Smokeless powder
  (i) Visual/microscopic examination.
  (ii) Ignition test

(iii) Chemical testing
(iv) FTIR to identify NG, NC
  (v) GC/MS to identify NG.

  Visual and microscopic examination of low explosive powders are highly informative, as explosive particles have distinct size, shape, color, heterogeneity, surface coatings and porosity. An ignition test provides a quick assessment of the general type of low explosive that is being examined. Then, appropriate methods are used to identify the inorganic or organic components in the material.

5.3.3.2 Postblast residue protocols

The identification of trace explosive residues is among the most challenging of the examinations conducted in a laboratory. In the absence of intact explosive particles, an examiner can attempt to identify postblast residues based upon the volatile components that are present, visible combustion products, or the extraction of invisible residues as shown in Figure 5.10.

  As shown previously, a critical procedure in the determination of trace explosive residues is that of sample preparation. In particular, if intact particles of explosive are not

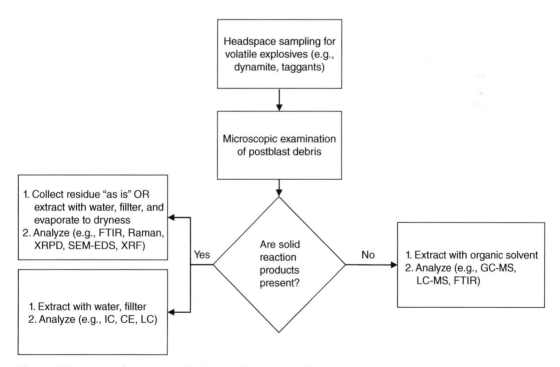

Figure 5.10 Example trace explosive residue protocol

found, extractions of the debris should be made. Extracts can be broken into three main categories:

(i) Aqueous extracts are analyzed using techniques capable of identifying trace anions and cations in a sample (e.g., IC, CE, and/or LC).

(ii) Evaporated aqueous extracts are analyzed using techniques capable of identifying solid inorganic materials (e.g., FTIR, XRD, EDX, Raman and PLM).

(iii) Organic extracts (e.g., acetonitrile, acetone, ethanol, methanol, methylene chloride and ethyl acetate) are analyzed using techniques capable of identifying low levels of organic explosives (e.g, LC-MS, GC-MS, GC-TEA, EGIS, IMS, GC-ECD, FTIR and TLC).

An ideal solvent will dissolve the explosive of interest but have little to no effect on the substrate. For most situations, acetonitrile, methanol, acetone or dichloromethane can be used. Dichloromethane is best used with substrates such as metal and glass. Plastic materials such as polystyrene or poly(vinyl chloride) should be extracted with methanol or acetonitrile to avoid dissolving the polymer. If substrate effects are not a concern, acetone is a convenient solvent for explosives because of their high solubility in this solvent. Following extraction with an organic solvent, the same location/item can be re-extracted with water to recover any inorganic residues. Taken together, the analysis of the organic and aqueous extract will typically give a clear indication of the explosive used.

Unlike in the analysis of intact explosives, the identification of a postblast residue often requires identifying combustion products of a particular explosive. In the case of inorganic low explosives, these are solid reaction products that are commonly found on device fragments. In the case of organic low explosives and high explosives, any combustion products are generally undetectable given the efficiency with which these materials decompose. Hence, trace residues of the original explosive must be identified.

5.3.3.3 Examination of control samples

Any analytical procedure that is intended to identify trace amounts of explosive requires a number of quality control elements to be in place in order to produce reliable data. The analytical techniques in question are typically those that involve mass spectrometry because of its high sensitivity and specificity (e.g., GC-MS, LC-MS and IC-MS). These techniques also have the advantage of being automated, but can only analyze one injected sample at a time.

The use of a control sample is particularly important when samples are not taken directly by a chemist and then analyzed. Even with routine analysis, any instrumental method that is intended to identify trace amounts of explosive must also include the following steps:

(i) Instrument blank (optional): The instrumental method is completed without any injection. This step verifies that the instrument is in a clean, stable state prior to beginning analysis.

(ii) Solvent blank: An aliquot of the same solvent that is used for the sample is injected. This verifies that the solvent is free of contaminants as well as the instrument being free of carry-over from previous injections.

(iii) Materials control: A sample of solvent is processed in parallel with a sample using the same filters, reagents, vials, and so on. This sample is a check against any systematic contamination of laboratory glassware or supplies.

(iv) Control sample: This is a "control" in the typical sense of the word in that it is the same material that was used to gather the sample (e.g., swab, gauze pad, filter, etc.). This control sample must be analyzed along with the sample to prove that there is no systematic contamination of the device that was used to gather the suspected residue.

5.3.3.4 Solid phase microextraction

An alternative sample preparation technique in explosives analysis is solid phase micro-extraction (SPME). SPME has the advantages of not requiring an extraction solvent and offering significant preconcentration of explosive residues onto the fiber prior to analysis by GC-MS or LC-MS. Headspace SPME has been used to identify volatile compounds emitted from intact explosives like smokeless powder (Joshi *et al.*, 2009; Chang *et al.*, 2014) as well as various high explosives (Lorenzo *et al.*, 2003; Brown *et al.*, 2004; Moore *et al.*, 2011; MacCrehan *et al.*, 2012; Kranz *et al.*, 2014). Although the use of SPME to extract explosive residues from forensic samples has also been demonstrated (Wu *et al.*, 1999; Furton *et al.*, 2000) it has not yet become a routine tool in post-blast cases.

5.3.3.5 Examination of device components

Broadly defined, an improvised explosive device (IED) is a non-military, non-commercial or modified explosive device designed and constructed by the bomb builder using available knowledge and materials. As such, IEDs come in a wide variety of forms, ranging from a simple pipe bomb to a modified military munition. Regardless of their complexity, however, all IEDs must have device components that serve as an initiation system, a container and, in some cases, shrapnel.

IED initiation systems can be triggered in three main ways: time (the device functions after a specified time period), action (an action by the victim causes the device to function) or command (an action by the perpetrator causes the device to function). The triggering mechanism will have a large effect on the actual device components that are used.

The nature of the initiation can also differ, to include non-electrical systems, electrical systems or a combination of the two. Non-electrical systems include mechanical initiators, where a physical action is required in order to initiate the device (e.g., a spring loaded striker impacts a primer). One of the most commonly encountered non-electrical initiators is a burning fuse (typically based on black powder). Finally, electrical initiating systems will contain a power source (e.g., a battery), conductor (e.g., wires), load (e.g., a detonator or other igniter) and a switch.

Other device components can be any one of a wide variety of materials. The most common include tape, adhesives, solder, welds and shrapnel (ball bearings, nails, tacks, etc.). For example, there are several examples of how careful visual/microscopic examinations coupled with chemical analysis can identify the manufacturer and/or brand of adhesive tapes that are found on IEDs (Keeler, 1938; Goodpaster *et al.*, 2007, 2009; Horacek *et al.*, 2008; Dietz *et al.*, 2012).

5.4 Chemical analysis of explosives

Many analytical methods are available for use in identification of intact explosives and their postblast residues. These techniques can be broken down into several broad categories(Box 5.1).

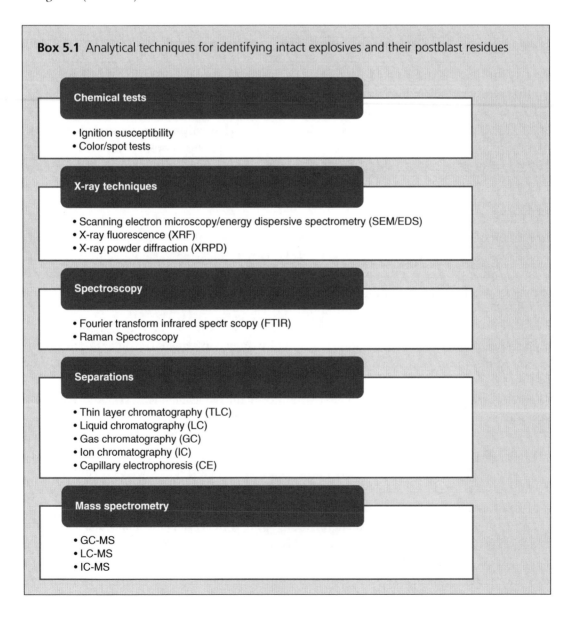

Box 5.1 Analytical techniques for identifying intact explosives and their postblast residues

Chemical tests

- Ignition susceptibility
- Color/spot tests

X-ray techniques

- Scanning electron microscopy/energy dispersive spectrometry (SEM/EDS)
- X-ray fluorescence (XRF)
- X-ray powder diffraction (XRPD)

Spectroscopy

- Fourier transform infrared spectr scopy (FTIR)
- Raman Spectroscopy

Separations

- Thin layer chromatography (TLC)
- Liquid chromatography (LC)
- Gas chromatography (GC)
- Ion chromatography (IC)
- Capillary electrophoresis (CE)

Mass spectrometry

- GC-MS
- LC-MS
- IC-MS

5.4.1 Consensus standards (TWGFEX)

Given the wide variety of techniques that are available for use in explosives analysis, a set of guidelines has been developed by the Standards & Protocols Committee of the Laboratory Explosion Group within the Technical Working Group on Fires and Explosions TWGFEX). These guidelines were based upon surveys and general consensus among explosive examiners in the United States. The documents are entitled *Recommended Guidelines for Forensic Identification of Intact Explosives* (TWGFEX, 2007a) and *Recommended Guidelines for Forensic Identification of Post-Blast Explosive Residues* (TWGFEX, 2007b), both of which are available on the TWGFEX web site (www.twgfex.org or www.swgfex.org).

As a part of these guidelines, the TWGFEX categorizes analytical techniques that can be used for either intact or post-blast analysis according to the amount of information they offer (Table 5.10).

The TWGFEX category for several common analytical techniques are listed in Table 5.11. Category 1 and 2 techniques are combined in this table. The rationale is that some Category 1 techniques are highly informative for some substances, but the same technique can offer only limited information about other substances. In this case, the same techniques would be considered to be Category 2. A familiar example is FTIR, which will yield highly specific spectra of organic explosives but cannot readily distinguish between inorganic oxidizers that differ only in their cation (e.g., sodium nitrate versus potassium nitrate).

The TWGFEX has made several key recommendations on protocols used to identify explosives, such as:

- Multiple techniques must be employed and supporting analytical data must be available for review (printed spectra, chromatograms, pictures, etc.).
- Suspected explosives should be inspected using macroscopic and microscopic techniques.
- When appropriate, a burn test should be carried out (e.g., for intact low explosives).

In addition, the analyst must identify the explosive compound itself or its key constituents according to the following criteria for choice of analytical method(s):

- A Category 1 technique is sufficient for identification.
- A Category 2 technique must be accompanied by an additional supporting technique.
- A Category 3 technique must be accompanied by two additional supporting techniques.
- A Category 4 technique must be accompanied by three additional supporting techniques.

Table 5.10 TWGFEX categories for analytical methods

TWGFEX category	Information provided
1	Significant structural and/or elemental information
2	Limited structural and/or elemental information
3	High degree of selectivity (e.g., separations)
4	Other (e.g., chemical spot tests)

Table 5.11 Analytical techniques sorted by TWGFEX category

Categories 1 and 2	Category 3	Category 4
Infrared spectroscopy (IR)	Gas chromatography (GC)	Burn test
Gas chromatography-mass spectrometry (GC-MS)	Gas chromatography Thermal energy analyzer (GC-TEA or EGIS)	Spot test
Energy dispersive X-ray analyzer (EDX)	Liquid chromatography (LC)	Melting point
Raman spectroscopy	Liquid Chromatography Thermal energy analyzer (LC-TEA)	
X-ray diffraction (XRD)	Ion chromatography (IC)	
Liquid chromatography-mass spectrometry (LC-MS)	Capillary electrophoresis (CE)	
	Thin layer chromatography (TLC) Ion mobility spectrometry (IMS) Polarizing light microscopy (PLM) Stereo light microscopy (SLM)	

Additional guidelines are:
- Ions must be identified using two techniques per ion.
- Chromatographic techniques that use different stationary and/or mobile phases can be counted as distinct Category 3 methodologies
- Polarized light microscopy (PLM) may be counted as two distinct Category 3 methodologies when two different identification tests are done, such as examination of the physical/optical properties plus a microcrystalline test

Particular analytical techniques can be mapped to intact explosives and several examples are shown in Table 5.12, in which analytical techniques from categories 1, 2 and 3 are shown for a given explosive. Provided that the TWGFEX guidelines are followed, this table indicates what analytical techniques are required in order to identify a given intact explosive.

The TWGFEX also recommends the use of reference collections of intact explosives complete with digital images or photographs. Some recommended exemplars are:
- Black powder
- Black powder substitutes (e.g., Pyrodex®/ Triple 7®)
- Perchlorate explosive mixtures
- Pyrotechnic compositions
- Smokeless powder (single base and double base)
- Dynamite
- Watergel/slurry
- Emulsion

- ANFO
- Binaries (e.g., Kinestick®)
- Plastic bonded explosives (e.g., Composition C-4, flexible sheet explosive)
- Single-component explosives (e.g., TNT, PETN, RDX, HMX)

5.4.2 Chemical tests
5.4.2.1 Ignition susceptibility test

An ignition susceptibility test (IST) or "burn test" is a fast and convenient means to determine if an unknown material is an explosive. In this test, a single grain or particle of the unknown is held in forceps and then placed in a flame. Careful observation of the ignition should include visual characteristics (i.e., flame color, sparks, smoke), sound (i.e., none, hiss, pop), smell (none versus sulfurous) as well as the type of residue that is generated (i.e., none versus solid reaction products).

Most explosives will ignite immediately and generate some combination of light, sound and smoke. Non-explosive materials will not burn at all or combust slowly. The results of an IST can be quite useful when looking at particulate low explosives, such as black powder, black powder substitutes, smokeless powders and flash powders as shown in Table 5.13.

The IST is less useful for high explosives. Although an IST will readily reveal that they are energetic, the burning characteristics of high explosives are all very similar.

5.4.2.2 Color tests

A number of colorimetric tests have been devised for anions that may indicate the presence of explosives. Some example tests and the species they indicate are the modified Greiss test (NO_3^-), the aniline sulfate test (ClO_3^-), the methylene blue test (ClO_4^-), Nessler reagent (NH_4^+) and pyridine/NaOH (sulfur) (Midkiff, 2002). Although color tests are fast, sensitive

Table 5.12 TWGFEX requirements for the identification of various intact high explosives

Material	GC-MS	LC-MS	IR	EDX	N	XRD	GC-TEA	GC	LC/TEA	LC	IC	CE	TLC	IMS	PLM
Ammonium nitrate			1	3	2	1				3	3				3
HMTD	1		1		2	1								3	
HMX	1	1	1		2	1	3		3			3	3		3
Nitromethane	1		1		2			3		3					
PETN	1	1	1		2	1	3		3			3	3	3	3
RDX	1	1	1		2	1	3		3			3	3	3	3
TATP	1		1		2	1								3	
Tetryl	2	1	1		2	1	3		3			3	3	3	3
TNT	1	1	1		2	1	3					3	3	3	3

Table 5.13 Typical IST results for various propellants

Explosive	Visual	Sound	Odor	Residue
Black powder	Fast burn Purple flash Grey smoke	yes	Sulfur	yes
Black powder substitute	Fast burn Yellow flash Grey smoke	yes	Sulfur (in Pyrodex)	yes
Flash powder	Fast burn White flash Grey smoke	yes	Sulfur (in some compositions)	yes
Pyrotechnic compositions	Fast burn Green or red flash Grey smoke	yes	Sulfur (in some compositions)	yes
Smokeless powder	Slow burn Yellow flash No smoke	no	no	No

and not prone to false negatives, they can react with materials other than the analyte of interest and the results cannot be easily documented. The advent of XRPD and IC techniques has largely replaced color tests for ions in most forensic science laboratories.

It is important to note that color tests are considered to be destructive, as they irreversibly react with explosives residue in order to generate colored products. Hence, samples may need to be collected from areas that have not been tested in this fashion. Furthermore, if presumptive tests are used on postblast debris, this can deposit chemical residues on the evidence that are undesirable from the point of view of the analytical chemistry laboratory.

5.4.3 X-ray techniques
5.4.3.1 Scanning electron microscopy-energy dispersive spectrometry

Scanning electron microscopy-energy dispersive spectrometry (SEM-EDS) is a powerful technique for the analysis of intact explosives that possess characteristic, but microscopic, structures. For example, emulsions are class of explosives that contain varying amounts of ammonium nitrate, sodium nitrate, water, emulsifier (e.g., sorbitol esters, sorbitan esters, oleates, and/or ethanolamides), oil/wax, oxidizer (calcium nitrate, sodium perchlorate, and/or aluminum) and glass microspheres or "microballoons".

Microspheres are designed as "physical sensitizers" in that they are incorporated into the emulsion and will crush under the pressure of a detonation wave. This creates localized "hot spots" in the explosive that allow the detonation wave to fully propagate through the material. Microspheres can be recovered from the suspected emulsion by performing a pentane extraction, which disrupts the wax/oil matrix. Observing the microspheres using SEM then confirms that the explosive is an emulsion (Figure 5.11).

Another explosive whose analysis benefits from a microscopic exam using SEM is intact dynamite. Nitroglycerin dynamite contains a variety of ingredients that differ by manufacturer. This includes nitroglycerin, ethylene glycol dinitrate (EGDN), sodium nitrate, ammonium nitrate, nitrocellulose, and various inert ingredients such as sawdust. Figure 5.12 shows nitroglycerin dynamite as well as an SEM image of sawdust particles found in the dynamite itself.

In addition to its powerful imaging capabilities, SEM-EDS can carry out elemental analysis and plays an important role in the identification of inorganic explosives. SEM-EDS can identify lighter elements such as carbon, nitrogen and oxygen; hence, organic explosives can be analyzed using this technique. However, the information obtained is less probative than for inorganic explosives that contain more characteristic elements, such as match heads (Glattstein *et al.*, 1991) or pyrotechnic compositions (Kosanke *et al.*, 2003;

Oxidizers: NH_4NO_3, $NaNO_3$
Fuels: Wax/oil, Al
"Sensitizers": Glass microspheres
alkylamine nitrates

Glass microspheres

Figure 5.11 An example of a commercial emulsion explosive and a SEM image of glass microspheres

Ethylene glycol dinitrate (EGDN)
Nitroglycerine (NG)

Sawdust filter

Figure 5.12 An example of commercial nitroglycerin dynamite with an SEM image of characteristic sawdust filler (*See insert for color representation of the figure.*)

Table 5.14 Explosives for which SEM-EDS yields significant elemental information

Explosive formulation	Component(s)	Elemental profile
Black powder	Potassium nitrate Sulfur Carbon	C, N, O, S, K
Pyrodex	Potassium nitrate Potassium perchlorate Dicyandiamide Sulfur Carbon	C, N, O, Na, S, Cl, K
Chlorate/perchlorate Explosive mixtures	Potassium perchlorate Metal fuel Sulfur	Al, S, Cl, K
Emulsion	Inorganic nitrate Oil/wax Emulsifiers Microballoons Aluminum	C, N, O, Si, Al
Water gel/slurry	Inorganic nitrate Gelling agent Microballoons Aluminum Alkylamine sensitizers	C, N, O, Si, Al

Vermeij *et al.*, 2009). Several intact explosives and the components of those explosives for which SEM-EDS is a Category 1 technique are listed in Table 5.14.

5.4.3.2 X-ray fluorescence

The use of X-ray fluorescence (XRF) to analyze explosives or explosive residues is not very uncommon. A key deficiency of the technique is that the low fluorescence yield of low Z elements precludes XRF from detecting elements lighter than sodium (e.g., carbon, nitrogen and oxygen). Hence, XRF cannot identify any organic material, including all organic explosives. Identification of heavier elements can and does play an important role in the identification of inorganic explosives, however. For example, the presence of elements due to some oxidizers (e.g., K + Cl for $KClO_3$ and $KClO_4$, K + Mn for $KMnO_4$), fuels (e.g., S) or pyrotechnic compositions (e.g., Al, Mg, Ba, Sr) can be readily determined using XRF (Midkiff, 2002). The identity of the compounds that these elements form must then be determined. It is important to note that many of these elements are quite uncommon in the environment and their identification can prove quite probative when they are found on the same (blast-damaged) object.

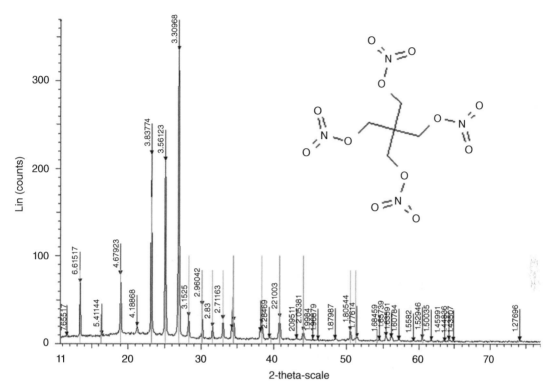

Figure 5.13 X-ray powder diffractogram of the base explosive recovered from a sample of flexible sheet explosive. Overlaid in green is the reference pattern for PETN

5.4.3.3 X-ray powder diffraction

X-ray powder diffraction (XRPD) finds routine use in the analysis of inorganic explosives and their postblast residues. This technique provides conclusive identification of most crystalline explosives and combustion products (Figures 5.13 and 5.14). XRPD is considered to be non-destructive and can analyze samples "as is" or as dried water extracts. However, XRPD is relatively insensitive to minor components in a mixture and *only* crystalline materials are detectable (Table 5.15).

5.4.4 Spectroscopy
5.4.4.1 Fourier transform infrared spectroscopy

The identification of explosives using Fourier transform infrared spectroscopy (FTIR) is a core analytical technique, particularly for intact explosives. For example, the IR spectrum of an inorganic explosive such as ammonium nitrate features a mixture of broad and sharp absorptions. But despite the relative simplicity of the molecule, its spectrum is suitably complex to uniquely identify this compound (Figure 5.15).

In organic explosives, the nitro functional group contributes characteristic absorptions depending upon its configuration relative to the rest of the molecule. For example, in

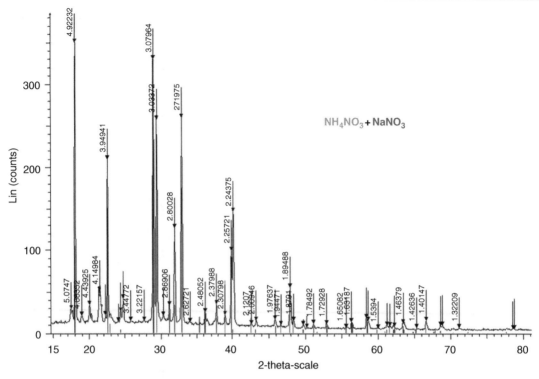

Figure 5.14 X-ray powder diffractogram of the water-soluble material recovered from a sample of a water gel explosive. The reference patterns for ammonium nitrate and sodium nitrate are overlaid on top of the experimental data

Table 5.15 Compounds in intact explosives and solid reaction products for which XRPD yields significant structural information

Compound	Formula	Explosive formulation(s)
Ammonium nitrate	NH_4NO_3	ANFO, dynamite, emulsions/water gels
HMTD	$C_6H_{12}N_2O_6$	Home made explosives, improvised detonators
HMX	$C_4H_8N_8O_8$	Shock tube
PETN	$C_5H_8N_4O_{12}$	Detonating cord, flexible sheet explosive
Potassium chlorate	$KClO_3$	Chlorate explosive mixtures, pyrotechnic compositions
Potassium nitrate	KNO_3	Black powder, Pyrodex/Triple Seven
Potassium perchlorate	$KClO_4$	Perchlorate explosive mixtures, Pyrodex/Triple Seven,
RDX	$C_3H_6N_6O_6$	Detonating cord, composition C-4
Sodium nitrate	$NaNO_3$	Dynamite, emulsions/water gels
Sulfur	S	Black powder, Pyrodex
TATP	$C_9H_{18}O_6$	Home made explosives, improvised detonators
TNT	$C_7H_5N_3O_6$	Cast explosives
Potassium chloride	KCl	Solid reaction product of chlorate/perchlorate based explosives (e.g., Pyrodex, Triple Seven, pyrotechnics)
Potassium sulfate	K_2SO_4	Solid reaction product of sulfur-containing explosives (e.g., black powder, Pyrodex)
Potassium carbonate	K_2CO_3	Solid reaction product of explosives with carbon-based fuels (e.g., black powder)

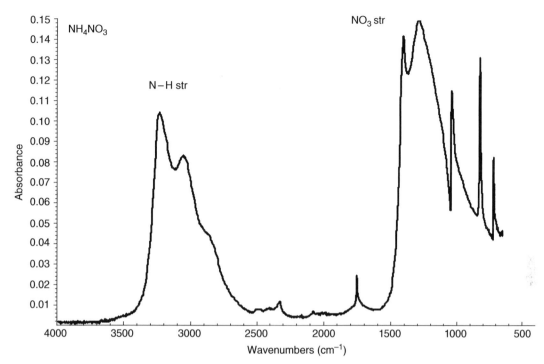

Figure 5.15 FTIR spectrum of pure ammonium nitrate acquired using an attenuated total reflectance (ATR) sampling accessory

nitroaromatics the NO_2 group is bound directly to an aromatic ring. The IR spectra of these compounds exhibit asymmetric and symmetric NO_2 stretching vibrations around 1510–1590 and 1320–1390 cm^{-1}. In contrast, nitrate esters contain a C–O–NO_2 functional group. The IR spectra of these compounds exhibit vibrations around 1640–1660 and 1270–1285 cm^{-1}. Finally, nitramines contain an N–NO_2 functional group. The IR spectra of these compounds exhibit vibrations around 1530–1590 and 1270–1310 cm^{-1} (Zitrin, 1998).

Although these various nitro vibrations are quite diagnostic, some closely related explosives (e.g., nitrate esters such as nitroglycerin, PETN and nitrocellulose) may be extremely difficult to distinguish by FTIR. In the case of nitroglycerin and nitrocellulose, only minor differences in the spectra around 1000 cm^{-1} differentiate the two compounds (Figure 5.16).

The main issue in situations such as this is that the group frequencies for some functional groups are very insensitive to the rest of the molecular structure. While this is very convenient for qualitative analysis and identification of compound classes, it can be problematic for identification of a particular member of a compound class.

The solution in the case of smokeless powder is to extract the sample first with dichloromethane in order to selectively remove and identify nitroglycerin in the exhibit. Then, the residue is extracted with acetone to remove and identify nitrocellulose. Examples of the FTIR spectra that result from this procedure are shown in Figure 5.17:

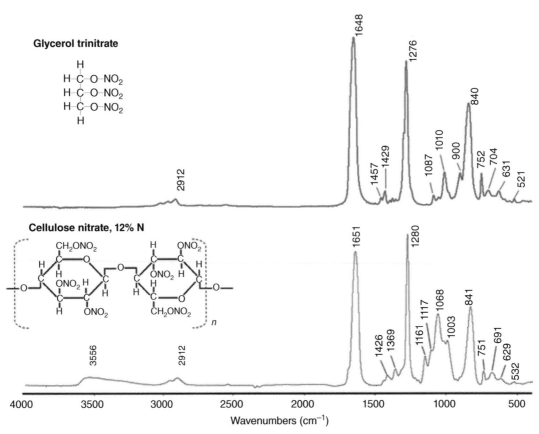

Figure 5.16 A comparison of the FTIR spectra of nitroglycerine and nitrocellulose

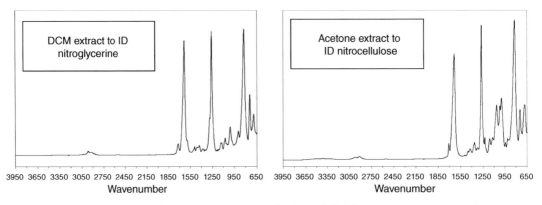

Figure 5.17 Comparison of the FTIR spectrum of the initial dichloromethane extract of a double-base smokeless powder (left) and the subsequent acetone extract (right)

Although it has been reported that FTIR can identify postblast residues of high explosives (Banas *et al.*, 2010), it is far more likely to use this technique in postblast cases involving smokeless powder. In these cases, nitroglycerin is identified by GC-MS after extracting debris with dichloromethane, then nitrocellulose is identified by FTIR after extracting the same debris with acetone.

Of course, the absence of the characteristic peaks due to the NO_2 group can in and of itself be diagnostic; for example, this may indicate the presence of a peroxide explosive such as TATP and HMTD, the IR spectra of which clearly differ from the more well-known explosives discussed above (Zitrin, 1998). In general, FTIR can be used to identify any explosive that contains ammonium nitrate as well as any single high explosive. FTIR is also used for the identification of intact low explosives such as Pyrodex and Triple Seven.

5.4.4.2 Raman spectroscopy

The forensic applications of Raman spectroscopy are as varied as those for FTIR. However, the technique has not reached the same broad acceptance and usage in the forensic community due to low sensitivity and difficulty in identifying unknowns. TWGFEX classifies Raman as a category 2 technique, in that it provides limited (rather than significant) structural information.

Raman spectroscopy can also be combined with other analytical techniques, such as SEM-EDS (Otieno-Alego, 2009). In this technique, high-resolution SEM images, elemental maps and Raman spectra provide a more complete picture of a sample, particular when a mixture of inorganic and organic material is present. Combining spectroscopy and imaging techniques (i.e., hyperspectral imaging) has allowed for images to be acquired in which spectral information is encoded. This technique was successful at detecting explosive residues on surfaces (Blake *et al.*, 2009).

5.4.4.3 Stand-off techniques

Several new stand-off spectroscopic techniques have been introduced and applied to explosives analysis. These techniques are largely focused on detection of explosives at long distances rather than conventional laboratory analysis. For example, photofragmentation followed by laser-induced fluorescence (PF-LIF) detection degrades an explosive sample into nitric oxide (NO), which can then be detected remotely (Cabalo and Sausa 2005; Arnold *et al.*, 2009). A related technique is laser induced breakdown spectroscopy (LIBS), which uses intense laser light to degrade an explosive sample into its constituent elements (carbon, hydrogen, nitrogen and oxygen), which are then detected (Gottfried *et al.*, 2008; De Lucia *et al.*, 2009) These techniques are largely geared towards detection of live explosive devices from a safe distance rather than the analysis of intact and postblast explosives.

5.4.5 Separations
5.4.5.1 Thin layer chromatography
Thin layer chromatography (TLC) has a number of applications to trace evidence, including the analysis of explosives. For example, TLC has historically been used to separate organic high explosives such as RDX, NG and PETN (Hiley, 1993). The Lawrence Livermore National Laboratory has developed a portable TLC kit for use on site (Reynolds *et al.*, 2006). The kit contains typical TLC supplies (i.e., TLC plates, solvents, pipets, spray reagents) as well as a digital camera/light box to capture an image of the processed TLC plate. The image is then loaded onto a laptop computer and a software program compares the R_f values of the sample to any explosives standards.

5.4.5.2 Liquid chromatography
One of the first official methods for explosives in environmental samples such as soil and water was EPA Method 8330 (Nitroaromatics and Nitramines by High Performance Liquid Chromatography (HPLC)), which was released in 1994. This method has since been updated in EPA methods 8330a and 8330b, which were released in 2006 and 2007 and now include PETN as an additional analyte. For example, the US Army Corps of Engineers has made extensive use of this method to monitor the distribution and fate of explosives at munitions storage facilities and bombing ranges (Pennington *et al.*, 2006).

HPLC is also an important means to identify the fuels in black powder substitutes like Pyrodex and Triple Seven. The organic fuels in these propellants are ionizable and, hence, water-soluble. The presence of nitrobenzoic acid in a sample is a clear indication of Triple Seven, as shown in Figure 5.18 for an IED constructed from copper tubing and a paper fuse.

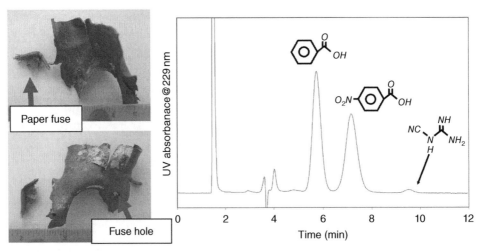

Figure 5.18 An example of the identification of benzoic acid, nitrobenzoic acid and dicyandiamide (DCDA) in the postblast residue of Triple Seven using the HPLC method described by Bender (Bender, 1998). Picture and data are courtesy of Lisa Lang, PhD, of the ATF Laboratory (*See insert for color representation of the figure.*)

Table 5.16 Common inorganic explosives and their characteristic ions

Explosive	Residues	Combustion products
Black powder	NO_3	CO_3^{2-}, SCN^-, SO_4^{2-}
Pyrodex	NO_3, ClO_4	Cl^-, SO_4^{2-}
Triple Seven	NO_3, ClO_4	Cl^-
Flash powder	ClO_4	Cl^-, SO_4^{2-}

5.4.5.3 Ion chromatography/capillary electrophoresis

The ability of ion chromatography (IC) and capillary electrophoresis (CE) to identify anions that are present in postblast residues is well established (Miller *et al.*, 2001; Johns *et al.*, 2008). Although anions such as nitrate, chlorate and perchlorate are usually the most probative, IC and CE systems can analyze cations and anions in postblast residues. In IC, two types of column will be needed – one for cation exchange and one for anion exchange. In CE, it is possible to separate and detect cations and anions in the same run, or two different methods can be developed.

The ions that are detected in the postblast residues of inorganic explosives originate either from residual explosive or its combustion products; they are listed in Table 5.16.

5.4.6 Gas chromatography

Methods utilizing GC and single-channel detectors (e.g., FID, ECD) have long been used to separate mixtures of explosives. Among these, the use of an electron capture detector (ECD) is a popular and relatively inexpensive way to detect nitrated explosives. Furthermore, a standardized GC-ECD method for explosives in soil or water also exists (EPA Method 8095).

Coupling a gas chromatographic system with a thermal energy analyzer (TEA) detector allows for both efficient separation and sensitive detection of complex mixtures containing explosives. A TEA pyrolyzes the eluent from the GC column, reacts it with ozone and thus forms chemiluminescent products that can be detected. A chemiluminescent reaction is one in which one of the chemical products is in an excited electronic state. This reaction product then relaxes to the ground state through the emission of a photon.

The EGIS system (manufactured by Thermo Electron Corporation) is based upon these concepts and includes a portable fast GC with chemiluminescence detection. In an EGIS system, a sample undergoes collection (thermal desorption), concentration (cryo-focusing), separation (via GC), conversion (pyrolysis and release of NO), detection (reaction with O_3 to produce NO_2 and light), and interpretation (qualitative analysis based on retention time, quantitative analysis based on peak area). Cryofocusing occurs in the EGIS at two "cold spots" located immediately after the thermal desorption unit that allow for preconcentration of the explosive vapors and their subsequent release into the separation system as a narrow band. The detection module consists of the reaction chamber, which reacts generated NO with O_3 to form NO_2 and light and a photomultiplier tube (PMT) that detects the emitted light after it passes through a red filter.

EGIS provides presumptive detection of common high explosives (e.g., EGDN, NG, PETN, DNT, TNT, RDX) and vapor phase taggants (e.g., dimethyldinitrobutane [DMNB]) with rapid results (Figures 5.19 and 5.20). EGIS is considered to be destructive as it requires extracting a sample with an organic solvent. EGIS is considered non-destructive

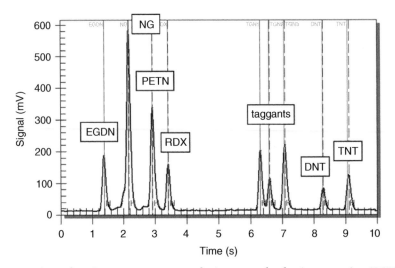

Figure 5.19 Separation of a nine-component explosives standard mixture using EGIS

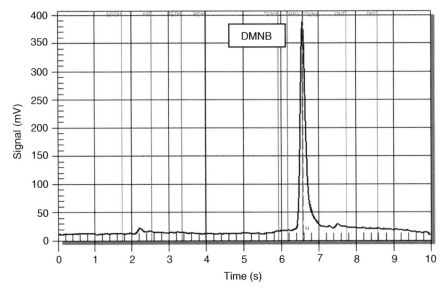

Figure 5.20 Detection of the taggant dimethyldinitrobutane (DMNB) in the headspace of a plastic-bonded explosive using EGIS

when it is used for headspace sampling, however. All EGIS results require confirmation via GC-MS.

5.4.7 Mass spectrometry

Mass spectrometry is one of the most powerful analytical tools for explosives analysis given its high selectivity, sensitivity and specificity. As explosives are inherently unstable, the process by which an explosive molecule is ionized will have a strong effect on the final results. For example, explosives do not typically exhibit a strong molecular ion under electron impact (EI) conditions due to their inherent instability in the gas phase. Therefore, when identifying nitrated compounds under EI conditions, analysts must look beyond the molecular ion to characteristic fragment ions such as m/z 30 (NO^+), m/z 46 (NO_2^+) and m/z 76 ($CH_2ONO_2^+$).

In the case of TNT, the loss of a hydroxyl radical is highly favored, so that the abundance of the molecular ion is very low (Figure 5.21). This phenomenon, known as the "ortho effect", is observed in the ortho isomers of nitrotoluenes and nitroanilines, yielding a $(M–OH)^+$ ion. There are also numerous small mass fragment ions present in the spectrum, including m/z 30 (NO^+).

In contrast, chemical ionization is a much "softer" ionization technique. Nitrated explosives are efficiently ionized in negative ion mode via an electron capture mechanism. An example of a mass spectrum using negative ion chemical ionization (NICI) with methane reagent gas is shown in Figure 5.22 for TNT. In this case, the neutral losses of OH or NO are still possible, but the molecular ion remains largely intact.

5.4.7.1 Gas chromatography-mass spectrometry

Gas chromatography-mass spectrometry (GC-MS) remains one of the most popular techniques for explosives analysis, as it combines high separation efficiency with the

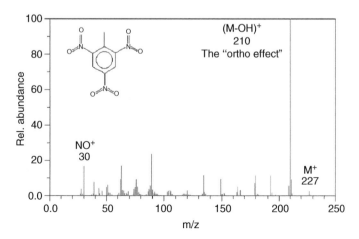

Figure 5.21 The EI mass spectrum of TNT

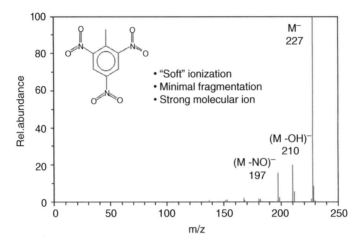

Figure 5.22 The NICI mass spectrum of TNT

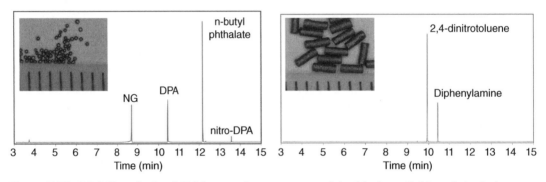

Figure 5.23 GC-MS analysis of dichloromethane extracts of double-base (left) and single-base (right) smokeless powders (NG = nitroglycerine, DPA = diphenylamine, nitro-DPA = nitrodiphenylamine). The GC-MS system consisted of a Gerstel liquid/headspace/SPME autosampler, split/splitless injector, Agilent GC and an Agilent MS with an EI source

sensitive and specific detection of a mass spectrometer. The general methodology for GC-MS analysis of explosives has also been largely standardized. For example, GC-MS using EI ionization is suitable for the separation and identification of components in smokeless powders (Figure 5.23).

GC-MS using EI ionization can also be used to identify compounds that have been derivatized prior to analysis, such as the organic fuels in black powder substitutes (Routon *et al.*, 2011).

For trace analysis, negative ion chemical ionization has the greatest sensitivity for nitrated high explosives (Sigman and Ma, 2001; Calderara *et al.*, 2004). Negative ion CI has impressive detection limits (on the order of picograms) due to the fact that explosives possess nitro functional groups. These nitro groups are strongly electron withdrawing and are, therefore, able to stabilize a negative charge. RDX, TNT (and other nitroaromatic species), and taggants can all be identified by NICI GC-MS. Nitrate esters (e.g., EGDN, NG, and PETN) give very similar (and simple) mass spectra consisting solely of fragment ions at m/z 46 (NO_2^-) and m/z 62 (NO_3^-).

5.4.7.2 Liquid chromatography-mass spectrometry

Liquid chromatography-mass spectrometry (LC-MS) is particularly well suited for the analysis of thermally sensitive and/or non-volatile explosives. Hence, the LC-MS results for a number of explosive materials, including nitroaromatics, nitramines, nitrate esters, oxidizers, and other samples of military, commercial or improvised origin, were reported soon after the technology became routinely available (Yinon, 2004). Given that detection limits for LC-MS lie in the parts-per-billion range, it is an appropriate technique for trace detection of the residual high explosive that can be found on post-blast debris.

The choice of either electrospray ionization (ESI) or atmospheric pressure chemical ionization (APCI) is an important one, as the ionization efficiency, and therefore sensitivity, depends upon it. It has been demonstrated that nitraromatics are best analyzed by APCI, while nitramine explosives ionize quite well under ESI conditions, particularly when a volatile buffer additive is present to form adduct ions. Nitrate esters perform well under either ionization technique (Yinon, 2004).

Research into LC-MS methods for explosives has increased steadily, and it is a powerful technique for the analysis of many types of explosives. Publications are available that describe LC-MS methods for black powder substitutes (Bottegal *et al.*, 2010; Cummins *et al.*, 2011), smokeless powder (Mathis and McCord, 2003), explosives residues on hand swabs (Perret *et al.*, 2008), as well as organic high explosives (Holmgren *et al.*, 2005).

5.4.7.3 GC-MS and LC-MS of improvised high explosives

Of the improvised explosives that have found use in criminal and terrorist bombings, triacetone triperoxide (TATP) and hexamethylene triperoxide diamine (HMTD) are the most well-known. The analysis of peroxide-based explosives such as TATP by GC-MS (Stambouli *et al.*, 2004) and LC-MS (Widmar *et al.*, 2002) has been reported and trends in the detection of peroxides have been reviewed (Burkes and Hage, 2009).

An additional improvised explosive that has captured significant attention is urea nitrate, which is both easy to make and powerful. As with the peroxide explosives, this explosive has been used frequently by terrorist organizations in the Middle East. There has been some research on color tests for urea nitrate. These are fast, sensitive and inexpensive, although not a means of identification (Almog *et al.*, 2005). Identification of this explosive using more definitive instrumental techniques, such as mass spectrometry, has also been reported (Tamiri *et al.*, 2009).

5.4.7.4 Ion chromatography-mass spectrometry

Coupling ion chromatography with mass spectrometry (IC-MS) is a more specialized technique for those laboratories that require identification of inorganic compounds in

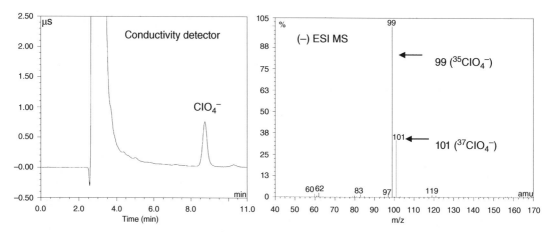

Figure 5.24 IC-MS results for a water extract of a fragment from a device containing a perchlorate-based explosive. The IC-MS consisted of a DIONEX DX500 system with an AS16 column. The system uses a CD-20 conductivity cell and a Thermo Electron MSQ detector (−ESI from m/z 30 to 130)

water samples (Barron and Gilchrist, 2014). IC-MS provides conclusive identification of all non-volatile inorganic explosives. However, it is considered destructive, as it requires extracting a sample with water.

Given that water samples are analyzed, IC-MS will detect preformed ions in solution and, hence, relies upon electrospray ionization. The detection of very low mass species such as chloride is more challenging, but those species that are the most diagnostic in a postblast residue (e.g., nitrate and perchlorate) are also among the ions with the best sensitivity.

As shown in Figure 5.24, perchlorate is readily detected using a conductivity detector. More importantly, the perchlorate mass spectrum is quite characteristic given the 3:1 ratio between the ^{37}Cl and ^{35}Cl isotopic peaks. Additional structural detail can be gained by increasing the fragmentation voltage on the ESI source, which will yield significant peaks at m/z 83 and 85, corresponding to the loss of oxygen from the parent compound, thus forming ClO_3^-.

5.4.7.5 Ambient ionization mass spectrometry

One of the biggest revolutions in the field of mass spectrometry has been the development of ambient sampling and ionization methods. These methods rely upon desorption of molecules from surfaces followed by identification using a mass spectrometer. The potential for such techniques to identify explosive residues on surfaces was immediately recognized. Specific techniques that have been discussed as a means for explosives detection are desorption electrospray ionization (DESI) (Cotte-Rodriguez *et al.*, 2005, 2008; Takats *et al.*, 2005; Soparawalta *et al.*, 2010), direct analysis in real time (DART) (Rowell *et al.*, 2012), and extractive electrospray ionization (EESI) (Gu *et al.*, 2010).

5.4.8 Provenance and attribution determinations

An important frontier in the area of explosive analysis is that of provenance and attribution. "Chemical provenance" is the chemical identification and source attribution of a sample. The extent to which the materials used in an explosive device can be accurately identified and traced can significantly impact an explosives investigation. Regardless of the circumstances of an incident, there are three general strategies for determining the source of an explosive, which are listed here in order of increasing complexity:

  (i) Brand identification: Identify the additives, binders and other components of an explosive formulation that can be brand and manufacturer specific.

 (ii) Impurity profiling: Determine the identity and amount of impurities related to the explosive. These compounds may indicate the synthetic route that was used, the level of sophistication of the manufacturer or the extent of degradation in the sample.

(iii) Isotopic analysis: Detailed characterizations of the explosive compound itself, to include its stable isotope ratios (SIR).

5.4.8.1 Brand identification

An immediate means of brand identifying commercial explosive samples is via manufacturer colors and markings. The maintenance of large exemplar collections and manufacturer information is vital to this effort. Fortunately, several commercial sources of explosive use distinct colorations to differentiate their product (Figure 5.25):

Another way to forensically differentiate plastic-bonded explosives (PBX) is by their plasticizers and binders (Table 5.17). These compounds are added to make the explosives malleable and able to be formed into any size and shape. It is also possible to perform high resolution fingerprinting of process oils in PBX such as Composition C-4 (Reardon and Bender, 2005).

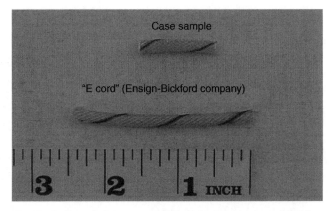

Figure 5.25 Comparison of an unknown sample of detonating cord with an exemplar of a commercial product (*See insert for color representation of the figure.*)

Table 5.17 Examples of plastic-bonded explosive (PBX) compositions

Product	Explosive	Binder	Plasticizer
Detasheet/Primasheet	PETN	Butyl rubber	Acetyl tributylcitrate (Citraflex)
Composition C-4	RDX	Polyisobutylene	Diocytladipate (DOA)
			Dioctylsebacate (DOS)
A-5	RDX	Stearic acid	N/A

PETN = pentaerythritol tetranitrate
RDX = cyclotrimethylenetrinitramine

5.4.8.2 Impurity profiling

Identification of compounds related to the explosive can differentiate high explosives made by different chemical synthetic pathways. These chemical syntheses may be used by legitimate (government or corporate) and non-legitimate (terrorist) organizations, respectively. Knowledge of the method used to synthesis an explosive can be important and can help with forensic investigations.

For example, it has been shown that a series of oligomeric peroxide compounds are formed when TATP is synthesized. These oligomers vary batch-to-batch and can still be detected postblast (Sigman *et al.*, 2009).

Another example of impurity profiling is the differentiation of RDX-based explosives. This can be done because RDX that is manufactured in the United States contains a small amount of HMX, which is a by-product of the synthesis. In contrast, RDX that is manufactured outside of the United States is made with a different synthetic method that does not generate HMX.

Impurity profiling is also useful for TNT, one of the most common and oldest explosives used for military or industrial explosives. TNT breakdown products will be present where military munitions are exploded frequently or where military munitions are buried as a method of disposal. The breakdown of TNT by photodecomposition depends on the wavelength of the light and duration of exposure. Therefore, these decomposition products can provide information on the age of the explosive.

5.4.8.3 Stable isotope ratios

The study of SIR concerns naturally occurring isotopes whose ratio is relatively constant. In practice, isotopic abundances are determined with isotope ratio mass spectrometry (IRMS). The application of IRMS as an analytical and forensic tool is well documented (Meier-Augenstein and Liu, 2004; Benson *et al.*, 2006). In IRMS, complex organic compounds are combusted to form simple molecules such as CO_2, H_2O and N_2 prior to analysis. These gases are then analyzed and the ratio of light and heavy isotopes is determined. To achieve the requisite level of mass resolution, single focusing magnetic sector mass spectrometers with fixed multiple detectors (one per isotope) are used.

The ability of SIR to elucidate the potential source of a sample is based upon the phenomenon of fractionation. Fractionation results in differences between the distribution of isotopes in a sample and a SIR standard material. Fractionation is based on differences in behavior that can be traced to differences in mass (e.g., evaporation favors lighter isotopes, condensation favors heavier isotopes).

SIR measurements are expressed in terms of the experimentally derived ratio of the isotopes (R_s) to that of a reference material (R_r), as per the equation:

$$\delta\%_{oo} = \left(\frac{R_s}{R_r} - 1\right) \times 1000$$

Negative values indicate that the sample has a lower heavy:light ratio than the standard. Reference materials include calcium carbonate ($CaCO_3$), atmospheric nitrogen and sea water.

The use of IRMS to attribute explosives to a particular source is a relatively new development and IRMS technology is still being validated for use in forensic applications. In the case of explosives, researchers have demonstrated source differentiation of Semtex (Pierrini *et al.*, 2007), ammonium nitrate (Benson *et al.*, 2009a) as well as TATP and PETN (Benson *et al.*, 2009b). The relationship between RDX and its hexamine starting material was also studied using IRMS (Lock and Meier-Augenstein, 2008). IRMS instrumentation and SIR methods for explosives have also been evaluated in an interlaboratory trial (Benson *et al.*, 2010a, 2010b).

One weakness of IRMS that has been acknowledged by several authors is that the SIR of an explosive and its postblast residues can be significantly different in unpredictable ways. The blast event perturbs SIR to a large extent, making association of postblast residues to the original explosive very difficult or even impossible. To date, the differentiating and sourcing of intact samples of explosive has been more successful. Finally, IRMS requires a specialized radiochemistry laboratory as well as the proper expertise and training. Hence, this technique will not be available to all forensic laboratories for explosives identification, due to the expense and specialized set up required.

5.5 Ongoing research

Research into new methods for identifying explosives and explosive residues regularly introduces increased sensitivity, selectivity and specificity for the explosives analyst. Keeping current with research can be challenging but there are several sources that regularly summarize new developments in the field, such as:
- Every three years, INTERPOL hosts the Interpol International Forensic Science Symposium. The Symposium is dedicated to all aspects of forensic science, including chemical criminalistics (e.g., fibers, firearms, tool marks, forensic geology, paint and

glass), drugs and toxicology, electronic evidence, fire, explosives and hazardous materials and identification sciences (e.g., biological evidence, document examination and fingerprints.

- The International Symposium on the Analysis and Detection of Explosives (ISADE) is also an important event in the area of explosives research. Offered every three years, the ISADE covers topics ranging from new analytical techniques, field sampling, incident reports, new and emerging technologies, and portable detection systems.
- Lastly, NATO has published a series of monographs on the detection, analysis, mitigation and disposal of explosives and improvised explosive devices. These monographs are based upon the proceedings of NATO Advanced Research Workshops at which international experts present their findings.

Acknowledgements

Portions of this chapter were adapted with permission from the author's contributions to the Trace Explosive Materials Reference Guide (TEMRG). The author acknowledges Dr Susan Jones and Mr Donald Houseman of Analytic Services, Inc. (ANSER) for their guidance in this effort. The TEMRG was sponsored by the Combating Terrorism Technical Support Office (CTTSO) and the Technical Support Working Group (TSWG) of the US Department of Defense.

The author also acknowledges his training and experience with the Bureau of Alcohol, Tobacco, Firearms and Explosives (ATF), which is reflected in many of the approaches described herein. The author would also like to acknowledge Dr Nicolas Manicke, Mr Clinton Carroll and Ms Dana Bors at IUPUI for providing additional references for this chapter.

References

Almog, J., Klein, A. Tamiri, T., *et al.* (2005) a field diagnostic test for the improvised explosive urea nitrate. *Journal of Forensic Sciences*, **50** (3), 582–586.

Arnold, B., Kelly, L., Oleske J.B. and Schill, A. (2009) Standoff detection of nitrotoluenes using 213-nm amplified spontaneous emission from nitric oxide. *Analytical and Bioanalytical Chemistry*, **395**, 349–355.

Banas, K., Banas, A., Moser, H.O., *et al.* (2010) Multivariate analysis techniques in the forensics investigation of the postblast residues by means of Fourier transform infrared spectroscopy. *Analytical Chemistry*, **82** (7), 3038–3044.

Barron, L. and Gilchrist, E. (2014) Ion chromatography-mass spectrometry: A review of recent technologies and applications in forensic and environmental explosives analysis. *Analytica Chimica Acta*, **806**, 27–54.

Benson, S., Lennard, C., Maynard, P., and Roux, C. (2006) Forensic applications of isotope ratio mass spectrometry – A review. *Forensic Science International*, **157** (1), 1–22.

Benson, S.J., Lennard, C.J., Maynard, P., *et al.* (2009a) Forensic analysis of explosives using isotope ratio mass spectrometry (IRMS) – Discrimination of ammonium nitrate sources. *Science & Justice*, **49**(2), 73–80.

Benson, S.J., Lennard, C.J., Maynard, P., *et al.* (2009b) Forensic analysis of explosives using isotope ratio mass spectrometry (IRMS) – Preliminary study on TATP and PETN. *Science & Justice*, **49**(2), 81–86.

Benson, S.J., Lennard, C.J., Hill, D.M., *et al.* (2010a) Forensic analysis of explosives using isotope ratio mass spectrometry (IRMS) – Part 1: Instrument validation of the DELTAplusXP IRMS for bulk nitrogen isotope ratio measurements. *Journal of Forensic Sciences*, **55** (1), 193–204.

Benson, S.J., Lennard, C.J., Maynard, P., *et al.* (2010b) Forensic analysis of explosives using isotope ratio mass spectrometry (IRMS) – Part 2: Forensic inter-laboratory trial: Bulk Carbon and nitrogen stable isotopes in a range of chemical compounds (Australia and New Zealand). *Journal of Forensic Sciences*, **55** (1), 205–212.

Bjella, K.L. (2005) *Pre-Screening for Explosives Residues in Soil Prior to HPLC Analysis Utilizing Expray*. US Army Corps of Engineers, Hanover, NH.

Blake, T.A., Kelly, J.F., Gallagher, N.B., *et al.* (2009) Passive standoff detection of RDX residues on metal surfaces via infrared hyperspectral imaging. *Analytical and Bioanalytical Chemistry*, **395**, 337–348.

Bors, D., Cummins, J., and Goodpaster, J. (2014a) The anatomy of a pipe bomb explosion: measuring the mass and velocity distributions of container fragments. *Journal of Forensic Sciences*, **59** (1), 42–51.

Bors, D., Cummins, J., and Goodpaster, J. (2014b) The anatomy of a pipe bomb explosion: The effect of explosive filler, container material and ambient temperature on device fragmentation. *Forensic Science International*, **234**, 95–102.

Bottegal, M., Lang, L., Miller, M.L., and McCord, B.R. (2010) Analysis of ascorbic acid based black powder substitutes by high-performance liquid chromatography/electrospray ionization quadrupole time-of-flight mass spectrometry. *Rapid Communications in Mass Spectrometry*, **24**, 1377–1386.

Brown, H., Kirkbride, K.P., Pigou, P.E., and Walker, G.S. (2004) New developments in SPME Part 2: Analysis of ammonium nitrate-based explosives. *Journal of Forensic Sciences*, **49** (2), 215–221.

Burkes, R.M. and Hage, D.S. (2009) Current trends in the detection of peroxide-based explosives. *Analytical and Bioanalytical Chemistry*, **395**, 301–313.

Cabalo, J.B. and Sausa, R.C. (2005) *Explosive Residue Detection by Laser Surface Photofragmentation–Fragment Detection Spectroscopy: II. In Situ and Real-time Monitoring of RDX, HMX, CL20, and TNT, by an Improved Ion Probe*. Defense Technical Information Center, Fort Belvoir, VA.

Calderara, S., Gardebas, D., Martinez, F., and Khong, S.-P. (2004) Organic explosives analysis using on column-ion trap EI/NICI GC-MS with an external source. *Journal of Forensic Sciences*, **49** (5), 1–4.

Chang, K.H., Yew, C.H., and Abdullah, A.F.L. (2014) Optimization of headspace solid-phase microextraction technique for extraction of volatile smokeless powder compounds in forensic applications. *Journal of Forensic Sciences*, **59** (4), 1100–1108.

Conkling, J.A. and Mocella, C.J. (2011) *Chemistry of Pyrotechnics: Basic Principles and Theory*. CRC Press, Boca Raton, FL.

Cooper, P. (1996) *Explosives Engineering*. Wiley-VCH, New York.

Cotte-Rodriguez, I., Takats, Z., Talaty, N., *et al.* (2005) Desorption electtrospray ionization of explosives on surfaces: sensitivity and selectivity enhancements by reactive desorption electrospray ionization. *Analytical Chemistry*, **77**, 6755–6764.

Cotte-Rodriguez, I., Hernandez-Soto, H., Chen, H., and Cooks, R.G. (2008) In-situ trace detection of peroxide explosives by desorption electrospray ionization and desorption atmospheric pressure chemical ionization. *Analytical Chemistry*, **80**, 1512–1519.

Cummins, J., Hull, J., Kitts, K., and Goodpaster, J.V. (2011) Separation and identification of anions using porous graphitic carbon and electrospray ionization mass spectrometry: application to inorganic explosives and their post blast residues. *Analytical Methods*, **3** (7), 1682–1687.

De Lucia, F.C., Gottfried, J.L., and Miziolek, A.W. (2009) Evaluation of femtosecond laser-induced breakdown spectroscopy for explosive residue detection. *Optics Express*, **17** (2), 419–425.

Dietz, M.E., Stern, L.A., Mehltretter, A.H., *et al.* (2012) Forensic utility of carbon isotope ratio variations in PVC tape backings. *Science & Justice*, **52** (1), 25–32.

Furton, K.G., Wu, L., and Almirall, J.R. (2000) Optimization of solid-phase microextraction (SPME) for the recovery of explosives from aqueous and post-explosion debris followed by gas and liquid chromatographic analysis. *Journal of Forensic Sciences*, **45** (4), 857–864.

Glattstein, B., Landau, E., and Zeichner, A. (1991) Identification of match head residues in post-explosion debris. *Journal of Forensic Sciences*, **36** (5), 1360–1367.

Goodpaster, J.V., Sturdevant, A.B., Andrews, K.L., and Brun-Conti, L. (2007) Identification and comparison of electrical tapes using instrumental and statistical techniques: I. Microscopic surface texture and elemental composition. *Journal of Forensic Sciences*, **52** (3), 610–629.

Goodpaster, J.V., Sturdevant, A.B., Andrews, K.L., *et al.* (2009) Identification and comparison of electrical tapes using instrumental and statistical techniques: II. Organic composition of the tape backing and adhesive. *Journal of Forensic Sciences*, **54** (2), 328–338.

Gottfried, J.L., De Lucia, F.C., Munson, C.A., and Miziolek, A.W. (2008) Strategies for residue explosives detection using laser-induced breakdown spectroscopy. *Journal of Analytical Atomic Spectrometry*, **23**, 205–216.

Gu, H., Yang, S., Li, J., *et al.* (2010) Geometry-independent neutral desorption device for the sensitive EESI-MS detection of explosives on various surfaces. *Analyst*, **135**, 779–788.

Hendricks, P.I., Dalgleish, J.K., Shelley, J.T., *et al.* (2014) Autonomous in situ analysis and real-time chemical detection using a backpack miniature mass spectrometer: concept, instrumentation development, and performance. *Analytical Chemistry*, **86**, 2900–2908.

Hiley, R.W. (1993) Investigations of thin layer chromatographic techniques used for forensic explosives analysis in the early 1970s. *Journal of Forensic Sciences*, **38** (4), 864–873.

Holmgren, E., Carlsson, H., Goede, P., and Crescenzi, C. (2005) Determination and characterization of organinc explosives using porous graphitic carbon and liquid chromatography-atmospheric pressure chemical ionization mass spectrometry. *Journal of Chromatography A*, **1099**, 127–135.

Horacek, M., Min, J.S., Heo, S., *et al.* (2008) The application of isotope ratio mass spectrometry for discrimination and comparison of adhesive tapes. *Rapid Communications in Mass Spectrometry*, **22** (11), 1763–1766.

Hutchinson, J.P., Johns, C., Breadmore, M.C., *et al.* (2008) Identification of inorganic ions in post-blast explosive residues using portable CE instrumentation and capacitively coupled contactless conductivity detection. *Electrophoresis*, **29**, 4593–4602.

Johns, C., Shellie, R.A., Potter, O.G., *et al.* (2008) Identification of homemade inorganic explosives by ion chromatographic analysis of post-blast residues. *Journal of Chromatography A*, **1182**, 2015–2214.

Johnson, C. (2006) The proper application of explosives detection kits. *The Detonator*, **May/June**, 1–4.

Joshi, M., Delgado, Y.,. Guerra, P., *et al.* (2009) Detection of odor signatures of smokeless powders using solid phase microextraction coupled to an ion mobility spectrometer. *Forensic Science International* **188** (1–3), 112–118.

Keeler, K. (1938) Comparison and identification of adhesive tape used in the construction of a bomb mechanism. *Journal of Criminal Law and Criminology*, **28** (6), 904–908.

Kosanke, K.L., Dujay, R.C., and Kosanke, B. (2003) Characterization of pyrotechnic reaction residue particles by SEM/EDS. *Journal of Forensic Sciences*, **48** (3), 531–537.

Kranz, W., Kitts, K., Strange, N., *et al.* (2014) On the smell of Composition C-4. *Forensic Science International*, **236**, 157–163.

Lock, C.M. and Meier-Augenstein, W. (2008) Investigation of isotopic linkage between precursor and product in the synthesis of a high explosive. *Forensic Science International*, **179** (2–3), 157–162.

Lorenzo, N., Wan, T., Harper, R.J., *et al.* (2003) Laboratory and field experiments used to identify *Canis lupus* var. *familiaris* active odor signature chemicals from drugs, explosives, and humans. *Analytical and Bioanalytical Chemistry*, **376**, 1212–1224.

MacCrehan, W., Moore, S., and Schantz, M. (2012) Reproducible vapor-time profiles using solid phase microextraction with an externally sampled internal standard. *Journal of Chromatography A*, **1244**, 28–36.

Mathis, J.A. and McCord, B.R. (2003) Gradient reversed-phase liquid chromatographic-electrospray ionization mass spectrometric method for the comparison of smokeless powders. *Journal of Chromatography A*, **988**, 107–116.

Meier-Augenstein, W. and Liu, R.H. (2004) Forensic applications of isotope ratio mass spectrometry, in J. Yinon (ed) *Advances in Forensic Applications of Mass Spectrometry*, CRC Press, Boca Raton, FL, pp. 149–180.

Midkiff, C.R. (2002) Arson and explosives investigation, in R. Saferstein (ed) *Forensic Science Handbook*. Prentice Hall, Upper Saddle River, NJ, pp. 4790150524.

Miller, M.L., Doyle, J.M., and Lee, R.A. (2001) Analysis of anions by capillary electrophoresis and ion chromatography for forensic applications. *Forensic Science Communications*, **3** (2) [online].

Moore, S., MacCrehan, W., and Schantz, M. (2011) Evaluation of vapor profiles of explosives over time using ATASS (Automated Training Aid Simulation using SPME). *Forensic Science International*, **212** (1–3), 90–95.

NFPA (National Fire Protection Association) (2011) *Explosives, in NFPA 921: Guide for Fire and Explosion Investigation*, National Fire Protection Association, Quincy, MA, Chapter 21.

Otieno-Alego, V. (2009) Some forensic applications of a combined micro-Raman and scannine electron microscopy system. *Journal of Raman Spectroscopy*, **40**, 948–953.

Oxley, J.C., Smith, J.L., Resende, E., *et al.* (2001) Improvised explosive devices: Pipe bombs. *Journal of Forensic Sciences*, **46** (3), 510–534.

Pennington, J.C., Jenkins, T.F., Ampleman, G., *et al.* (2006) *Distribution and Fate of Energetics on DoD test and Training Ranges*. US Army Engineer Research and Development Center, Vicksburg, MS.

Perret, D., Marchese, S., Gentili, A., *et al.* (2008) LC-MS-MS Determination of Stabilizers and Explosives Residues in Hand Swabs. *Chromatographia*, **68**, 517–524.

Pierrini, G., Doyle, S., Champod, C., *et al.* (2007) Evaluation of preliminary isotopic analysis (C-13 and N-15) of explosives A likelihood ratio approach to assess the links between semtex samples. *Forensic Science International*, **167** (1), 43–48.

Reardon, M.R. and Bender, E.C. (2005) Differentiation of Composition C-4 based on the analysis of the process oil. *Journal of Forensic Sciences*, **50** (3), 564–570.

Reynolds, J., Nunes, P., Whipple, R., and Alcaraz, A. (2006) On-site analysis of explosives, in H. Schubert and A. Kuznetsov, *Various Matrices. Detection and Disposal of Improvised Explosives*, Springer, Dordrecht, pp. 27–32.

Routon, B.J., Kocher, B.B., and Goodpaster, J.V. (2011) Discriminating Hodgdon Pyrodex (R) and Triple Seven (R) using gas chromatography-mass spectrometry. *Journal of Forensic Sciences*, **56** (1), 194–199.

Rowell, F., Seviour, J., Lim, A., *et al.* (2012) Detection of nitro-organic and peroxide explosives in latent fingermarks by DART- and SALDI-TOF-mass spectrometry. *Forensic Science International*, **221** (1–3), 84–91.

Sigman, M.E. and Ma, C.-Y. (2001) Detection limits for GC/MS analysis of organic explosives. *Journal of Forensic Sciences*, **46** (1), 6–11.

Sigman, M.E., Clark, C.D., Painter, K., *et al.* (2009) Analysis of oligomeric peroxides in synthetic triacetone triperoxide samples by tandem mass spectrometry. *Rapid Communications in Mass Spectrometry*, **23**, 349–356.

Soparawalta, S., Salazar, G., Sokol, E., *et al.* (2010) Trace detection of non-uniformly distributed analytes on surfaces using mass transfer and large-area desorption electrospray ionization (DESI) mass spectrometry. *The Analyst*, **135** (8), 1953–1960.

Stambouli, A., El Bouri, A., Bouayoun, T., and Bellimam, M.A. (2004) Headspace-GC/MS detection of TATP traces in post-explosion debris. *Forensic Science International*, **1468**, 8191–8194.

Takats, Z., Cotte-Rodriguez, I., Talaty, N., *et al.* (2005) Direct, trace level detection of explosives on ambient surfaces by desorption electrospray ionization mass spectrometry. *Chemical Communications*, 1950–1952.

Tamiri, T., Rozin, R., Lemberger, N., and Almog, J. (2009) Urea nitrate, an exceptionally easy-to-make improvised explosive: studies towards trace characterization. *Analytical and Bioanalytical Chemistry*, **395**, 421–428.

TWGFEX (2007a) Recomended Guidlines for Forensic Identification of Intact Explosives. Standards & Protocols Committee of the Laboratory Explosion Group within the Technical Working Group on Fires and Explosions (TWGFEX). http://swgfex.org/; last accessed 14 June 2015.

TWGFEX (2007b) Recomended Guidelines For Indetification of Post-Blast Explosives Residue. Standards & Protocols Committee of the Laboratory Explosion Group within the Technical Working Group on Fires and Explosions (TWGFEX). http://swgfex.org/; last accessed 14 June 2015.

Vermeij, E., Duvalois, W., Webb, R., and Koeberg, M. (2009) Morphology and composition of pyrotechnic residues formed at different levels of confinement. *Forensic Science International*, **186** (1–3), 68–74.

Widmar, L., Watson, S., Schlatter, K., and Crowson, A. (2002) Development of an LC/MS method for the trace analysis of triacetone triperoxide (TATP). *Analyst*, **127**, 1627–1632.

Wu, L., Almirall, J.R., and Furton, K.G. (1999) An improved interface for coupling solid-phase microextraction (SPME) to high performance liquid chromatography (HPLC) applied to the analysis of explosives. *Journal of High Resolution Chromatography*, **22** (5), 279–282.

Yinon, J. (2004) Analysis of explosives by LC/MS, in J. Yinon (ed.) *Advances in Forensic Applications of Mass Spectrometry*, CRC Press, New York, pp. 243–286.

Zitrin, S. (1998) Analysis of explosives by infrared spectrometry and mass spectrometry, in A. Beveridge (ed.), *Forensic Investigation of Explosions*, Taylor & Francis, London, pp. 267–314.

Further reading

A single book chapter cannot hope to treat all topics in the field of explosives analysis with the depth that is available in all other sources. Hence, the following references are recommended for continued study in the areas of explosives investigation and chemical analysis of explosives:

- Akhavan, J., (2006) *The Chemistry of Explosives*, 2nd edn. RSC Paperbacks.

- Beveridge, A. (ed.) (2012) *Forensic Investigation of Explosives*, 2nd edn. Taylor and Francis, Boca Raton, FL.
- Cooper, P.W. and Kurokski, S.R. (1996) *Introduction to the Technology of Explosives*. Wiley-VCH, New York.
- National Fire Protection Association (2011) *NFPA 921: Guide for Fire and Explosion Investigations*. National Fire Protection Agency, Quincy, MA.
- Marshall, M. and Oxley, J.C. (eds) (2009) *Aspects of Explosives Detection*. Elsevier, New York.
- Meyer, R., Kohler, J., and Homburg, A. (2002) *Explosives*, 5th edn. Berlage GmgH/Wiley-BHS, Weinheim.
- Midkiff, C.R. (2002) Arson and explosives investigation, in R. Saferstein (ed.) *Forensic Science Handbook*, 2nd edn, Prentice Hall, Upper Saddle River, pp. 479–524.
- Seigel, J.A. and Suakko, P.J. (ed) (2013) *Encyclopedia of Forensic Sciences*, 2nd edn. Elsevier/Academic Press, Boston, MA.

CHAPTER 6

Analysis of glass evidence

Jose Almirall and Tatiana Trejos

Department of Chemistry and Biochemistry and International Forensic Research Institute, Florida International University, USA

6.1 Introduction to glass examinations and comparisons

When glass breaks a number of small fragments can be readily transferred to a person or to other objects in the vicinity of the breaking glass. If the breaking of the glass is related to criminal activity, glass evidence may provide information about how and when the events happened and associate a person or an object to the breaking event. Broken glass is commonly encountered as trace evidence in cases of burglaries, assaults, car accidents such as hit-and-runs, shootings and bombings (Almirall and Trejos, 2006; Almirall, 2001b). Small glass fragments can be retained on the breaking object (i.e., hammer, bat, rock, bullet, and brick), other objects in the vicinity, and on the individual(s) involved in the breaking activity (hair, clothing and/or shoes). These fragments can be collected and submitted to the forensic laboratory labeled as the "recovered glass" or as a "questioned sample". In addition, if the source of the broken glass is known or found, the glass is submitted to the laboratory for comparison purposes labeled as a "control glass" or as the "known sample".

Forensic examiners then conduct glass examinations to determine whether there is an association between the questioned and the known glass fragments, measuring physical, optical and chemical properties to then infer contact or vicinity. An example is shown in the following hit-and-run case scenario: *A vehicle strikes a bicyclist and the body of the cyclist is lifted onto the windshield with such force that the windshield is broken and glass fragments are transferred to the cyclist's clothing. The vehicle leaves the scene of the accident and the car is found abandoned a few miles away from the scene. After serving the vehicle owner a warrant to search her house, the owner claims that her friend borrowed her car. She denies to driving the vehicle at the time of the accident and that her friend never returned the car. The crime scene investigators search her house and person and find multiple glass fragments on the suspect's hair, clothing and*

Forensic Chemistry: Fundamentals and Applications, First Edition. Edited by Jay A. Siegel.
© 2016 John Wiley & Sons, Ltd. Published 2016 by John Wiley & Sons, Ltd.

home. The glass recovered from the suspect's home is analyzed and compared to the glass known to originate from the windshield to establish a link between the car owner and the accident. Any broken glass fragments found on the cyclist or on the ground at the scene of the accident can also be linked to the vehicle.

Materials such as glass found as **trace evidence** usually exhibit class characteristics. The value and significance of the evidence will depend on different factors, such as the discrimination power of the techniques used for its comparison. The value of evidence can also be enhanced by the **cross-transfer** of evidence between victim and suspect, the transfer of multiple items of evidence and/or the rarity of the characteristics of the material(s) transferred. Glass can only provide individual characteristics in the case where there is a perfect fit of two glass fragments along the edges. Such a perfect "fracture match" is rarely found in actual cases (Almirall, 2001a).

Improvements in quality assurance during the manufacture of glass have required glass examiners to use techniques to compare glass samples that have higher sensitivity and better discrimination power with minimum sample manipulation, destruction and sample amount.

The scientific literature related to forensic glass examinations spans more than 50 years, including the photographic documentation and transfer studies from breaking glass published by Nelson and Revell from the ESR in New Zealand in 1967 (Nelson and Revell, 1967) and very early elemental analysis and comparison studies of window glass by Coleman and Goode from the Atomic Weapons Research Laboratory (AWRL) in Aldermaston, Berkshire (UK) in 1973 using neutron activation for the analysis of 25 elements (Coleman and Goode, 1973). Many other groups and individual researchers in government research laboratories, operational forensic laboratories and academic research laboratories have contributed to the body of knowledge with developments and recommendations on analysis/comparison tools as well as interpretation tools. More recently, some coordinated efforts to standardize analytical methods have resulted in very good analytical methods that have helped the forensic scientist to improve the quality of the results **and** the conclusions derived from glass examinations and comparisons.

The Federal Bureau of Investigation funded Scientific Working Group on Materials (SWGMAT) created a glass subgroup in the mid-1990s in order to improve glass analysis through writing of recommended procedures of analysis (analytical guidelines). The SWGMAT glass subgroup of users contributed to the development of a refractive index standard method of analysis entitled *Standard Test Method for the Automated Determination of Refractive Index of Glass Samples Using the Oil Immersion Method and a Phase Contrast Microscope* (E1967), originally published by ASTM in 1998 and then modified and reapproved several times since then (ASTM, 2011). Similar groups in Europe, including the European Network of Forensic Science Institutes (ENFSI) and the European-funded Natural Isotopes and Trace Elements in Criminalistics and Environmental Forensics (NITECRIME), have also contributed research efforts and standard operating procedures that have improved the state-of-the-art in glass examinations, including the development of guides for trace evidence analysis in the case of ENFSI and research on analysis of glass using LA-ICP-MS that led to a publication (Latkoczy *et al.*, 2005).

More recently, a National Institute of Justice funded research effort (Elemental Analysis Working Group or EAWG) comprising 31 forensic scientists and researchers representing 23 organizations participated in a series of interlaboratory analytical exercises that evaluated the performance and ruggedness of μ-XRF (X-ray fluorescence), LIBS, ICP-MS and LA-ICP-MS for the analysis of glass. This effort resulted in three publications (Ernst *et al.*, 2012; Trejos *et al.*, 2013a, 2013b) describing the optimized methods and the results of the interlaboratory trials. In addition, the EAWG authored standard methods of analysis for the use of both μ-XRF and LA-ICP-MS in the analysis and comparison of glass (ASTM, 2013a, 2013b).

Described in this chapter are the progression and maturation of forensic analysis of glass from individual research efforts through interlaboratory trials and, ultimately, to the adoption of consensus methods with known precision, accuracy and limits of detection – the figures of merit needed to evaluate the scientific foundation of an analytical technique. Many laboratories around the world now routinely use these standard methods for refractive index (RI), μ-XRF and LA-ICP-MS thanks to the efforts of the scientists willing to participate and contribute their time and talent. The newly established Organization for Scientific Area Committees (OSAC) by the US National Institute of Standards and Technology (NIST), as part of the National Commission on Forensic Science, aims to further raise awareness of the scientific underpinnings that justify the use of forensic methods by recognizing those methods that include the necessary scientific foundations. The NIST OSAC publishes a registry of recognized methods that are approved by the forensic scientific community at large and, in the case of the three standard methods described for glass, this recognition is in addition to the rigorous standard development process of ASTM. All of these efforts have aided the forensic community and the judicial system by providing excellent tools to better characterize and compare glass and to give proper weight to the interpretation of the evidence.

Glass evidence has a special role within trace evidence in that it represents a "model matrix" due to the following characteristics of this evidentiary material:

- It is a common type of trace evidence as a consequence of its fragile nature and wide use in society.
- It is easily transferred from a broken source to the scene/victim(s)/suspect(s).
- It is easily recovered from a scene/object.
- It can persist after transfer but its chemical composition does not degrade or vary over time after transfer.
- The sizes of the fragments that are typically recovered from breaking events are normally sufficient for analysis by a variety of methods.
- Sensitive methods for the determination of the optical and chemical properties of glass currently exist in the form of standard methods of analysis and suitable reference materials also exist for calibration and/or bias determination.

- Its chemical composition is relatively homogeneous within a single pane or sheet of glass.
- Despite the continuous improvements in the manufacturing processes from automation and other developments, source variations (and temporal variations within the same source) are detected when sensitive methods are used for analysis and comparison of glass.
- The physical–chemical measurements result in numerical values that can be subjected to statistical analysis methods to aid the interpretation of the results. These data can also be used to assess frequency of occurrence of RI and/or elemental composition for a given population of glass.
- The scientific literature is very extensive (over 100 peer-reviewed publications and including four ASTM standard methods), providing the necessary scientific foundation to the forensic examination of glass evidence.

In order to evaluate the value of glass as evidence it is necessary to understand the nature and chemistry of this material, the transfer and persistence of small fragments and the advantages and limitations of the analytical techniques used for its examination. This chapter covers these topics to offer the reader a broad perspective of the significance of glass evidence.

6.2 Glass, the material

6.2.1 Physical and chemical properties

Glass is defined as an inorganic product of fusion that has been cooled to a rigid condition without crystallization (ASTM, 2010). This material is composed of a mixture of inorganic materials that are responsible of its different physical properties.

The inorganic materials may be present in the final product at high, minor or trace levels. Some components determine the glass structure; others are added intentionally to decrease the cost of manufacture or to provide desired properties such as color, viscosity, heat resistance and safety.

Glass can be classified in different groups according to its intended use as: flat glass (for architecture and automobiles), containers (bottles and jars), fiberglass (for insulation) and specialty glass (fiber optic, semiconductors and optical). It can also be classified by its chemical composition as soda-lime glass (containers and windows), leaded glass (housewares and decorations) and borosilicate glass (industrial, lamps and cookware) (Almirall, 2001a).

The essential raw materials used for the manufacture of soda-lime glasses are sand (SiO_2), soda ash (Na_2CO_3) and limestone (CaO). Sand, the major source of silica, has a tetrahedral pyramid shape with silicon linked symmetrically to four oxygen atoms. During glass making, the high melting point of silica is reduced by adding sodium carbonate and calcium oxide. Calcium and sodium atoms enter the silicon–oxygen network, reducing the melting point and the viscosity of silica. The amount of soda ash,

Table 6.1 Common contaminants in quartz sand (Tooley, 1984; Koons *et al.*, 2002)

Contaminant	Chemical formula	Effect on glassmaking
Feldspars	$[(K, Na)AlSi_3O_8-CaAl_2Si_2O_8]$	Add color to glass and alter furnace temperatures
Iron	Fe	Add color to glass and alter furnace temperatures
Ilmenite	$FeTiO_3$	Non-glassy impurities
Corundum	Al_2O_3	Non-glassy impurities
Chromite	$FeCr_2O_4$	Non-glassy impurities
Spinel	$MgAl_2O_4$	Non-glassy impurities

Table 6.2 Classification of components of glass according to their principal manufacturing function (Almirall, 2001a; Koons *et al.*, 2002)

Function	Common components
Formers (primary)	SiO_2, B_2O_3
Intermediate formers	Al_2O_3, ZrO_2, V_2O_5, Sb_2O_5, PbO, ZnO, TiO_2
Modifiers	Na_2O, CaO, MgO, Li_2O, BaO, SrO
Colorants	Fe_2O_3, Cr^+, Se^+
Decolorants	As_2O_3, MnO_2, CoO, CeO_2
Refining agents	As_2O_3, $CaSO_4$

limestone and minor and trace components varies by manufacturer and depends on the eventual use of the glass.

Borosilicate glass contains boron for thermal and physical resistance; "leaded" glasses, as the name implies, use lead to replace some of the calcium atoms to increase the refractory properties, desired in ornamental tableware and lamps for example (Almirall, 2001a).

Sand requires certain characteristics in order to be employed in the manufacture of glass. Small impurities present in the sand could produce undesired optical properties in the final product. Some of the common impurities in sand and how they affect the product are summarized in Table 6.1 (Tooley, 1984; Koons *et al.*, 2002).

The components of glass are classified according to their function as: formers, fluxes, modifiers, stabilizers, colorants, decolorants, accelerants, and refining and opaliser agents. **Former** agents are products that generally form the framework of the glass structure; when cooled quickly after melting they solidify without crystallizing. **Fluxes** are components that are added to the formers to lower the melting temperature and to reduce cost of production. **Stabilizers** are added to offer chemical resistance to the glass, while decolorants are used to clarify the glass. Refining agents are also important components of glass that help to remove trapped bubbles from the molten glass during its production. Table 6.2 shows some of the main components of glass and their principal function (Almirall, 2001a; Koons *et al.*, 2002).

The use of recycled glass or cullet is commonly employed in the manufacture of glass to decrease the melting temperature and to reutilize the broken glass, reducing the cost of the manufacturing process. Most of the cullet used in sheet glass is recycled within the plant while some container plants use recycled consumer glass, which adds some heterogeneity between batches originating from the same manufacturing plant. This is favorable from a forensic point of view because the elemental profiling will differ widely amongst different sources or different batches.

Although technological advances in the manufacture of glass have led to standardized products, minor variations in the physical properties and chemical composition of the glass remain between manufacturing sources, due to the inherent trace contaminations within the raw materials as well as any differences in manufacturing (e.g., change of color production). In addition, for float glass, as the inner surfaces of the furnace age, the tendency of some elements to leach from the bricks into the molten glass is greater. Two of the trace contaminants that could leach from the furnace "corundum-zirconium oxide refractory" bricks are aluminum and zirconium (Koons *et al.*, 2002). These trace "contaminants" incorporate an elemental "signature" into the glass product that is detectable and distinctive, allowing discrimination between glass fragments that originate from different sources (i.e., glass manufactured in different plants or at the same plant at different time intervals). A corollary proposition to the above hypothesis is that when comparing two glass fragments and no difference in trace element content is found using analytically sensitive methods, it suggests that the glass samples originated from the same plant and were manufactured at around the same time.

6.2.2 Manufacturing

Glassmaking has a long history, with the first formulations reported in 1500 BC by the Egyptians for use mainly in the manufacture of vessels and jewelry. Later, press molded glass was formulated in Alexandria in 400 BC and the Syrians introduced the first examples of flat glass for use in windows in 200 BC. The Romans, who were known for their contributions in oven technology, improved the quality of glass manufacture further and developments in optical glass and heat-resistant glass for use in thermometers and cooking glassware were reported in Germany in the 19th century (Koons *et al.*, 2002). Since then, the manufacture of glass has been in constant change in order to improve the physical and optical properties through the incorporation of automated processes and, ultimately, to decrease costs of manufacturing.

Today, glass is one of the most used materials in society for many reasons. Glass production ranges from simple glass containers to advanced microcomponents. The manufacturing of glass usually follows five steps: (i) manipulation of raw materials (storage, weighing and mixing), (ii) melting (refining and homogenizing), (iii) forming, (iv) annealing and (v) warehouse or secondary processing. The melting process takes place at furnaces that are resistant to high temperatures (>1500°C), where a continuous flow of molten glass is fed into automatic forming machines (Copley, 2001). During the refining process, the elimination of bubbles from the molten glass takes place and thermal and mechanical stirring ensures homogenizing of the glass, which is very important to offer uniform RI in the product.

The forming procedure is then followed by gradual changes in viscosity that allow molten glass to be converted into different products; the process will depend on the final product (container, blowing, flat glass or fiberglass). After the forming step, the annealing stage takes place and the glass is cooled at a controlled rate to solidify without crystallization.

Some glass products require secondary processing such as tempering, coating and coloring or decolorizing. Tempered glass is glass that has been subjected to a tempering process (preferential cooling on the surfaces) to impart additional strength and safety characteristics. Tempered glass is widely used in the manufacturing of automobile side and back windows as well as glass doors and panels in architectural applications. During the tempering process, the surfaces of the glass are cooled more rapidly than the center, forcing the edges into a state of compression. This process can be used for flat glass or some curved screens; however, it cannot be used for containers (Copley, 2001).

The coating method is used for decoration, protection or strengthening of the glass. This is a usual method in the manufacture of containers to improve the handling of the material. The coatings are applied by spray or as a vapor, first between the forming step and the annealing step (hot end) and then just after the annealing (cold end). Hot end coatings are mainly made of titanium or tin and the cold ends are organic waxes or fatty acids. Some flat glass products are also coated.

As mentioned, some impurities from the raw materials can produce a color in the glass. For some products, a clear appearance is desirable and, therefore, additional quantities of elements such as selenium and cobalt are added to counterbalance the green or yellow color caused by iron. On the other hand, some products are colored intentionally for decorative or technical reasons. Common colorants are iron (green, brown or blue), manganese (purple), cobalt (blue, green, pink), titanium (purple, brown), cerium (yellow) and gold (red) (Copley, 2001).

There are many different glass compositions but in general glass that may come to the attention of a forensic scientist as part of a case normally falls into one of the following types: flat glass (architectural and automobile), containers (bottles and jars), domestic glass (tableware), technical glass (laboratory glassware) and fiberglass (insulators), and now, much less frequently, borosilicate glass from headlamps, since current automobile headlamps are polymer based. Each end product has a different use and composition. Therefore, the manufacturing process differs from one product to the other. The following sections describe the main processes for manufacturing glass.

6.2.2.1 Flat glass

In flat glass manufacture there are two main glass-forming processes: the float process and the rolling process. The rolling process is used to produce patterned or textured glass for decorative purposes. In this method, a ribbon of glass is passed through water-cooled metallic rollers, which transport the glass horizontally into an annealing oven and, finally, the glass is cut to size (Copley, 2001). Adjusting the gap between the rollers controls the thickness of the glass. The rollers can also impress a final pattern into the glass, if required. When the glass is further ground and polished, the resulting flat glass is called "polished plate".

The float glass method is the preferred manufacturing method for flat glass. It was introduced commercially in 1959 in England. In this process, the raw materials are introduced into one end of a furnace and are melted and fused into molten glass. The molten glass emerges from the furnace into a chamber that contains a pool of liquid tin. This chamber is kept at a strictly controlled temperature in an atmosphere free of oxygen in order to avoid oxidation of the tin. At the entrance of the chamber the glass temperature is approximately 1000°C and at the exit the temperature is 600°C. Metal knobbed wheels on both sides of the 12-foot-long sheet (or ribbon) of glass are used to pull the glass over the surface of the smooth liquid tin surface. The final thickness of the glass is controlled by the speed at which the glass ribbon is drawn over the tin bath resulting in very flat, smooth and polished surfaces on both sides of the glass. The glass is then cut into 12 × 12 foot squares for use in different applications. The glass can also be coated by online chemical vapor deposition to improve the optical properties of the glass with a chemical coating less than a micron of chemical thickness. The glass can be annealed after it is cut to the final size and shape, since cutting annealed glass results in breaking.

6.2.2.2 Containers

Containers are typically produced using a blowing or flowing process. In the first method, the glass is placed in a mold and the glass is blown to expand and take the desired shape of the mold; then the container is reheated and annealed to prevent cracking of the product.

In the flowing method, the homogeneous molten glass flows and drops into the mold, where it is shaped by pressing (wide neck jars) or blowing (bottles) before being transported to the blow mold. The containers usually undergo surface treatments as they pass through the annealing process. The flowing method is used in automatic production of containers. Most manufacturers of containers use recycled glass or "cullet" as raw materials, adding more heterogeneity to the glass.

6.2.2.3 Fiberglass

Fiberglass is widely used in composite materials, reinforcement of plastics, gypsum and as a thermal insulator. From a forensic point of view, these materials are readily encountered in houses and buildings and, therefore, are potential trace evidence that can be transferred to clothing in crime scenes (Koons *et al.*, 2002).

In the manufacture of fiberglass there are two main products; continuous glass filament and glass wool. The main difference in the manufacturing process is that in the first one, the filaments of molten glass are drawn mechanically downwards from the bushing of a set diameter, while in the glass wool process the glass flows by gravity from the bushing and the filaments are random in diameter (Copley, 2001).

6.2.2.4 Manufacturing trends

The forensic examiner must be aware that glass manufacturing is dynamic and formulations and manufacturing processes may change over time as a product of market-driven requirements and environmental restrictions. For instance, trends in the automotive industry are towards addition of more glass in cars for cosmetic and aerodynamic reasons. Some modern vehicles now have expanded glass visual openings around the corners to reduce blind spots, and some windshields and back windows are extended into the roof. These trends require changes in glass formulation to facilitate the bending properties. Safety, driver comfort and fuel savings have also driven changes in the thickness of laminated glass to make it thinner and lighter. New coatings on laminated and tempered glass are being formulated to be IR and UV reflective to reduce the interior heating of the car (Clawson, 2009). Alternative raw materials, such as lithium feldspars, have been explored as a potential replacement for limestone to reduce melting temperature and energy costs (Merivale, 2008).

The 2010 Interpol review papers (Daiéd, 2010) reported new trends in glass manufacturing, including the production of a light emitting glass (Planilum), which consists of four layers of tin-doped indium oxide conducting glass, and the "self cleaning" glass, which is coated with a proprietary hydrophobic coating that repels organic contaminants on its surface. The flat glass manufacturing industry is also rapidly growing in other countries, such as China, Korea and Brazil, and reaching the United States' market, which directly affects studies involving databases and statistical interpretations, including frequency of occurrence statements (Wray, 2008).

6.2.3 Fractures and their significance

The study of glass fractures may help to determine the direction of the force that produced the fracture. There are a few publications on the nature of glass fractures and their use in forensic examinations (McJunkins and Thornton, 1973; Thornton and Cashman, 1979, 1986; Mencik, 1992; Haag, 2012; Tulleners *et al.*, 2013). However, the interpretation of the fractures still relies on the experience of the examiner.

Glass is broken by tension rather than compression; the surface that breaks first is the opposite side of the impact zone. Figure 6.1 shows the process involved in the breaking of flat glass. A force applied to a surface will produce compression to the side facing the force and tension on the opposite side. The tension state on the forward side creates radial fractures that will produce, at the same time, subsequent compression of that side of the glass. This compression will produce further tension on the side originally exposed to the force, creating concentric fractures.

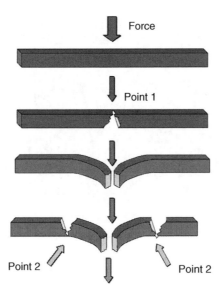

Figure 6.1 Diagram of the breakage phenomena of a glass pane produced by a low-velocity impact. After the application of the force, the glass bulges on the side opposite to the impacting force and radial fractures begin at Point 1. Later, concentric fractures will form at Points 2

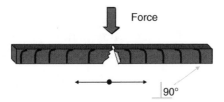

Figure 6.2 Diagram of typical ridges on a radial fracture

Figure 6.2 shows that radial fractures propagate away from the point of impact and generates ridges that tend to be perpendicular to the opposite side of the impact and parallel to the side of impact.

Fractures on tempered glass have to be studied separately, since tempered glass surfaces are, by default, under a state of compression while the central region is under tension.

When an object at high velocity impacts a glass surface (i.e., a bullet), it tends to produce a cone or crater, with the opening having a larger diameter at the exit side. Examiners could look at these features to determine if a bullet was shot from inside or outside a vehicle. Figure 6.3 shows a picture of glass fragments moving toward and backward from the direction of a projectile. Nevertheless, it has to be taken into account that bullets moving at low velocities may not pass through the glass pane, leaving a larger cone toward the point of impact.

Figure 6.3 High speed photography of a bullet penetrating a glass pane

The nature of glass fractures is not discussed in detailed in this chapter. However, good references on this topic can be found in the books *Forensic Examination of Glass and Paint* (Thornton, 2001) and *Forensic Science Handbook* (Koons *et al.*, 2002).

6.2.4 Forensic considerations: Transfer and persistence of glass
6.2.4.1 Transfer and persistence
Understanding the phenomena of transfer and persistence of glass fragments is fundamental for interpreting and assessing the significance of the evidence. For instance, some glass fragments, such as broken containers, are more commonly found in streets than fragments originating from architectural windows, and glass is more commonly found on footwear than on clothing of the general population. For this reason, it is important to evaluate the findings in a specific set of circumstances to assess the correct value of the evidence.

Numerous studies have been conducted to investigate these aspects. These studies have been widely used to enable estimations of the likelihood ratio (LR) of obtaining glass fragments on a suspect that "matches" a control piece of glass, as well as other Bayesian network approaches for comparison and characterization of glass (Curran *et al.*, 1997, 1998; Daéid, 2001; Zadora, 2007, 2009; Zadora and Neocheous, 2009; Zadora and Ramos, 2010; Ramos and Zadora, 2011). The LR is estimated using Bayes theorem given by the probability of the evidence "if the prosecution proposition is true" divided by the probability of the evidence "if the defense proposition is true" (Cook *et al.*, 1998). In the United States, classical statistical methods (otherwise known as the frequentist approach) are more commonly employed in courtrooms rather than the Bayesian approach. Nonetheless, the implications of transfer and persistence of glass fragments are also taken into account to judge the value of the findings.

When glass is broken, fragments are expelled not only with the direction of the breaking force but also backward. This is known as backward fragmentation. Glass has been reported to travel more than 3 m from the breaking source and has been recovered from persons standing up to 1 m away from the broken object (Daéid *et al.*, 2009).

According to experimental data, the number and size of fragments transferred and retained in the recipient is related to several factors, such as type of clothes and garments (Luce *et al.*, 1991, Hicks *et al.*, 1996), distance from the breaking window (Hicks *et al.*, 1996; Allen and Scranege, 1998), wet versus dry clothing (Allen *et al.*, 1998a), type and thickness of glass (Hicks *et al.*, 1996), the breaking force and the number of blows and the object that broke the window (Pound and Smalldon, 1977; Locke and Scranage, 1992; Allen *et al.*, 1998b; Hicks *et al.*, 2005). Some of the studies found that glass fragments are more likely to transfer to wet clothing than to dry clothing; garments such as jumpers and socks retain more fragments than trousers; the number of fragments transferred decreases exponentially with increases in distance from the source; the rate of decrease of fragments versus distance is even more pronounced for fragments larger than 1mm.

Hicks *et al.* (2005) reported that it is possible to predict with statistical methods the number of fragments transferred into a garment at the moment of the breaking from the number and distribution of fragments found on the ground. These estimations depend on the type and thickness of the window and how the window was broken. This was the first study to investigate these behaviors when a firearm is used to break the glass. Some studies suggest that a large number of fragments with original surfaces may be indicative of backward fragmentations versus contact with already broken glass (Zoro, 1983; Luce *et al.*, 1991; Underhill, 1997).

It has also been determined that glass fragments usually range from 0.25 to 1 mm in size and that large fragments (>5 mm) are more likely to be easily lost. Although glass tends to fall out from clothing, hair and shoes over time, it was demonstrated that glass fragments could persist on the recipient for at least eight hours after the breaking event (Allen *et al.*, 1998c; Hicks *et al.*, 1996).

Other authors have estimated the frequency of the occurrence of glass based on surveys of some populations. In general, most of the studies showed that few glass fragments, if any, were found in clothing and footwear of a randomly selected population not involved in breaking of glass (Lambert *et al.*, 1995; Lau *et al.*, 1997; Petterd *et al.*, 1999; Coulson *et al.*, 2001; Roux *et al.*, 2001). It was also reported that the prevalence of glass is higher in the soles of shoes than in the upper area (Roux *et al.*, 2001). Therefore, for real casework, it is also important to document the areas where the glass is collected from as well as its appearance ("fresh" or clean surfaces versus scratched and dirty surfaces). Caution should be taken, however, in interpreting the location of the glass fragments in the clothing, since glass fragments can migrate from pockets to the surface of the clothing and to the evidence bag during the handling of the evidence prior to analysis (O'Sullivan *et al.*, 2011).

All these studies support the hypothesis that if a suspect activity involves glass breaking, the significance of finding glass in the clothing or hair of a suspect is very high, particularly when multiple fragments are recovered.

Nonetheless, law enforcement personnel must be aware of possible secondary transfer. Secondary transfer is the transfer of glass fragments to an object or individual that was not directly involved in the breaking of the glass. Some researchers have confirmed that secondary transfer of glass fragments between objects and people may occur (Allen *et al.*, 1998a, 1998c; Daéid *et al.*, 2009; Cooper, 2013).

The general consensus is that the transfer and prevalence of glass in the general public by secondary transfer is very small. However, for persons in contact with broken glass, including law enforcement personnel, the indirect transfer of multiple glass fragments could be significant. For instance, recent studies of secondary transfer stress the importance of taking precautions when an officer attends a crime scene and shortly afterwards (<1 hour) arrests a suspect (Daiéd *et al.*, 2009; Cooper, 2013).

A good review of transfer and persistence studies can be found in the book entitled *Forensic Interpretation of Glass Evidence* (Curran *et al.*, 2000). The authors reported ten major factors that could affect the transfer and persistence of glass:

(i) Size of the window.
(ii) Type and thickness of the glass.
(iii) How the window is broken.
(iv) The position of the offender relative to the window.
(v) Whether or not entry was gained to the premises.
(vi) The nature of the clothing worn by the suspect.
(vii) The activities of the suspect between the time of the incident and the arrest.
(viii) The length of time between the arrest and the clothing being taken.
(ix) The way in which the clothing was obtained from the suspect.
(x) The weather at the time of the incident.

All of these findings about transfer and persistence could support the scientific conclusions that strengthen either the prosecution or the defense proposition and must be considered when weighing the significance of an association between samples of glass from a known source and fragments recovered from a suspect or victim (questioned).

6.2.4.2 Recovery of glass

At the crime scene, examiners and investigators are encouraged to be meticulously thorough in deciding how to sample glass fragments. Some recommendations about recovering glass fragments are:

- Sample as many fragments as practical from the known source because it is important to thoroughly characterize the heterogeneity of the source for later comparisons of their physical and chemical properties.
- Fragments originating from laminated glass, for example, windshields, should be collected separately and labeled to indicate the outer or the inner panel.
- Pieces removed from a frame should be collected and packaged separately from broken fragments found on the ground at the scene.
- Whenever it is possible, identify the outer and inner surface of a glass fragment.

- Due to the nature of glass transfer, the clothing of suspects and victims, as well as their hair are important sources of glass fragments. Combing the hair and recovering the debris on a clean paper is recommended.
- Tools or materials suspected of having a direct impact on a glass surface may also contain traces of glass (bats, rocks, metallic tools, bullets).
- Glass fragments from different areas should be packaged separately and labeled properly.
- Do not shake the garments during collection from the suspect or victim nor during the packaging, since glass fragments can be lost.
- It is necessary to collect and submit all items as soon as possible to the laboratory, since the length of time affect the efficiency of recovery.
- It is important to be aware of secondary transfer and **avoid any cross-contamination** during the collection of glass evidence from the scene, suspects and victims.

At the laboratory, there are different techniques available for collection of glass fragments. These include hand picking with tweezers, scrapping, vacuuming, taping and shaking a garment over a large metal cone (Daéid *et al.*, 2009).

In our opinion, the first method of choice for recovery of glass fragments from clothing is the removal by means of sharp plastic tweezers, under strong light and magnification, if necessary, followed by shaking. However, the method selected depends upon the experience and criteria of the examiner as well as the nature of the sample.

The scrapping method is also used in many laboratories and involves using a spatula to scrap the surface of the clothing and recovering the debris on a clean sheet of paper that is later inspected under a stereomicroscope. This technique is useful for particulate materials; nonetheless, trace examiners must consider whether or not this method could increase the risk of contamination of other trace evidence, such as fibers.

Pound and Smalldon (1977) reported a high rate of recovery of glass fragments using the shaking method. Daiéd *et al.* (2009) reported the recovery of glass fragments from the examiner's fleeces; therefore, the use of gloves and disposable laboratory coats are recommended to minimize secondary transfer or contamination.

6.3 A brief history of glass examinations

An optical property – the refractive index (RI) – of a transparent material such as glass can be measured with high precision, accuracy and in an automated manner. Therefore, RI, which ranges between 1.47 for borosilicate glass and 1.55 for leaded glass, is routinely measured and compared in forensic casework. Improvements in quality control during the manufacture of glass have reduced the range of variation for RI values of glass as a population (Koons and Buscaglia, 2001), thereby reducing the "informing power" of RI as a discrimination tool for glass fragments. As a consequence, it has become necessary to use additional techniques, such as elemental composition analysis, to enhance the informing power of some comparisons between fragments (Almirall, 2001a; Duckworth *et al.*, 2002). Refractive index, however, is still the screening tool of choice for forensic examinations of glass.

Extensive research has been carried out on the use of elemental analysis of glass by radiochemical, spectroscopic and mass spectrometric techniques (Stoecklein and Becker, 2001). These include atomic absorption (Hughes *et al.*, 1976; Catterick and Wall, 1978), X-ray fluorescence (Koons *et al.*, 1991; Buscaglia, 1994), neutron activation (Coleman and Weston, 1968; Coleman and Goode, 1973), scanning electron microscopy (Kuisma-Kursula, 2000), inductively coupled plasma-atomic emission spectrometry (ICP-AES) (Koons *et al.*, 1988, 1991) and inductively coupled plasma-mass spectrometry (ICP-MS) (Zurhaar and Mullings, 1990; Parouchais *et al.*, 1996; Duckworth *et al.*, 2000). Each technique has its own advantages and shortcomings but ICP-MS has been shown to be the most effective analytical method for the comparison of trace elements in small glass fragments (Duckworth *et al.*, 2002). Some of the advantages of ICP-MS over other analytical techniques include its multi-element capability, excellent sensitivity, high sample throughput and the capability to provide quantitative and isotopic information (Montero *et al.*, 2003). The isotope dilution (ID) method applied to ICP-MS analysis usually provides the best accuracy and precision when the sample size is limited but this technique is quite cumbersome and, therefore, rarely used in operational laboratories (Smith, 2000).

Although conventional solution ICP-MS using external calibration (EC) has been shown to be an excellent tool for elemental analysis of glass, this method is time consuming and has the disadvantage of requiring the dissolution of the sample, thereby destroying the sample prior to its introduction into the ICP-MS. Fortunately, other sample introduction techniques, such as laser ablation (LA-ICP-MS and LIBS), overcome these limitations (Watling *et al.*, 1997; Russo, 1998, 2011; Watling, 1999; Trejos *et al.*, 2003; Latkoczy *et al.*, 2005; Almirall and Trejos, 2006; Bridge *et al.* 2006, 2007; Naes *et al.*, 2008; Rodriguez-Celis *et al.*, 2008; Cahoon and Almirall, 2010). Alternatively, µ-XRF has also been adopted by many forensic laboratories for elemental analysis and, while XRF does not provide equal discrimination to ICP-MS or LA-ICP-MS analysis, it is suitable and fit-for-purpose for forensic examination and comparison of glass (ASTM, 2013a).

6.4 Glass examinations and comparison, standard laboratory practices

Glass examiners often measure the physical and optical properties of glass, such as color, thickness, density, RI, and also, if necessary, they conduct elemental analysis to enhance the value of an association. Figure 6.4 shows the typical scheme for the forensic analysis of glass fragments.

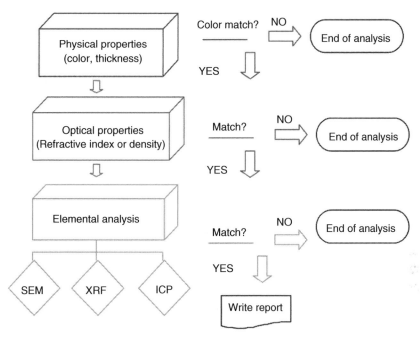

Figure 6.4 Basic scheme for the forensic examination of glass (*See insert for color representation of the figure.*)

6.4.1 Physical measurements

The first step in any glass analysis is to identify the glass by physical and optical properties such as hardness, amorphous structure and isotropism. Physical observations such as color, thickness, curvature, flatness, presence of manufacture markings, fluorescence and fracture characteristics are made in the preliminary stages of the analysis. Forensic examiners should be meticulous during the physical examination of glass as it can address important information such as:

- type of impact that caused the fracture of glass, for example, a fracture caused by a gunshot or by a hard object such as a baseball bat;
- direction of the force, for example, to establish if a window was broken from inside or from outside;
- type of material, for example, a flat architectural window, a tempered glass, a headlamp, an eyeglass lens or a colored bottle;
- source of origin of the glass fragment, for example, to establish that two or more pieces of glass were originally joined.

6.4.1.1 Physical fit

A perfect physical fit or physical "match" is the best-case scenario that a forensic examiner could have during the physical evaluation of glass, but it is rarely found in real cases. For comparison purposes, the fragments are inspected visually in order to determine whether or not there is a "fracture match" between any of the recovered fragments to any of the source fragments. Such a match requires the edges of one fragment to perfectly fit into the corresponding edges of another, much like a jigsaw puzzle (Almirall, 2001a). The physical match of the broken edges of glass is three-dimensional and the analyst should document the details by photography and written documentation. The match can also be observed under the microscope in order to determine microscopic marks (conchoidal or hackle marks). Other methods, such as shadowgraphy and laser interferometry, may also be used (von Bremen, 1975; Thornton and Cashman, 1979; Welch *et al.*, 1989).

6.4.1.2 Thickness

Sometimes glass can be recovered from the scene as pieces that present the full thickness of the source of origin. In such cases, it may be useful to measure the thickness of the glass, as it can provide information about the type of object from where it came, for example, a vehicle side window, tempered glass, or beverage bottle. For comparison purposes thickness measurements provide limited information when the thickness is found to be the same but may be used to exclude a fragment as originating from a source that clearly presents a different thickness.

6.4.1.3 Density

Density comparisons of glass can be accomplished using the sink-float method. The comparison of density has the disadvantages of involving toxic liquids and requiring at least 5 mg of sample. Another limitation of density measurements is that measurements of small, irregular or dirty fragments of glass may be inaccurate (Koons *et al.*, 2002). Density measurements have been mostly replaced by RI measurements, which provide good optical property information with the advantage of being faster, more accurate and more precise. Density and RI are reported to be highly correlated, so the measurement of density does not significantly add to the information collected from the RI measurements (Almirall, 2001a).

6.4.2 Optical measurements

Optical properties of glass, such as isotropism, are typically determined in forensic laboratories as part of the screening test to determine whether or not a fragment is actually glass. The fragment is observed under crossed polarized light using a microscope (polarized light microscope) and only isotropic specimens, such as glass, would remain dark in any orientation when it is placed between the crossed polars while rotating the sample stage. Anisotropic or birefringent substances, such as quartz or other minerals, become bright if the specimen is rotated 45° from its extinction position.

The foremost optical property measured on glass is the "refractive index"; it has been the most commonly used method in forensic laboratories for examination of this material. Refractive index is defined by the Snell's law as the ratio of the wave velocity in a vacuum to the wave velocity in the transparent medium:

$$RI = V_{vacuum} / V_{Glass}$$

In 1892, Becke observed a bright line inside the edge of a mineral specimen that had a higher RI than its surroundings (Koons *et al.*, 2002). When he raised the objective of the microscope, he noticed that this line moved into the object that had a higher RI. This line is known as the "Becke line" and it was used to measure the RI of glass specimens. The method consisted of immersing a glass fragment in oil with a higher RI and observing the Becke line moving away from the edge of the glass. The RI of the oil was then modified with a miscible liquid until the RI of the glass equaled the RI of the liquid, at that point the Becke line disappeared. The RI of the oil mixture was then measured with a refractometer. However, this method was very time consuming and not very accurate (Hamer, 2001).

Another method employed for measuring RI was the immersion method that was based on the technique described in 1926 by Winchell and Emmons (Koons *et al.*, 2002). In this method, a fragment of glass is immersed in suitable oil and heated until the glass observed under a microscope disappears. At this temperature, it is assumed that the RI of the glass is the same as the RI of the oil. The RI of the oil at that temperature is measured, providing an indirect measurement of the RI of the glass (Hamer, 2001). The method is still in use in some laboratories but it has been almost completely replaced since the 1980s by an automated method V usually referred to as GRIM, due to the model name given by the company that produces the instrument (Foster and Freeman, Evesham, UK), which stands for "Glass Refractive Index Measurement". ASTM has published a standard method for the automated determination of RI for the forensic comparison of glass fragments (ASTM, 2011). The advantages of this method include its speed, that it is less tedious for the operator and produces more precise and accurate data, and that it is automated and thus results in reduced operator error.

The automated determination of the RI of glass fragments using the oil immersion method with phase contrast microscope is described in the ASTM method for glass samples as small as 300 μg. In this method, the glass fragments should be cleaned with deionized water and ethanol, dried and then crushed into small pieces of approximately less than 150 μm in size. A small sample is then deposited onto a slide and immersed in a microdrop of oil and covered with a cover slip. This slide is placed into a hot stage under a phase contrast microscope. The magnified image (~160×) of the glass particles immersed in the oil is observed using a charge-coupled device (CCD) camera. The temperature of the oil is changed at a controlled rate using the hot stage. The temperature at which the glass disappears (minimum contrast between the glass and the oil) is recorded electronically and the average of the match temperature from both the heating and the cooling cycle is converted to RI value according to a calibration curve of the oils (Figure 6.5). This calibration curve is obtained from the standardization of oils using

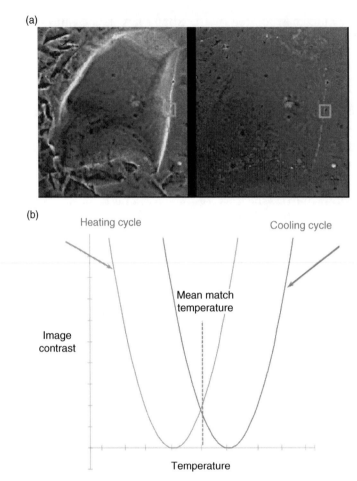

Figure 6.5 (a) Glass fragment above the matching temperature (left) and at the matching temperature (right). (Curran *et al.*, 2000. Reproduced by permission of Taylor and Francis Group, LLC, a division of Informa plc.) (b) Heating and cooling cycle used for refractive index determination (*See insert for color representation of the figure.*)

reference glass with certified RI values to construct the calibration curves. ASTM recommends a minimum of three standard glasses for the calibration of each oil, in addition to a separate glass standard to verify the calibration curve.

A hot-stage control precision of 0.1°C is required by ASTM for interlaboratory comparisons and the method reports a standard deviation in RI of 0.00002 during a five hour period of operation and a standard deviation of 0.00003 over a five day period. There are different factors that could affect the precision of RI measurements, such as natural heterogeneity within a glass object. For instance, there are studies that reported larger variations within patterned glass as well as within float versus non-float surfaces (Hamer, 2001). Suzuki *et al.* (2000) reported a variation of RI values of as much as 0.00010 for some containers. Newton (2011) reported the effect of debris contamination on the precision of RI measurements, the potential of false exclusions when glass fragments from surfaces are compared with non-surface fragments (Newton *et al.*, 2004) and the effect of edge morphology on the precision of the measurements (Newton and Buckleton, 2008). In order to account for the inherent heterogeneity of the source, it is important to measure several replicates from several edges and fragments. For comparisons, it is recommended to conduct as many replicates of the known sample as is practical (Garvin and Koons, 2011).

The automated method for measuring RI has the advantages of improved precision as well as improved accuracy, as it minimizes the variation between different operators and laboratories. Over the years, the original model, GRIM1, has been replaced by newer models such as GRIM2 and the newly available GRIM3, which have some technical improvements but retain the fundamentals of the technique. The new GRIM3 allows for the simultaneous measurement of several edges, which provides the examiner with more data points for the same amount of time of analyses.

6.4.2.1 Annealing of glass fragments

Several authors (Locke *et al.*, 1982, 1984; Marcouiller, 1990; Curran *et al.*, 2000; Newton *et al.* 2005; Zadora, 2009; Newton, 2011) have reported the use of annealing methods to distinguish tempered glass from non-tempered glass and as a tool to improve the discrimination power from RI measurements. The annealing method uses a furnace to heat both control and suspect samples, which are then cooled in a controlled manner (Hamer, 2001). The temperature is high enough to remove stress in the glass but below deformation. A typical heating cycle starts at room temperature and then is heated up to approximately 600°C. The cooling cycle cools the glass at a very slow rate down to approximately 400°C followed by cooling to room temperature.

Refractive index of glass specimens is measured before and after the annealing process and the refractive index difference (ΔRI) is used either to classify glass as tempered or not tempered, or to conduct further comparisons.

Locke and Hayes (1984) reported that a ΔRI value of 0.00120 is typical of tempered glass, while values below 0.00060 classifies the glass as non-tempered. Nevertheless, there is still an overlapping range between some non-tempered specimens that could reach values of ΔRI up to 0.000140.

Annealing of glass is also used as an additional discriminating tool for glass comparisons, mainly for tempered samples. The RI of two specimens (recovered and known) can be compared before and after the annealing. Usually, the annealing process contributes to decreasing the relative standard deviation of the RI measurements and, therefore, the informing power is enhanced.

6.4.3 Chemical measurements: elemental analysis
6.4.3.1 SEM-EDS

X-ray detection methods such as scanning electron microscopy-energy dispersive spectroscopy (SEM-EDS) are also used in forensic laboratories for elemental analysis of glass. SEM is used for imaging the microstructural characteristics of solid objects by using electrons produced when a focused beam, at a given accelerating voltage, interacts with the surface of a material (Brozel-Mucha and Zadora, 1998). This beam produces a number of electron products that are picked up by detectors capable of analyzing the specific electron, photon or X-ray. The created X-rays are detected and identified by EDS for multi-elemental analysis.

Glass is refractory and non-conductive in nature, requiring a coating process that is usually made with carbon. The coating step prevents the sample from "charging", otherwise the sample will build a charge in the interaction volume, affecting the ability of the SEM to properly image the sample.

Reeve *et al.* (1976) first reported the use of SEM-EDS to further discriminate glass fragments that were undistinguishable by RI and density measurements. The authors found that the discrimination power was enhanced by measuring the elemental intensity ratios of nine different elements to calcium (Ti/Ca, Mn/Ca, Fe/Ca, Cu/Ca, Zn/Ca, As/Ca, Rb/Ca, Sr/Ca and Zr /Ca). Two years later, Andrasko and Maehly (1978) reported the use of the concentration ratios Na/Mg, Na/Al, Mg/Al, Ca/K and Ca/Na to better characterize glass evidence. In 1986, Ryland (1986) reported a classification scheme for container and sheet glass samples using the intensity ratios of Ca/Mg (by SEM/EDS) and Ca/Fe (by XRF). Zadora *et al.* (Zadora, 2007, 2009; Zadora and Ramos, 2010) have reported the use of SEM-EDS data of major and minor elements (O, Na, Al, Mg, Si, K, Ca and Fe) and statistical methods such as Naïve Bayes Classifiers and Support Vector Machines to improve the classification of glass based on its end use and the application of this data to databases that facilitate such classification. Although the classification of glass fragments into categories is important for forensic examinations of glass, the ability to discriminate between glasses of the same type is even more valuable for comparisons. In this sense, SEM-EDS has some limitations due to the lack of sensitivity for trace elemental analysis. Its detection limits allow the identification of minor and major elements only (>0.1%). In addition, precision and accuracy is generally poor, quantitative analysis is usually not possible and the amount of elements that can be detected is limited in comparison to other methods.

SEM-EDS has some advantages in that it is non-destructive and can be used in the forensic analysis of glass debris in bullets, where the sample size may not be suitable to other elemental analysis methods. Nonetheless, glass examiners are encouraged to use analytical methods with superior sensitivity to fully characterize the elemental signature of glass exhibits. SEM-EDS is not recommended for elemental analysis in forensic case-work unless the fragment size does not allow for other elemental analysis methods, such as cases involving small glass debris on bullet surfaces.

6.4.3.2 XRF

The X-ray fluorescence methods (µ-XRF) use similar fundamental principals to SEM-EDS, with the main difference being the excitation source used (Figure 6.6). During XRF analysis, a primary X-ray excitation source from an X-ray tube strikes a sample and is either absorbed by the atom or scattered through the material. The process in which the

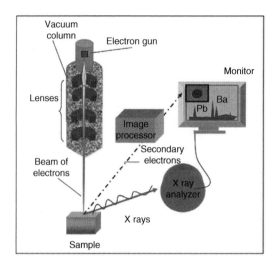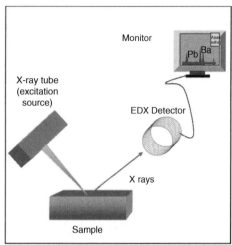

Figure 6.6 Comparison of the schematic diagram of SEM-EDS and XRF instruments (*See insert for color representation of the figure.*)

X-ray is absorbed by the atom by transferring all of its energy to an innermost electron is called the "photoelectric effect". During this process, if the primary X-ray has sufficient energy, electrons are ejected from the inner shells, creating vacancies. These vacancies present an unstable condition for the atom. As the atom returns to its stable condition, electrons from the outer shell are transferred to the inner shells and in the process give off characteristic X-rays with an energy difference specific to the binding energies between the corresponding shells. Because each element has a unique set of energy levels, each element produces X-rays at different energies, allowing the characterization of the elemental composition of a sample. The process of emitting characteristic X-rays is called "X-ray fluorescence" or XRF.

XRF has been incorporated into forensic laboratories for the analysis of glass fragments and has similar advantages to those obtainable by SEM-EDS: it is considered to be non-destructive, rapid and relatively sensitive (Howden *et al.*, 1978; Curran *et al.*, 2000; Krüsemann, 2001; Roedel *et al.*, 2002; Hicks *et al.*, 2003). Nonetheless, XRF has the advantage over SEM of being more sensitive (~100 ppm vs ~1000 ppm for SEM-EDS), especially for elements with higher atomic numbers. As a consequence, XRF is more discriminating than SEM, allowing not only the classification of glasses into categories but also a better discrimination among glasses of the same type (Ernst *et al.*, 2012).

Although XRF is limited to semi-quantitative analysis, it has been demonstrated that it provides useful information and can be used as a complementary tool for discrimination of glasses. In 1976, Reeve *et al.* (1976) were able to distinguish 97.5% of the glass sources under study. In 1980, Dudley *et al.* (1980) were able to distinguish 98% of the glasses originating from a set composed of 50 pairs of glasses, including window and non-window glasses.

In 1991, Koons *et al.* (1991) analyzed a total of 81 samples that originated from different tempered sheet glasses in order to evaluate the discrimination capabilities of RI, XRF and ICP-AES. Both methods for elemental analysis (XRF and ICP-AES) offered improved discrimination when combined with RI. In 2011, Ryland (2011) reported the improved discrimination of flat glass samples measured by μ-XRF with similar RI. In 2013, the Elemental Analysis Working Group performed a series of interlaboratory tests that compared the analytical performance of μ-XRF, ICP-MS and LA-ICP-MS for the analysis of glass fragments manufactured in the same plant at short time intervals (Trejos *et al.*, 2013a). A standard method (ASTM, 2013a) is now available to forensic examiners for the examination of glass fragments by μ-XRF as a product of these studies.

6.4.3.3 ICP-AES and ICP-MS

Inductively coupled plasma methods (ICP-AES and ICP-MS) are also currently used as standard methods for the analysis of glass samples in forensic laboratories. Since the early 1980s, numerous scientists have demonstrated the relevance of applying ICP methods to conduct elemental analysis of glass samples. (Koons *et al.*, 1991; Parouchais *et al.*, 1996; Duckworth *et al.*, 2000, 2002; Suzuki *et al.*, 2000; Montero, 2002; Almirall and Trejos, 2006).

In general terms, ICP instruments are composed of three main parts: the sample introduction system, the ionization source and the detector. The most common sample introduction system introduces liquid samples into the torch with the help of two key components: the nebulizer and the spray chamber. The nebulizer produces an aerosol of liquid particles that are then selected according to their size in the spray chamber. Only the liquid particles that are small enough will pass from the spray chamber to the torch and the rest will be drained to the waste.

The plasma is produced in an ionization source that comprises a torch assembly and load coil. The torch, a device that contains and shapes the plasma, consists of three concentric quartz tubes through which argon gas flows. The outermost tube serves to isolate the heat from the outer components of the torch; the intermediate tube is the one responsible for transporting the stream of gas that will generate the plasma; and the innermost tube serves as carrier gas for the sample. Plasma is generated by injecting electrons from a Tesla coil into an argon mixture in the presence of a radio frequency field. After the plasma is generated, a process known as "inductive coupling" sustains it. The plasma reaches high temperatures (5000–10,000K) and the sample within it becomes desolvated, atomized, excited and ionized. Characteristic emission lines from excited species relaxing back to the ground energy state are detected with a spectrometer and reported as wavelength in the ICP-AES experiment. In the ICP-MS experiment, ions formed in the plasma are directed into a mass analyzer with magnetic focusing optics; these ions are measured and using their mass to charge ratio are reported as mass from 7 amu (Li) to 238 amu (U).

During glass analysis by ICP-AES or ICP-MS, glass fragments are rinsed and dried, then crushed and weighed (approximately 5–8 mg for AES and 2 mg for MS for each of at least three replicates). The glass samples are then digested using a mixture of acids (HF, HCl and HNO$_3$). Ultrasonic baths are used to aid the digestion process and then the sample isdried at 80°C. Samples are then reconstituted in nitric acid in order to be analyzed and quantified. The sample preparation scheme for glass analysis by ICP is time consuming and the sample is destroyed, which are considered the main drawbacks of the technique. Nonetheless, it has been shown that its multi-elemental capabilities, precision and sensitivity made ICP-AES worth it, since the discrimination power is significantly improved (Koons *et al.*, 1991).

The main advantages of ICP-AES are the relatively low cost and complexity, large linear dynamic range and the ease of use. The main advantages of ICP-MS are high increased multi-element capability and isotopic information.

The use of ICP-MS for forensic examinations of glass has been the focus of several investigations (Zurhaar and Mullings, 1990; Suzuki *et al.*, 2000; Duckworth *et al.*, 2002). This technique combines the strengths of ICP-AES with the added advantages of providing isotopic information, superior selectivity and superior sensitivity, which reduce the sample consumption during the analysis and allow element detection at ultra-trace levels. ICP-MS is regarded as the most powerful technique for elemental analysis of glass. A standard method for analysis of glass by digestion ICP-MS was first published in 2004 and has since been updated (ASTM, 2012).

The main disadvantage of conventional (digestion) ICP methods is that it they destroy the sample and are time consuming. Fortunately, these limitations have been overcome by alternative sample introduction systems.

6.4.3.4 LA-ICP-MS

A laser ablation (LA) system enables the introduction of the products from the direct sampling of solids into the plasma. A typical LA-ICP-MS set-up consists of a laser, an ablation cell and the ICP-MS, which is used as an ionization source and analyzer. A solid sample is placed inside the ablation cell and a laser beam is focused on the surface of the sample (Figure 6.7). When the laser is fired, the high-energy interaction between the laser and the sample surface produces a cloud of very small particles and microdroplets. These particles are removed from the sampling cell by a carrier gas, usually argon or helium, and are swept into the ICP plasma for atomization, ionization and subsequent analysis (Russo, 1998).

The ablation cell is equipped with a quartz window that allows the laser pulse to enter the closed cell that is mounted on a translation stage, providing X-Y-Z positioning control for laser targeting on the sample, and is under computer control. The Z-axis of the translation stage is used to focus the laser using a CCD camera viewing system.

LA-ICP-MS has many advantages over solution ICP methods. When the analysis is carried out using laser ablation, minimum sample quantities are consumed during the

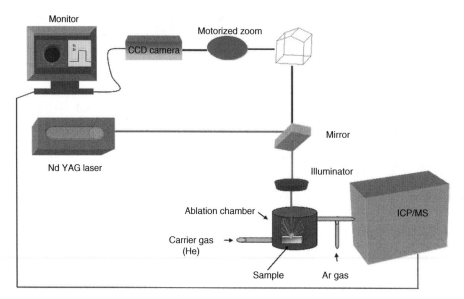

Figure 6.7 Schematic diagram of a typical LA-ICP-MS set-up (*See insert for color representation of the figure.*)

analysis (approximately 300 ng). Fragments as small as 0.2–0.4 ug (~0.1–0.4 mm) can be analyzed in several replicates using this method (ASTM, 2013b).

In addition, LA does not require the use of concentrated acids and minimizes the spectrochemical interferences of oxides or hydride formation, mainly due to water. The background signal for a "dry" plasma is lower than the conventional HNO_3 blank used in solution analyses.

Since little or no chemical preparation is required, the time of analysis is reduced and, more importantly, the potential for contamination from reagents and airborne particulate material is also greatly reduced. This advantage is particularly significant when determining low analyte concentrations, as such contamination can potentially mask a unique feature that may facilitate differentiation. The elimination of the digestion and solution step not only reduces the cost of high purity reagents and standards but also eliminates the health risks associated with handling hazardous materials such as HF.

Several authors have reported the analysis of glass samples by LA-ICP-MS. Features of forensic interest, such as true quantitative analysis, discrimination power, accuracy, precision and reproducibility of LA-ICP-MS, have been documented to justify its acceptance by courts in several countries in Europe, Australia, South Africa, Canada and the United States. Performance of the method has been compared to the traditional ICP-MS solution methods of external calibration and isotope dilution and to µ-XRF, in terms of time and ease of analysis, repeatability, reproducibility, precision, accuracy and discrimination power (Moenke-Blankenburg *et al.*, 1992; Becker *et al.*, 2001, 2003; Almirall *et al.*, 2003; Trejos *et al.*, 2003, 2013a; Bajic *et al.*, 2005; Latkoczy *et al.*, 2005; Trejos and Almirall, 2005a, 2005b; Almirall and Trejos, 2006, 2010; Berends-Montero

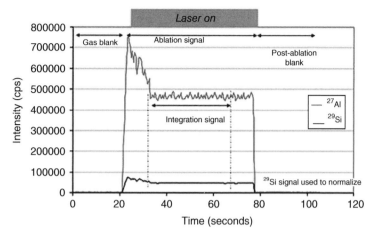

Figure 6.8 A strategy for the quantitative analysis of glass using LA-ICP-MS involves the use of a low natural abundance isotope of Si (^{29}Si) to normalize the analyte signal (in this case ^{27}Al) for differences in ablation yield as the internal standard, as it is present in a very high concentration (~70% as SiO$_2$) in glass. The intensity signal (cps) is converted to concentration using calibration standards such as the NIST 6XX series or the FGS standards (Latkoczy, 2005; ASTM, 2013b) (*See insert for color representation of the figure.*)

et al., 2006; Cahoon and Almirall, 2010; Trejos and Almirall, 2010; Koch and Gunther, 2011; Weis *et al.*, 2011).

Quantitative analysis with LA-ICP-MS can be achieved using the approach first described by the members of the NITECRIME network and based on collaborative trials between operational forensic scientists, government researchers and academics (Latkoczy, 2005). Figure 6.8 shows an example for the time-resolved (transient) signal for one analyte isotope of interest in glass (^{27}Al) while using a low natural abundance isotope of Si (29) to ***normalize*** for any differences in ablation yield throughout the 500 laser shots that create the steady stream of particles that make up the signal for ^{27}Al in this example. In a sense, ^{29}Si is used as an inherent internal standard because it is always present in float glass samples at very high concentrations (typically 70% as SiO$_2$). The signal is integrated at some interval during the ablation after it has stabilized following the initial laser–material interaction. A background is subtracted from a gas signal before the initiation of laser ablation and the area from the integration signal is used to calculate the concentration when compared to one or more calibration standards of glass samples with reported concentration values (i.e., NIST 6XX series or the FG series developed by the NITECRIME effort and reported by Latkoczy). The ASTM E2927-13 method expands on the NITECRIME effort by providing all the analytical figures of merit necessary for a Standard Method of Analysis and also reports interlaboratory results from several participating laboratories.

More than 35 forensic laboratories around the world in every continent now routinely employ the use of LA-ICP-MS for materials characterization. The recently approved international ASTM standard method for the forensic analysis and comparison of glass by LA-ICP-MS (ASTM, 2013b) can assist any forensic laboratory in the world to adopt LA-ICP-MS into routine use for glass analysis and comparisons.

6.4.3.5 Laser Induced breakdown spectrometry

An emerging analytical tool that can provide a cost effective alternative to LA-ICP-MS methods is laser induced breakdown spectrometry (LIBS).

Advantages of LIBS over LA-ICP-MS are: extremely fast measurements (<1 s each and usually 30 s for multiple shot analysis, twice as fast as LA-ICP-MS); elemental analysis of elements that are difficult to analyze by ICP-MS, such as Li, Fe, K, Ca and S, can be detected by LIBS; potential for portability; increase versatility; and lower instrument cost and maintenance. Finally, LIBS data are informationrich and amenable to automated interpretation.

Nevertheless, analytical LIBS is not yet as mature as LA-ICP-MS in the forensic laboratory. The analytical performance of LIBS in terms of selectivity, sensitivity and reproducibility is superior to μ-XRF but not quite as good as LA-ICP-MS. However, its analytical figures of merit are fit-for-purpose for glass examinations, allowing the detection of major, minor and trace elements. Research has demonstrated the utility, reliability and cost efficiency of LIBS in glass examinations (Bridge *et al.*, 2006, 2007; Naes *et al.*, 2008; Rodriguez-Celis *et al.*, 2008; Naes, 2009; Cahoon and Almirall, 2010; Russo, 2011). Continuation of these research efforts and future interlaboratory validation studies will facilitate the standardization of methods of analysis and the acceptance of LIBS in the courtroom for the forensic examinations of glass.

The main components of a LIBS system are the pulsed laser, focusing mirrors/lenses, the ablation stage, the light collection system (lens, mirror or fiber optic), the detection system composed of the spectrometer that filters or disperses the light, a detector and, finally, a computer that processes the data (Cremers and Radziemski, 2006).

A low energy pulsed laser is used in LIBS, typically of the order of 10–100 mJ/pulse; it is focused into the bulk or on the surface to generate a microplasma. That microplasma produces vaporization of small amounts of sample and excitation of species present in the material. A portion of that plasma light is emitted by excited atoms and ions. The light is then collected and dispersed by a spectrometer. The detector records the signal of the emitted species, which is documented in the form of a spectrum of intensity versus wavelength. The generated spectrum can be used as a fingerprint of the chemical composition of the emitting species (Cremers and Radziemski, 2006). Figure 6.9 shows an example of a typical LIBS spectrum taken with a UV laser of a float glass sample with concentrations of Sr of 89 ppm and Ba of 32 ppm (insets). Note the excellent signal-to-noise ratios for both Sr and Ba.

The generated microplasma is temporal and typically last a few microseconds. The spectrum changes as the plasma evolves and, therefore, the time for collection and

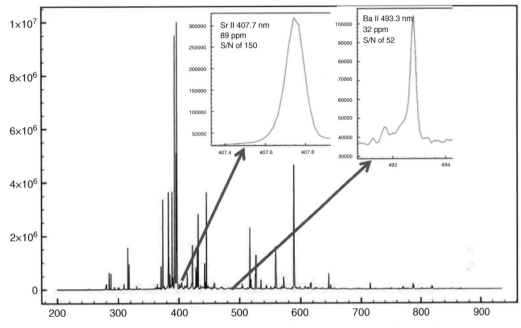

Figure 6.9 LIBS spectrum of NIST 1831 float glass with 89 ppm Sr and 32 ppm Ba peaks shown. [Experimental set-up: 266 nm Q-switched Nd:YAG NewWave laser (27 mJ/5 ns), Andor Mechelle 5000 spectrometer with an ICCD detector, laser pulse frequency of 0.67 hz. Experimental parameters: single pulse ablation at 27 mJ in atmospheric pressure using an Ar gas bath (900 mL/min), delay after laser pulse of 1.50 μs, acquisition time 12 μs, 100 pulses, using the latter 50 shots for accumulation and analysis.]

detection of the species of interest is a fundamental factor that has to be optimized for each particular matrix (Russo, 1998). In a typical 5–10 ns pulse laser, the ionization of ions and atoms is initially very high. After a few hundred nanoseconds, a recombination of electrons and ions occurs and neutral moieties and molecules form. During the plasma lifetime there will also be a continuum background formed by photons emitted by electrons accelerated or decelerated by collisions. The continuum decays more quickly than the analytical spectral lines (Cremers and Radziemski, 2006) and a signal is usually observed after 1 μs (1000 ns).

As a result, the detector is typically gated to look at the signal once the continuum has decreased while the signals of interest become more pronounced. The time between the initiation of the laser pulse and the opening of the detector window is called the gate delay. The detector will collect data for a specific time, usually 1–10 μs. This is referred to as the gate width (Cremers and Radziemski, 2006).

There are different ways of improving the LIBS signal, all of which are well documented in the literature, including spectral resolution, gating, average of signals from many microplasmas, the use of double pulse lasers (either collinear or orthogonal), and the ambient gas (Horn *et al.*, 2001; Gonzalez *et al.*, 2002; Mokgalaka and Gardea-Torresdey, 2006; Wen *et al.*, 2007; Evans *et al.*, 2010; Gornushkin and Panne, 2010; Koch and Gunther, 2011). As a result, the appropriate selection of laser parameters is critical for getting good sensitivity and precision by LIBS.

6.5 Interpretation of glass evidence examinations and comparisons

6.5.1 Defining the match criteria

Several methods for data analysis are used in many areas of forensic science to assist in the interpretation of evidence. Some examinations for the identification and comparison of glass may generate discrete or qualitative data, such as color or appearance. By contrast, examinations such as elemental analysis and RI generate quantitative or semi-quantitative data that permit the application of statistical tools for a better characterization of the evidence, to measure associations between variables, to calculate confidence intervals, estimate systematic or random errors, to assign discrimination values and to present the data in a more understandable manner (Almirall, 2001a; Trejos *et al.*, 2013b).

There are many statistical software packages – SYSTAT, Excel, .JMP, R and Minitab, to mention some – that greatly simplify the statistical analysis of data.

6.5.2 Descriptive statistics

For data reduction, it is very useful to describe the sample sets by the arithmetic mean, the standard deviation and relative standard deviation. The arithmetic mean for a set of data can be estimated using the equation (Miller and Miller, 2000):

$$\overline{X} = \frac{\sum\limits_{i=1}^{n} X_i}{n} \tag{6.1}$$

The standard deviation for the entire set or population is calculated as:

$$s = \sqrt{\frac{1}{n}\sum\left(X_i - \overline{X}\right)^2} \tag{6.2}$$

For practical reasons a small representative sample is used instead of the entire population and the standard deviation is approximated using the equation:

$$s = \sqrt{\frac{1}{n-1}\sum\left(X_i - \overline{X}\right)^2} \tag{6.3}$$

The variance is another important value that measures the dispersion of the data about the mean. It is defined as the squared of the standard deviation:

$$s^2 = \frac{1}{n-1}\sum\left(X_i - \bar{X}\right)^2 \qquad (6.4)$$

The relative standard deviation and the percentage of relative standard deviation is a value commonly employed to estimate the precision of a set of measurements and is calculated using the standard deviation and mean values. The only difference between RSD and %RSD is that the latter is multiplied by 100:

$$\%RSD = \frac{s}{\bar{X}} \times 100 \qquad (6.5)$$

These descriptive statistics are commonly used in forensic examinations as part of the quality control protocol to verify the performance of standards and the reproducibility of the measurements. Precision, bias, limits of detection and limits of quantitation are typical figures of merit that should be documented in forensic examinations.

6.5.3 Match criteria for refractive index measurements

Garvin and Koons (2011) reported the type I error rates (how often one would incorrectly exclude samples of originating from the same source) when using a variety of match criteria for RI data. These authors compared different intervals including fixed criteria (+/– 0.00020 and 0.00010, +/– 1 and 2 standard deviations (SDs) of the mean of the controls, a t-value times the same SD to create an interval (not the t-test) and a test based on a range defined by the maximum and minimum values observed from known (K) values. The first conclusion reached from this study was that the type I error rate is greatly improved when three measurements of the questioned (Q) glass are taken, in comparison to one measurement. A second conclusion was that the type I error decreases considerably for all the above tests when 10 measurements of the known (K) glass are taken. The t-value multiple of the SD produced the lowest type I error rate (Garvin and Koons, 2011). One of the advantages of a true t-test is that the user selects the type I error rate for the test, normally 5% (p-value of 0.05). The Garvin study did not address the type II error rate (the frequency of incorrectly associating samples that did not originate from the same source) because this study did not include sufficient different samples of known source to conduct the necessary pairwise comparisons. While a p-value of 0.01 would serve to lower the type I error rate to approximately 1%, this approach potentially increases the type II error rate to an unacceptable level. The study served to demonstrate the importance of taking at least three measurements of the Q samples and, perhaps, at least 10 measurements of the K samples for a RI comparison, in order to avoid very high type I error rates.

For multivariate comparison purposes, the match criteria should be selected based on the analytical method used to take the measurements. For instance, the ASTM method for μ-XRF analysis (ASTM, 2013a) recommends the use of spectral overlay followed by range overlap or ±3s for the ratios used in the multivariate comparisons, while a wider match criteria is recommended when more sensitive and reproducible methods are used for comparisons, such as LA-ICP-MS. In LA-ICP-MS, a criterion of ±4s (with minimum of 3% RSD) for each element is recommended as the match criteria (Weis *et al.*, 2011; ASTM, 2013b; Trejos *et al.*, 2013b).

Each of these multivariate comparison methods is described in more detail in the following sections.

6.5.3.1 Range overlap

Individual comparisons of a pair of glass samples can be performed using different approaches. A simple match criteria criterion would determine if the overall range of the control and recovered samples overlap. If so, the samples cannot be distinguished. For example, for μ-XRF data of glass, a semi-quantitative measurement of the area under each element peak can be obtained by integration in the software. Ratios of elements with similar energy are then calculated for the known and the questioned glass replicates. Because standard deviations are not calculated, this statistical measure does not directly address the confidence level of an association. If the ranges of one or more elements in the questioned and known specimens do not overlap, it may be concluded that the specimens are not from the same source.

6.5.3.2 Match interval

Another method consists of calculating mean and standard deviations of both control and recovered samples; if the mean value of the recovered sample is within the mean value of the control ±3 or ±4 standard deviations of the control replicates, they are considered to match or originate from a common source. If the mean concentration of one (or more) element(s) in the recovered fragment falls outside the match interval for the corresponding element in the known fragments, the element(s) does not "match" and the glass samples are considered distinguishable. Because LA-ICP-MS can generate reproducibility between measurements lower than 3% RSDs, a minimum of 3–5% RSD is recommended in such circumstances to minimize false exclusions (Weis *et al.*, 2011; Trejos *et al.*, 2013b).

6.5.3.3 Analysis of variance

When more than two samples (or two means) need to be compared, a pairwise comparison (multiple pairs) should be carried out. A method used for multiple comparisons is the analysis of variance (ANOVA) (Miller and Miller, 2000). The results of an ANOVA, however, only indicate whether these multiple means differ significantly without identifying which of the means are significantly different. Tukey's *post hoc* test is useful to determine which pairs of means differ significantly. This test is very useful for estimating the discrimination power of a technique, where sets of a large number of pair comparisons are required. Tukey's test defines confidence values based on the mean square error (MSE) within "k" groups of "n" replicates. A further explanation of this statistical tool can be encountered in the literature (Kleinbaum and Kupper, 1978; Duckworth *et al.*, 2000).

6.5.3.4 Multivariate analysis

There are some multivariate analyses tools, such as principal component analysis (PCA) and cluster analysis, that can be used for data reduction and visualization of the multivariate space. PCA is a method of identifying patterns in data, and expressing the data in such a way that could serve to highlight their similarities and differences. The principal objective of a PCA is to reduce the amount of data when there is correlation between variables (Miller and Miller, 2000). Cluster analysis fits together data into classes according to their similarities. Therefore, data with similar characteristics will cluster into groups that are closer to data that exhibit very different characteristics. One of the advantages of these types of multivariate analyses is that it allows for a graphical display of the groupings of data that can be easily understood by a jury. Nonetheless, these methods are used to describe data and patterns rather than the more strict "match" and "no-match" typically employed in the two-stage frequentist approach to forensic comparisons as described in the ASTM methods for μ-XRF, digestion ICP-MS and LA-ICP-MS analysis.

6.5.3.5 Bayesian approach

Several authors have reported the use of the Bayesian approaches to assist the interpretation of evidence (Evett and Buckleton, 1990; Walsh *et al.*, 1996; Curran *et al.*, 2000; Taroni *et al.*, 2004; Hicks *et al.*, 2005; Zadora, 2007, 2009; Ramos and Zadora, 2011). The Bayesian approach consists of assessing the LR, which considers the probability of the evidence, with respect to at least two possible explanations or hypotheses. A simple example of some explanations might be "The suspect was present at the crime scene when the window was broken" and "The suspect was not present at the crime scene when the window was broken". The evaluation of these probabilities takes into account information such as the strength of the match between the crime scene source and fragments recovered from the suspect, and the relative rarity (or frequency) of glass with similar characteristics in the population. In addition, information related to the event, such as probability of transfer fragments and persistence, and the presence of non-matching sources of glass, may be incorporated into the calculations. The point of this approach is that it explicitly accounts for, and includes, explanations other than commission of the crime for the matching glass. This method is particularly popular within the forensic community in Europe and Australasia. It represents an alternative and/or complementarty approach to conventional methods to incorporate details of transfer and persistence in the assessment of the significance of an association or an exclusion.

6.5.4 Informing power of analytical methods, forming the opinion

Advanced manufacturing processes and stringent quality control in glassmaking has led to less variability of RI in glasses produced within a single plant or even between different plants. As a result, the evidential value of RI measurements has been limited and glass examiners have found it necessary to complement physical and optical measurements with elemental analysis.

An effective way of estimating the discrimination potential of elemental analysis is using ANOVA together with Tukey's tests in order to determine the occurrence of glass samples that have originated from different sources having the same elemental profiles. Duckworth *et al.* (2000, 2002) investigated the overall discriminating power of ICP-MS and the individual discriminating potential of elements with the purpose of using them for the development of glass databases.

An interesting survey of the variation of RI and elemental composition in a single plant for a period of 4.5 years, as well as variations between 36 different manufacturing plants within the United States, showed that the variation of RI is very limited while elemental analysis can distinguish small differences in glass composition, even from glass originating from the same plant over short periods of time (Montero, 2002).

Another study showed that for a set of 45 headlamps originating from different sources, RI was able to discriminate only 90% of the headlamps, while the combination of RI with elemental analysis enhanced those values to 100% (Trejos *et al.*, 2003). The literature referring to survey studies supports the hypothesis that the elemental profile of glass offers a very discriminating tool for comparison of glass fragments (Koons *et al.*, 1991, 2002; Buscaglia, 1994; Parouchais *et al.*, 1996; Montero *et al.*, 2003; Weis *et al.*, 2011). The conclusions from these studies are in agreement that elemental analysis, when sensitive methods are used, can differentiate glass samples that originate from different plants and also glass samples that originate from the same plant but that are manufactured at different times. The corollary to this conclusion supports the hypothesis that when no differences are found between elemental profiles between two samples, there is very strong support for the fact that the two samples originated from the same manufacturing source.

In 2009, the National Institute of Justice (US Department of Justice) sponsored the creation of the Elemental Analysis Working Group to coordinate an effort of a group of forensic glass examiners to evaluate the relative performance of elemental analysis methods used for the comparison of elemental profiles of glass and to generate standard methods of analysis. The group consisted of 34 members representing 23 different local, state, federal, private and academic laboratories from the USA, Canada, Mexico, Germany and the Netherlands. The group performed a series of interlaboratory studies to evaluate the analytical performance and informing power of current methods in use for elemental analysis of glass. SEM-EDS was excluded from the study, since its sensitivity does not allow for the detection of trace elements that are essential for forensic comparisons of glass. These interlaboratory studies allowed for a direct comparison between four of the most sensitive methods currently available for the forensic elemental analysis of glass samples (LA-ICP-MS, solution ICP-MS, LIBS and µ-XRF). The methods were compared in terms of analytical performance and discrimination capability.

ICP-based methods (ICP-MS and LA-ICP-MS) are the most sensitive, with sub-ppm detection limits in the solid material. Advantages of these methods are that they are fairly standardized among participant laboratories, they are currently used in forensic laboratories and they have been accepted in court. A standardized ASTM method already exists for the digestion and analysis by ICP-MS (ASTM, 2012) and for LA-ICP-MS (ASTM, 2013b). Both methods are fairly mature, with several publications previously reporting the evaluation of their capabilities and limitations. In addition, laser ablation sampling has unique advantages over digestion-based methods, such as reducing the sample consumption from milligrams to just a few hundred nanograms, reducing the time for analysis and eliminating the use of hazardous digestion reagents. Interlaboratory comparisons of glass reference standard materials demonstrated that ICP-methods provide accurate and precise quantitative data with deviations lower than 10% for nearly all elements measured in the studies.

Micro-XRF (µ-XRF) methods provided consistent data among participants after normalization with a reference standard material such as SRM NIST 1831. The EAWG used the experience gained from these interlaboratory tests to draft an ASTM method (ASTM, 2013a). Limits of detection are 2–3 orders of magnitude higher (poorer) in comparison to ICP-based methods; therefore, the number of trace elements typically detected in glass samples is more limited. Nevertheless, good performance was also observed among µ-XRF laboratories. The measurement of limit of detection (LOD) provided a better understanding of the capabilities of the technique and permitted a means of quantitatively comparing the performance of different instrument configurations.

The interlaboratory studies also revealed that a wide variety of match criteria are currently employed by forensic laboratories to conduct statistical comparisons of elemental composition data. Extensive discussions between the group members led to the design of additional interlaboratory tests to address the interpretation of evidence and the systematic selection of match criteria for elemental comparisons of glasses, based on simultaneously minimizing the frequency of both false exclusions (type I) and false inclusions (type II) errors.

Based on results obtained in the interlaboratory tests, it was concluded that the match criteria for comparisons of elemental composition of glass fragments should be carefully selected based on the technique used for analysis as well as the number of replicates that are conducted to characterize the variability of the known and questioned samples as previously explained.

In terms of interpretation of elemental comparisons of glass, it can be concluded from the study that glass samples that are manufactured in different plants, or even at the same plant years apart, are clearly differentiated by elemental composition when μ-XRF or ICP-based methods are used for analysis. Samples produced in the same plant over time intervals of weeks to months may also be differentiated. This level of differentiation can be used to add significance to an association, when one is found, and to assist in assigning recovered fragments to a source when selecting among several potential sources.

6.5.5 Report writing and testimony

The job of a glass examiner does not end with the completion of the analysis of evidence exhibits. Report writing and expert testimony remain a challenge for forensic examiners. It is essential to develop special skills to be able to combine the amount of data generated during the analysis and the examiner's expertise into brief and understandable statements that reflect the nature of the evidence without overstating or understating the value of the evidence. Forensic scientists should apply their expert opinion to assess the correct weight to their conclusions. In order to do that, they have to be aware of the advantages and limitations of the analytical techniques used during the analysis of the submitted samples, as well as their discrimination potential. Quality control policies at the forensic laboratory must be followed properly to support the findings and the expert opinion must be based on the facts of the case; it is strongly recommended that objective match criteria are used in order to reach the correct conclusion. Finally, it is important to focus the opinion on the physical evidence (known and questioned samples) rather than on a suspect or victim.

To date, there are no standard guidelines for the reporting of conclusions for glass examinations. However, as a response to the NRC report (2009), the OSAC committees are currently working towards the standardization of reporting language for glass examinations. One example interpretation scale that is being promoted in Germany and in some states in the United States incorporates different levels of association, ranging from elimination of the unknown all the way to identification. This language scale is shown in Figure 6.10. On the left of the scale is the typical language that can describe the levels and on the right of the scale is the equivalent LR (proposed by the authors of this chapter) as an example of how the frequentist approach to evidence interpretation can be related to a LR approach to evidence interpretation. The language highlighted in blue is offered as the language used in both a non-matching elemental profile determination "Very strong evidence of no association" and also the matching elemental profile determination (Type 2 Association: Very strong evidence). The corresponding LR is provided as a guide to start to think about the significance of this evidence in comparison to other types of evidence such as matching fingerprint or matching DNA profiles. The LR approach provides a continuous scale to consider and describe the relative significance using a number rather than the language currently being proposed by several operational laboratories. A justification for the LR values given stems from database searches of over 1000 glass samples from different sources, as reported in the surveys listed in the previous section.

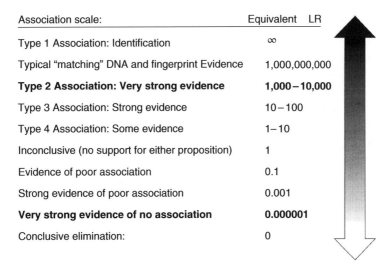

Association scale:	Equivalent LR
Type 1 Association: Identification	∞
Typical "matching" DNA and fingerprint Evidence	1,000,000,000
Type 2 Association: Very strong evidence	**1,000 – 10,000**
Type 3 Association: Strong evidence	10 – 100
Type 4 Association: Some evidence	1 – 10
Inconclusive (no support for either proposition)	1
Evidence of poor association	0.1
Strong evidence of poor association	0.001
Very strong evidence of no association	**0.000001**
Conclusive elimination:	0

Figure 6.10 A proposed interpretation scale using language that communicates the value of an association using elemental analysis comparisons of glass (left) and using the corresponding likelihood ratio (right). The highlighted text (in bold) and LR values (also in bold) correspond to a "non-match" for elemental analysis as well as the "match" for an elemental profile, when sensitive analytical methods are used for the comparison using the ASTM methods described

6.6 Case examples

The case examples presented here are provided to exemplify the utility of glass as forensic evidence under different circumstance. Some specific details about the cases are not provided to protect anonymity of the suspect or victim.

6.6.1 Case 1: Hit-and-run case

Case scenario

Early in the morning, a woman was hit by a vehicle. She was crossing the street while coming home from the supermarket. The victim impacted the windshield and died at the scene; the driver left the scene of the accident. Witnesses described the vehicle involved in the hit-and-run as a black BMW. Investigators conducted a search and found a black BMW vehicle abandoned in a parking lot not far from the accident with severe damage to the windshield. The investigators obtained a search warrant and visited the vehicle owner's house. Upon arrival to the suspect's home, the owner of the vehicle claimed that his vehicle must have been stolen the previous night. The investigators executed a search warrant and found several glass fragments in the living room of the house, the shower, bathroom sink and on some clothing found in the bedroom (pants and shirt). These recovered fragments were submitted for analysis to the forensic laboratory. Investigators also collected control glass (known fragments) from the windshield and labeled them separately as inner and outer pane of the windshield. The recovered and control samples were analyzed and compared by physical examination, RI and LA-ICP-MS.

Summary of forensic analysis results and the opinion
Microscopic and instrumental examination of the recovered fragments submitted for analysis revealed that these were indistinguishable from the glass fragments submitted as control samples with respect to their physical, optical and chemical characteristics. In this case, the expert can indicate in their opinion that this association was "very strong evidence" that the recovered glass (from the shower, the living room, the shirt and the pants) originated from the same source as the broken windshield that was submitted as the control, given the following:
- There was a large amount of fragments recovered from four different objects associated with the suspect (clothing, shower, sink, and living room).
- Some fragments had full thickness consistent with non-tempered flat glass.
- The samples were analyzed in several replicates with very sensitive methods (LA-ICP-MS).
- There were enough glass fragments from the control standard to characterize the physical and chemical properties.

Moreover, given the existence of the scientific background information about the variability and discrimination of the elemental composition for flat glass when analyzed by the methods used here, the expert can include in their opinion that "there is very strong support for the hypothesis that the recovered glass originated from the windshield submitted as a standard or another broken windshield manufactured in the same manufacturing plant and at approximately the same time period". The glass evidence found in the suspect's house is very strong evidence of association of someone residing in the house being present during the hit-and-run accident.

6.6.2 Case 2: Multiple transfer of glass in breaking-and-entry case
Case scenario
Fifteen vehicle windows were broken on the same day at an airport parking lot in an effort to gain entry into the vehicles. The crime scene investigators collected fragments of glass from the windows of all fifteen vehicles and sent them as known (control) samples for analysis. Some time later, a suspect wearing overalls with several pockets was questioned by the police related to the event. A total of 42 glass fragments were recovered from the suspect's overall and from the police vehicle seat where the suspect traveled (Montero *et al.*, 2003).

Summary of forensic analysis results and the opinion
Recovered and control samples were analyzed by physical examination, density, RI and solution digestion ICP-MS. Following the analytical protocol, all the known samples (originating from 15 different vehicles) were differentiated from each other.

Moreover, seven of those known samples were associated with several (twelve) of the questioned glass fragments.

Given the fact that a sensitive elemental analytical method (ICP-MS) was used for the associations of the twelve questioned glass with a particular known sample, the examiner may render an opinion that the recovered glass originated from the windows submitted as standards or from another broken window manufactured in the same manufacturing plant and at approximately the same time period. The odds of a person not involved with the breaking activity to have such a large number of fragments on their clothing and to have twelve fragments matching the distinctive physical and chemical properties of seven separate vehicles are extremely low.

This case represents a good example where the evidence supports the prosecution proposition (the suspect was involved in multiple breaking activities) rather than the defense proposition (the suspect was not involved in the breaking activities).

In this particular case, existing scientific research published on transfer and persistence provides solid support for the significance of finding large amounts of glass on an individual's clothing and the extreme rarity of finding so many "matching" fragments from multiple broken sources on a person not involved in the breaking activities.

6.7 Conclusions

It has been demonstrated that glass fragments generated and collected at crime scenes can provide very useful scientific evidence to assist in the resolution of a case. There are numerous analytical techniques that can be used to conduct glass analysis. Refractive index, using automated systems such as the GRIM series and the ASTM 1967 standard, is the screening tool of choice. For samples found to be indistinguishable by RI, it is recommended that elemental analysis is the next step in the forensic examination.

The literature supports the proposition that μ-XRF or ICP methods are excellent tools for the discrimination of glass samples originating from different manufacturing sources. Of these ICP methods, LA-ICP-MS provides a faster and less destructive alternative that offers good accuracy, good precision, and good discrimination power, making this technique one of the most complete tools that a forensic glass examiner may have. LIBS is an emerging and promising alternative for forensic examination of glass and the authors believe it will become one of the methods of choice for elemental analysis of glass in the near future, but no standard methods currently exist.

Good sampling techniques, proper preservation of the evidence and the correct interpretation of the results is of great importance throughout the evaluation of glass evidence. Consequently, glass examiners must always consider the overall picture of each individual case in order to reach the right conclusions and communicate this conclusion effectively.

References

Allen, T.J.; Scranage, J.K. (1998) The transfer of glass – part 1: Transfer of glass to individuals at different distances. *Forensic Science International*, **93**, 167–174.

Allen, T.J.; Hoefler, K.; Rose, S.J. (1998a) The transfer of glass – part 2: A study of the transfer of glass to a person by various methods. *Forensic Science International*, **93**, 175–193.

Allen, T.J.; Locke, J.; Scranage, J.K. (1998b) Breaking of flat glass – part 4: Size and distribution of fragments from vehicle windscreens. *Forensic Science International*, **93**, 209–218.

Allen, T.J.; Cox, A.R.; Barton, S.; *et al.* (1998c) The transfer of glass – part 3: The transfer of glass fragments from the surface of an item to the person carrying it. *Forensic Science International*, **93**, 201–208.

Almirall, J.R. (2001a) Elemental analysis of glass fragments. In *Forensic Examination of Glass and Paint* (ed. Caddy, B.), Taylor & Francis, London, pp. 65–80.

Almirall, J.R. (2001b) Manslaughter caused by a hit-and-run: glass as evidence of association. In *Mute Witnesses: Trace Evidence Analysis* (ed. Houck, M. M.), Academic Press, San Diego, CA, pp. 139–155.

Almirall, J.R.; Trejos, T. (2006) Advances in forensic analysis of glass fragments with a focus on refractive index and elemental analysis. *Forensic Science Review*, **18** (2), 74–96.

Almirall, J.R.; Trejos, T. (2010) Forensic applications of mass spectrometry. In *Encyclopedia of Mass Spectrometry*, 1st edn, Vol. **5** (eds Beauchemin, D.; Matthews, D.) Elsevier, pp. 705–717.

Almirall, J.R.; Trejos, T.; Hobbs, A.; Furton, K.G. (2003) Trace elemental analysis of glass and paint samples of forensic interest by ICP-MS using laser ablation solid sample introduction. In *Sensors, and Command, Control, Communications, and Intelligence (C31) Technologies for Homeland Defense and Law Enforcement II*, Proceedings of the International Society for Optical Engineering (SPIE) (ed. Carapezza, E.M.), Orlando, FL, 21 Apr 2003, **5071**, pp. 193–196.

Andrasko, J.; Maehly, A.C. (1978) The discrimination between samples of window glass by combining physical and chemical techniques. *Journal of Forensic Sciences*, **23**, 250–262.

ASTM (2010) Standard C162-05, Standard Terminology of Glass and Glass Products. In *Annual Book of ASTM Standards*, **Vol. 15.02**, ASTM International, West Conshohocken, PA.

ASTM (2011) Standard E1967-11a, Standard Test Method for the Automated Determination of Refractive Index of Glass Samples Using the Oil Immersion Method and a Phase Contrast Microscope. In *Annual Book of ASTM Standards*, **Vol. 14.02**, ASTM International, West Conshohocken, PA.

ASTM (2012) Standard E2330-12, Standard Test Method for Determination of Trace Elements in Glass Samples Using Inductively Coupled Plasma Mass Spectrometry (ICP-MS). In *Annual Book of ASTM Standards*, **Vol. 14.02**, ASTM International, West Conshohocken, PA.

ASTM (2013a) Standard E2926-13, Standard Test Methods for Forensic Comparison of Glass Using Micro X-ray Fluorescence (μ-XRF) Spectrometry. In *Annual Book of ASTM Standards*, **Vol. 14.02**, ASTM International, West Conshohocken, PA.

ASTM (2013b) Standard E2927-13, Standard Test Method for Determination of Trace Elements in Soda-Lime Glass Samples Using Laser Ablation Inductively Coupled Plasma Mass Spectrometry for Forensic Comparisons. In *Annual Book of ASTM Standards*, **Vol. 14.02**, ASTM International, West Conshohocken, PA.

Bajic, S.J.; Aeschliman, D.B.; Saetveit, N.J.; *et al.* (2005) Analysis of glass fragments by laser ablation-inductively coupled plasma-mass spectrometry and principal component analysis. *Journal of Forensic Sciences*, **50**, 1123–1127.

Becker, S.; Gunaratnam, L.; Hicks, T.; *et al.* (2001) The differentiation of float glass using refractive index and elemental analysis: Comparisons of techniques. *Problems of Forensic Sciences*, **47**, 80–92.

Becker, S.; Watzke, P.; Dücking, M.; Stoecklein, W. (2003) Laser ablation in forensic glass analysis: The use of matrix matched standards for quantitative float glass analysis. *Forensic Science International*, **136**, 361.

Berends-Montero, S.; Wiarda, W.; de Joode, P.; van der Peijl, G. (2006) Forensic analysis of float glass using laser ablation inductively coupled plasma mass spectrometry (LA-ICP-MS): validation of a method. *Journal of Analytical Atomic Spectrometry*, **21**, 1185–1193.

Bridge, C.M.; Powell, J.; Steele, K.L.; *et al.* (2006) Characterization of automobile float glass with laser-induced breakdown spectroscopy and laser ablation inductively coupled plasma mass spectrometry. *Applied Spectroscopy*, **60**, 1181–1187.

Bridge, C.M.; Powell, J.; Steele, K.L.; Sigman, M.E. (2007) Forensic comparative glass analysis by laser-induced breakdown spectroscopy. *Spectrochimica Acta, Part B: Atomic Spectroscopy*, **62**, 1419–1425.

Brożek-Mucha, Z.; Zadora, G. (1998) Differentiating between various types of glass using SEM-EDX elemental analysis: A preliminary study. *Problems of Forensic Sciences*, **37**, 68–89.

Buscaglia, J. (1994) Elemental analysis of small glass fragments in forensic science. *Analytica Chimica Acta*, **288**, 17–24.

Cahoon, E.M.; Almirall, J.R. (2010) Wavelength dependence on the forensic analysis of glass by nanosecond 266 nm and 1064 nm laser induced breakdown spectroscopy. *Applied Optics*, **49**, C49–C57.

Catterick, T.; Wall, C.D. (1978) Rapid analysis of small glass fragments by atomic-absorption spectroscopy. *Talanta*, **25**, 573–577.

Clawson, M. (2009) Trends in Automotive Glass. Proceedings of the Trace Evidence Symposium, 3–7 August 2009. http://projects.nfstc.org/trace/2009/presentations/4-clawson-glass.pdf; last accessed 18 June 2015.

Coleman, R.F.; Goode, G.C. (1973) Comparison of glass fragments by neutron activation analysis. *Journal of Radioanalytical Chemistry*, **15**, 367–372.

Coleman, R.F.; Weston, N.T. (1968) A case concerning neutron activation analysis of glass. *Journal of Forensic Sciences Soc.*, **8**, 32–33.

Cook, R.; Evett, I.W.; Jackson, G.; *et al.* (1998) A model for case assessment and interpretation. *Science and Justice*, **38**, 151–156.

Cooper, G. (2013) The indirect transfer of glass fragments to a jacket and their subsequent persistence. *Science and Justice*, **53**, 166–170.

Copley, G. J. (2001) The comparison and manufacture of glass and its domestic and industrial applications. In *Forensic Examination of Glass and Paint* (ed. Caddy, B.), Taylor & Francis, London, pp. 27–46.

Coulson, S.A.; Buckleton, J.S.; Gummer, A.B.; Triggs, C.M. (2001) Glass on clothing and shoes of members of the general population and people suspected of breaking crimes. *Science and Justice*, **41**, 39–48.

Cremers, S.A.; Radziemski, L.J. (2006) *Handbook of Laser-Induced Breakdown Spectroscopy*. John Wiley & Sons Ltd, Chichester, UK.

Curran, J.M.; Triggs, C.M.; Almirall, J.R.; *et al.* (1997) The interpretation of elemental composition measurements from forensic glass evidence: I. *Science and Justice*, **37**, 241–244.

Curran, J.M.; Triggs, C.M.; Buckleton, J.S.; *et al.* (1998) Assessing transfer probabilities in a Bayesian interpretation of forensic glass evidence. *Science and Justice*, **38**, 15–21.

Curran, J.M.; Hicks, T.N.; Buckleton, J.S. (2000) *Forensic Interpretation of Glass Evidence*. CRC Press, Boca Raton, FL.

Daéid, N.N. (2001) Statistical interpretation of glass evidence. In *Forensic Examination of Glass and Paint* (ed. Caddy, B.), Taylor & Francis, London, pp. 85–94.

Daéid, N.N. (2010) Review paper, 87–107, 16th International Forensic Science Symposium, Interpol-Lyon, 5–8 October 2010.

Daéid, N.N.; McColl, D.; Ballany, J. (2009) The level of random background glass recovered from fleece jackets of individuals who worked in Law enforcement or related professions. *Forensic Science International*, **191**, 19–23.

Duckworth, D.C.; Baynes, C.K.; Morton, S.J.; Almirall, J.R. (2000) Analysis of variance in forensic glass analysis by ICP-MS: Variance within the method. *Journal of Analytical Atomic Spectrometry*, **15**, 821–828.

Duckworth, D.C.; Morton, S.J.; Baynes, C.K.; *et al.* (2002) Forensic glass analysis by ICP-MS: A multi-element assessment of discriminating power via analysis of variance and pairwise comparisons. *Journal of Analytical Atomic Spectrometry*, **17**, 662–668.

Dudley, R.J.; Howden, C.R.; Taylor, T.J.; Smalldon, K.W. (1980) The discrimination and classification of small fragments of windowand non-window glasses using energy-dispersive x-ray fluorescence spectrometry. *X-Ray Spectrometry*, **9**, 119–122.

Ernst, T.; Berman, T.; Buscaglia, J.; *et al.* (2012) Signal-to-noise ratios in forensic glass analysis by micro X-ray fluorescence spectrometry. *X-Ray Spectrometry*, **43**, 13–21.

Evans, E.H.; Day, J.A.; Palmer, C.; Smith, C.M.M. (2010) Advances in atomic spectrometry and related techniques. *Journal of Analytical Atomic Spectrometry*, **25**, 760–784.

Evett, I.W.; Buckleton, J. (1990) The interpretation of glass evidence: A practical approach. *Journal of the Forensic Science Society*, **30**, 215–223.

Garvin, E.J.; Koons, R.D. (2011) Evaluation of match criteria used for the comparison of refractive index of glass fragments. *Journal of Forensic Sciences*, **56**, 491–500.

Gonzalez, J.; Mao, X.L.; Roy, J.; *et al.* (2002) Comparison of 193, 213 and 266 nm laser ablation ICP-MS. *Journal of Analytical Atomic Spectrometry*, **17**, 1108–1113.

Gornushkin, I.B.; Panne, U. (2010) Radiative models of laser-induced plasma and pump-probe diagnostics relevant to laser-induced breakdown spectroscopy. *Spectrochimica Acta, Part B: Atomic Spectroscopy*, **65**, 345–359.

Haag, L.C. (2012) Behavior of expelled glass fragments during projectile penetration and perforation of glass. *American Journal of Forensic Medicine & Pathology*, **33**, 47–53.

Hamer, P.S. (2001) Microscopic Techniques for Glass Examination. In *Forensic Examination of Glass and Paint* (ed. Caddy, B.), Taylor & Francis, London, pp. 47–63.

Hicks, T.; Vanina, R.; Margot, P. (1996) Transfer and persistence of glass fragments on garments. *Science and Justice*, **36**, 101–107.

Hicks, T.; Monard Sermier, F.; Goldmann, T.; *et al.* (2003) The classification and discrimination of glass fragments using non destructive energy dispersive X-ray μ fluorescence. *Forensic Science International*, **137**, 107–118.

Hicks, T.; Schütz, F.; Curran, J.M.; Triggs, C.M. (2005) A model for estimating the number of glass fragments transferred when breaking a pane: Experiments with firearms and hammer. *Science and Justice*, **45**, 65–74.

Horn, I.; Guillong, M.; Günther, D. (2001) Wavelength dependant ablation rates for metals and silicate glasses using homogenized laser beam profiles – implications for LA-ICP-MS. *Applied Surface Science*, **182**, 91–102.

Howden, C.R.; Dudley, R.J.; Smalldon, K.W. (1978) The analysis of small glass fragments using energy dispersive X-ray fluorescence spectrometry. *Journal of the Forensic Science Society*, **18**, 99–112.

Hughes, J.C.; Catterick, T.; Southeard, G. (1976) The quantitative analysis of glass by atomic absorption spectroscopy. *Forensic Science*, **8**, 217–227.

Kleinbaum, D.G.; Kupper, L.L. (1978) *Applied Regression Analysis and Other Multivariable Methods*. Duxbury Press, Belmont, CA.

Koch, J.; Günther, D. (2011) Review of the state-of-the-art of laser ablation inductively coupled plasma mass spectrometry. *Applied Spectroscopy*, **65**, 155–162.

Koons, R.D.; Buscaglia, J. (2001) Distribution of refractive index values in sheet glasses. *Forensic Science Communications*, **3** (1).

Koons, R.D.; Fiedler, C.; Rawalt, R.C. (1988) Classification and discrimination of sheet and container glasses by inductively coupled plasma-atomic emission spectometry and pattern recognition. *Journal of Forensic Sciences*, **33**, 49–67.

Koons, R.D.; Peters, C.A.; Rebbert, P.S. (1991) Comparison of refractive index, energy dispersive X-ray fluorescence and inductively coupled plasma atomic emission spectrometry for forensic characterization of sheet glass fragments. *Journal of Analytical Atomic Spectrometry*, **6**, 451–456.

Koons, R.D.; Buscaglia, J.; Bottrell, M.; Miller, E.T. (2002) Forensic glass comparisons. In *Forensic Science Handbook*, 2nd edn (ed. Saferstein, R.), Prentice Hall, Upper Saddle River, NJ, pp. 161–214.

Krüsemann, H. (2001) SEMs and Forensic Science. *Problems of Forensic Sciences*, **47**, 110–121.

Kuisma-Kursula, P. (2000) Accuracy, precision and detection limits of SEM-WDS, SEM-EDS and PIXE in the multi-elemental analysis of medieval glass. *X-Ray Spectrometry*, **29**, 111–118.

Lambert, J.A.; Satterthwaite, M.J.; Harrison, P.H. (1995) A survey of glass fragments recovered from clothing of persons suspected of involvement in crime. *Science and Justice*, **35**, 273–281.

Latkoczy, C.; Becker, S.; Dücking, M.; *et al.* (2005) Development and evaluation of a standard method for the quantitative determination of elements in float glass samples by LA-ICP-MS. *Journal of Forensic Sciences*, **50**, 1327–1341.

Lau, L.; Beveridge, A.D.; Callowhill, B.C.; *et al.* (1997) The frequency of occurrence of paint and glass on the clothing of high school students. *Canadian Society of Forensic Science Journal*, **30**, 233–240.

Locke, J.; Hayes, C.A. (1984) Refractive index variations across glass objects and the influence of annealing. *Forensic Science International*, **26**, 147–157.

Locke, J.; Scranage, J.K. (1992) Breaking of flat glass – Part 3: Surface particles from windows. *Forensic Science International*, **57**, 73–80.

Locke, J.; Sanger, D.G.; Roopnarine, G. (1982) The identification of toughened glass by annealing. *Forensic Science International*, **20**, 295–301.

Luce, R.J.W.; Buckle, J.L.; McInnis, I. (1991) A study on the backward fragmentation of window glass and the transfer of glass fragments to individual's clothing. *Canadian Society of Forensic Science Journal*, **24**, 79–89.

Marcouiller, J.M. (1990) A revised glass annealing method to distinguish glass types. *Journal of Forensic Sciences*, **35**, 554–559.

McJunkins, S.P.; Thornton, J.I. (1973) Glass fracture analysis. A review. *Forensic Science*, **2**, 1–27.

Mencik, J. (1992) Strength and fracture of glass and ceramics. In *Glass Science and Technology* (ed. Mecholsky, J.J., Jr.), Elsvier, Amsterdam, Vol. **12**, pp. 99–152, 219–229.

Merivale, C. (2008) The Batch Solution Opportunity. The Glass Researcher. *American Ceramics Society Bulletin*, **87** (11), 37–42.

Miller, J.N.; Miller, J.C. (2000) *Statistics and Chemometrics for Analytical Chemistry*, 4th edn. Pearson Education/Prentice Hall, Harrow, U.K.

Moenke-Blankenburg, L.; Schumann, T.; Günther, D.; *et al.* (1992) Quantitative analysis of glass using inductively coupled plasma atomic emission and mass spectrometry, laser micro-analysis inductively coupled plasma atomic emission spectrometry and laser ablation inductively coupled plasma mass spectrometry. *Journal of Analytical Atomic Spectrometry*, **7**, 251–254.

Mokgalaka, N.S.; Gardea-Torresdey, J.L. (2006) Laser ablation inductively coupled plasma mass spectrometry: Principles and applications. *Applied Spectroscopy Review*, **41**, 131–150.

Montero, S. (2002) *Trace Elemental Analysis of Glass by Inductively Coupled Plasma-Mass Spectrometry (ICP-MS) and Laser Ablation-Inductively Coupled Plasma-Mass Spectrometry (LA-ICP-MS).* Florida International University, Miami, Florida,.

Montero, S.; Hobbs, A.L.; French, T.A.; Almirall, J.R. (2003) Elemental analysis of glass fragments by ICP-MS as evidence of association: Analysis of a case. *Journal of Forensic Sciences*, **48**, 1101–1107.

Naes, B.E. (2009) *Elemental Analysis of Glass and Ink by Laser Ablation Inductively Coupled Plasma Mass Spectrometry (LA-ICP-MS) and Laser Induced Breakdown Spectroscopy (LIBS).* Florida International University, Miami, Florida.

Naes, B.E.; Umpierrez, S.; Ryland, S.; *et al.* (2008) A comparison of laser ablation inductively coupled plasma mass spectrometry, micro X-ray fluorescence spectroscopy, and laser induced breakdown spectroscopy for the discrimination of automotive glass. *Spectrochimica Acta, Part B: Atomic Spectroscopy*, **63**, 1145–1150.

Nelson, D.F. and Revell, B.C. (1967) Backward fragmentation from breaking glass. *Journal of the Forensic Science Society*, **7**, 58–61.

Newton, A.W.N. (2011) An investigation into the variability of the refractive index of glass: Part II – The effect of debris contamination. *Forensic Science International*, **204**, 182–185.

Newton, A.W.N.; Buckleton, J.S. (2008) An investigation into the relationship between edge counts and the variability of the refractive index of glass: Part I – Edge morphology. *Forensic Science International*, **177**, 24–31.

Newton, A.W.N.; Curran, J.M.; Triggs, C.M.; Buckleton, J.S. (2004) The consequences of potentially differing distributions of the refractive indices of glass fragments from control and recovered sources. *Forensic Science International*, **140**, 185–193.

Newton, A.W.N.; Kitto, L.; Buckleton, J.S. (2005) A study of the performance and utility of annealing in forensic glass analysis. *Forensic Science International*, **155**, 119–125.

NRC (2009) *Strengthening Forensic Science in the United States: A Path Forward.* Committee on Identifying the Needs of the Forensic Sciences Community, National Research Council. National Academic Press, Washington, DC.

O'Sullivan, S.; Geddes, T.; Lovelock, T.J. (2011) The migration of fragments of glass from the pockets to the surfaces of clothing. *Forensic Science International*, **208**, 149–155.

Parouchais, T.; Warner, I.M.; Palmer, L.T.; Kobus, H. (1996) The analysis of small glass fragments using inductively coupled plasma mass spectrometry. *Journal of Forensic Sciences*, **41**, 351–360.

Petterd, C.I.; Hamshere, J.; Stewart, S.; *et al.* (1999) Glass particles in the clothing of members of the public in south-eastern Australia – a survey. *Forensic Science International*, **103**, 193–198.

Pound, C.A.; Smalldon, K.W. (1977) *The Efficiency of Searching for Glass on Clothing and the Persistence of Glass in Clothing and Shoes.* Forensic Science Service.

Ramos, D.; Zadora, G. (2011) Information-theoretical feature selection using data obtained by scanning electron microscopy coupled with and energy dispersive X-ray spectrometer for classification of glass traces. *Analytica Chimica Acta*, **705**, 207–217.

Reeve, V.; Mathiesen, J.; Fong, W. (1976) Elemental analysis by energy dispersive x-ray: a significant factor in the forensic analysis of glass. *Journal of Forensic Sciences*, **21**, 291–306.

Rodriguez-Celis, E.M.; Gornushkin, I.B.; Heitmann, U.M.; *et al.* (2008) Laser induced breakdown spectroscopy as a tool for discrimination of glass for forensic applications. *Analytical and Bioanalytical Chemistry*, **391**, 1961–1968.

Roedel, T.C.; Bronk, H.; Haschke, M. (2002) Investigation of the influence of particle size on the quantitative analysis of glasses by energy-dispersive micro x-ray fluorescence spectrometry. *X-Ray Spectrometry*, **31**, 16–26.

Roux, C.; Kirk, R.; Benson, S.; *et al.* (2001) Glass particles in footwear of members of the public in south-eastern Australia – a survey. *Forensic Science International*, **116**, 149–156.

Russo, R.E. (1998) Laser ablation. In *Focus on Analytical Spectrometry: A Compendium of Applied Spectroscopy Focal Point Articles (1994–1997)* (eds Holocombe, J. A.; Hieftje, G. M.; Majidi, V.), Society for Applied Spectroscopy, Frederick, MD, pp. 41–55.

Russo, R.E.; Suen, T.W.; Bolshakov, A.A.; *et al.* (2011) Laser plasma spectrochemistry. *Journal of Analytical Atomic Spectrometry*, **26**, 1596–1603.

Ryland, S.G. (1986) Sheet or container? Forensic glass comparisons with emphasis on source classification. *Journal of Forensic Sciences*, **31**, 1314–1329.

Ryland, S.G. (2011) Discrimination of flat (sheet) glass specimens having similar refractive indices using micro X-ray fluorescence spectrometry. *Journal of the American Society of Trace Evidence Examiners*, **2**, 2–12.

Smith, D.H. (2000) Isotope dilution mass spectrometry. In *Inorganic Mass Spectrometry* (eds Barshick, C. M.; Duckworth, D. C.; Smith, D. H.), Dekker, New York, pp. 223–230.

Steocklein, W.; Becker, S. (2001) Paint and glasses. *Proceedings of the 13th INTERPOL Forensic Science Symposium*, Lyon, France, 16–19 October 2001.

Suzuki, Y.; Sugita, R.; Suzuki, S.; Marumo, Y. (2000) Forensic discrimination of bottle glass by refractive index measurement and analysis of trace elements with ICP-MS. *Analytical Science*, **16**, 1195–1198.

Taroni, F.; Biedermann, A.; Garbolino, P.; Aitken, C.G.G. (2004) A general approach to Bayesian networks for the interpretation of evidence. *Forensic Science International*, **139**, 5–16.

Thornton, J.I. (2001) Interpretation of physical aspect of glass evidence. In *Forensic Examination of Glass and Paint* (ed. Caddy, B.), Taylor & Francis, London, pp. 97–119.

Thornton, J.I.; Cashman, P.J. (1979) Reconstruction of fractured glass by laser beam interferometry. *Journal of Forensic Sciences*, **24**, 101–108.

Thornton, J.I.; Cashman, P.J. (1986) Glass fracture mechanism – A rethinking. *Journal of Forensic Sciences*, **31**, 818–824.

Tooley, F.V. (1984) *The Handbook of Glass Manufacture*, 3rd edn, Vol. **1**. Ashlee Publishing Co., New York.

Trejos, T.; Almirall, J.R. (2005a) Sampling strategies for the analysis of glass fragments by LA-ICP-MS Part I. Micro-homogeneity study of glass and its application to the interpretation of forensic evidence. *Talanta*, **67**, 388–395.

Trejos, T.; Almirall, J.R. (2005b) Sampling strategies for the analysis of glass fragments by LA-ICP-MS Part II: Sample size and sample shape considerations. *Talanta*, **67**, 396–401.

Trejos, T.; Almirall, J.R. (2010) Laser ablation inductively coupled plasma mass spectrometry in forensic science. In *Encyclopedia of Analytical Chemistry* (ed. Meyers, R. A.), John Wiley & Sons Ltd, Chichester, UK, pp. 1–26.

Trejos, T.; Montero, S.; Almirall, J.R. (2003) Analysis and comparison of glass fragments by laser ablation inductively coupled plasma mass spectrometry (LA-ICP-MS) and ICP-MS. *Analytical and Bioanalytical Chemistry*, **376**, 1255–1264.

Trejos, T.; Koons, R.; Becker, S.; *et al.* (2013a) Cross-validation and evaluation of the performance of methods for the elemental analysis of forensic glass by μ-XRF, ICP-MS, and LA-ICP-MS. *Analytical and Bioanalytical Chemistry*, **405**, 5393–5409.

Trejos, T.; Koons, R.; Weis, P.; *et al.* (2013b) Forensic analysis of glass by μ-XRF, SN-ICP-MS, LA-ICP-MS and LA-ICP-OES: evaluation of the performance of different criteria for comparing elemental composition. *Journal of Analytical Atomic Spectrometry*, **28**, 1270–1282.

Tulleners, F.A.; Thornton, J.; Baca, A.C. (2013) Determination of Unique Fracture Patterns in Glass and Glassy Polymers. Award No: 2010-DN-BX-K219; US Department of Justice, Davis, CA. https://www.ncjrs.gov/pdffiles1/nij/grants/241445.pdf; last accessed 18 June 2015.

Underhill, M. (1997) The acquisition of breaking and broken glass. *Science and Justice*, **37**, 121–127.

von Bremen, U. (1975) Shadowgraphs of bulbs, bottles, and panes. *Journal of Forensic Sciences*, **20**, 109–118.

Walsh, K.A.J.; Buckleton, J.S.; Triggs, C.M. (1996) A practical example of the interpretation of glass evidence. *Science and Justice*, **36**, 213–218.

Watling, R.J. (1999) Novel application of laser ablation inductively coupled plasma mass spectrometry in forensic science and forensic archaeology. *Spectroscopy*, **14**, 16.

Watling, R.J.; Lynch, B.F.; Herring, D. (1997) Use a laser ablation inductively coupled plasma mass spectrometry for fingerprinting scene of crime evidence. *Journal of Analytical Atomic Spectrometry*, **12**, 195–203.

Weis, P.; Dücking, M.; Watzke, P.; *et al.* (2011) Establishing a match criterion in forensic comparison analysis of float glass using laser ablation inductively coupled plasma mass spectrometry. *Journal of Analytical Atomic Spectrometry*, **26**, 1273–1284.

Welch, A.; Rickard, R.; Underhill, M. (1989) The observation of banding in glass fragments and its forensic significance. *Journal of the Forensic Science Society*, **29**, 5–13.

Wen, S.B.; Mao, X.; Greif, R.; Russo, R.E. (2007) Expansion of the laser ablation vapor plume into a background gas. I. Analysis. *Japanese Journal of Applied Physics*, **101**, 023114.

Wray, P. (2008) China: Ceramics world continues eastward shift. *American Ceramics Society Bulletin*, **87** (10), 26–30.

Zadora, G. (2007) Glass analysis for forensic purposes – a comparison of classification methods. *Journal of Chemometrics*, **21**, 174–186.

Zadora, G. (2009) Evaluation of evidence value of glass fragments by likelihood ratio and Bayesian Network approaches. *Analytica Chimica Acta*, **642**, 279–290.

Zadora, G.; Neocleous, T. (2009) Likelihood ratio model for classification of forensic evidence. *Analytica Chimica Acta* , **642**, 266–278.

Zadora, G.; Ramos, D. (2010) Evaluation of glass samples for forensic purposes – An application of likelihood ratios and an information–theoretical approach. *Chemometrics and Intelligent Laboratory Systems*, **102**, 63–83.

Zoro, J.A. (1983) Observations on the backward fragmentation of float glass. *Forensic Science International*, **22**, 213–219.

Zurhaar, A.; Mullings, L. (1990) Characterization of forensic glass samples using inductively coupled plasma mass spectrometry. *Journal of Analytical Atomic Spectrometry*, **5**, 611–617.

CHAPTER 7

The forensic comparison of soil and geologic microtraces

Richard E. Bisbing

Richard E. Bisbing LLC, USA

Like other organisms, a soil possesses special characteristics by means of which it can be recognized and classified.

The Study of Soil in the Field (1936) (G.R. Clarke)

Dust is a result of the divisibility of matter.

The Kingdom of Dust (1912) (J. Gordon Ogden)

7.1 Soil and geologic microtraces as trace evidence

Soil and geologic microtraces from a crime scene can be forms of associative evidence, that is, by comparing them with apparently similar material located on the perpetrator, their clothing, or in their home or automobile, some connection or association can be established between crime scene and criminal. Paul L. Kirk, Professor of Criminalistics, University of California, Berkeley, summarized the value of soil as associative evidence in 1953:

> "Soil is useful precisely to the extent that it can be employed to determine whether or not a person has been at a certain place. This will be possible to the extent that the soil at different points varies in some detectable manner. It would be expected that the soil at two different spots would invariably be different to some degree but only careful experiment will determine whether the differences that exist are sufficient to detect. Present indications are that it will be a rare circumstance when the soils from two different places are indistinguishable. In fact, differences are often perceived in the soil composition from spots only a few feet or even inches apart." (Kirk, 1953)

Fortunately, there are well established analytical procedures available from the geological sciences to classify soils, to detect differences between soils and to compare soil and geologic microtraces.

Forensic Chemistry: Fundamentals and Applications, First Edition. Edited by Jay A. Siegel.
© 2016 John Wiley & Sons, Ltd. Published 2016 by John Wiley & Sons, Ltd.

Soil is a three-dimensional, dynamic natural body of accumulated clastic mineral grains, decaying organic solids, living organisms, water, and gas-filled pores, occupying the earth's surface, isolated from its parent material below, of variable depth, differentiated into horizons, sometimes in places man-modified with earthly materials. Soil has an individuality resulting from the integrated effects of climate and living organisms acting over time upon parent material, as conditioned by topography. A soil sample is a volume of soil, an accumulation of mineral, ecological, and anthropogenic particles, separated from the natural soil body, which, if unconsolidated, may have a dusty, sandy, stony or cobbly texture, or, if consolidated, may have a granular, blocky, or platy structure. Soil samples sometimes are in the form of crumbs, lumps, or clods.

Geologic microtraces are unconsolidated accumulations of small bits and pieces of geological materials; if the motes are relatively large, the traces can be called debris and if the particles are small, the microtraces can be called dust. When transferred from a crime scene, the microtraces can serve as reminders that someone or something was at the site previously; in other words, they are used to associate someone or something to an event at a site of interest. Geologic microtraces can originate from three possible sources: (i) rocks, sediments, sand, and soil (geological materials); (ii) products made from geological materials, such as concrete, safe insulation, and building materials; or (iii) dust produced from the geological materials or products.

Soil is the most common geological material used as trace evidence. A soil sample of unknown origin might be on a shovel used to dig a grave, on shoes worn through the mud at the crime scene, in a vehicle's wheel-well used to maliciously destroy a golf green, or in a car trunk used to move marihuana plants. The usual question for the forensic scientist is whether the recovered soil could have originated from the crime scene; therefore, known samples from the crime scene are compared with the questioned soil. The goal is to determine whether or not the known and questioned soil samples are similar in all respects and, therefore, in the case of an association (match), cannot be distinguished one from another in any significant way. Normally, the comparisons are accomplished using forensic microscopy (Chapter 11).

7.2 Comparison process

From an historical perspective, a useful forensic comparison process is revealed in Georg Popp's cases, Walter McCrone's legacy, and from Ontario's Centre of Forensic Sciences' research and practice initiated by Oded Frenkel. An understanding of their methods and experiences provides a model for the forensic comparison of soil and geologic microtraces – a *morphomicroscopical contextual analysis process* (MCAP).

Georg Popp, in 1908, was the moving spirit of the Popp–Becker Chemico-technical and Hygiene Institute in Frankfurt, Germany, an independent public chemical laboratory for food studies, an experimental laboratory for technical aspects of fire, and a microscopy laboratory for trace evidence. His reputation as a forensic microscopist developed after a Frankfurt newspaper highlighted his work under headlines such as "The Microscope as Detective." He often lectured on the potential of chemistry, mineralogy, and botany as forensic sciences (Thorwald, 1967). The following case illustrates his approach and the process described in this chapter.

When Margarethe Filbert was murdered, the suspect, a local farmer named Andreas Schlicher, denied having anything to do with the crime, including walking through the field where the crime occurred. Having heard that, Popp wrote to the investigators offering his services:

> "Since Schlicher denies having been at the place where the body was found and at the presumptive crime scene, this assertion might possibly be checked by comparison of the soil found on the boots he wore that day. We might therefore compare his boots with samples of soil from these places, as well as considering the soil from the field he allegedly visited and the path he allegedly took. I therefore recommend that the samples of soil from the shoes not be removed, but left for me to remove during my studies, so that the succession of layers and the characteristic stratification may be as far as possible recorded photographically. It would be useful if I myself could collect the necessary soil samples from the locales of interest." (Thorwald, 1967)

The investigators agreed and Popp collected soil samples from the various scenes associated with the crime and suspect, and studied them with help from a geologist. They found: (a) the soil from Schlicher's property and nearby street contained greenish goose droppings; (b) the soil from Schlicher's fields was on average very dark and heavily mixed with broken-down porphyry rocks, milk-quartz, and mica, along with root fibers, weathered straw, and some leaves; (c) the soil near the crime scene was different, a reddish sandy earth, a decomposition product of red sandstone mixed with splintery quartz granules, mica, and a large amount of ferriferous red clay but with very little vegetable matter; and (d) the soil from the path was similar to the crime scene except it contained decaying beech leaf fragments from the trees that lined the path and, as expected, contained some horse manure.

Examining the suspect's shoes, Popp noted thickly caked and layered soil stuck to the heel. He reasoned that the layers represented a sequential deposit, with the earliest material deposited directly on the leather. The initial comparisons between the soil from Schlicher's shoes and the exemplars revealed the soil from the shoes had a strikingly reddish color similar to the soil from the crime scene (c). Popp moistened the soil on the shoe then sliced the sample down to the leather with a razor blade, so producing neat cross-sections of the different layers. The layer closest to the instep contained goose droppings, such as Popp had found in Schlicher's farmyard (a). Next came a mud layer containing grass fragments, as if Schlicher had crossed the meadow. Then there was a layer of red, sandy earth, such as found at the crime scene (c). As for the alibi, though Schlicher had claimed he had walked in his own fields on that day (b), no porphyry fragments or milky quartz were found on the shoes. Finally, in order to separate plant components from the soil, Popp suspended several slices in a little water. To his surprise, he found tiny fibrous elements in the red strata that presumably came from the crime scene, including tiny wool and cotton fiber fragments similar to the fibers comprising the victim's skirt and petticoat (Thorwald, 1967). By systematically recovering each individual soil layer in context and identifying mineralogical, ecological, and anthropogenic constituents microscopically, Popp was able to retrace the steps, literally, taken by the suspect on the day of the murder.

Walter C. McCrone was a student of the chemical microscopist Émile Monnin Chamot, senior author of the *Handbook of Chemical Microscopy* (1930). In 1956, he founded McCrone Associates, Inc., Chicago (now located in Westmont, IL), a laboratory dedicated to microscopy and microanalysis of materials, and, in 1960, McCrone Research Institute, Chicago, a not-for-profit company devoted to teaching light microscopy.

Throughout his life, McCrone promoted the use of the light microscope for materials analysis. He lectured, conducted hands-on workshops, and wrote articles for trade journals extolling chemical microscopy. Over the years, McCrone's methods became the gold standard for the identification of geological materials including asbestos (McCrone, 1974) and artist's pigments (McCrone, 1981). He is best known to the general public for his Shroud of Turin study where the particle approach, with microscopical observation of the problem, morphological analysis, isolation of individual constituents for polarized light microscopy and ultra-microanalysis led to a most interesting conclusion (McCrone, 1996). In 1967, McCrone summarized his philosophy regarding particle identification in *The Particle Atlas*:

> "The microscopist should be able to identify any particle he can see and describe. Often he can identify it on sight, just as we all identify macroscopic objects, e.g. people, buildings, birds and flowers. In some cases he may have to measure additional properties such as refractive index, solubility, density and x-ray diffraction data. He has the further advantage over other analysts that he identifies the actual compound in the existing physical phase and he may even be able to differentiate between particles of the same substance from different sources." (McCrone *et al.*, 1967)

McCrone's particle approach is in accord with the most effective approach to soil and geologic microtrace comparisons. McCrone compared soil with regard to particle size distribution, mineral content with modal analysis, and any unusual particles that might be unique to the site (McCrone, 1982). Based on his teachings, today's leaders in trace evidence and soil comparisons were either protégés, colleagues or students of this remarkable man. Mentored by Walter McCrone, they used his methods throughout their successful careers, opened their own laboratories using McCrone's business model, and taught their colleagues and students how to identify geologic particles in dust and compare soils. For example, Thom Hopen, of the US Bureau of Alcohol, Tobacco, Firearms and Explosives, Atlanta, GA, who acknowledges Walter McCrone as his mentor, describes several cases where the methods he learned from his friend, like those described here, were used to compare soil and geologic microtraces, demonstrating once again that soil evidence is valuable associative evidence (Hopen, 2004).

Oded J. Frenkel, a geologist, joined the Centre of Forensic Sciences in Toronto, Canada, in the early 1960s. The Attorney General's Laboratory was created in 1951 with H. Ward Smith as Director; the laboratory was renamed the Centre of Forensic Sciences in 1966 and Doug Lucas was made the new director. He recently shared his memories of the soil research with me. As a geologist, Frenkel recognized the potential value of soil comparisons. Before he joined the laboratory, they did virtually nothing with soil evidence. He quickly realized the paucity of data available about soil for forensic purposes and,

Pollen and Pigments – Turin, Italy

The Shroud of Turin is a linen cloth bearing the image of a man who appears to have suffered wounds in a manner consistent with crucifixion. It is believed by some to be the burial shroud of Jesus of Nazareth. The negative image was first observed in 1898 on the photographic plate of amateur photographer Secondo Pia, who was allowed to photograph it while on exhibit in Turin. The shroud is kept in the royal chapel of the Cathedral of Saint John the Baptist in Turin, Italy, but it was made available for scientific study, first in 1969 and 1973 by a committee appointed by Cardinal Michele Pellegrino and then again in 1978 by the Shroud of Turin Research Project (STURP).

Max Frei-Sulzer, Zurich Police Department Crime Laboratory, has been credited with first recommending collecting microtraces completely invisible to the naked eye by pressing ordinary scotch tape to places where they were suspected to be present. In 1973 he used transparent sticky tapes pressed into the linen of the Turin Shroud to collect small particles for microscopical study. He found 58 different pollen species, 45 from the Jerusalem area, 6 from the eastern Middle East, one growing exclusively in İstanbul, and two from Edessa, Turkey. He concluded that the Turin Shroud must have originated in the Middle East.

In 1979, Walter McCrone analyzed the sticky tape samples he was given by STURP, using polarized light microscopy, microchemical tests, electron microscopy, and Raman spectroscopy. He demonstrated that the image is actually made up of millions of submicrometer pigment particles and concluded the apparent bloodstains were tempera paint tinted red with hematite and vermillion. His results are still controversial; for example, Alan D. Adler, Western Connecticut University, claimed the reddish stains are blood and interpreted the iron oxide as a natural residue of hemoglobin.

In 1988, very small samples from the shroud's linen were radiocarbon dated by accelerator mass spectrometry (AMS) in laboratories at Arizona, Oxford and Zurich Universities. The results provided conclusive evidence that the linen of the Shroud of Turin is medieval, AD 1260–1390, which corresponds to when the relic surfaced on the antiquities market.

therefore, conducted numerous field studies to better understand the comparison process. Based on his research and the work of Carol I. Dell, of the Ontario Research Foundation, Toronto (Dell, 1959a, 1963), Frenkel initiated a mineralogical approach to soil analysis and devised the concept of a rapid counting technique for modal analysis. After a heavy liquid separation, Dell identified the relative percentages of each mineral, mainly by the use of the petrographic microscope for the coarse silt and fine sand, and the binocular microscope for the medium and coarse sand. She described the distinguishing characteristics of about 35 minerals in Ontario soils (Dell, 1959b). Bill Graves soon thereafter joined the Centre's staff and Frenkel trained him in the procedures he had developed. From there the Centre's reputation for soil comparisons grew. With Frenkel's procedure, soil samples are separated and cleaned by wet sieving and polarized light microscopy is used to classify individual mineral grains. The types and numbers of mineral grains in the questioned and known samples are tallied and the frequencies compared – modal analysis. In conjunction with color comparisons, a large number of soil samples from southern Ontario, separated by no more than the size of a footprint, could be distinguished; and, in most instances, soil samples within a footprint-size area were indistinguishable (Graves, 1979).

7.3 Developing expertise

This chapter describes a practical forensic practice primarily from experiences in North America. It is not a review or critique of all possible ways to make comparisons. Kenneth Pye and Debra Croft have edited a special publication of *The Geological Society* containing articles describing numerous other techniques for soil analysis (Pye and Croft, 2004). Neither is this chapter a detailed primer or atlas describing all the various constituents in soil and the many materials that could comprise geologic microtraces. There are adequate resources elsewhere, including the web sites shown in Box 7.1 and textbooks in the suggested reading. Rather, this chapter describes a comparison process for soil and geologic microtraces that has been shown to distinguish soils in close proximity, using

---

Box 7.1 Useful web sites

Rocks and Soil
http://highered.mheducation.com/sites/0072530634/student_view0/web_links.html
www.geology.com
www.landfood.ubc.ca/soil200/index
www.minerox.com
www.munsellstore.com
www.nrcs.usda.gov/wps/portal/nrcs/site/soils
www.sandatlas.org
www.soils.org
www.soils.org.uk

Minerals
www.ces.ncsu.edu/plymouth/programs/drees2
http://consorminex.com/atlas.htm
http://funnel.sfsu.edu/courses/geol426/Handouts/mintable.pdf
http://www.hmag.gla.ac.uk/John/teaching/mintable.htm
http://www.mindat.org/
http://www.mineralatlas.com/General%20introduction/Introduction.htm
http://www.minerals.net/MineralMain.aspx
www.minsocam.org/MSA/collectors_corner/id/mineral_id_keyi1
www.webmineral.com

Pollen and Phytoliths
http://apsa.anu.edu.au/
http://www.geo.arizona.edu/palynology/polonweb.html
http://gepeg.org/enter_PCORE.html
http://www.homepages.ucl.ac.uk/~tcrndfu/phytoliths.html
http://pollen.usda.gov/

simple techniques performable by most analysts in forensic laboratories and in a cost effective way (Junger, 1996). Figure 7.1 outlines the process in a general way. Ray Murray, University of Montana (retired), has often recounted the history of forensic geology and summarized some seminal soil cases including several investigated by former colleagues and students of Walter McCrone. A few are described in text boxes throughout this chapter to illustrate practical application of the process.

Young scholars interested in forensic geology may benefit from studying and practicing the comparison process presented here. An understanding of physical geology (petrology and mineralogy) and pedology (soil science) assists in realization of the complexity and, therefore, the value of soil and other geological materials as trace evidence. Therefore, study how rocks and soil vary from place to place. Practice collecting soil samples, comparing soil color, and separating soils into size and density fractions; isolate small particles for analysis. Learn to recognize mineral, ecological, and anthropogenic constituents found in soil and dust. Mineral grains comprise most of the soil and dust from geological materials. Therefore, practice mineral identification using optical mineralogy; and, conduct counting procedures to determine the relative abundance of the constituents. Most importantly, make comparisons and test yourself with matching tests – tests prepared by others where the results (whether from the same source or not) are known only to the preparer. This chapter briefly summarizes each topic.

7.4 Genesis of soil

The work of water, wind, living organisms, and changing temperatures reduce rocks comprised of minerals to regolith, sand, soil, and dust. Igneous rocks are of volcanic origin formed from molten rock (magma), which cools and solidifies below the surface (intrusive rocks) or from lava, which cools and solidifies above the surface (extrusive rocks), in each case forming minerals. Rocks exposed at the surface are slowly broken down into smaller particles by mechanical and chemical weathering. As they accumulate in layers, along with plant and animal remains, the particles are sometimes compressed and cemented into sedimentary rocks, which are subsequently themselves slowly weathered. Metamorphic rocks are formed from other rocks at depths under the earth's surface in regions of great heat and pressure, thereby changing the minerals; when later exposed to the environment, metamorphic rocks also weather. This cycle, depicted in Figure 7.2, produces rock bodies containing a variety of minerals, beginning the spatial differentiation essential to forensic comparisons. Table 7.1 lists some common rock forming and accessory minerals. The net effect of weathering is that the rocks ultimately spall and shed fragments containing minerals that are sooner or later swept away by wind and water to be deposited elsewhere. Through that process, the clastic grains that accumulate in the soil, debris, and dust have widely varying compositions depending on from where they came and what they picked up along the way.

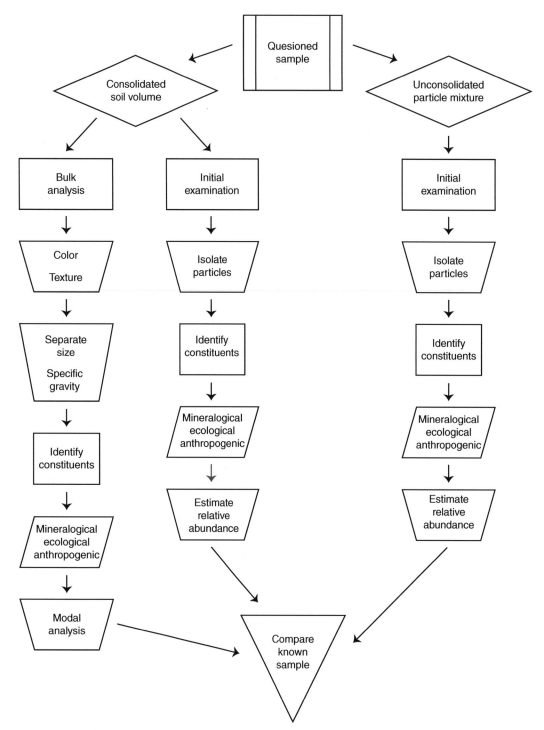

Figure 7.1 Soil and geologic microtrace comparison process. The flow chart illustrates a morphomicroscopical contextual analysis process (MCAP) modeled after Georg Popp's historic cases, Walter McCrone's legacy and Oded Frenkel's research and practice

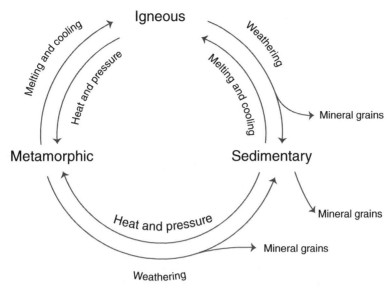

Figure 7.2 Rock cycle as the source of mineral grains. Mineral grains are eroded from rocks to ultimately constitute sediments, sand, soil, and geologic microtraces

Pedology, a component of soil science, is the study of the soil body in its natural position, including its geography and morphology. In the late 19th and early 20th centuries, it was recognized how soils spatially differentiate, providing individuality to soil samples. Although climate is a driving factor of soil formation, parent material (rocks and sediments) is a key soil-forming factor, especially at the regional scale. Topography is a key soil-forming factor at the landscape level; and, organisms play a key role in soil development from a microscopic scale. Furthermore, through natural processes, over time, the soil body changes. For example, wind (aeolian-turbation) causes the winnowing of fines, leaving well sorted top soils like the loess covering the Great Plains of Nebraska, Kansas, and Colorado; freezing and thawing (cryo-turbation) pushes the soil apart causing lateral movement; and bioturbation by flora and fauna acts to displace and mix sediment particles. Along with nature, man moves and contaminates soil by cultivating, excavating, mine-dumping, and filling space, creating made land that further differentiates the soil spatially. As a consequence, there are numerous types of soils possible at a site; they can be undisturbed natural soils with horizons or disturbed where the surface and underlying horizons have been mixed, exposed, inverted, or transported to another location. Usually these changes cause the structure, texture, color, and mineralogy of soil to vary markedly within short distances over the landscape (horizontally) and downward into the depths (vertically). The spatial differentiation can be experienced easily by observing soil in road cuts, and over the landscape across cultivated fields. The importance of seeing soil in the field cannot be overemphasized.

Table 7.1 Common rock minerals

Igneous		Sedimentary		Metamorphic		Gangue
Rock-forming	Accessory	Mechanical	Chemical	Regional	Contact	Accessory
Augite	Apatite	Feldspars	Anhydrite	Actinolite	Chalcopyrite	Aragonite
Biotite	Corundum	Kaolinite	Calcite	Chlorite	Diopside	Barite
Hypersthene	Hornblende	Montmorillonite	Dolomite	Epidote	Graphite	Fluorite
Microcline	Ilmenite	Quartz	Gypsum	Garnet	Sphalerite	Opal
Muscovite	Magnetite	Talc	Halite	Kyanite	Spinel	Rhodochrosite
Olivine	Pyrite		Hematite	Serpentine	Tremolite	Siderite
Orthoclase	Pyroxenes		Limonite	Sillimanite	Wollastonite	Topaz
Phlogopite	Rutile		Magnesite	Staurolite		Tourmaline
Plagioclase	Sphene					
Sodalite	Zircon					

Note: In order to minimize redundancy and save space, many of the listed minerals are not found exclusively in the rocks under which they are listed; many of the minerals are found in several different types of rocks.

Root balls – California

The theft of $40,000 worth of palm trees from a San Diego nursery provided soil scientists Brad Lee, Tanja Williamson, and Robert Graham an opportunity to apply their expertise to solving the crime. The owner had raised the exotic trees from seeds using a special potting soil. Soil samples were collected from root balls left in the owner's yard and from root balls in the suspect's yard. A comparison of carbonates, color, particle size distribution, ratio of light to heavy minerals and identification of the heavy minerals associated the root balls in the suspect's yard with the root balls in the owner's yard (Lee et al., 2002).

Sand occurs in slightly different circumstances from soil. Unconsolidated when compared to soil, normally sand samples do not contain enough decaying organic solids, living organisms, water, silt, and clay to glue the grains together. Instead, a handful of sand from a beach, dune, or riverbed contains billions of loose, relatively large (between 50 μm and 2 mm) mineral grains or skeletal remains that are mute witnesses of its origin. Sand is mainly found in rivers, beaches, dunes, deserts, alluvial fans, deltas, swamps, and bayous. Like soil, the composition of sands is remarkably diverse.

Mineral sands (abiogenic sands) are formed as rocks disintegrate through weathering and erosion. For example, volcanic islands, such as the Hawaiian Islands, are made of basalt, a dense, black rock that contains olivine, a heavy green mineral. On the shores of Hawaii (The Big Island), the green beaches are formed from the shattering of basaltic lava rich in olivine. The nearby black sand beaches are dominated by shattered grains of black volcanic glass. The continental landmasses are composed mostly of granite; therefore, mineral sands, as on most American beaches, are formed by the breakdown of granite and usually contain mostly quartz and some feldspar. On the other hand, in the Canadian Rockies, the limestone rock is ground down and washed away leaving glacial flour (silt-size calcite) along the edges of streams. The white sand dunes of New Mexico are comprised of gypsum weathered from the rocks of nearby mountains. Mineral beach sands might also contain ecological particles (diatoms and foraminifera) and anthropogenic particles (slag and fly ash).

The size distribution of the sand grains can reveal its source; for example, beach sand is likely well sorted with a relatively normal size–frequency curve. In comparison, the sand of glacial till will likely be unsorted and the size–frequency curve will be bi- or trimodal. The grains may be further characterized based on their sphericity, roundness, and microscopic surface texture (smoothness, polish or gloss and surface markings). Finally, like soil, heavy mineral analysis is useful for comparing sands.

Balloon bombs WWII

During the Second World War, the Japanese sent thousands of incendiary bombs attached to balloons released into the easterly wintertime jet stream winds to float cross the north Pacific to the US mainland, the world's first intercontinental weapon. The balloons were relatively ineffective as weapons; despite their low success, the authorities thought they might get lucky with a balloon carrying biowarfare agents.

The balloons were crafted from mulberry paper, glued together with potato flour and filled with expansive hydrogen. The Japanese programmed the balloons to release hydrogen if they ascended to over 38,000 feet and to drop pairs of sand-filled ballast bags if the balloon dropped below 30,000 feet. Three-dozen sand-filled ballast bags were hung beneath the balloon, along with the bomb.

The sand in some of the sandbags was analyzed by the US Geological Survey (USGS). It was immediately clear that the ballast was beach sand, but from where? They quickly eliminated North America as a source of the sand. The sand contained over 100 species of diatoms but no coral. In Japan coral grows along the coast of the main island of Honshu only as far north as Tokyo Bay. They also found foraminifera that had been described in Japanese geologic papers from beaches north of Tokyo on the eastern shore of Honshu. Heavy mineral analysis narrowed the source area to the northerly thousand miles of Japan's eastern coasts. Slag in the sand probably originated from a nearby blast furnace.

They determined that the sand samples likely came from either of two locations: a northerly site along the great beach at Shiogama, close to Sentai, Japan, and/or the Ninety-nine League Beach at Ichinomiya, Japan – an actual launch site, as learned later. (Mikesh, 1973; Rogers, 2014)

The skeletal remains of plants and animals are a second source of sands. Most biogenic sands are composed of fragments of corals, coralline algae, mollusks, and other resistant biological fragments, such as sea urchin spines, sponge spicules, and fossil remains such as tiny teeth; but, some contain skeletal remains of entire organisms, such as the micromollusks or the single-celled foraminifera. Caribbean coral beach sands sometimes contain anthropogenic sea glass from dumping of bottles in a nearby bay.

7.5 Genesis of geologic microtraces

Geologic microtraces are contained in the microscopic debris and dust that accumulates from the disintegrated components of our environment. Hans Gross, Professor of Criminology at the University of Prague, in his book *Handbuch fur Untersuchungsrichter als System der Kriminalistik*, published in the last quarter of the 19th century, described dust as "the eroded vestige of the more easily apprehended world of normal-scaled matter – our environment or surroundings in miniature" (Burney, 2013). Dust has three distinguishing characteristics that make it useful as trace evidence: (i) it is found as bits of materials in a pulverized state; (ii) its compositional and morphological features derive from its material origin, and (iii) its microscopic size facilitates its transfer as trace evidence.

Airborne dust from geological sources and from industrial activities is commonly encountered and accumulates in places where it can subsequently transfer to our houses, cars, and clothing. For example, dust in agricultural areas contains soil minerals; dust in the neighborhoods surrounding quarries contains rock fragments from the quarry; and dust in the neighborhoods surrounding coal piles contains coal dust. Therefore, as the French Forensic Microscopist Edmond Locard of Lyon explained in 1930, "The microscopic debris that covers our clothes are the mute witnesses of all our movements and all our encounters" (Locard, 1930).

Unlike soil samples where a volume of soil is compared with another volume, most dust samples are unconsolidated mixtures of particles accumulating over time. Therefore, a direct comparison of the percentage of each component (modal analysis) might be meaningless. The more productive approach is to compare principal constituents that are either the most in number or the most unusual with possible sources. In other words, within the context of the case, the comparisons are made between relevant particles in the dust and possible source materials without ignoring the potential value of the extra dimension that the particle suite as a whole represents. As described by Chesterene Cwiklik, of Cwiklik and Associates, Seattle, WA, the debris found on an item reflects the item's individual history and, therefore, the comparison of the suite of particles provides information for comparison that is far more specific than would otherwise be possible by comparison of individual microtraces (Cwiklik, 1999).

In a similar way, Dave Stoney and Andy Bowen, of Stoney Forensics, Chantilly, VA, students and colleagues of Walter McCrone, describe their approach to comparing geologic microtraces. "We are continually exposed to a 'soup' of these very small particles. Populations of these particles, as adsorbed on surfaces, will reflect cumulative exposures and conditions, which are expected to be highly characteristic of the local environment and their presence, identity, and relative quantities represent an untapped source of individuality for conventional trace evidence" (Bowen and Stoney, 2013). Put another way, "Fine dust particles, adhering to virtually any object and within virtually any product, are the result of a history of exposure. Routinely, such dusts contain a tremendous variety of particles, including those of mineral, botanical, zoological, microbial, and anthropogenic character. The large number of particles in these dusts, and their variety, provide an extremely rich source of potential information – translating into a powerful inferential tool – when the particles are [appropriately collected], appropriately analyzed, and appropriately interpreted" (Stoney *et al.*, 2011).

When comparing geologic microtraces, it is important to determine which particles are foreign to the object on which they are found, that is, those that originate from something or someplace other than those particles that originate from habitual activities or the object's usual place. Therefore, appropriate and adequate samples are required, as are required for soil comparisons. "Samples of debris from the sites relevant to the event under investigation must be available, as well as debris standards from the usual environments of the people [and objects] involved" (Cwiklik, 1999). In order to use geologic microtraces for association, there must be enough particles, enough types, and the correspondences must significantly outweigh the non-correspondences (Cwiklik, 1999).

The comparison of geologic microtraces is best accomplished using the particle approach and chemical microscopy practiced by Walter McCrone. Hans Gross explained in 1893, "The recognition of the almost infinite variety of mineral, biological, and manufactured constituents of dusts calls for a considerable amount of patience and involves a number of different methods of treatment and study which may demand the collaboration of the chemist, the mineralogist and the biologist" (Jackson, 1962). That is why Skip Palenik while at McCrone Associates in 1979 listed the many topics requiring study by a forensic microscopist including: chemistry, physics, biology, mineralogy, polarized light microscopy, microchemical analysis, microscopy of hairs and fibers, wood identification, and pollen identification (Palenik, 1979). They both understood that dust particles are identified from a vast knowledge base by a gestalt of contextual analysis, stereomicroscopy, and polarized light microscopy by observing the *morphomicroscopic* features listed in Table 7.2. These observations, coupled with experience, generally enable

Table 7.2 Particle features observed by polarized light microscopy

Ordinary light	Plane polarized light	Crossed polarized light
Orthoscopically	**Orthoscopically**	**Orthoscopically**
Cleavage; parting; fracture	Dispersion staining	Anomalous polarization colors
Color: transmitted and reflected	Pleochroism	Birefringence: retardation colors
Diaphaneity: transparent; translucent; opaque	Refractive indices	Extinction: complete, incomplete, absent, undulose, symmetrical; angle
Elastomericity	Relief	Isotropy or anisotropy
Hardness; brittleness		Sign of elongation: length fast or length slow
Magnetic susceptibility		Twinning
Morphology (shape)		
Size (micrometry)		
Surface features		
Tendency to dissolve		
Texture, sculpting, luster		

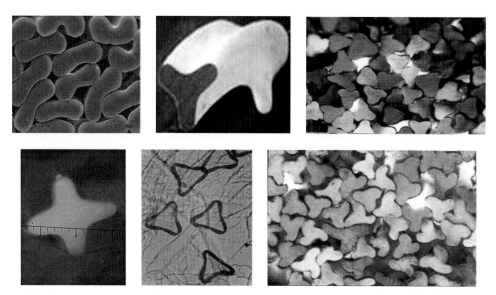

Figure 2.2 The cross-section of a manufactured fiber relates to its end use. About 500 cross-sections are used in the textile industry for various fibers types

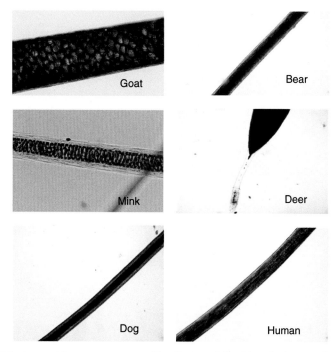

Figure 2.7 Animal hairs and humans are readily distinguished by their morphology, coloration, and medulla size

---

Forensic Chemistry: Fundamentals and Applications, First Edition. Edited by Jay A. Siegel.
© 2016 John Wiley & Sons, Ltd. Published 2016 by John Wiley & Sons, Ltd.

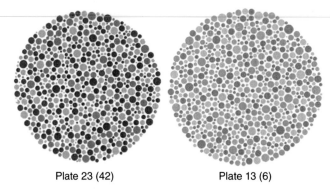

Plate 23 (42)   Plate 13 (6)

Figure 2.15 Two plates from the Ishihara color vision test. The correct answers are shown in parentheses. (Source: Wikimedia Commons.)

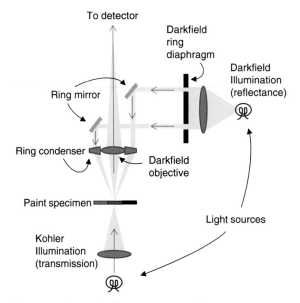

Figure 3.14 Schematic diagram of a typical microspectrophotometer

Figure 5.1 Explosion of 20,000 pounds of Composition C-4 inside a small truck. Note that the shock wave is visible due to its extreme effect on the refractive index of the surrounding air. The "thermal event" occurs within the volume defined by the shock wave. Photo courtesy ATF laboratory

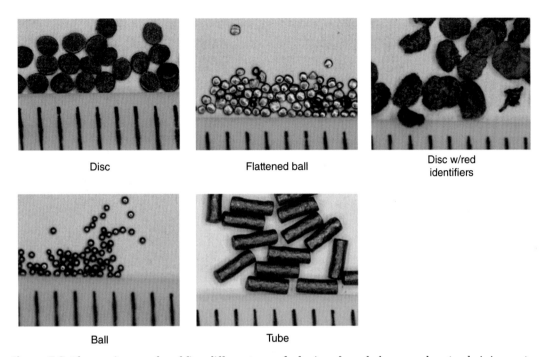

Disc

Flattened ball

Disc w/red
identifiers

Ball

Tube

Figure 5.3 Photomicrographs of five different morphologies of smokeless powders (scale is in mm)

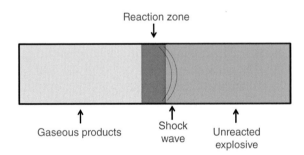

Reaction zone

Gaseous products

Shock
wave

Unreacted
explosive

Figure 5.4 Depiction of a detonation wave propagating through a high explosive

Figure 5.7 Conducting a line search during a postblast training exercise

Ethylene glycol dinitrate (EGDN)
Nitroglycerine (NG)

Sawdust filter

Figure 5.12 An example of commercial nitroglycerin dynamite with an SEM image of characteristic sawdust filler

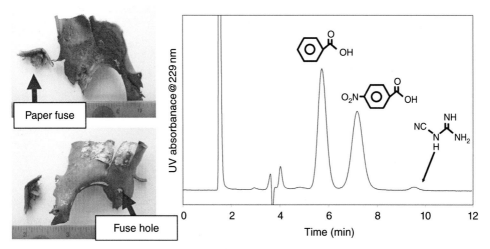

Figure 5.18 An example of the identification of benzoic acid, nitrobenzoic acid and dicyandiamide (DCDA) in the postblast residue of Triple Seven using the HPLC method described by Bender (Bender, 1998). Picture and data are courtesy of Lisa Lang, PhD, of the ATF Laboratory

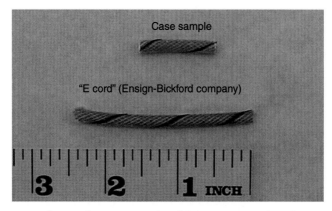

Figure 5.25 Comparison of an unknown sample of detonating cord with an exemplar of a commercial product

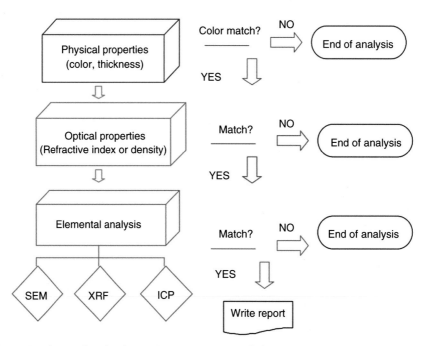

Figure 6.4 Basic scheme for the forensic examination of glass

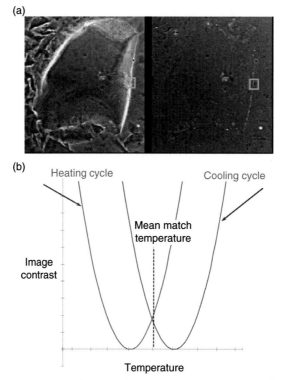

Figure 6.5 (a) Glass fragment above the matching temperature (left) and at the matching temperature (right). (b) Heating and cooling cycle used for refractive index determination

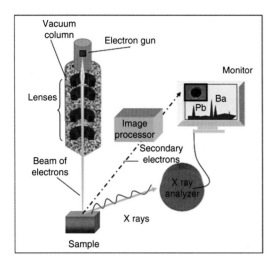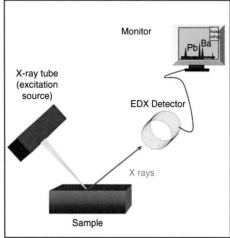

Figure 6.6 Comparison of the schematic diagram of SEM-EDS and XRF instruments

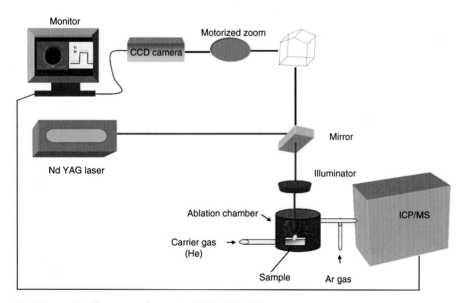

Figure 6.7 Schematic diagram of a typical LA-ICP-MS set-up

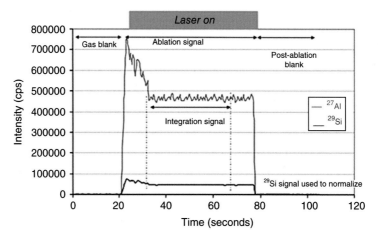

Figure 6.8 A strategy for the quantitative analysis of glass using LA-ICP-MS involves the use of a low natural abundance isotope of Si (^{29}Si) to normalize the analyte signal (in this case ^{27}Al) for differences in ablation yield as the internal standard, as it is present in a very high concentration (~70% as SiO$_2$) in glass. The intensity signal (cps) is converted to concentration using calibration standards such as the NIST 6XX series or the FGS standards (Latkoczy, 2005; ASTM, 2013b)

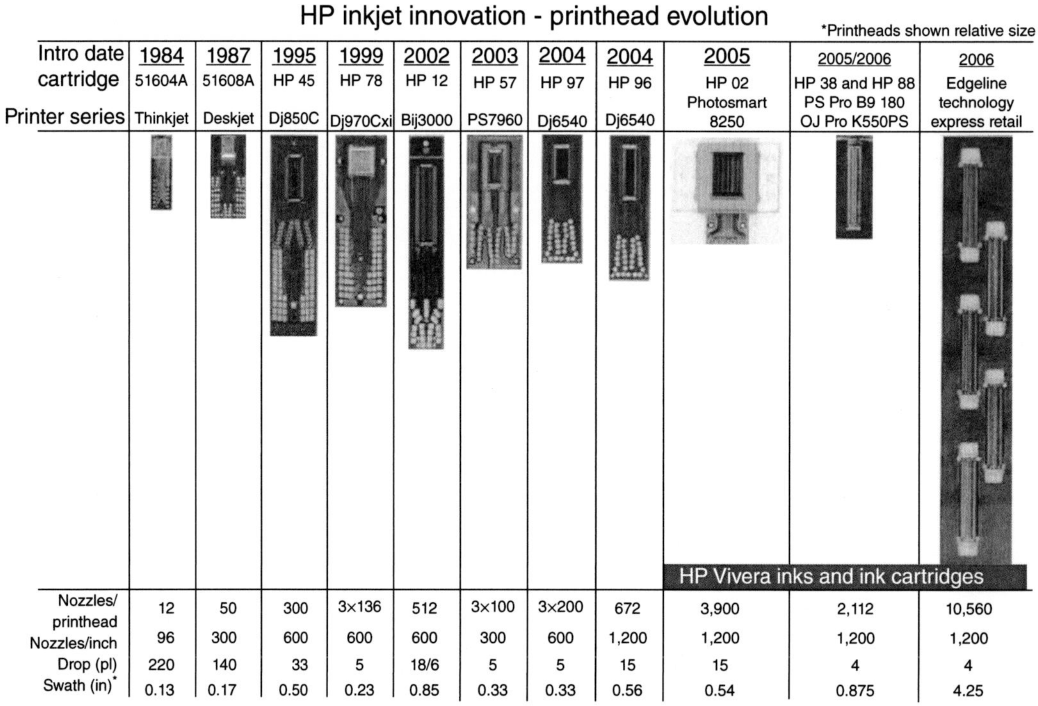

HP inkjet innovation - printhead evolution

*Printheads shown relative size

Intro date	1984	1987	1995	1999	2002	2003	2004	2004	2005	2005/2006	2006
cartridge	51604A	51608A	HP 45	HP 78	HP 12	HP 57	HP 97	HP 96	HP 02 Photosmart 8250	HP 38 and HP 88 PS Pro B9 180 OJ Pro K550PS	Edgeline technology express retail
Printer series	Thinkjet	Deskjet	Dj850C	Dj970Cxi	Bij3000	PS7960	Dj6540	Dj6540			
Nozzles/ printhead	12	50	300	3×136	512	3×100	3×200	672	3,900	2,112	10,560
Nozzles/inch	96	300	600	600	600	300	600	1,200	1,200	1,200	1,200
Drop (pl)	220	140	33	5	18/6	5	5	15	15	4	4
Swath (in)*	0.13	0.17	0.50	0.23	0.85	0.33	0.33	0.56	0.54	0.875	4.25

HP Vivera inks and ink cartridges

Figure 8.1 Demand for rapid quality inkjet printing gave rise to an evolution of printhead technology in inkjet printers. These technological advances resulted in changes to the chemistry of inks and new inkjet formulations because of the necessity for inks to be compatible with the printhead speed and droplet size requirements

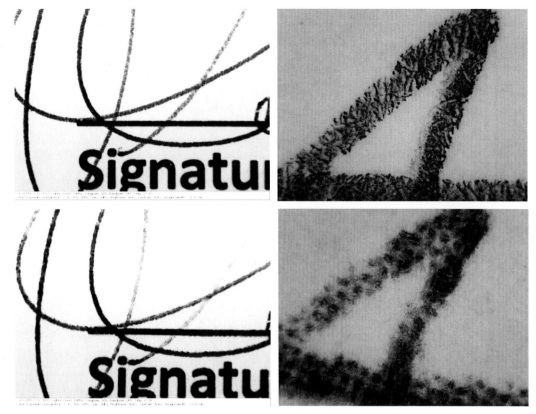

Figure 8.5 The top left image is a portion of a signature created with a blue ballpoint pen and the corresponding top right image was captured at 70× magnification with a digital microscope. The bottom left image is color toner from a copy of the same signature and the corresponding bottom right image was captured at 70× magnification with a digital microscope. The signature with the pen cannot be differentiated from the signature with the toner until visualized with a microscope

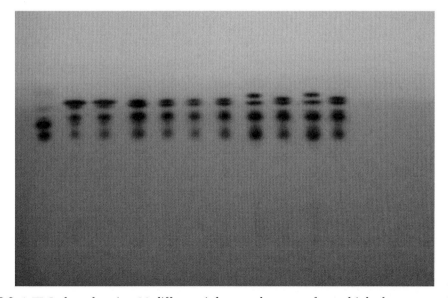

Figure 8.8 A TLC plate showing 11 different inks tested as part of a multiple document submission. The colored spots, and their relative migration positions, represent different dyes used in black point inks

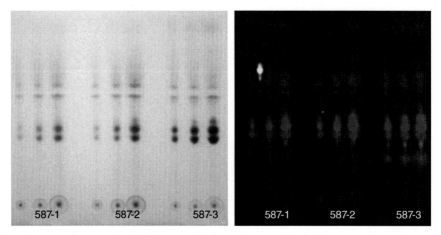

Figure 8.9 The TLC plate on the left is a single formulation of ink (587). Lanes 1 through 3 represent different concentrations of batch 587-1; lanes 4 through 6 represent different concentrations of batch 587-2; and lanes 7 through 9 represent different concentrations of batch 587-3. The TLC plate on the right is the same plate visualized with a red filter using an ALS set at 515 nm. No differences are observed when the TLC plate is visualized with visible light, but differences are readily apparent with an ALS and the appropriate filter

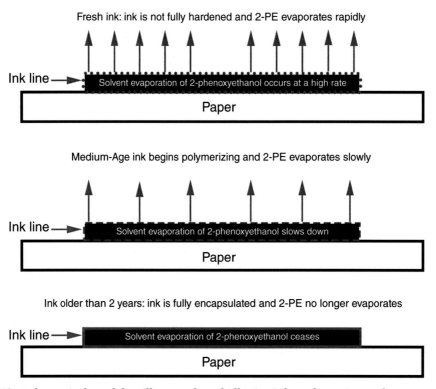

Figure 8.10 A theoretical model to illustrate how ballpoint inks polymerize as they age and the effect on the solvent loss rate of 2-PE

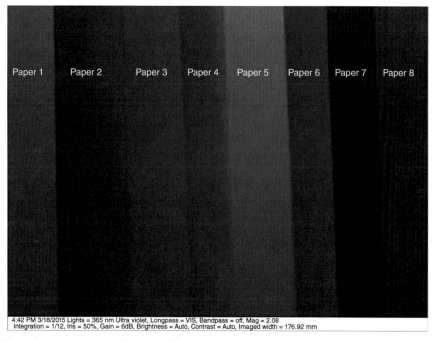

4:42 PM 3/18/2015 Lights = 365 nm Ultra violet, Longpass = VIS, Bandpass = off, Mag = 2.08
Integration = 1/12, Iris = 50%, Gain = 6dB, Brightness = Auto, Contrast = Auto, Imaged width = 176.92 mm

Figure 8.12 A demonstration of eight different pieces of paper visualized using a 366 nm UV source. Each piece of paper has a different UV fluorescent property

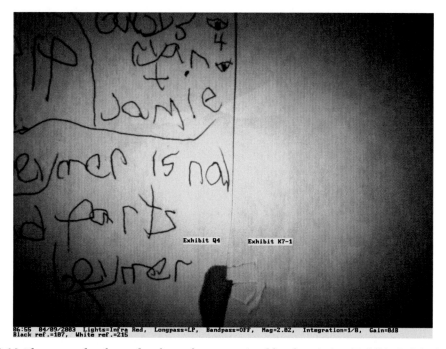

06:55 04/09/2003 Lights=Infra Red, Longpass=LP, Bandpass=OFF, Mag=2.02, Integration=1/8, Gain=8dB
Black ref.=107, White ref.=215

Figure 8.14 The torn edge from the threat letter received by the victim (Exhibit Q4) is shown to align with a torn page from a notepad that was seized from the suspect (Exhibit K7-1). This method of association is sometimes referred to as a fracture match

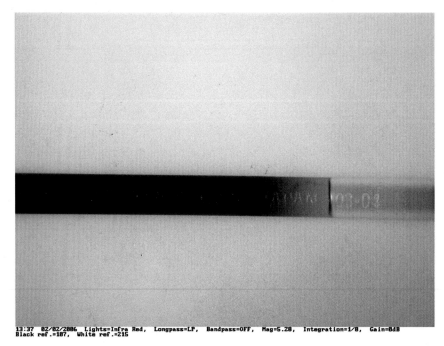

13:37 02/02/2006 Lights=Infra Red, Longpass=LP, Bandpass=OFF, Mag=5.28, Integration=1/8, Gain=0dB
Black ref.=107, White ref.=215

Figure 8.15 The ink used to produce a signature and date found on a financial agreement of substantial value purported to be created in 1999, was found to match an ink formulation produced by the Pilot Corporation. Once the information was provided to the investigator, a Pilot pen was seized from the suspect. The seized pen cartridge contained a date code indicating that the cartridge was filled with ink in April of 2003 (highlighted area). When confronted with the additional information, the suspect proceeded to confess that the financial agreement was created in 2004

(a)   (b)

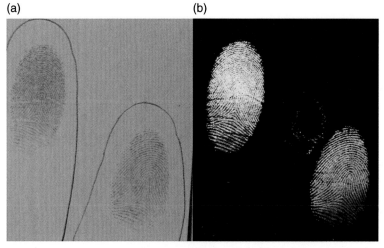

Figure 9.6 Fingermarks on white copy paper developed with 1,2-indanedione, viewed (a) under ambient lighting conditions and (b) through a 550 nm barrier filter with illumination at 505 nm

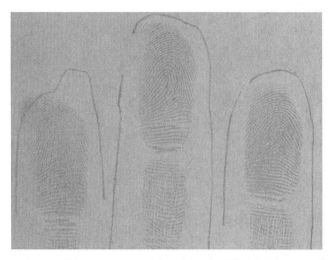

Figure 9.10 Fingermarks on white copy paper developed with ORO in propylene glycol

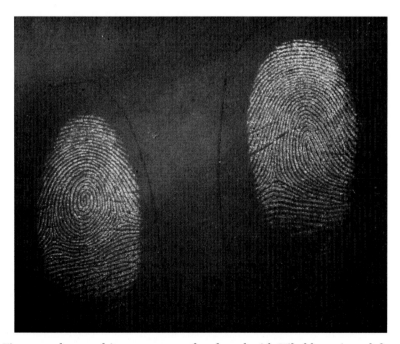

Figure 9.14 Fingermarks on white copy paper developed with Nile blue, viewed through a 550 nm barrier filter with illumination at 505 nm

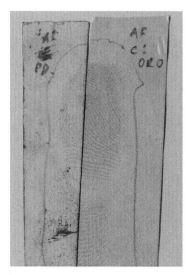

Figure 9.15 Example of a split fingermark on white copy paper, treated with PD (left) and ORO (right)

(a)   (b)

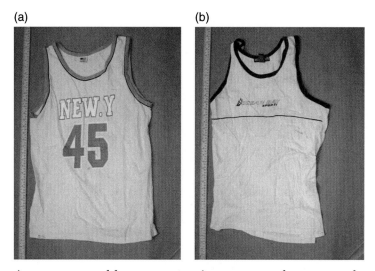

Figure 11.1 Two jerseys recovered from separate crime scenes as they appeared upon receipt at the laboratory

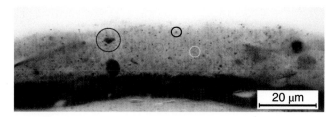

Figure 11.2 Image of a thin paint section (~2 μm thickness) in which effect pigments and some larger agglomerations of pigments can be resolved. However, the finer organic pigments cannot be resolved and are manifest simply as a mass tone color

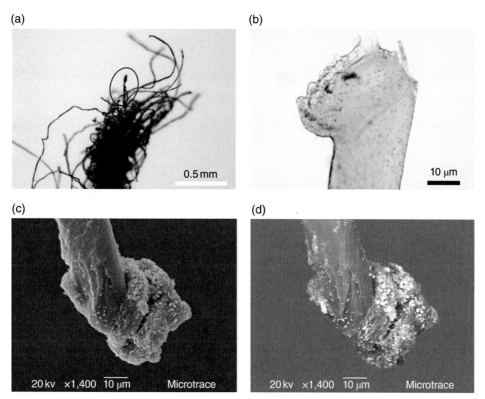

Figure 11.3 (a) Image showing a mass of fibers partially fused together as the result of a bullet severing them. (b) Appearance of a melted, globular end, by transmitted light microscopy. Appearance of a globular end by (c) secondary electron imaging and (d) backscatter electron imaging, showing the appearance of metal particles embedded when the fiber was molten

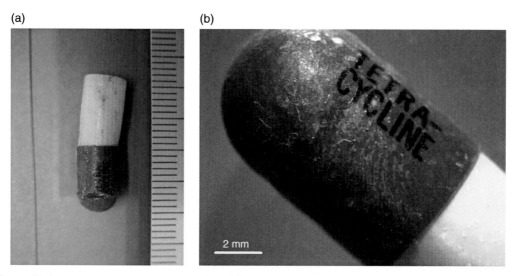

Figure 11.6 (a) A macroscopic image of a tablet that is qualitatively consistent with its visually observed color (green and yellow). (b) The same tablet as it appears in an image that was illuminated using a "white" light LED and a digital camera

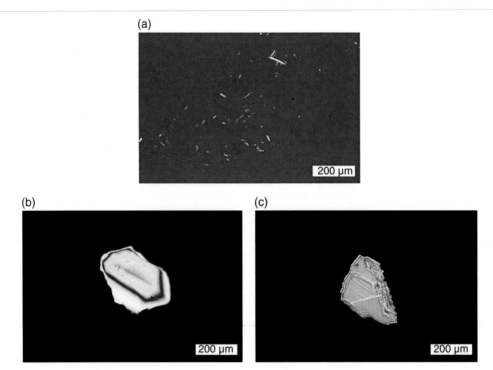

Figure 11.10 Example of crystals observed between crossed polarizers with varying levels of birefringence. (a) Glass is isotropic and has no birefringence, (b) quartz is uniaxial and shows low order interference colors, (c) calcite shows high order interference colors

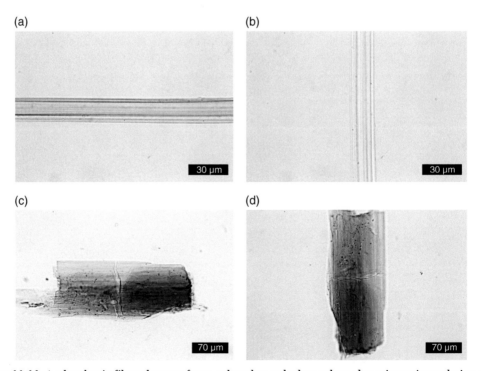

Figure 11.11 A pleochroic fiber changes from colored to colorless when the orientation relative to the polarizer changes from (a) length parallel to (b) length perpendicular. Certain minerals are also pleochroic, as noted by the change in this glaucophane grain when the mineral is rotated from an orientation in which the length is (c) parallel to the polarizer to one that is (d) perpendicular to it

the identification of most particles in a given dust sample. Sometimes, when particles cannot be identified by microscopical observation alone, the particles can be removed from the preparation and identified by other microanalytical techniques, such as infrared microspectroscopy, Raman spectroscopy, cathodoluminescence, electron microprobe analysis, and scanning or transmission electron microscopy. The relative quantity of each component in the dust is estimated by arriving at a volume percentage for each component by inspecting numerous fields of view.

7.6 Collecting questioned samples of unknown origin

Questioned soil samples are collected, either at the crime scene or in the laboratory, from objects of interest, such as shovels, shoes, floors, or vehicles. The questioned samples may be unconsolidated dust, thin clay films, small individual soil crumbs, layered and stratified clods or relatively large bulk samples. The collection process is critical because the examination of soil and geologic microtraces on tools and clothing, for example, not only provides associative evidence, it tells a story that can be told only by evaluating every sample's context in the following dimensions: temporal, spatial, depositional, and typological (Taupin and Cwiklik, 2011). Its context is its position at a site or on the tool or shoe, including its physical and temporal association with other objects and other facts. Therefore, in all cases, the questioned sample should be preserved in the context and condition as originally observed; the samples should be left intact as much as possible until they can be systematically separated for analysis in the laboratory.

More than 75 years ago, Edmond Locard described the systematic recovery and description of mud stains in Chapter IV of his *Manual of Police Techniques*:

> "In a great number of criminal trials, the identification of mud stains, supplemented by an analysis of dust is useful in determining where a suspect has been just prior to the time of his arrest. The first priority is to act fast to remove the mud. The best thing to do is to seize the suspect's shoes and clothing. Only in cases where seizing these items proves to be impossible should one resort to scraping samples with a knife or gathering dust samples with a brush. If such samples are taken in a police lab, it is best to remove the dried mud in successive layers, noting the order of the layers and numbering the samples. The individual may, in effect, be carrying traces of various terrains. First, one will describe the microscopic aspects of the various layers. For each layer, one will then proceed with a microscopic examination. In a series of preparations, one will research the various crystals in the samples and set up a system. One will note the degree of frequency of the various types in order to be able to differentiate those which determine the types of dirt and those that are accidentally present." (Locard, 1939)

When it is time to remove the questioned samples from the object, while observing in good light and possibly with magnification, separate and pick up crumbs of soil or bits of other microtraces with a spatula or forceps and transfer them to a small container; nudge or scrape samples into a clean folded aluminum foil or paper packet (bindle); lift smaller particles with tape and place on a transparent backing; or vacuum dust through an appropriate filter trap. Skip and Chris Palenik, of Microtrace LLC, Elgin, IL, describe in great detail in one of Saferstein's *Forensic Science Handbook*s (Saferstein, 2005), appropriate collection methods for microtraces, such as picking, scraping, lifting with tape, microvacuuming, and washing followed by moving and isolating particles of interest into appropriate preparations using tungsten needles (Palenik and Palenik, 2005).

7.7 Collecting soil samples of known origin

Unfortunately, my experiences and anecdotal evidence continue to suggest that difficulties with forensic soil comparisons revolve mainly around problems of sample selection and collection. The goal is to find the actual source (locus) from which to collect a primary sample, the exact location from which the questioned soil was actually transferred. Too often the known samples are not from the locus. After comparison, when questioned soil samples differ from known samples, it is possible the questioned soil was transferred from some other location, the known sample used for comparison did not adequately represent the locus, or the questioned soil is a mixture of soils from multiple sources.

In order to collect the correct primary known sample, follow a strategy that considers empathy for the criminal activity, an understanding of the soil site, prediction of the likely transfer mechanisms, and any other evidence available including witness statements. The more information available about the questioned soil sample, the more likely it is that the location from where the questioned soil originated will be found and the primary sample collected from where the questioned soil actually originated. For example, if the questioned sample has a particular color, structure, consistency, or contains vegetation, choose a known sample from a locus with similar features.

In addition to primary samples dictated by the case, collect alibi samples dictated by the accused or the investigator. Alibi locations are suggested as an excuse for why the questioned soil was found. Given the suggested location, thoroughly search for soil likely to match the questioned soil. Alibi samples are of paramount importance because a soil comparison is essentially an exclusionary process; when all other possible sources of the questioned soil are eliminated, the probative value of any association is increased markedly. Finally, collect delineating samples (survey samples) to assist in demonstrating the variability across the site and the relative uniqueness of the soil at the locus.

These known samples are usually bulk samples consisting of a single soil volume collected from a single locus. Soil samples are part of a soil body, which by definition is an accumulation of heterogeneous particles; therefore, comparison samples must be a coherent, consolidated volume of soil that represents the accumulated particles at the locus. Surface samples are most common; collect spoon-size samples from either the top layer of a soil or to a depth of any existing impressions. In locations where soil has been disturbed or removed from below the surface, including holes and ruts, collect soil from any visually distinct areas separately. Burial sites are special situations, as shown in

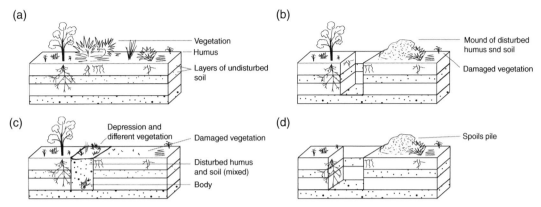

Figure 7.3 Mixing of soil from burial site: (a) cross-section of site before burial; (b) cross-section of site after digging hole for burial; (c) cross-section of site after burial and refilling hole; (d) cross-section of site after excavation to remove buried object

Figure 7.3. Before the burial (Figure 7.3a) the soil will likely be undisturbed, supporting plant growth, with natural horizons. As the hole is dug (Figure 7.3b) before the object is buried, the plants, top soil, and soil from the horizons will be mixed and thrown into a spoils pile, damaging the plants on the surface next to the hole. The object is then buried and the hole refilled (Figure 7.3c) primarily with soil from the spoils pile, further mixing

---

Shovel mark – Michigan

In 1959, while investigating a breaking and entering, a Michigan State Police Trooper was shot and killed and buried in a shallow grave. When the suspect finally led detectives to the burial site, Charlie Meyers and his Crime Laboratory colleagues carefully excavated the body and collected soil evidence. Meyers describes the evidence. "At this site, a couple of clods of dried soil were noted to be relatively intact and displayed a shape indicative of falling off a shovel blade. One clod in particular displayed numerous striated markings, caused by imperfections in the tip of the shovel blade." When the suspect's shovel was recovered, soil samples from the shovel were associated with the crime scene. "In this case, striated markings on the clod from the scene and test markings made with the shovel blade were strikingly similar and there could be no doubt that this shovel had been used to bury the officer's gear (Meyers, 2004)."

---

it, after which it is allowed to settle into a depression of compacted soil where new and different plants will grow. Finally, the burial is discovered, excavated using archaeological techniques (Figure 7.3d), including sifting the soil to recover trace evidence, leaving a sifted soil pile, after which the object is recovered. The spoils pile and backfill are normally mixed soil and often mottled in color and texture. It is always possible that questioned soil from a shovel or boots could be like some of the soil in the backfill. During the excavation, look for any impressions (shoe, shovel) in the backfill and sidewalls of the hole. Finally, collect column samples from the excavated hole where the best-lighted face is troweled clean and vertical revealing the entire soil profile (Figure 7.3d). Cut blocks of soil from each visually distinct (color or texture) layer. Preserve any layer structure during the collection, packaging, and submission to the laboratory.

7.8 Initial comparisons

When it is time to make comparisons, before disturbing the structures use a stereomicroscope to visually compare the overall nature of the known and questioned samples with regard to structure, color, texture, and particle types therein. If the questioned and known samples are undeniably different, no further analysis is necessary; report the samples as dissimilar. If the samples cannot be distinguished, separate any rock fragments and botanical components such as seeds, leaf fragments, twigs, and grass; separate any intact soil clumps and note their size, shape, color, and any layers; separate manufactured materials such as paint chips, glass fragments, and fibers; and use a magnet to separate any susceptible particles. If helpful, pass the sample through a sieve or place the sample in water to allow the constituents to separate. Compare the color and morphology of any sand grains. Compare any botanical and manufactured portions with known samples in the ways commonly used for trace evidence analysis. If the two samples still cannot be distinguished and if bulk soil samples of sufficient quantity are available, begin a more systematic comparison of their color and texture. Treat small-consolidated soil samples, natural and persistent aggregates (peds) with structure (blocky, prismatic, platy, or granular), or artificially layered soil fragments and clods as individual samples.

7.9 Color comparison

Although soil has many features, including structure, texture (proportions of sand, silt, and clay), and chemical composition useful to pedologists, color is the most useful feature for characterizing and comparing soils in a forensic case. Soil color reflects its genesis including: climate, drainage, age, parent materials, geochemistry, organic matter accumulation, weathering, leaching, erosion, accidental and purposeful additions, and cultural activities. As shown in Figure 7.4, the colors usually are a mixture of white, black, red, and yellow. The color within a soil sample may vary with layers and mottling. Describe the number and thickness of layers along with each color; describe mottles with regard to abundance, color, size, and contrast.

In addition to describing the soil color in colloquial terms, as in Figure 7.4, colors may be described in more systematic terms by using a Munsell® Soil Color Chart, where the

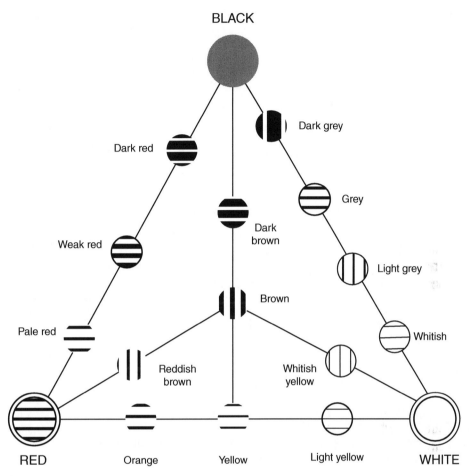

Figure 7.4 Soil color triangle. Soil color is a mixture of black, white, red, and yellow

hue (dominant spectral color), value (relative lightness) and chroma (purity and strength of color) are estimated by comparing the soil with color chips in the chart. Figure 7.5 illustrates the arrangement of hue, value, and chroma in the Munsell color space along with the common names associated with the hue card 10R, as an example. The soil color can be named in two complementary ways: color names or Munsell notation. Although more systematic and using Munsell notation seems more precise, the notation remains somewhat subjective when selecting the appropriate hue card, determining colors that are intermediate between the hues in the chart and distinguishing between value and chroma, where chromas are strong. To improve objectivity, it is also possible to use a colorimeter to measure soil color.

The Munsell notations are for the sole purpose of recording observations made during the sample examination. The notations and names are often useful when asked, "What color was the soil?" In no case should the names or Munsell notations be used for the

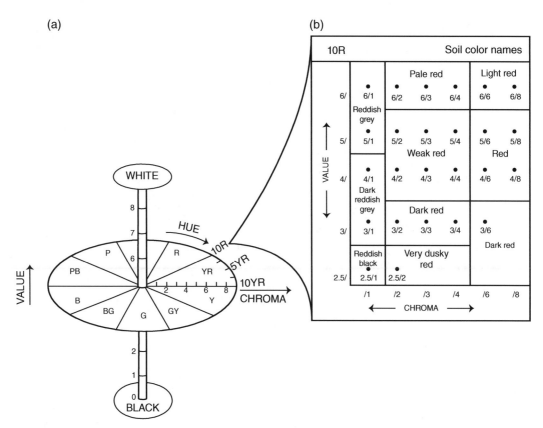

Figure 7.5 (a) The colors of Munsell Soil Color Hue 10R; (b) The Munsell color space and Munsell 10R soil names. The soil color can be named in two complementary ways: Munsell notation or color names

actual forensic comparison of soil samples. In the end, visually compare the questioned and known soil samples side-by-side under like conditions.

When comparing soil color, the conditions must be carefully controlled in order to ensure that comparisons are made between representative samples prepared and observed in the same way. How the soils are separated or blended, how they are moistened, dried, or treated, and in what light they are compared all must be carefully controlled. Prepare representative known soil samples in a way that mimics the nature of the questioned samples: clods, crumbs, dust, or smears, for example.

When sufficient soil is available, after observation and documentation of structures, a portion may be homogenized by lightly grinding; the samples can be split, mixed, blended, and quartered to prepare more manageable representative samples for analysis and comparison. If desired, vary the moisture content, organic content, and chemistry. For example, the samples can be wetted to field capacity (saturated to excess water), dried at room temperature (~20°C), oven dried (~100°C), ashed to remove organic materials (~500°C) or ignited to remove carbonates (~1000°C). Separate different size

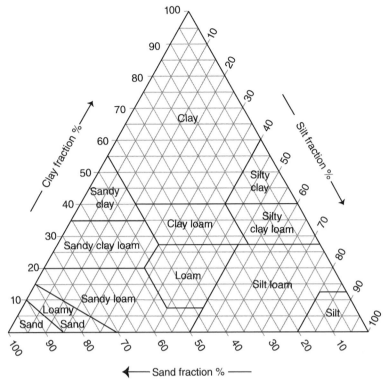

Figure 7.6 Soil texture classification triangle. To classify a soil, measure the relative percentages of clay, silt, and sand. From any two values, follow the lines within the ternary diagram until they intersect; the soil name will be within the box at the intersection

fractions by sieving or sedimentation and compare the color of each size fraction, using corresponding fractions and preparations. In all cases, compare the colors with the same type of light source: incandescent, fluorescent, or daylight from a north window. If the soils samples cannot be distinguished by color, continue the process by comparing the texture followed by comparison of the soil constituents.

7.10 Texture comparison

Texture is a fundamental soil property; when sufficient soil is available, texture is a useful way to describe and compare soils. Soil texture is named by the relative proportions of sand, silt, and clay as indicated in the ternary diagram in Figure 7.6. The conventional particle size limits in soil and the corresponding sieves are listed in Table 7.3.

When sufficient soil in an individual sample is available, after observation and documentation of structures, a portion may be homogenized and split as needed. In order to accurately measure the particle sizes of individual soil grains, the soil must be disaggregated and dispersed in some way into ultimate particles so that each grain is

Table 7.3 Sieve sizes and soil particle size grades

Mesh	Aperture (mm)	Aperture (µm)	Grade
6	3.36	3360	GRAVEL
7	2.83	2830	
8	2.38	2380	
10	2.00	2000	
12	1.68	1680	
14	1.41	1410	
16	1.19	1190	
18	1.00	1000	COARSE SAND
20	0.84	840	
25	0.71	710	
30	0.59	590	
35	0.50	500	
40	0.42	420	MEDIUM SAND
50	0.30	297	
60	0.25	250	
70	0.21	210	FINE SAND
80	0.18	177	
100	0.15	149	
120	0.13	125	
140	0.11	105	
170	0.09	88	
200	0.07	74	
230	0.06	62	
270	0.05	53	
325	0.04	44	SILT
400	0.04	37	
		CLAY	

separate from all others. The particle size distribution of sand and soil is normally determined by sieve analysis.

After disaggregation, pour the soil sample into the top sieve, either dry or wet. If dry, shake the stack of sieves for a time to allow the grains to pass down through the stack. If wet, pass a stream of water through the stack to wash the grains down through the screens. Wet sieving has the advantage of also cleaning the sand grains of adhering silt and clay. If wet sieved, dry in an oven at approximately 100°C. Weigh the fraction on each sieve screen. Inspect each fraction with a stereomicroscope for any unusual particles of interest; transfer the motes to a separate container. Using a geologist hanging pendulum pivot magnet or pocket magnet pickup tool with the magnet covered with plastic wrap

$$V = \frac{d^2\, g\, (\rho_p - \rho_s)}{18\eta}$$

Figure 7.7 Stokes law of sedimentation. V = terminal settling velocity of a solid particle; d = diameter of settling particle; g = gravitational acceleration; ρ_p = density of settling particle; ρ_s = density of liquid through which particle is settling; η = dynamic viscosity. The larger the particle, the faster it falls

---

Bentonite mud – Colorado

A wife arranged to kill her husband while he camped near a pond. Jacqui Battles, Colorado Bureau of Investigation, Denver, identified bentonite clay on the wife's clothing and associated it with the pond near the husband's camp. The wife claimed the mud on her clothing was from a bog or pond near her own camp, but the mud was inconsistent with the two alibi locations (Murray, 2011).

---

collect any grains susceptible to the magnet, such as metal particles. Select a fraction for further analysis; for example, fine sand is usually selected for mineral identification.

Fine-grained sediments (silt and clay) are disaggregated, dispersed, and separated by sedimentation (settling) in water where the particles settle in accordance with Stokes's Law, a formula for determining the rate of sedimentation (Figure 7.7). The rate can be very slow for particles whose density is close to that of the liquid, for particles whose diameter is small, or where the viscosity is high. Sand grains settle out more rapidly than silt and silt more rapidly than clay. Shake the sample in a tube of water; larger particles settle faster and the smallest clay particles normally remain suspended in the water. After a period of time, compare the relative abundance of the particle sizes by layer thickness and select a size range for analysis as needed. The third, less common method for measuring particle size distribution is by using a microscope to measure the diameters of thousands of individual grains, most often by automated image analysis.

To separate particles by specific gravity, use a watch glass and swirl a sample in water to separate plant fragments and other light particles from the heavier mineral grains (like panning for gold). To separate heavy minerals from the light minerals in the fine sand fraction, use a high-density liquid such as bromoform or polytungstate solution in a 30 mL separating funnel (Figure 7.8); enlarge the hole in the Teflon stopcock to allow sand grains to pass. For smaller samples, ultrasonicate the soil in a heavy-walled glass centrifuge tube with the heavy liquid, freeze the bottom of the tube by dipping in liquid nitrogen, and decant the light minerals with a stream of acetone while the heavy minerals are frozen in the heavy liquid (Figure 7.9); after thawing, mount the heavy minerals for microscopy. Table 7.4 lists a range of light and heavy liquids that can be used with caution for separating particles from soil. The specific gravities of common soil minerals and whether they sink or float in bromoform are listed in Table 7.5. The minerals in the questioned and known samples are then identified and compared.

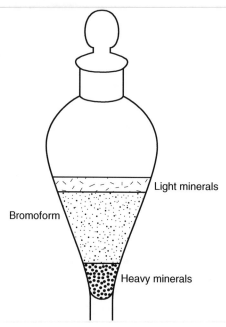

Figure 7.8 Mineral separation in heavy liquid using separating funnel. First, allow the heavy minerals to pass the stopcock, followed by the excess bromoform and light minerals

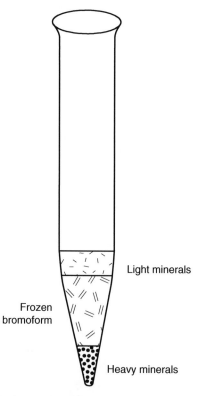

Figure 7.9 Mineral separation in heavy liquid using centrifuge tube. Agitate the selected size fraction in the bromoform and allow grains to settle. Dip the end of the centrifuge tube in a Dewar flask containing liquid nitrogen (LN_2). Decant the light minerals off the top; allow the bromoform to melt and collect the heavy minerals from the bottom

Table 7.4 Density of liquids for soil particle separations

Liquid	Density (g/cm³)	Refractive index	Use
Ethanol	0.79	1.36	
Acetone	0.79	1.36	Rinse bromoform from grains
Xylene	0.87	1.51	
Water	1.00	1.33	Used for panning
Nitrobenzene	1.20	1.55	
Dichlorobenzene	1.29	1.55	
Chloroform	1.48	1.44	
Bromobenzene	1.50	1.56	
Carbon tetrachloride	1.58	1.46	Separate light minerals from plant debris
Iodonaphthalene	1.73	1.70	
Dibromobenzene	2.26	1.57	
Bromoform	2.89	1.60	
Sodium polytunstate solution	2.45 2.89	1.56	Water soluble and non-toxic
Tetrabromomethane	2.96	1.64	
Methylene iodide	3.33	1.73	
Clerici solution	4.25	1.70	Dilute with water to lower density

7.11 Mineral comparison

Soil and geologic microtraces contain mostly mineral grains; therefore, accurate mineral identification is an essential step in forensic comparisons. Fortunately, the sciences of petrology and mineralogy provide a solid foundation for mineral identification. Petrology is the study of rocks, their origin, forms, texture (grain size), mineral content, and geographic distribution. Rocks are comprised of minerals; mineralogy is the study of minerals. Minerals are naturally occurring solid substances with definite chemical composition, primarily formed by inorganic processes, with highly ordered atomic arrangement in a crystalline form that is a homogeneous solid that cannot be physically subdivided. The rocks of the mountain, the sand on the beach, the soil in the garden and many of the products we use daily are completely or in large part made from minerals. In the 18th and 19th century, scholars standardized mineral nomenclature and demonstrated how crystals were formed from chemical compounds.

In 1828, William Nicol invented the Nicol polarizing prism and built a polarizing microscope, which allowed the study of optical properties of minerals. By 1850, Henry Sorby used polarized light microscopy to study minerals in thin rock sections and became known as the Father of Modern Petrographic Methods. Although the microscopic study of minerals in thin section initiated by Sorby in the mid-19th century is still an essential tool in the study of geological materials, the establishment of the immersion method, the only method for determining the

Table 7.5 Mineral specific gravity

	Heft	Specific gravity	Mineral
Floats in bromoform	Very light	<2	Borax
	Light	2–2.4	Gypsum Halite Sodalite
	Average	2.5–2.8	Calcite Dolomite Feldspar Muscovite Quartz Talc
Sinks in Bromoform	Above average	2.9–4	Apatite Biotite Epidote Fluorite Hornblende Kyanite Limonite Olivine Pyroxene Sillimanite Staurolite Tourmaline
	Heavy	4–5	Barite Corundum Garnet Rutile Sphalerite Zircon
	Very heavy	5–10	Galena Hematite Magnetite Pyrite
	Extremely heavy	>10	Gold Platinum Silver

optical properties of minute, translucent solid particles, required in the study of soil and geologic microtraces, evolved during the early 20th century, primarily after Winchell and Winchell's *Elements of Optical Mineralogy* was published in 1909 (Winchell and Winchell, 1909).

For purposes of soil and geologic microtraces comparisons, two approaches to mineral identification are useful. First, develop the necessary skills to understand and use optical mineralogy to identify minerals. The principles, techniques, and descriptions can be found in numerous optical mineralogy reference books; some are suggested in the recommended reading at the end of this chapter. Second, learn how to identify common minerals found in soils, geologic products and dust using features that are learned from practice so that minerals are rapidly recognized in a single mountant.

Microslide preparations are necessary in order to study the morphological and optical properties of minerals with a microscope. Rocks are usually studied in petrographic thin sections where slabs are cut to approximately 30 μm in thickness and then polished to facilitate their microscopic study. Likewise, consolidated soil samples can be collected from soil horizons, impregnated with a resin, cut, and polished like a rock slab. The mineral composition, weathering, pore space, and so on of these micromorphology samples are then studied. Unconsolidated lithic, gravel, and sand grains from soil can also be dispersed in a resin, cut into 30 μm thick sections, polished, and mounted for polarized light microscopy. The advantage is that now each grain is approximately 30 μm thick and can be characterized with the features learned for petrography.

Most often, soil and other geologic microtraces are prepared as grain mounts in the following way. Scatter a portion of the whole sample or a useful size fraction (from sieving) in a mounting medium of known refractive index on a clean microscope slide; then tip a coverslip onto the preparation. If in a liquid, roll the grains with the coverslip as needed to disperse them and to observe optical properties in different orientations; wave a magnet over the preparation to wiggle magnetic minerals in order to assist in the identification of opaque minerals; and, observe their reflected light color and luster with epi-illumination. Commonly, for rapid identification of soil minerals and modal analysis, the grains are mounted in an index liquid or other mountant with a refractive index between 1.53 and 1.55; heavy minerals are better dispersed in a mountant with a refractive index of 1.66.

The polarizing microscope is used to observe a plethora of optical characteristics useful for mineral identification. Unfortunately, a thorough discussion of the techniques of optical mineralogy is clearly impossible in a few paragraphs. Nevertheless, this brief description of the optical properties of minerals and how they are measured can provide a basis for further reading, training and practice. Optical mineralogy textbooks will provide more theory, detailed instruction regarding these techniques and descriptions of the optical properties of specific minerals. Images and diagrams of optical properties and videos illustrating these techniques can be found on the Internet; search Google images and follow the links to web pages and YouTube. The following microscopical skills must be learned and practiced until proficient in order to use the textbooks to approach the problem of identifying unknown minerals.

Adjust the microscope for maximum resolution. Center the stage, objectives, condenser, and field diaphragm to ensure Köhler illumination. The microscope should be equipped with analyzer and polarizer in order to use both plane polarized light (uncrossed polars) and crossed polarized light with orthoscopic illumination, in the way all compound microscopes are used. It should also have a Bertrand lens to view the image at the back focal plane of the objective; if none is present, remove an eyepiece to view the objective back focal plane and observe interference figures with conoscopic illumination.

Observe isotropy and anisotropy. Isotropic minerals, such as garnet, are those that are not doubly refracting; they have a crystalline structure that passes light through all directions at the same velocity. Therefore, with crossed polarized light (crossed polars), there is no retardation and no interference colors – they go to black. As an illustration, compare isotropic ground glass with anisotropic conchoidally fractured quartz. Anisotropic minerals are those where light passes through different mutually perpendicular paths at different velocities; they are doubly refracting, causing retardation and producing interference colors when the polars are crossed. While viewing with crossed polars, normally, an anisotropic grain becomes dark, called extinction, at four positions each time the stage is rotated 360 degrees, usually at 90 degrees; but, subtle inclination angles are characteristic of some minerals. The retardation and interference colors of a grain using crossed polars will depend on the material's birefringence (difference in refractive indices in two different orientations) and thickness through which the light passes. The colors and relationships between birefringence and thickness are illustrated in a Michel–Lévi Interference Color Chart.

Observe relief using plane polarized light. The darkness and width of the contours of a mineral grain increase with increasing difference between the refractive index of the particle and that of the mounting medium – the relief. The difference or relative refractive index is determined by the Becke line test. With the condenser diaphragm closed down and while focusing away from the specimen, usually by lowering the microscope stage, observe whether the Becke line's bright halo travels away from the grain or toward it; the Becke line always moves toward the medium of higher refractive index. Compare quartz and corundum in a mountant with a refractive index of 1.66, as an example. When the grain has the same refractive index as the mounting medium, it will disappear.

Observe color and pleochroism. Pleochroism is due to different light absorption in different optic directions of anisotropic minerals. As an example, observe the color of hornblende using plane polarized light and observe any color change during stage rotation.

Observe any twinning. Crystal twins are an orderly and symmetrical intergrowth of two crystals of the same mineral. Observe tartan twinning, a pattern consisting of crisscrossed horizontal and vertical bands, in microcline or the alternating light and darks bands in albite with crossed polars.

Determine sign of elongation. Using crossed polars, observe the sign of elongation, length slow or length fast, by orienting an elongated crystal's prominent face or cleavage plane to a SW–NE position (assuming the polarizer is oriented E–W); then insert the first-order red wave plate (530 nm compensator) and observe color changes along the grain. If the grain has a low birefringence and the interference colors are light gray, the grain will turn either blue or yellow: blue indicates a positive elongation, or length slow; yellow indicates a negative elongation, or length fast.

Table 7.6 Optical properties of minerals

Plane polarized light	Crossed polarized light	
Orthoscopic	Orthoscopic	Conoscopic
Color	Birefringence	Uniaxial
Pleochroism	Extinction	Biaxial
Refractive indices	Isotropy or anisotropy	Optic sign
Relief	Sign of elongation Twinning	Optic axial angle (2V)

Observe interference figures. Learn to recognize and interpret interference figures and find the optic axes. An optic axis is an isotropic direction of light travel in an otherwise anisotropic mineral; one is present in uniaxial minerals, and two are present in biaxial minerals. Locate grains using crossed polars at lower magnifications; optic axis figures are found in a grain that remains dark gray during stage rotation. View interference figures using crossed polars with conoscopic illumination (Bertrand lens inserted into the light path) at high magnifications. Compare the interference figures of uniaxial quartz or zircon and biaxial epidote or olivine. Determine the optic sign using a first-order compensator superimposed on an optic axis interference figure. For example, with a uniaxial mineral, given a centered optic axis figure (black cross with "arms" (isogyres) pointing N–S and E–W), observe the color change in the northeast quadrant near where the two isogyres cross (the melatope), which marks the point where the optic axis emerges; if blue the optic sign is positive and if yellow, negative. Compare the angle between optic axes (2V) using thin basal cleavage flakes of biotite and muscovite (both biaxial micaceous minerals).

In order to identify minerals, compare the optical properties of questioned mineral grains, as determined from the techniques described here and listed in Table 7.6, with the descriptions of likely minerals and the identification tables in an optical mineralogy textbook, such as William Nesse's *Introduction to Optical Mineralogy* (Nesse, 2004), or on a reliable web site (Box 7.1). For example, if a biaxial mineral cannot be identified readily, using the necessary refractive index liquids with the Becke line test, determine the refractive index perpendicular to the optic axis ($n\beta$), where the grain is in an isotropic orientation (dark gray interference colors), and use an identification table of biaxial minerals sorted by increasing beta to get started. Confirm the mineral identification by verifying physical characteristics and other optical properties. Learn the techniques by practicing with correctly identified comminuted minerals.

7.12 Modal analysis

Once the minerals and other particles in the sample have been identified, the measurement of the relative number of each mineral or particle type is the means by which to compare with other samples. The modal analysis of soils is understood to be the determination of the mineralogical composition. If two soil samples are from the same source, the relative abundance of the constituents should be similar. Therefore, the relative abundance is either estimated or measured by modal analysis. Some might choose to use bulk chemical analyses to compare the compositions of samples, but they are indirect means and do not actually determine the morphology and phases of the minerals or the nature of other ecological and anthropogenic constituents so valuable to a forensic comparison of soil and geologic microtraces.

Normally, modal analysis involves tallying the different minerals by counting grains from the soil's fine sand fraction. Generally, 300 or more grains must be counted in order to provide the number percentage with some statistical significance; with less than 300, the counting error increases rapidly, whereas with more than 300 it decreases slowly. However, when comparing heavy mineral suites, less than 300 grains may be sufficient; a count of 100 grains has commonly been used. Normally, a single size fraction is used for modal analysis in order to produce a number percentage of minerals in each sample (Galehouse, 1969).

The identification of 300 mineral grains in each sample can be an arduous task if the microscopist must stop at each grain and study it orthoscopically and coniscopically as normally required in optical mineralogy. A more practical approach is to first study the mineral suite, identify the types of minerals using the techniques of optical mineralogy as needed. The types may include not only the mineral species but also the variety of colors and forms of the minerals. Then, by memorizing the morphomicroscopic features and key optical properties of the minerals in the preparation, rapidly scan the preparation again and count each of the featured grains. For example, feldspars can be rapidly classified according to whether the grain's refractive index is greater or less than 1.53–1.54, as in Table 7.7. Be careful not to misidentify or ignore unusual minerals in the preparation. Finally, count the oddities and other ecological and anthropogenic particles. Tables 7.8 and 7.9 are examples of schemes for rapidly and systematically classifying mineral grains; Graves provided another approach (Graves, 1979).

Table 7.7 Refractive indices of feldspars

	Name	Formula	Refractive index	
			Lowest (α)	Highest (γ)
Alkali Feldspars	Sanidine	$(K, Na)AlSi_3O_8$	1.518	1.531
	Orthoclase		1.518	1.530
	Microcline		1.517	1.530
	Anorthoclase		1.519	1.536
	Adularia	$KAlSi_3O_8$	1.518	1.526
Plagioclase Feldspars	Albite	$NaAlSi_3O_8$	1.528	1.542
	Oligoclase		1.533	1.552
	Andesine		1.543	1.562
	Labradorite		1.554	1.573
	Bytownite		1.563	1.583
	Anorthite	$CaAl_2Si_2O_8$	1.572	1.588

Table 7.8 Rapid mineral identification by polarized light microscopy: fine sand grains mounted in medium with refractive index of 1.53

Group	Class	Mineral	Transmitted color	Reflected color	Relative refractive index	Interference color	Form	Fracture cleavage	Distinguishing feature
I	Tectosilicate	Quartz	Transparent	Transparent	>1.53	gray & central tints	Equant	Concoidal	Fusi
		Plagioclase Feldspar	Transparent	Coating	>1.53	colors & tan center	Flattened	Terraced	
		Alkali Feldspar	Transparent	Coating	<1.53	colors & tan center	Tabular	Pitted	
		Microcline	Transparent	Coating	<1.53	colors & tan center	Tabular	Pitted	Cross hatch twinning
II	Nesosilicate	Garnet	Color	Transparent	>>1.53	Isotropic	Irregular	Concoidal	Opaque inclusions
		Sphene	Transparent	Transparent	>>1.53				Ultra-blue at extinction
		Olivine	Olive green	Pitted	>>1.53	High			
		Zircon	Transparent	Transparent	>>1.53	High	Prismatic	Fracture	
III	Pseudo-silicates	Rutile	Transparent	Transparent	>>>1.53	High	Prismatic	Concoidal	
		Glass	Transparent	Transparent	~1.53	Isotropic			Bubbles
IV	Inosilicate	Pyroxene	Green & brown	Transparent	>1.53	Colors			Augite
		Amphibole	Blue	Transparent	>1.53	Colors (+)	Prismatic	56/124°	Hornblende pleochroic (+) sign
V	Sorosilicate	Epidote	Yellow	Transparent	>1.53	High anomalous	Angular		Pleochroic & twinning

(Continued)

Table 7.8 (*Continued*)

Group	Class	Mineral	Transmitted color	Reflected color	Relative refractive index	Interference color	Form	Fracture cleavage	Distinguishing feature
VI	Phyllosilicate	Phylogopite	Tan	Transparent	>1.53	Black	Thin plates	Basal	Birds-eye
		Biotite	Tan	Transparent	>1.53	Black	Thin plates	Basal	Birds-eye
		Chlorite	Yellow	Transparent	>1.53	Gray anomalous	Platy		
		Serpentine	Green	Transparent	>1.53	Gray anomalous	Platy		
		Muscovite	Transparent	Transparent	~1.53	High	Splintered	Basal	Sparkling
		Talc	Transparent	Transparent	~1.53	High	Splintered	Basal	Sparkling
VII	Cyclosilicate	Tourmaline	Brown	Transparent	>1.53	High (−)	Irregular	Striations	Pleochroic (−) sign
VIII	Carbonates	Calcite	White	Transparent	>1.53	High	Macro/micro crystal	Rhombohedra	Twinkling
		Dolomite	White	Transparent	>1.53	High	Macro/micro crystal	Rhombohedral	Twinkling
IX	Sulfates	Gypsum	Transparent	Transparent	<1.53	Low	Thin plates	Fibrous	Pitted
X	Opaque	Pyrite	Black	Brassy			Polygonal		
		Siderite	Black	Black			Round		
		Spheruliticpyrite	Black	Black			Round		Non-magnetic
		Magnetite	Black	Black			Round		Magnetic
XI	Aggregates	Hematite	Red	Red	>>1.53	Low	Aggregate		
		Limonite	Brown	Yellow	>>1.53	Low	Aggregate		

Table 7.9 Rapid mineral identification by polarized light microscopy: mineral grains mounted in medium with refractive index of 1.66

Group	Refraction	Transmitted color	Relative refractive index	Mineral	Fracture cleavage	Axi	Sign
I	Isotropic	colorless	<<1.66	Fluorite	Octahedral		
				Opal	Conchoidal		
				Pumice	Vesicular		
II		colored	>1.66	Garnet	Conchoidal		
				Sphalerite	Excellent		
III	Anisotropic	Colorless	>1.66	Clinozoisite	Flakes	Biaxial	+
				Corundum	Rounded	Uniaxial	−
				Kyanite	Square blades	Biaxial	−
				Sphene	Diamond shaped	Biaxial	+
				Zircon	Rounded	Uniaxial	+
IV		Colored	>1.66	Augite		Biaxial	+
		Green		Biotite	Micaceous	Biaxial	−
		Brown		Diopside		Biaxial	+
		Green Pleochroic		Epidote		Biaxial	−
		Green Pleochroic		Hematite	Conchoidal	Uniaxial	−
		Red		Hornblende	Flate blades	Biaxial	+
		Green Pleochroic		Hypersthene		Uniaxial	−
		Green Pleochroic		Malachite		Biaxial	−
		Green		Rutile	Conchoidal	Uniaxial	+
		Yellow Pleochroic		Staurolite		Biaxial	+
		Yellow					

(Continued)

Table 7.9 (*Continued*)

Group	Refraction	Transmitted color	Relative refractive index	Mineral	Fracture cleavage	Axi	Sign
V		Colorless	<1.66	Calcite	Rhombohedral	Uniaxial	−
				Dolomite	Rhombohedral	Uniaxial	−
				Apatite	Conchoidal	Uniaxial	−
				Tourmaline	Blades	Uniaxial	−
				Barite	Tabular	Biaxial	+
				Sillimanite	Prisms and needles	Biaxial	+
				Topaz	Conchoidal	Biaxial	+
VI			<<1.66	Anorthite	Tabular and twinned	Biaxial	−
				Beryl	Conchoidal	Uniaxial	−
				Muscovite	Micaceous	Biaxial	−
				Quartz	Conchoidal	Uniaxial	+
VII			<<<1.66	Albite	Albite twinning	Biaxial	+or−
				Gypsum	Arrowhead twins	Biaxial	+
				Microcline	Crossed lamellar twin	Biaxial	−
				Orthoclase		Biaxial	−

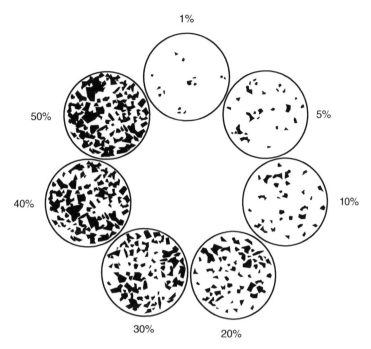

Figure 7.10 Estimating relative abundance of grain types. By observing several fields of view, use these illustrations to assist in estimating the relative abundance of a particular particle type in the preparation

A systematic counting method for determining the relative abundance of various minerals in grain mounts was introduced by Fleet in 1926 (Fleet, 1926). Until then, the abundance was visually estimated more subjectively. For example, the illustration in Figure 7.10 is sometimes used to assist with estimating the relative abundance of the various minerals in a petrographic section or sand sample.

Fleet's method requires that the selected size fraction be mounted and all the grains identified and counted. More commonly, in order to improve efficiency, only a random sample of the preparation is counted, as indicated in Figure 7.11. Instead of counting all the grains on the entire slide, count only grains in a representative area using a reticle to define the area; move the stage randomly and count each type of mineral within the square until 300 or so grains are tallied. Similarly, using the ribbon method, move the slide along the stage in a straight line and count all the grains that pass between two horizontal lines. In order to keep the scan in a line and to prevent counting grains more than once, control the slide with a mechanical stage, moving it in only one direction, or slide the preparation between two microscope slides taped to the stage. In these ways, the stage can be rotated as needed to observe optical properties of the mineral grains, while maintaining the linear movement of the slide.

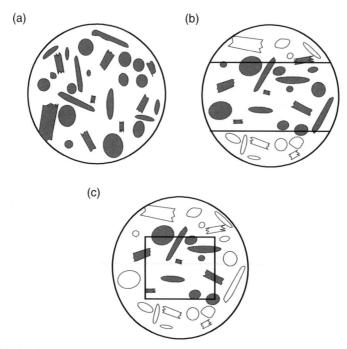

Figure 7.11 Methods of counting grain types: (a) Fleet method – all the grains are counted in each field of view; (b) ribbon method – all the grains between the two lines are counted as the slide traverses the stage; and (c) area method – all the grains within a selected area counted as the field of view is randomly moved from place to place

7.13 Automated instrumental modal analysis

The identification of hundreds of mineral grains in a grain mount can be the most difficult and time consuming task when comparing soils. Automated instrumental methods might be a more practical way to identify and quantify the minerals in the sample. For example, powder X-ray diffraction (XRD) is a classic method for mineral identification and can be used to quantitatively measure the relative abundance of minerals in a bulk sample. On the other hand, XRD can only identify crystalline materials and cannot identify amorphous particles such as phytoliths, diatoms, or pollen in the sample, let alone paint and glass particles.

Consistent with the particle approach to soil and geologic microtrace comparisons, particle characterization by X-ray microanalysis can be used to assist with the identification of individual mineral grains and to quantify (count) their relative abundance in a sample. Unattended particle-by-particle automated analyses can be achieved with a scanning electron microscope (SEM) (Pirrie, 2004). Energy dispersive X-ray spectrometry (EDS) is most often used, as it is more readily available and has been validated for other common trace evidence analyses, such as paint comparisons and gunshot residue (GSR) identification. Thousands of particles can be analyzed unattended in short order providing a sound statistical base. The more difficult task is then classifying the particles by composition in order to compare the inventory with other samples.

In soil comparisons, the task is essentially one of identifying each mineral grain, one by one, with a polarizing microscope, and tallying the number of each mineral type present in the sample. Doing the same job by automated SEM-EDS requires that each optically identifiable mineral has a distinct elemental composition that can be recognized by the software. In practice, not all mineral types can be identified in this way, as crystallographic forms that may be optically distinguished can have identical elemental compositions, such as quartz and opal or calcite and shell. On the other hand, SEM-EDS can easily distinguish between some minerals, such as rounded quartz and untwined plagioclase feldspars (oligoclase), that are not rapidly distinguished in grain mounts by polarized light microscopy, without observing interference figures. Likewise, SEM-EDS can readily distinguish some opaque heavy minerals (McVicar and Graves, 1997).

Modal analysis requires the automated SEM-EDS to briefly stop and analyze all particles on the stub without regard to their composition. The SEM detects a particle by contrast between the particle and background or other particles; therefore, ideally, a sample should be prepared so that all particles are separated from each other. Particles that are touching each other, sitting on each other or in some way closely juxtaposed will likely be mistaken by the SEM for a single particle, thereby confusing the interpretation of the EDS spectrum and preventing the correct identification. To minimize overlaps, the Centre of Forensic Sciences developed reverse sieving for automated SEM-EDS (McVicar and Graves, 1997). Particles in the fraction on a single sieve's screen have a diameter range between the larger sieve opening directly above and the smaller mesh they rest on; smaller particles pass through. To reverse sieve, the particles on the lower sieve (smaller openings) are passed back through the sieve above (larger openings) upon which is stuck, on the reverse side of the screen, a SEM pin stub specimen mount with an adhesive conductive-carbon tab. As each grain passes through the screen, single grains adhere to the stub in a regular array (Figure 7.12).

Although soil and geologic microtraces contain mostly mineral particles and their identification and modal analysis is paramount, they also will likely contain other particle types from the environment, some indistinguishable chemically from minerals (diatoms) and others organic (pollen) that may be elementally indistinguishable from background in the SEM. These other constituents are equally useful for further differentiating samples and for confirming an association.

Figure 7.12 Fine sand grains reversed sieved onto SEM stub. As the size fraction is passed through the next larger sieve openings, the grains line up in a pattern like the screen

7.14 Ecological constituents

In addition to mineral grains, soil and geologic microtraces contain particles from the ecological environment of both geologic and biologic origins. Biologic grains are of particular value for comparisons because of their remarkable durability in soils and sediments. For example, animals leave microfossils (diatoms, foraminifera, and radiolaria) in aquatic sediments, and plants leave macroflora (seeds, wood, and leaf fragments) and microflora (pollen, starch grains, and bioliths such as phytoliths) in soil where the plants grow, die, and decompose. Diatoms, foraminifera and radiolaria are microscopic unicellular organisms with a silica skeleton that are taxonomically classified based on their morphology, structure, and ornamentation. Phytoliths are biosiliceous particles formed in plant tissues, usually composed of hydrated silica; the greatest variety of shapes are found in grasses, sedges, and cereals. Morphomicroscopical analysis of phytoliths can be an effective way to discriminate soil samples that experienced different land use, even though indistinguishable by mineral composition (Yanai and Marumo, 1986).

Like phytoliths, pollen might also be found in the soil and dust. The forensic microscopist can recognize them in airborne dust and separate them from the soil for comparison in order to help distinguish samples that are otherwise similar. These palynomorph assemblages include modern and fossil pollen, spores, and other acid-resistant plant materials originating from plants present at and near the site. Pollen is an ideal microtrace because the grains are small, highly variable, and found on almost any item that comes in contact with the out-of-doors. The wide variety of shapes, sizes, and surface characteristics of pollen grains are easily recognized and compared. Nevertheless, although routinely used in Great Britain, Australia, and New Zealand, where they have come to appreciate the full potential of pollen analysis in their investigations, unfortunately, pollen analysis has seen extremely limited use in the United States forensic science community, probably because of a lack of expertise in most crime laboratories (Korejwo *et al.*, 2014).

Although the fine sand fraction is most often used for mineral identification, other fractions might contain pollen, starch, diatoms, or phytoliths. The biosiliceous particles are sometimes best discovered by treating a water suspension of the soil, after the sand has settled, with 30% hydrogen peroxide (H_2O_2) and separating them from the soil by flotation in a liquid with a density of about 2.3. The separation of pollen from soil sometimes requires treatment with hydrofluoric acid (HF) followed by acetolysis.

Shells – California

A mother and her children accepted a ride from an acquaintance who then raped and attempted to kill the mother. The suspect was identified and arrested; although he showered and cleaned his clothes, he forgot to clean his shoes. The mother led the police to the crime scene and was able to pinpoint the place on the riverbank where the assault had taken place. Marianne Stam, California Department of Justice, Riverside, visited the crime scene and collected soil samples from the spot of the assault and surrounding area. While at the scene, Stam observed the various types of soil at the scene and observed a different windblown deposit composed of mostly quartz and small shells of a fresh water species. A combination of the soil's color, texture, mineral grain shapes, and heavy and light mineral composition along with the shells associated the soil on the suspect's shoes with the crime scene (Murray, 2011).

Pollen and building materials – Maryland

A serial rapist's modus operandi included covering the victim's head with a shirt after which he left the shirts at the crime scenes. Microscopical study of dust vacuumed from the shirts revealed finely divided drywall particles, paint overspray particles and oak pollen. This allowed Skip Palenik, Microtrace, LLC, Elgin, IL, to conclude: "Both shirts were worn by the same person. The subject was employed working indoors as a drywall installer and finisher. He worked on large-scale commercial projects and was, therefore, probably a professional. He wore one of the shirts in the early spring while he was working in the vicinity of oak trees." After the information was released to the public, a suspect driving a commercial drywall installation truck was stopped for questioning on another matter and confessed to the rapes even before any questions were asked (Ballou *et al.*, 2013).

7.15 Anthropogenic constituents

In addition to minerals and ecological constituents, anthropogenic materials are often found in the soil and dust. Nearby industries may have dumped slag on the ground or spewed fly ash into the air. Landfills (made land) can contain almost anything. Building materials will be found in soil and dust nearly anywhere people are found. Soil near houses might contain shingle stones from a weathered roof, paint fragments from the pealing siding or brick and mortar particles from the crumbling face brick. In another example, urban street soil often contains: reflective glass beads, road paint, asphalt, tire rubber, auto paint, metal parts, or window glass. Of course, trace evidence from the people at the crime scene can sometimes end up in the soil and dust as well, such as hairs and fibers, or DNA.

Paint, asphalt and shingle stones – Washington

The body of a laborer was found in his own van on a busy residential street. Police noticed soil caked in the wheel wells and noted where some had fallen onto the street directly below the van. The suspect's yard contained numerous muddy areas and debris littered the front and back yards. Soil samples were collected from the driveway, front and back yards and from the surrounding neighborhood. Using stereomicroscopy and polarized light microscopy, Bill Schneck, Washington State Patrol, demonstrated the soil from the van and the yard to be similar in color and mineralogy. In addition, he observed fiberglass coated paint particles of various colors, asphalt, and shingle stones in both questioned and known samples (Murray, 2011).

7.16 Reporting comparison results

When the comparisons are done, the findings need to be communicated in a report. The report should contain analytical results and conclusions, and the basis for an interpretation of the significance of the evidence. Make it perfectly clear what is meant by the words used; explain the meaning of terms and use enough words so that everyone understands the relative certainty of the conclusions and the real significance of the evidence. Although each case is different and requires careful evaluation within the context of all the evidence, with soil and geologic microtraces, as with most associative evidence, the conclusions can be expressed in some form of the ways in Table 7.10.

Table 7.10 Report concepts and definitions

Associations			Non-associations	
Source identity	**Strong association**	**Conventional association**		
Source identity or source attribution is used when broken soil clods are physically matched along an irregular surface demonstrating that the clods were once part of the same object.	Strong association with unusual characteristics is used when two soils cannot be differentiated based on color, texture and observed constituents, including modal analysis and/or chemical composition and in addition contain comparable unusual, unexpected mineralogical, ecological or anthropogenic constituents.	Association with conventional characteristics is used when samples cannot be differentiated based on color, texture, observed constituents including modal analysis and/or chemical composition and where it is likely other soils somewhere share the same properties.		
Inconclusive	**Dissimilar**			**Elimination**
Inconclusive is used when the possibility that the soils originated from the same source can neither be confirmed nor eliminated due to limited sample or the inability to identify particles of interest.	Dissimilar or non-association is used when samples are dissimilar in observed or measured properties, indicating different sources; however, the samples might share enough general corresponding characteristics that, due to site variations that could reasonably be expected, the source itself cannot be eliminated.			Elimination or exclusion is used when the soil samples are different enough to demonstrate that they could not originate from the same source.

A slice of bread – Ohio

A burglar had stepped on a slice of soft, white bread found on the floor at the crime scene. While comparing the shoes from a suspect with the footwear impression on the slice of bread, Mike Trimpe, Hamilton County Coroner's Office Crime Laboratory, Cincinnati, Ohio, found a crumb of soil on the bread which proved similar to soil in the grooves of the sole of the suspect's shoes. In addition to the pattern match and soil comparison, the irregularly-shaped crumb could be physically refit into a missing section of soil trapped in one of the grooves, ultimately confirming a source identity.

7.17 Future directions and research

The continued acceptance of soil and geologic microtraces as reliable evidence will depend, in part, on continued validation of processes and analytical methods to determine their suitability for their intended use of associating crimes with criminals, with particular attention to the question, "What is the possibility of wrongly associating samples when they do not originate from the same source?" Validate the methods with laboratory experiments, using samples that mimic actual questioned samples commonly encountered in real cases that test the reproducibility, ruggedness and robustness of the process under varying sample sizes and circumstances. Evaluate new and novel methods with due consideration for their use with small samples (microtraces) and the use of discriminating target substances that do not change when they are removed from the soil body and transferred to an object or turned to dust. Test the methods with local laboratory surveys to study how often soil samples from different sites are indistinguishable. In all instances, the goal should be to understand the uncertainty in the comparison process and to proffer criteria for associations and exclusions.

Unfortunately, describing a process for soil and geologic microtrace comparisons will be for naught without more awareness about geologic trace evidence by investigators, crime scene investigators, criminalists, geologists, lawyers, and judges. Better connections must be established between experts in the forensic comparison of soil and geologic microtraces and law enforcement agencies, research institutions, and law schools.

Acknowledgments

Thanks go out to my friends Jay Siegel for asking me to contribute this chapter, Ray Murray for his encouragement, and Mike Trimpe for reading the manuscript and offering some valuable suggestions. I also appreciate the help from Jennifer Bisbing with copy-editing and from Emma Bilyk with illustrations. As usual, my wife Bonnie is always helpful with some really good ideas.

References

Ballou, S., Houck, M., Siegal, J.A., *et al.* (2013) Criminalistics: the bedrock of forensic science. In: Ubelaker, D.H. (ed.) *Forensic Science: Current Issues, Future Directions*, John Wiley & Sons Ltd, Chichester, pp.29–101.

Bowen, A. and Stoney, D.J. (2013) A new method for the removal and analysis of small particles adhering to carpet fiber surfaces. *Journal of Forensic Sciences*, **58**, 789–796.

Burney, I. (2013) Our environment in miniature: dust and the early twentieth-century forensic imagination. *Representations (Berkeley)*, **121**, 31–59.

Cwiklik, C. (1999) An evaluation of the significance of transfers of debris: criteria for association and exclusion. *Journal of Forensic Sciences*, **44**, 1136–1150.

Dell, C.I. (1959a) A study of the mineralogical composition of sand in southern Ontario. *Canadian Journal of Soil Science*, **39**, 185–196.

Dell, C.I. (1959b) Methods of study of sand and silt from soils. *Canadian Mineralogist*, **6**, 363–371.

Dell, C.I. (1963) A study of the mineralogical composition of sand in northern Ontario. *Canadian Journal of Soil Science*, **43**, 189–200.

Fleet, W.F. (1926) Petrological notes on the old red sandstone of the West Midlands. *Geological Magazine*, **63**, 505–516.

Galehouse, J.S. (1969) Counting grain mounts: number percentage vs. number frequency. *Journal of Sedimentary Petrology*, **39**, 812–815.

Graves, W.J. (1979) A mineralogical soil classification technique for the forensic scientist. *Journal of Forensic Sciences*, **24**, 323–338.

Hopen, T.J. (2004) The value of soil evidence. In: Houck, M.M. (ed.) *Trace Evidence Analysis: More Cases in Mute Witnesses*, Elsevier Academic Press, London, pp.105–122.

Jackson, R.L. (1962) *Criminal Investigation: A Practical Textbook for Magistrates, Police Officers and Lawyers*, 5th edn (adapted from the *System Der Kriminalistik* of Dr Hans Gross). The Carswell Company Ltd, Toronto, ON.

Junger, E.P. (1996) Assessing the unique characteristics of close-proximity soil samples: just how useful is soil evidence? *Journal of Forensic Sciences*, **41**, 27–34.

Kirk, P.L. (1953) *Crime Investigation: Physical Evidence and the Police Laboratory*. Interscience Publishers, Inc., New York.

Korejwo, D.A., Webb, J.B., Willard, D.A., and Sheehan, T.P. (2014) *Pollen Analysis: an Underutilized Discipline in the U.S. Forensic Science Community*. http://projects.nfstc.org/trace/docs/Trace%20Presentations%20CD-2/Korejwo_paper.pdf (last accessed 20 June 2015).

Lee, B.D., Williamson, T.N., and Graham, R.C. (2002) Identification of stolen rare palm trees by soil morphological and mineralogical properties. *Journal of Forensic Sciences*, **47**, 190–194.

Locard, E. (1930) The analysis of dust traces. *The American Journal of Police Science*, **1**, 276.

Locard, E. (1939) *Manual of Police Techniques*, 3rd edn. Payot, Paris, France.

McCrone, W.C. (1974) Detection and identification of asbestos by microscopical dispersion staining, *Environmental Health Perspectives*, **9**, 57–61.

McCrone, W.C. (1981) The microscopical identification of artists' pigments. *J. IIC-CG: Journal of the International Institute for Conservation, Canadian Group*, **7** (1 & 2), 11–34.

McCrone, W.C. (1982) Soil comparison and identification of constituents. *The Microscope*, **30**, 17–25.

McCrone, W.C. (1996) *Judgement Day for the Turin Shroud*. Microscope Publications, Chicago, IL.

McCrone, W.C., Draftz, R.G., and Delly, J.G. (1967) *The Particle Atlas*. Ann Arbor Science Publishers, Ann Arbor, MI.

McVicar, M.J. and Graves, W.J. (1997) The forensic comparison of soils by automated scanning electron microscopy. *Canadian Society of Forensic Sciences Journal*, **30**, 241–261.

Meyers, C. (2004) *Silent Evidence: Cases from Forensic Science*. Parkway Publishers, Inc., Boone, NC.

Mikesh, R.C. (1973) *Japan's World War II Balloon Bomb Attacks on North America*. Smithsonian Institution Press, Washington, DC.

Murray, R.C. (2011) *Evidence from the Earth: Forensic Geology and Criminal Investigation*, 2nd edn. Mountain Press, Missoula, MT.

Nesse, W.D. (2004) *Introduction to Optical Mineralogy*. Oxford University Press, New York.

Palenik, S. (1979) Microscopy and the Law. *Industrial Research and Development*. Microtrace LLC, Elgin, IL.

Palenik, S. and Palenik, C. (2005) Microscopy and microchemistry of physical evidence. In: Saferstein, R. (ed.) *Forensic Science Handbook*, Vol. **II**. Pearson Education, Prentice Hall, Upper Saddle River, NJ, pp. 175–230.

Pirrie, D., *et al.* (2004) Rapid quantitative mineral and phase analysis using automated scanning electron microscopy (Qemscan): potential applications in forensic geoscience. In: Pye, K. and Croft, D.J. (eds) *Forensic Geoscience: Principles, Techniques and Applications*. Special Publication 232, The Geological Society, London, pp. 123–136.

Pye, K. and Croft, D.J. (eds). (2004). *Forensic Geoscience: Principles, Techniques and Applications*. Special Publication 232, The Geological Society, London.

Rogers, J.D. (2014) *How Geologists Unraveled the Mystery of Japanese Vengeance Balloon Bombs in World War II*. http://web.mst.edu/~rogersda/forensic_geology/japenese%20vengenance%20bombs%20new.htm (last accessed 20 June 2015).

Saferstein, R. (2005) *Forensic Science Handbook*, Vol. **II**. Pearson Education, Prentice Hall, Upper Saddle River, NJ.

Stoney, D.A., Bowen, A.M., Bryant, V.M., *et al.* (2011) Particle combination analysis for predictive source attribution: tracing a shipment of contraband ivory. *Journal of the American Society of Trace Evidence Examiners*, **2**, 13–72.

Taupin, J.M. and Cwiklik, C. (2011) *Scientific Protocols for Forensic Examination of Clothing*. CRC Press, Taylor and Francis Group, Boca Raton, FL.

Thorwald, J. (1967) *Crime and Science: The New Frontier in Criminology*. Harcourt, Brace & World Inc., New York.

Winchell, N.H. and Winchell, A.N. (1909) *Elements of Optical Mineralogy*. D. Van Nostrand Company, New York.

Yanai, H. and Marumo, Y. (1986) Morphological analysis of opal phytoliths for soil discrimination in forensic science investigation. *Journal of Forensic Sciences*, **31**, 1039–1049.

Further reading

Armstrong, H.A. and Brasier, M.D. (2005) *Microfossils*, 2nd edn. Blackwell Publishing, Malden, MA.

Bergslien, E. (2012) *An Introduction to Forensic Geoscience*. John Wiley & Sons Ltd., Chichester, UK.

Bigham, J.M. and Ciolkosz, E.J. (eds) (1993) *Soil Color*. SSSA Special Publication Number **31**. Soil Science Society of America, Inc., Madison, WI.

Bloss, F.D. (1999) *Optical Crystallography*. Mineralogical Society of America, Washington, DC.

Buol, S.W., Southard, R.J., Graham, R.C., and McDaniel, P.A. (2011) *Soil Genesis and Classification*, 6th edn. John Wiley & Sons Ltd., Chichester, UK.

Deer, W.A., Howie, R.A., and Zussman, J. (2013) *An Introduction to the Rock Forming Minerals*, 3rd edn. Mineralogical Society, London.

Dyar, M.D., Gunter, M.E., and Tasa, D. (2008) *Mineralogy and Optical Mineralogy*. Mineralogical Society of America, Chantilly, VA.

Faegri, K., Kaland, P.E., and Krzywinski, K. (1989) *Textbook of Pollen Analysis*, IV edn. John Wiley & Sons, Inc., New York, NY.

Greenberg, G. (2008) *A Grain of Sand: Nature's Secret Wonder*. Voyageur Press, Minneapolis, MN.

Gribble, C.D. and Hall, A.J. (1992) *Optical Mineralogy: Principles and Practice*. CRC Press, New York.

Hester, T.R., Shafer, H.J., and Feder, K.L. (2009) *Field Methods in Archaeology*, 7th edn. Left Coast Press, Inc., Walnut Creek, CA.

Kapp, R.O., Davis, O.K., and King, J.E. (2012) *Ronald O. Kapp's Pollen and Spores*, 2nd edn, 3rd printing. AASP Foundation – The Palynological Society, Dallas, TX.

Klein, C. (2002) *The 22nd Edition of the Manual of Mineral Science*. John Wiley & Sons, Inc., New York, NY.

Lewis, D.W. and McConchie, D. (1994) *Analytical Sedimentology*. Chapman & Hall, New York.

Murray, R.C. and Tedrow, J.C.F. (1992) *Forensic Geology*. Prentice Hall, Upper Saddle River, NJ.

Piperno, D.R. (2006) *Phytoliths: A Comprehensive Guide for Archaeologists and Paleoecologists*. AltaMira Press, Lanham, MD.

Plummer, C.C., Carlson, D.H., and Hammersley, L. (2013) *Physical Geology*, 14th edn. McGraw-Hill, New York.

Pye, K. (2002) *Geological and Soil Evidence: Forensic Applications*. CRC Press, New York.

Raymond, L.A. (2002) *Petrology: The Study of Igneous, Sedimentary, and Metamorphic Rocks*, 2nd edn. Waveland Press, Inc., Long Grove, IL.

Round, F.E., Crawford, R.M., and Mann, D.G. (2007) *The Diatoms: Biology and Morphology of the Genera*. Cambridge University Press, New York.

Ruffell, A. and McKinley, J. (2008) *Geoforensics*. John Wiley & Sons Ltd., Chichester, UK.

Schaetzl, R. and Anderson, S. (2013) *Soils: Genesis and Geomorphology*. Cambridge University Press, New York.

Singer, M.J. and Nunns, D.N. (1999) *Soils: An Introduction*, 4th edn. Prentice Hall, Upper Saddle River, NJ.

Tarbuck, E.J. Lutgens, F.K., and Tasa, D. (2014) *Earth: An Introduction to Physical Geology*, 11th edn. Pearson Education, Inc., Upper Saddle River, NJ.

Torrence, R and Barton, H. (eds) (2006) *Ancient Starch Research*. Left Coast Press. Inc., Walnut Creek, CA.

Traverse, A. (2007) *Paleopalynology*. 2nd edn. Springer, Dorddrecht, The Netherlands.

Ulery, A.L. and Drees, L.R. (eds) (2008) *Methods of Soil Analysis: Part 5 – Mineralogical Methods*. Soil Science Society of America, Inc., Madison, WI.

Welland, M. (2009) *Sand: The Never Ending Story*. University of California Press, Berkeley, CA.

CHAPTER 8

Chemical analysis for the scientific examination of questioned documents

Gerald M. LaPorte

Office of Investigative and Forensic Sciences, National Institute of Justice, Department of Justice, USA

The information ascertained from the examination of questioned documents can be critical to an investigation and, subsequently, used to help resolve disputes in civil litigation and enforce the law in criminal court. Documents can present themselves in numerous formats, including letters, envelopes, packages, calendars, diaries, currencies, identification cards, financial documents, contracts, wills, and business records. Once submitted for forensic analysis, the item is considered a questioned document, which is defined as "...any material containing marks, symbols, or signs that convey meaning or message" (Lindblom, 2006). The forensic examination of questioned documents encompasses a broad spectrum of areas that extend beyond the realm of traditional handwriting analysis (Osborn, 1929; Conway, 1959; Hilton, 1982; Ellen, 1997). However, for the purpose of this chapter, the author has chosen to focus on applications of chemistry related to the materials used to construct a document, such as writing inks, paper, printing inks, and toner.

Generally, requests for the chemical analysis of documents comprise three types. The first is to determine if a portion of a document has been altered or if all of the entries were created contemporaneously. As an example, the procedure for this type of examination would involve the analysis and comparison of writing and/or printing inks from various areas of the document to evaluate whether the entries are consistent with being prepared at or around the same time. If the document is composed of multiple pages then the forensic document examiner will ascertain if a page(s) was substituted through the analysis of the paper, printing inks, or writing inks. The second form of request is to establish where a questioned document originated. Often anonymous letters, such as kidnapping notes, extortion letters, or threatening correspondence, become the focus of an investigation. This could include an analysis of the components of a questioned document to determine the manufacturing source or comparisons with seized materials,

Forensic Chemistry: Fundamentals and Applications, First Edition. Edited by Jay A. Siegel.
© 2016 John Wiley & Sons, Ltd. Published 2016 by John Wiley & Sons, Ltd.

such as writing instruments, paper, and printing devices. The third, and most challenging, request is to determine if a document is authentic with respect to the purported date of preparation. As discussed later, some writing inks have well documented aging properties that allow for estimating their age.

Utilizing an analytical approach for the scientific examination of documents has a long recorded history that dates back over the past century. In 1911, C. Ainsworth Mitchell first published *Science and the Criminal*, where he wrote about the use of the microscope for forged documents, how to distinguish inks in handwriting, and how the scientific examination of documents was used in some notable forgery trials. In 1922, Mitchell published *Documents and Their Scientific Examination* and wrote, "The scientific examination as a branch of forensic chemistry has been strangely neglected, and hitherto there has been no work dealing fully with the subject from all points of view." With a special reference to the chemistry involved in cases of suspected forgery, indeed this was a glimpse into the future of how information could be gleaned from a document through the application of the physical and chemical sciences to expose the truth about whether a document is fraudulent.

While the history of the chemical analysis of documents has been critical to the underpinnings of forensic document examination, the purpose of this chapter is not simply to discuss the evolution of analytical tests used to examine documents, but instead, to provide an overview of current methods – including strengths, limitations, and the interpretation of findings. However, it is first necessary to acknowledge the significant contributions of Richard Brunelle and Antonio Cantu with respect to their efforts in paving a way to the modern era of analytical approaches used to examine questioned documents. Brunelle and Reed (1984) wrote the first comprehensive textbook dedicated to the forensic examination of ink and paper, covering the applications of physical and chemical analysis to the examination of questioned documents. Brunelle and Crawford (2003) provided an update on the forensic analysis and dating of writing ink.

Cantu was the first to introduce forensic scientists to two different approaches of using analytical methods for the dating of documents to include the examination of items in or on documents such as inks, papers, and their components. The first of these is the static approach, which generally applies to methods based on the comparison of components, or ingredients of materials such as ink and paper, to a standard reference collection to determine the first date of production (Cantu, 1995). This enables the determination of the time when items in or on a document first came into existence. The second approach is referred to as the dynamic approach (Cantu, 1996), which incorporates procedures that measure the physical and/or chemical properties of materials that change with time. The changes that occur over a given period can generally be referred to as aging characteristics. It is important that both static and dynamic approaches are considered for any examination when attempting to ascertain the aging parameters because there are limitations to both of these approaches.

8.1 Static approach

The primary advantage of using the static approach is that when a forensic document examiner is able to obtain results about the first introductory date, a conclusion with a high degree of confidence can be reached. The use of chemical analysis is often perceived by law enforcement, attorneys, and jurors as a means to establish unequivocal facts, which is sometimes not the case. Although chemical examinations are often based on fundamental principles of chemistry, like most types of forensic examinations, there is human interpretation required, which does not always result in unequivocal conclusions. Nevertheless, the information gleaned from these types of examinations can still be used to help establish facts and corroborate other findings. There has been little published on the interpretation of evidence with respect to chemical examinations used for document examination, so interpretation is addressed throughout this chapter and is followed with a more extensive discussion at the end of the chapter.

The static approach can be used for any components of a document, such as writing inks, printer inks, paper, stamp pad inks, and even staples. Manufacturers have historically changed ingredients used in document materials because of cost savings, improvements, and changes in consumer demand. Therefore, the materials used to compose a document can change over time. The complexity of using static factors and information about changes will vary. For example, there is historical knowledge about the introduction of ballpoint inks (circa 1943), felt tip pens (early 1960s), roller ball pens (early 1970s), erasable inks (circa 1979), and gel inks (circa 1984). Punctuated changes in the evolution of new products are generally well documented, but great caution still must be exercised in how this information is interpreted. Often, historical information will provide introduction dates based on first commercial availability, but it is possible a manufacturer test marketed a new product before mass commercialization. Because gradations in introduction are not always well defined, it is necessary to consider the gap between the purported date of a document and the first introductory date of a product. For example, gel ink is a water-based ink that was first created in 1984 by Sakura Color Products Corporation of Japan, but the first gel inks did not arrive in the United States until the late 1980s (Wilson *et al.*, 2004). In addition, the manufacturer may have test marketed this new gel ink in 1983 in Japan.

Three different comparative scenarios can be considered to illustrate how a conclusion may vary depending on the circumstances. Assume that an examiner is provided with three different documents in three different cases, all written with gel ink and generated in the United States, and the documents are purported to have been dated in 1979, 1983, and 1984. Indeed, a conclusion with high confidence can be reached in the case of the 1979 document, since gel ink was reportedly released in 1984. The likelihood of the 1983 document being produced with gel ink seems to be quite low, but the confidence in an unequivocal conclusion will decrease because there might be some likelihood, albeit very small, that a gel ink was available because it was test marketed before mass commercialization. When considering the 1984 ink, it is possible that gel inks were obtained in Japan and brought into the United States in 1984, which will increase the likelihood of availability when compared to the 1983 ink. All of this information must be considered in a forensic examination and it will be critical to express these limitations when reporting and testifying.

Changes to an ingredient or component, in contrast to the introduction of new products as discussed above, is another level of consideration, but requires verifiable information. There are well documented introduction dates of specific components, such as the change from oil to glycol-based ballpoint inks or the use of optical brightening agents in paper first detectable using an ultraviolet source around 1950. The development of new dyes and pigments, and knowing when they were first used by the industry, can be used, too. Also, changes in technology can sometimes result in changes to materials such as inks, toners, and paper. For example, the rapid technological advances in inkjet printing over the course of 20 years brought about differences in print head technology in order to meet the demand for speed and quality. The quality of an inkjet droplet, once it is ejected onto a substrate, is largely due to the chamber size, nozzle thickness, and the technology of a print head. These technological changes resulted in changes to the chemistry of inks because of the critical relationship between ink composition, droplet size, the speed at which the ink is ejected, and the drying time for the ink once placed on paper. Figure 8.1 shows the evolution of print head technology over a period of 20 years. A significant change was made to the inks used in Hewlett-Packard printers in 2005, resulting in a markedly different formulation of ink.

Using patent literature can be helpful but should always be interpreted with caution, since it is possible for companies to apply for patents after a new product is introduced or ingredients are changed. In addition, it may be a number of years after a patent is obtained before a new product is commercialized. For example, Hewlett-Packard patented a process for inkjet printing known as black fortification on 13 October 1998 (Berge and Pathak, 1998). Printers using black fortification first disperse color ink dots onto a selected region, which is then followed by black ink printed over the color ink dots. According to Hewlett Packard, "[t]he color ink interacts chemically with the black ink to cause the pigment to fall out of solution almost instantaneously. When the black drop falls on top of a color drop, the pigment is deposited quickly on the surface of the paper where the black drop landed and does not soak into the paper." (Hewlett-Packard Development Company, 2014). The process results in a vibrant black text and allows the black ink to dry faster to prevent smudging. In addition, the black pigment is a smaller dot and, therefore, does not scatter as easily.

There are specialized ink tags in the form of fluorescent compounds or rare earth elements that were added to some writing inks from about 1970 until 1994. Factors have precluded some ink manufacturers from participating in such a program, including, but not limited to, insufficient resources, low priority, and/or disagreement about the type of tag utilized. Formulabs Incorporated (no longer in business) used four fluorescent

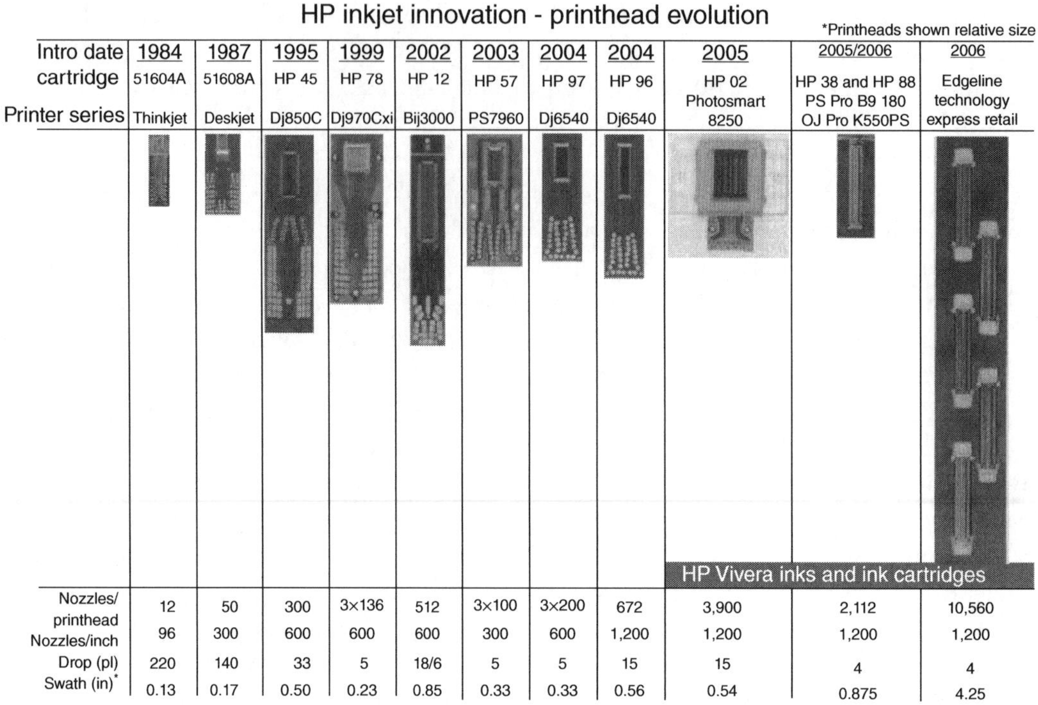

Figure 8.1 Demand for rapid quality inkjet printing gave rise to an evolution of printhead technology in inkjet printers. These technological advances resulted in changes to the chemistry of inks and new inkjet formulations because of the necessity for inks to be compatible with the printhead speed and droplet size requirements (*See insert for color representation of the figure.*)

compounds, referred to as Tag A, B, C, and D, from 1981 through to 1994. The four tags were placed in batches of ballpoint ink formulations in various permutations each year to identify the year of manufacture. Tag D was re-introduced into ballpoint inks by a major manufacturer in November 2002 (LaPorte, 2004). As depicted in Figure 8.2, the fluorescent tags are detected using thin layer chromatography (TLC) followed by visualization with an ultraviolet source using the standardized procedure for extraction and solvent development (ASTM, 2005).

Another powerful tool that can be used for helping establish the authenticity of a document is the creation and development of a database. From the perspective of forensic discriminatory value, there are numerous writing ink formulations that have been, and continue to be, developed, which are available to the general public. Personal preferences (e.g., fountain, ballpoint, roller, and gel pens), changes in technological features, cost considerations, and the availability of raw materials are some examples of why so many writing inks exist. Furthermore, most ink companies spend substantial resources on research and development. LaPorte *et al.* (2006) outlined the importance of utilizing a database of standards for forensic casework and show that as more standards

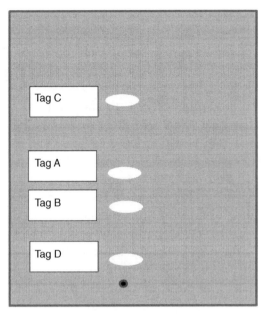

Figure 8.2 Schematic representation of a thin layer chromatography plate showing the respective Rf values of four fluorescent tags that were commonly used in various combinations depending on the year of manufacture of the ink. The tags are visible when viewed with an ultraviolet source

(and corresponding information about the specimens) are collected, there is greater likelihood of matching an ink from a questioned document with a known standard following series of analyses. As defined in ASTM E1422:05, a "match" is defined as the inability to distinguish between ink samples at a given level of analysis. Great caution must be taken when using this terminology because even when all observable aspects of all the techniques are considered, and two inks cannot be differentiated, the results would not be sufficient to support a definite conclusion of common origin. That is, one cannot conclude that the two inks are the same and have the same source.

The United States Secret Service (USSS) maintains the largest known forensic collection of writing inks in the world. The collection includes more than 10,000 samples of ink that date back to the 1920s and have been obtained from various manufacturers throughout the world. For many years, pen and ink manufacturers were contacted on an annual basis and requested to submit any new formulations of inks, along with appropriate information, so that the new standards could be chemically tested and added to the reference collection. In addition, writing pens are obtained on the open market and compared with the library of standards to identify additional inks that may not have been formally submitted by a manufacturer. Maintenance of the library is a formidable task that obviously requires significant resources and is not often practical for most forensic laboratories. The Bavarian State Bureau of Investigation in Munich, Germany, also maintains a reference collection of over 6000 writing inks.

8.2 Dynamic approach

Since the dynamic approach to document analysis incorporates procedures that measure the physical or chemical properties that change with time, it is critical that all methods undergo rigorous testing and are shown to be accurate and reliable. Of the utmost importance is to understand that there will be limitations on the test methods and interpretation of results. Not only are there various types of writing inks, such as ballpoint, gel inks, markers, and fountain pens, but each of these classes will also vary because manufacturers change the types and ratios of ingredients, which will have an impact on their drying characteristics. Since inks are composed of complex mixtures of compounds and there are numerous formulations available, drying rates of writing inks will vary significantly. Other factors, such as substrate (paper), environmental conditions, and variations in pen pressure, line thickness, and unequal distribution of ink, can have a substantive effect on the measured parameters. Before proceeding into a more extensive discussion about the dynamic methods used to approximate age of a document, it is important to understand some of basic chemistry of inks.

8.3 Ink composition

Generally, inks mainly consist of colorants (dyes and/or pigments) and vehicles (solvents and resins), as well as a variety of other ingredients such as antioxidants, preservatives, and trace elements, which when combined are collectively considered the ink formulation. These characteristics are applicable to writing inks, stamp inks, inkjet inks, typewriter inks, and commercial printing processes that utilize inks, such as offset lithography, flexography, and intaglio. However, this description does not apply to dry toner – printing material commonly used in photocopiers, laser printers, and some facsimile machines. Dry toner has a different composition to facilitate a markedly different process for printing a document and is discussed later. Thermal transfer printing applies to printing processes that utilize heat to produce an image by either physical or chemical means or by a combination of both. This technology is has evolved in the past two decades with the advent of bar coding, retailer receipts, fax machines, event tickets, and the use of high mega pixel digital photography and plastic identification cards. Some thermal transfer inks can be analyzed using TLC, but must be extracted differently than writing inks and developed in a specialized solvent system (LaPorte *et al.*, 2004).

For writing inks to perform properly, their chemical composition must be very specific and meet several requirements. Inks must have a high tinctorial strength (ability to impart color) and the solvents should have a low volatility to prevent evaporation and clogging at the tip of the writing instrument. The viscosity of writing inks will vary, but must be specific to a certain pen type to prevent too much flow of ink at the tip while writing. In addition, inks must have the appropriate pH and cannot be corrosive to any metal parts of the pen. Most importantly for users, writing inks are intended to create permanence of a written record and should be resistant to fading after prolonged exposure to light (lightfast) and resistant to chemical breakdown when exposed to water (water fast).

Writing inks are often classified into ballpoint and non-ballpoint (e.g., roller ball, felt tip, gel, fountain pens). This general categorization has developed because writing mechanisms (e.g., the tip of the pen) vary and, therefore, require a specific ink composition to ensure the proper flow of ink onto a substrate. Non-ballpoint inks are primarily aqueous based and more fluid, while most ballpoint inks are thicker and more viscous, and contain glycols and other solvents. Therefore, ballpoint and non-ballpoint inks have different aging parameters and must be extracted differently given the disparities in their general chemical compositions.

Another major difference in inks is whether they are composed of dyes, pigments, or a combination – often referred to generically as colorants. Dyes are generally considered to be compounds with highly conjugated resonance structures that are non-planar and non-symmetrical, making them easily soluble. Their molecular weights can vary from the low hundreds to the high thousands. The Colour Index, published by The Society of Dyers and Colourists and the American Association of Textile Chemists and Colorists (SDC/AATCC, 1971), is a compilation of the various classes of colorants. The Color Index divides dyes into a series of large groups: acid dyes, azoic dyes (monoazo-, diazo-, triazo-), basic dyes, developers dyes, direct dyes, disperse dyes, fluorescent brighteners, food dyes, ingrain dyes, leather dyes, mordant dyes, natural dyes, oxidation bases, phthalocyanine dyes, reactive dyes, reducing agents, solvent dyes, sulfur dyes, and vat dyes. These colorants are used in many industries to yield colors on an array of substrates. However, solvent dyes and pigments are the most commonly used colorants in writing instruments. Acid dyes and reactive dyes are also used, but less frequently.

The major distinguishing feature between dyes and pigments is that the latter consist of fine particles of insoluble material that are suspended in the vehicle. Generally, pigments are planar and symmetrical molecules that are considered stabile, lightfast, and water-fast. Their color is often derived from a metal-centered complex and is generally less vibrant than dyestuffs. There are five categories of pigments: organic pigments, toners, lakes, extended pigments, and inorganic pigments.

The most common dyes used in inkjet inks are acid, direct, and reactive dyes. Acid dyes (e.g., azo, anthraquinone, triphenylmethane, azine, and xanthene) are water soluble and display a wide gamut of hues. Direct dyes (e.g., phthalocyanine) have a variety of bright hues and some of them are fluorescent, as well as possessing relatively good light- and water-fastness properties. Additionally, reactive dyes have been added to some formulations because they contain desirable chromophores and are absorbed very strongly by cellulosic fibers in paper.

Weyermann *et al.* (2009) studied the degradation of crystal violet, a common dye used in some ballpoint inks. Over a period of 2–3 years, the authors compared the degradation pathways of the pure dye in water and ethanol using UV-Visible spectrophotometry and laser desorption ionization. They observed that degradation reactions were quenched by the presence of another dye due to competitive absorption and that the thickness of a stroke (concentration of ink) influenced the degradation process. In the absence of light, only one ballpoint pen showed slight degradation. Ultimately, the authors concluded there were significant differences in the products and the kinetics of the degradation. The study by Weyermann and her colleagues confirms similar findings by other researchers who found that once a writing ink is placed on paper, the solvents begin an evaporation process, leaving mainly the dyes behind (Grim *et al.*, 2001). Based on subjecting inks to UV irradiation as a means of accelerated aging, Grim *et al.* (2001) confirmed that dye degradation products are formed via an oxidative demethylation process.

The vehicle portion of writing inks can be divided into two broad classes: resins and solvents. Resins are natural or synthetic compounds, which are used in inks to adjust the viscosity and serve as a binding agent between the colorants and vehicles. Once applied to a substrate, resins allow for a bond to be created between the ink and the substrate as the ink dries and is best described as follows, "Resins start hardening as soon as the ink has been placed on paper. Hardening ('solidifying') of ink resins is a complex physical and chemical age-transforming process that can include cross-linking, polymerization, decreasing of intermolecular distances (this leads to decrease of solubility) due to solvent evaporation, and so forth" (Aginsky, 1996).

There has been little published on the introductory dates of resins, if they have changed over time, and the frequency of use for specific resinous components in ballpoint inks. Bügler *et al.* (2005) reported on the use of thermal desorption and gas chromatography/ mass spectrometry (GS-MS) to characterize solvent and resinous components of writing inks, including acetophenone-formaldehyde (AF) resin, cyclohexanoneformaldehyde (CF) resin, and various alkyd resins in 121 ballpoint inks. In a subsequent study, 25 different inks with known compositions of resins and solvents were studied to assess aging their parameters (Bügler *et al.*, 2008). The authors characterized the inks as "slow aging" (aging period extends several months) and "fast aging" (aging period lasts a few weeks). Interestingly, 12 of the inks were categorized as "slow aging," and 11 of those inks contained AF resin. The remaining one ink, classified as fast aging, contained CF and alkyd resins in combination. Only one of the 13 "fast aging" inks contained the AF resin, but also contained a combination of 2-phenoxyethanol (2-PE) and benzyl alcohol solvents – all "fast aging" inks contained PE and benzyl alcohol.

These studies indicated that resins affect the aging parameters of an ink and have shown that the chemical analysis of resins can be used to help characterize and classify inks. Yet there have not been in studies to date showing that quantitative analysis of resins (e.g., the decay or decomposition rate) can be used as a dynamic parameter to estimate the age of an ink once placed on paper. There has not been significant research to determine if resins, either in their current form or derivatized, could be used as a stable measurement component relative to other ingredients that change with time.

The other major component of an ink vehicle is the solvent(s), which aids in the application of the ink to paper. Glycols, alcohols and water are the most commonly used solvents in ballpoint pen ink today. The choice of solvents in an ink formulation must take into consideration the environment the ink will be stored in, type of writing instrument, cartridge composition, and region of sale. Two types of cartridges can be used in ballpoint pens, plastic based (e.g., poly(vinyl chloride), polyethylene, polypropylene) or metal (e.g., brass, stainless steel), and the properties of the solvents used must be taken into consideration when producing an ink formulation. It is well established that solvent evaporation is the first process to occur once ink is placed on a document and the quantity of solvent will decrease over time. The analysis of ink solvents, most commonly using gas chromatography coupled with mass spectrometry (GC-MS), has been studied and reviewed in the literature for over two decades by numerous authors (Beshanishvily *et al.*, 1990; Aginsky, 1993a, 1996, 1998; Andrasko, 2001; Gaudreau and Brazeau, 2002; LaPorte *et al.*, 2004; Bügler *et al.*, 2005, 2008; Brazeau and Gaudreau, 2007).

These authors have laid the groundwork for a dynamic approach that involves the analysis of 2-phenoxyethanol (2-PE), a common volatile organic compound found in most ballpoint writing inks. 2-PE is a glycol ether, which is used as the principal solvent in many ballpoint ink formulations. It is a colorless, slow evaporating, viscous liquid with a faint aromatic odor and is used in most ballpoint ink formulations because it is stable in the presence of acids and alkalis, and is non-hygrosopic (does not absorb water, making it amenable to hot, humid climates), non-hazardous, economical, and especially good at solubilizing resins. It is catalogued as Chemical Abstracts Service (CAS) number 122-9-6 and has a molecular weight of 138.17 with a boiling point of 245.2°C. Figure 8.3 shows the chemical structure of 2-PE.

Another well documented finding is that some inks contain the solvent benzyl alcohol. Weyerman *et al.* (2007) noted that "Benzyl alcohol (61%) and propylene glycol (54%) were also found in several ballpoint pen inks. These substances are more volatile than the ones previously mentioned. Therefore, they disappear faster from the stroke and would not perform as adequately in the possible dating of ink." Also noted in another study (Bügler *et al.*, 2008), the authors stated the following after testing 85 ballpoint inks that did not exceed 1.5 years of age: "… the analysis of ink components … revealed that more than 95% of all ballpoint inks contain PE as a major solvent (11). Similar results were obtained by LaPorte, *et al.* (12). Other solvents include benzyl alcohol, 2-ethylhexanol, N-methylpyrrolidone, alkylglycols and alkylethers, organophosphates, and phthalates … other common constituents of ballpoint inks were found not to be as useful as PE for ink dating. For example, diethylene glycol evaporates quickly, labile compounds such as

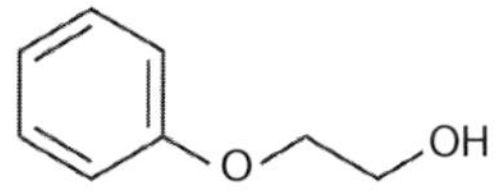

Figure 8.3 Chemical structure of ethanol, 2-phenoxy (2-phenoxyethanol)

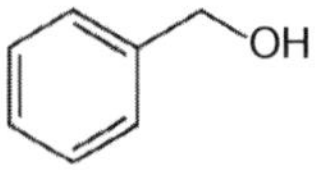

Figure 8.4 Chemical structure of benzyl alcohol

benzyl alcohol are subject to oxidation…" That is, benzyl alcohol is a labile compound, which means it breaks down and becomes undetectable at sufficient levels in inks. Therefore, observing high levels of benzyl alcohol from an ink on a document purported to be several years old may be used as parameter to help determine whether a document is as old as the purported date. Benzyl alcohol is catalogued as CAS number 100-51-6 and has a molecular weight of 108.14, with a boiling point of 205°C; its chemical structure is shown in Figure 8.4.

The overall ink composition plays a pivotal role in the aging characteristics of an ink formulation, so there is no generally accepted single ink aging method applicable to all types of inks. However, qualitative and quantitative analysis of ballpoint ink solvents has been studied extensively, and therefore can be used as a dynamic method to estimate the age of a ballpoint ink. A common procedure for this type of analysis, the limitations of the testing, and interpretation of results are discussed in the following section. Given variations in dye decomposition and the relative stability of resins, many authors have concluded that using dye decomposition and relative decay rate is not an accurate and reliable method to estimate the age of writing inks (Aginsky, 1993a, 1996; Andermann and Neri, 1998; Andrasko, 2001).

8.4 Examinations

When attempting to render conclusions regarding the authenticity of questioned documents, it is critical to conduct a comprehensive examination and to consider the results collectively. The importance of clearly describing all aspects of a document that are examined is critical because the failure to consider the results from all examinations deviates from the scientific method upon which the questioned document discipline is based. As stated by Ordway Hilton in the *Scientific Examination of Questioned Documents*: "Everyone knows that a document is the product of a combination of several materials, such as ink and paper, put together by means of certain common instruments […]. Almost everyone, however, fails to appreciate fully that each of these materials and instruments has its individual and class characteristics, which in combination help to personalize and identify the document's source and history […]. It thus becomes necessary to discover and evaluate correctly each of these factors to reconstruct as much as the document's history as possible" (Hilton, 1993).

Forensic experts employ a variety of physical, optical, and chemical methods to determine an analytical profile of each component of a document, which can then be compared with reference standards, materials from other documents, or seized materials from a suspect source. Although the emphasis of this chapter is on the chemical analysis of documents, it is critical to understand that physical and optical examinations are requisite before conducting chemical examinations. Therefore, a brief description of the examination process is necessary.

8.4.1 Physical examinations

Physical examinations include non-destructive methods for inspecting documents visually with an appropriate light source, taking measurements, identifying macroscopic features, and viewing the questioned document with a stereomicroscope. A stereomicroscope is a binocular microscope capable of blending both eyepiece images, allowing the examiner to discern depth of field. At this stage of the analysis, there are numerous materials and features of a questioned document that should be considered. In some cases, features such as alterations, deletions, or obliterations might be observed.

One of the first steps of any physical examination is to determine how a questioned document was produced and whether any written entries are original. If there are handwritten entries then the forensic document examiner must determine the type of writing instrument used to produce them. Writing inks can often be classified into ballpoint and non-ballpoint based on their unique microscopic characteristics, which result from the combination of their differential chemical composition and interactions with paper. Therefore, the physical examination of writing inks is a critical first step, as ballpoint and non-ballpoint inks must be extracted differently given that they are vastly different in their general chemical compositions. Determining the type and color of a writing ink is commonly reported following a physical examination and is further described in American Society for Testing and Materials (ASTM) International E1422: *Standard Guide for Test Methods for Forensic Writing Ink Comparison* (ASTM, 2005). However, ASTM standard documents, once created under the auspices of the Scientific Working Group for Forensic Document Examination (SWGDOC), are no longer published through ASTM International. Instead SWGDOC now maintains all standard documents, which are available on their website (http://www.swgdoc.org/).

The text, format, and/or images on documents can be printed using various methods. These methods of production are referred to as printing processes and are identifiable using a magnifying device, such as a stereomicroscope, with an appropriate light source. The most common types of home and office machines utilize toner (e.g., photocopiers, laser printers, and some facsimile machines) or inkjet technology (e.g., inkjet printers and some types of multifunction machines capable of scanning, copying, faxing, and printing). Documents submitted for forensic analysis are often created with machine printing and, in many cases, there are no accompanying handwritten entries, especially in cases involving threatening correspondence, extortion letters, and multiple page contracts (e.g., only the last page is signed). Inkjet ink absorbs into the paper and appears planar, or flat, when visualized with a microscope. Dry toner consists of a particulate material and sits on top of the paper, which appears to exhibit a three-dimensional effect when observed with a stereomicroscope. Both of these technologies are capable of printing in black and/or color. In some instances, the printed material on a document may appear black to the naked eye, but is actually composed of a mixture of colors. Figure 8.5 illustrates a signature written with a blue ballpoint pen that cannot be differentiated from a copy of the same signature created with color toner until viewed with a microscope.

Printers and copiers, both inkjet and toner-based systems, use at least four different ink colors: cyan (C), magenta (M), yellow (Y), and black (K). In addition, some inkjet printers also incorporate other inks to create a larger color palette such as light cyan (c), light magenta (m), orange, green, and "photo" black. The CMY components are

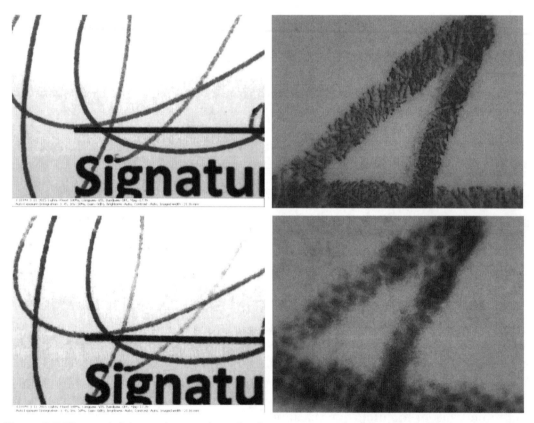

Figure 8.5 The top left image is a portion of a signature created with a blue ballpoint pen and the corresponding top right image was captured at 70× magnification with a digital microscope. The bottom left image is color toner from a copy of the same signature and the corresponding bottom right image was captured at 70× magnification with a digital microscope. The signature with the pen cannot be differentiated from the signature with the toner until visualized with a microscope (*See insert for color representation of the figure.*)

primary subtractive colors which, when combined, form a colored image or text. Therefore, prior to conducting any chemical examinations, a microscopic examination is necessary to ensure that the region of interest contains representative quantities of each color. The black text found on documents can be produced with black ink or a combination of CMY (i.e., composite black). It is highly recommended that separate chemical analyses be conducted on the CMY and K areas when using non-chromatographic techniques, but the analysis of each of the components on a document can sometimes be difficult, as one or more colors overlay each other in half tone patterns to create an image or text.

In some cases, when the text found on a document is produced from an office machine, they may contain printing imperfections, commonly referred to as trash marks. The identification and comparison of individual and class characteristics for linking multiple items is a generally accepted principle to link questioned documents with each other and/or with documents from a known source. Printing defects, or imperfections, sometimes occur on machine printed documents and can be used to help prove that two or more items originated from a common source. Printing defects can be found on documents produced by office machines (e.g., photocopiers, inkjet printers, typewriters) and commercial printing presses (e.g., offset lithography, gravure, flexography). With respect to photocopiers and some printers, printing imperfections can be imparted onto a document as a result of marks, scratches, dust or dirt on the glass platen, cuts on the delivery belt, dirt on the lens, scratches on the drum, and/or problems with the fusion system. These circumstances can arise as a result of normal wear and tear, abuse, or poor care and maintenance of the office machine. Therefore, when documents are produced contemporaneously, they may contain the same trash marks. Moreover, when documents are created at different intervals from the same machine, new trash marks may begin to appear over time. The trash marks created on documents can be class or individual characteristics, but as more trash marks are identified in similar constellations then the likelihood of uniquely identifying a suspect machine will increase.

Printing imperfections can also occur in documents printed from inkjet printers. Figure 8.6 is an example of printing defects that can be used to associate a questioned document with a suspect printer. In this case, a victim was poisoned, later revived at a local hospital, and then admitted she did not attempt to commit suicide. An alleged suicide note was found at the crime scene (see top image) and compared with a document created and produced on a printer seized from the suspect (middle image) and the victim (bottom image). The physical examination in this case was supplemented with the presence of other common printing defects found in the exemplar document seized from the suspect, which were not found in the document obtained from the victim's printer. Supplementing the physical examination with chemical testing of the ink, toner, and perhaps even the printing imperfections should always be considered when trying to determine source attribution.

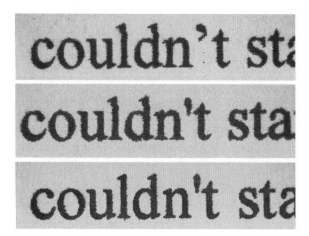

Figure 8.6 An example of how printing defects can be used to associate a questioned document with a suspect printer. The top image is a portion of text from an alleged suicide letter with a defect in the top portion of the l and d in the word "couldn't". The middle image is a portion of the same text from an exemplar document created and produced on a printer seized from a suspect. The bottom image is a portion of the same text from an exemplar document created and produced from a victim who was poisoned, but later revived

8.4.2 Optical examinations

Optical examinations, also referred to as filtered light examinations, are non-destructive and can provide valuable insight regarding the overall composition of inks and paper. Ink and paper are made from components that respond differently to different wavelengths of light, sometimes in regions of the electromagnetic spectrum beyond that which the human eye is capable of seeing. The presence of colorants and other materials will directly affect the manner in which inks and paper absorb, reflect, and transmit light. Ultraviolet (UV), infrared reflectance (IRR) and infrared luminescence (IRL) illumination are energy sources that can be used to evaluate the properties of an ink. When illuminating a sample, there are two ways the sample can generate light. The first involves light from the illuminant being reflected to a detector. The second way to detect light from a sample is through IRL. The greatest advantage in examining the IRL characteristics of a sample is the ability to filter out the incident energy. When the incident energy is filtered out, only the energy that has been re-radiated at a longer wavelength can be observed with the appropriate equipment.

Forensic document examiners commonly use a video spectral comparator (VSC) for this type of examination. This instrument is used to examine ink and paper. It is fitted with an infrared sensitive camera and light sources for producing both ultraviolet and infrared light. The camera and light sources are both equipped with filters that allow the examiner to use many combinations. These different responses to light allow items to be discriminated from each other if the items are composed of different components. While controlling both the wavelength of light being used and the wavelength or region

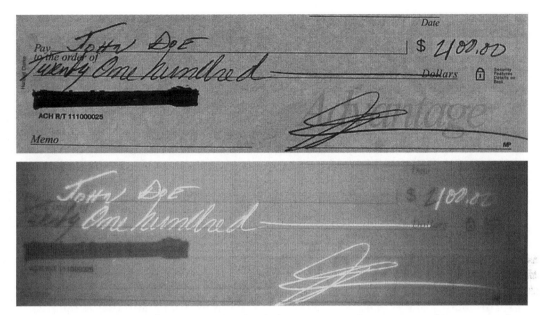

Figure 8.7 The top image is a portion of a check that appears to be written in the amount of $2100 with no signs of forgery. The bottom image, viewed using infrared luminescence, shows that the check was originally written for $100 and changed to $2100 using a different ink

being viewed with the aid of the camera, inks and paper can be characterized. This part of the examination is also used to determine if the entries on a document are heterogeneous, especially in cases when there are numerous pages of writing. Since it is impractical to sample and chemically analyze every aspect of a written entry, the optical examination is a critical step to sample representatively from homogenous areas of the document. How an optical examination can be used to distinguish different inks based on their IRL properties is shown in Figure 8.7.

It is the combination of all of the ingredients in inks that allow forensic examiners to differentiate inks using optical methods. An extensive study using a filtered light examination (FLE), TLC, and reflectance visible microspectrophotometry (MSP) was conducted on a total of 1,091 black and blue ballpoint inks and the authors concluded that "The power of the individual techniques to discriminate inks between and within brands, models and batches varied, the most informative techniques being TLC>FLE>MSP" (Roux *et al.*, 1999). They showed that FLE of blue inks (n = 549) and black inks (n = 542) resulted in a discriminating power (DP) of 0.83 and 0.96, respectively.

8.4.3 Chemical examinations

TLC, GC-MS, scanning electron microscopy coupled with energy dispersive spectroscopy (SEM-EDS), Fourier transform infrared spectroscopy (FTIR), and Raman spectroscopy are some examples of the analytical techniques that can be utilized to (i) determine if a document was prepared on its purported date, (ii) associate seized material from a suspect with questioned evidence from a crime scene, and (iii) link questioned documents originating from multiple sources.

TLC, an effective and efficient method of examining inks, is considered to be one of the most widely used methodologies for separating and characterizing colorants to help characterize an ink formulation. TLC is an inexpensive separation technique that offers several key benefits. It is relatively inexpensive compared to other instrumental methods, very little sample is required for analysis, the analysis is not significantly time intensive, and the interpretation is often quite objective and easy to convey to a non-scientific audience. TLC has achieved acceptability within the scientific community, which is essential for admissibility of analyzed evidence in civil and criminal courts.

TLC, for the purpose of ink examinations, has been discussed extensively in the literature (Witte, 1963; Brunelle and Pro, 1972; Kelly and Cantu, 1975; Brunelle and Reed, 1984; Aginsky, 1993b; Brunelle and Crawford, 2003; LaPorte, 2004; LaPorte *et al.*, 2006). Analysis of ink using TLC begins with the removal of ink from a document. Like most forensic examinations, a minimal amount of questioned ink should be removed due to the destructive nature of the analysis. A hypodermic needle-like apparatus or micropunch can be used to remove ink plugs with a diameter of 0.5–2.0 mm. Obtaining 3–5 hole punches with a 1 mm hole punch and extracting with 3–5 microliters of solvent is usually more than sufficient to achieve a sufficient chromatographic profile. The extract from the sample should be applied to a TLC plate (e.g., glass or plastic surface), which is coated with a thin layer of silica gel to act as a stationary phase. After applying the extract to the TLC plate's loading area and waiting for it to completely dry, the plate should be placed into a TLC bath containing the solvent system that has reached vapor equilibrium. Common solvent systems that are employed for forensic ink analysis include ethyl acetate:ethanol:water (70:35:30) and butyl acetate:butanol:water:acetic acid (10:41:17:32). The solvent system will act as a mobile phase, carrying the complex mixture up the plate by capillary action. As the solvent front migrates vertically up the plate, the mixture of components in the ink will separate. This separation is due to the interactions between the silica gel of the stationary phase and the mixture of components found within the ink being carried by the mobile phase. The polar compounds will interact more with the polar stationary phase causing their migration to be retarded. The non-polar compounds within the ink will migrate further up the plate. Following separation, the plate should be removed from the TLC bath and dried while the solvent evaporates off the plate.

Examination of the plate will reveal distinct bands of color at different heights on the TLC plate. Some colored bands may be attributed to the mixture of dyes used to produce the color of the writing ink. The color and the retention factor (Rf) can be compared against a known sample of ink within a reference library. Additional visualization techniques should include shortwave ultraviolet light (254 nm) or longwave ultraviolet light (366 nm). Figure 8.8 is an image of a TLC plate showing how the colorant profiles can be used to compare and evaluate writing inks.

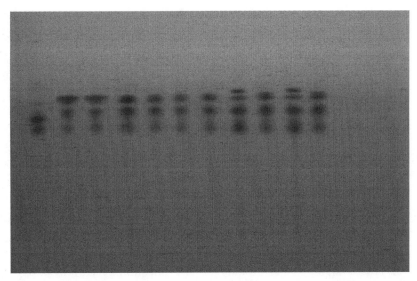

Figure 8.8 A TLC plate showing 11 different inks tested as part of a multiple document submission. The colored spots, and their relative migration positions, represent different dyes used in black point inks (*See insert for color representation of the figure.*)

Filtered light examinations using an alternate light source (ALS), similar to an examination with a video spectral comparator, can be used for the evaluation of inks analyzed using TLC. The alternate light source is used to illuminate samples under various wavelengths. Houlgrave *et al.* (2011) showed that evaluating TLC plates, coupled with the appropriate filter, can provide additional discrimination following visualization with ambient light. In addition, filtered light examinations can be used to detect variations in batches of inks by identifying different components (possibly unintended) in the same formulation of ink, but produced in different batches. Figure 8.9 shows how an alternate light source, with the appropriate filter, can be used to view and discriminate different batches of inks of the same ink formulation.

GC-MS is routinely used for ink analysis to compare the non-colorant organic ingredients such as resins, volatiles, and semi-volatile components. Although TLC is an excellent method to characterize the colorant components in an ink formulation, the colorants are only a fraction of the total ink formulation. Although GC-MS can be used as a supplemental test to compare questioned inks, its utility when comparing to a known standard is minimized. As discussed previously, when ink is placed on a document, some of the components change as the ink ages. Therefore, comparing a questioned ink of unknown age with a standard ink from a database will likely not be probative in all cases unless the objective of the testing is to identify specific resins or other compounds. From the perspective of estimating the age of an ink, GC-MS is ideal and can be used to measure these changes. In addition, reference standards are not required for comparison and "irrespective of an ink formula, ink solvents age on paper for two years" (Aginsky, 2002).

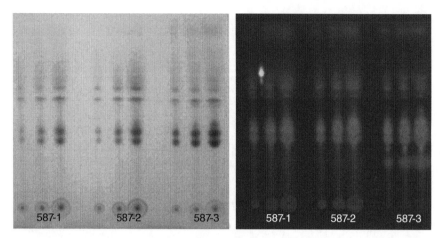

Figure 8.9 The TLC plate on the left is a single formulation of ink (587). Lanes 1 through 3 represent different concentrations of batch 587-1; lanes 4 through 6 represent different concentrations of batch 587-2; and lanes 7 through 9 represent different concentrations of batch 587-3. The TLC plate on the right is the same plate visualized with a red filter using an ALS set at 515 nm. No differences are observed when the TLC plate is visualized with visible light, but differences are readily apparent with an ALS and the appropriate filter (*See insert for color representation of the figure.*)

In 2002, Gaudreau and Brazeau presented their findings from an extensive study on solvent loss once an ink is placed on paper and concluded that the "... phenoxyethanol in ink evaporates at a high rate during the first six to eight months following its application on paper. The rate of evaporation stabilizes over a period of six to eighteen months. This process is no longer significant after a period of about two years." That is, 2-PE evaporates very quickly when an ink is first placed on paper and then eventually slows, but continues to evaporate in the 24 months after the ink has been placed on the document. After 24 months, 2-PE no longer evaporates at a significant or measurable rate. Figure 8.10 is a theoretical illustration to show how an ink polymerizes as it ages and the effect on the solvent loss rate of 2-PE.

For the purpose of ink dating, GC-MS is used to measure differences in the concentration of 2-PE when samples of the questioned ink are heated and unheated. In this method, samples of the questioned ink are removed from the document and one set of the samples is then heated and the other set is not, such that the quantity of 2-PE from the heated samples can be compared with the quantity of 2-PE from the unheated samples. A greater concentration of 2-PE will evaporate from fresh ink compared to older ink when the samples are heated at a temperature of 70°C (Bügler *et al.*, 2008). The basis of the theory is that if an ink is fresh then it will be much easier to evaporate 2-PE from the ink stroke, resulting in a significant difference in the amount of 2-PE when the unheated sample is compared with the heated sample. If more than 25% of the 2-PE is lost after heating a sample then the ink is said to be less than two years. There are factors that may affect the concentration of 2-PE prior to

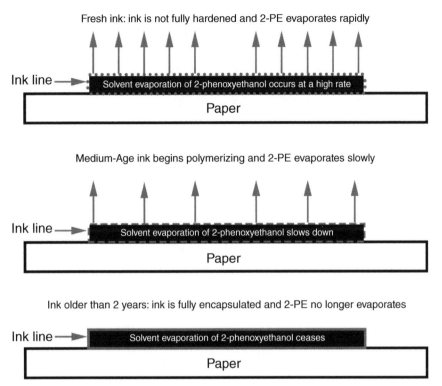

Figure 8.10 A theoretical model to illustrate how ballpoint inks polymerize as they age and the effect on the solvent loss rate of 2-PE (*See insert for color representation of the figure.*)

testing, such as storage in extreme cold, which slows the ink drying process, or extreme heat, which hastens the ink drying process, but none of these factors would be expected to cause an increase in the level of 2-PE.

This method requires very careful precision when removing samples and selecting samples to be heated and unheated – this cannot and should not be a randomized process. First, the micro-hole punches used for this type of analysis are 0.5 mm and should be carefully centered within the ink stroke. Second, the punches should be removed from areas in the ink line that are consistent with respect to ink deposition and volume following a microscopic examination. Third, although relative differences in the quantity of unheated and heated ink will eventually be determined, samples of ink selected for the heated and unheated experiments should be removed alternatively and within close proximity to each other. That is, sample 1 should be removed and placed into vial 1 (unheated samples) and sample 2 (taken in the same area as sample 1) should be placed in vial 2 (heated samples). This process should be continued for additional samples. Figure 8.11 is a microscopic image of blue ballpoint ink showing two pairs of punches measuring 0.5 mm in diameter. One punch from a pair will be used for the unheated testing set and the other micro hole punch will be used as part of the set heated at 70°C.

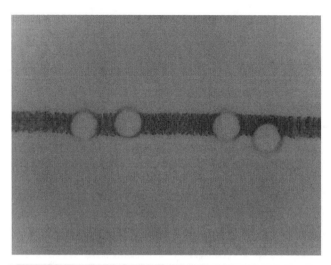

Figure 8.11 A microscopic image of blue ballpoint ink showing two pairs of micro-hole punches measuring 0.5 mm in diameter. The pair of hole punches on the right is unacceptable for chemical analysis, as the hole punch on the far right is off-center. Using the pair of hole punches on the left, one micro-hole punch will be used as part of a set for analyzing the amount of 2-PE without subjecting them to heat. The other micro-hole punch will be used as part of set subjected to 70°C for 60 minutes

The right pair is unacceptable for testing, as the punch to the far right is off-center and, therefore, missing a significant amount of ink.

However, when an ink does not lose more than 25% of the 2-PE after heating, one cannot conclude the ink is older than two years. Bügler *et al.* (2008) showed that 15 out of 60 inks (25%) dried at an extremely fast rate within the first two weeks from the time the ink is placed on paper and, hence, were classified as "fast aging." Therefore, if an ink loses less is than 25% of 2-PE after heating at 70°C then it is not an indication that the ink is older than 2 years – the ink may have "fast aging" characteristics.

It has been well established that the use of absolute measurements for 2-PE cannot be used to estimate the age of an ink or determine that the level of 2-PE is "elevated." One reason is that the amount of pressure applied and the amount of ink deposited in different areas of written notations will vary significantly, and so will the levels of 2-PE. Therefore, depending on where samples are taken from, the results for the amount of 2-PE will vary significantly within the same written entry. The other reason is that the amount of 2-PE will vary significantly in different ink formulations. Bügler *et al.* (2008) proved that making absolute measurements of 2-PE will undoubtedly result in errors when attempting to interpret the data. The authors estimated that the error introduced by sampling resulted in a maximum difference of 800% between ink strokes written with minimum and maximum force. The authors also evaluated 25 different pen types and determined that the solvent content of these ballpoint inks ranged from 0.3 up to 25 ng/mm for the ink samples with an age of one week, and from 0.1 up to 15 ng/mm for the samples with an age of 22 weeks.

8.4.3.1 Toner analysis

Chester Calson was the first to use a photoconductive material to make an image in 1938, commonly referred to as electrophotography. Electrophotography is a non-impact process whereby toner is affixed to a substrate to create an image. Generally, the toner adheres to a drum based on image recognition via software and the fact that the toner carries an opposite charge. The toner is then transferred to a substrate and is affixed by heat and/or pressure. Although not classified as a printing ink, toners are generally composed of colorants (primarily pigments), binders (e.g., styrene-based compounds, methacrylate polymers, and epoxy resins), and other inorganic materials (e.g., iron, calcium, barium, manganese, chromium, sulfur, silicon).

Toners can be dry or liquid, but dry toners are the most prevalent. In addition to having different microscopic properties, dry toners are prepared as homogeneous particles, while liquid toners are colloidal dispersions of pigmented or dyed resin particles. There are a number of authors who have published various chemical techniques used to analyze toners, but many of these techniques have not been implemented into standardized protocols for database development in forensic laboratories throughout the United States (Lile, 1976; Meng and Wang, 1999; Andrasko, 1994; Merrill *et al.*, 1996; Tandon *et al.*, 1997; Munson, 1999; Totty, 1999). Analytical methods such as pyrolysis-GC-MS, FT-IR, SEM-EDS, and X-ray fluorescence (XRF) can be used to discriminate toners, but their use for dating has not been established. In fact, there are no generally accepted methods for estimating the age of toner once it has been placed and fused onto a document. This does not preclude the use of toner analysis to compare questioned and known sources of toner to determine if they have a common origin or to ascertain if a page substitution may have occurred in a multiple page document. TLC analysis can be used initially to analyze both black and color toners (Tandon *et al.*, 1995; LaPorte and Ramotowski, 2003), followed by XRF or SEM-EDS for elemental composition, FT-IR for resins and polymers, or GC-MS for other organic constituents.

8.4.4 Paper examinations

Paper consists of various types of fibers (e.g., wood, seed hair, bast, grass), binders (e.g., starch), fillers (e.g., clay, calcium carbonate, titanium dioxide), sizings (e.g., rosin), optical brightening agents, or OBAs (e.g., stilbenes), colorants and additives, making it an extremely complex material. A discussion about the history, manufacturing, purposes of the various components, and types and grades of papers is far beyond the scope of this chapter. There are comprehensive writings dedicated to understanding the fundamentals of paper and the papermaking process (Hunter, 1974; Browning, 1977; Mead, 1993). Therefore, the focus here is on the forensic analysis of paper, but these examinations can range in complexity and may require diverse expertise depending on the nature of the request.

If the request is to compare a questioned document with other questioned or known documents then the objective is to determine if the papers originated from a common source, which would incorporate physical, optical, and chemical testing. This type of examination is quite common in cases when one or more pages are suspected to have been substituted in a multipage document. The dimensions, thickness, color, and general type of paper should be assessed. For paper thickness, at least six measurements using a micrometer should be taken from various areas and compared using statistical analysis, such as Student's t-test, to ascertain whether any differences are statistically significant. The presence of hole punches, embedded materials (e.g., security fibers), pre-printed lines, and margins can be used to assess common origin. Although paper fibers can be identified and characterized to help determine where the pulp or paper originated, this type of examination is highly specialized and not often utilized in forensic laboratories.

UV and IR sources may also be used to evaluate the optical properties of paper, such as paper brightness, opacity, subtle differences in color, and the overall interrelationships of the paper components. The "UV brightness" of papers can vary and one method to record differences is to assign subjective values such that zero represents no fluorescence and four represents extremely bright fluorescence; however, paper with known brightness (e.g., 96 brightness represents highly UV fluorescent paper) should be used as a calibration standard. Figure 8.12 shows the UV fluorescent (365 nm) characteristics of eight pieces of paper. When differences are observed in the UV fluorescence of paper, one study concluded that 30% of the reams of paper examined with a UV source revealed differences in paper brightness within the same ream (Green, 2012). The differences in paper can result from using different batches of paper that get intermingled during paper production. Therefore, great caution must be exercised when interpreting findings from differences in UV fluorescence. Overall, the physical and optical characteristics should be interpreted collectively. TLC (Wallace *et al.*, 1967; Gupta *et al.*, 1981) and GC-MS can be used to analyze the organic components within paper, such as OBAs and dying agents, while SEM-EDS (Polk *et al.*, 1977) or inductively coupled plasma mass spectrometry (ICP-MS) (Spence *et al.*, 2000) can be employed for elemental constituents. When using more sensitive instrumental analysis techniques like GC-MS and ICP-MS, caution should be taken to ensure that contaminants on the paper are not interpreted as components of the paper.

A second type of examination is to establish if the document was produced on its purported date. There are no generally accepted dynamic methods to establish the age of paper with any specificity, and static methods are very limited. The first step in this process is to determine if a watermark is present. If a watermark is identified then the information can be used to determine the manufacturer, and, in some cases, the manufacturer will provide information regarding the first use of the watermark. Also, some watermarks contain a date code, which is usually demarcated by using a small rectangular line placed over a certain letter. The manufacturer will place the demarcation over different letters in following years.

In some cases, documents purported to be several years old (e.g., 15 years old) might have been created more recently using new paper. Observations regarding the general condition of the paper should always be recorded. As discussed previously, modern

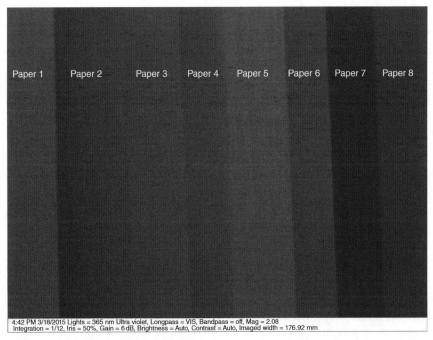

Paper 1   Paper 2   Paper 3   Paper 4   Paper 5   Paper 6   Paper 7   Paper 8

4:42 PM 3/18/2015 Lights = 365 nm Ultra violet, Longpass = VIS, Bandpass = off, Mag = 2.08
Integration = 1/12, Iris = 50%, Gain = 6 dB, Brightness = Auto, Contrast = Auto, Imaged width = 176.92 mm

Figure 8.12 A demonstration of eight different pieces of paper visualized using a 366 nm UV source. Each piece of paper has a different UV fluorescent property (*See insert for color representation of the figure.*)

papers contain OBAs to enhance the white appearance of the paper, which then absorb some wavelengths of light in the UV region. Crable (2011) noted that "the world of white papers changed tremendously in 2005 … In that year, International Paper increased the TAPPI directional brightness from 84, on average, to 92." Although OBAs have been used since the 1950s (Messier, 2005), it was not until the early 2000s when there was a paradigm shift to use stronger brightening agents. There is very little published on when certain OBAs were first used in paper, and there is no clear punctuated time for forensic document examiners to use as a reference. For that reason, the results from a paper examination should always be used with the results from other aspects of the document examination.

One other scenario that is not too uncommon is the presentation of documents that are alleged to be several years old but may have in fact been artificially aged. Paper naturally ages through a combination of processes whose effects vary depending on paper composition and storage conditions. The most significant deterioration of paper is caused by acid hydrolysis, a process in which the cellulose fibers split into smaller and smaller fragments in the presence of sulfuric acid. On the contrary, "artificial aging" is used as a legitimate technique used by hobbyists to produce a document with an antique or older appearance. However, these same methods can be used to create a fraudulent document with the appearance that it has aged significantly. One of the more common methods used to create an antiquated appearance involves staining the paper with various liquids, such as coffee, tea, vinegar, lemon juice, and milk. In cases where documents are exposed to coffee and tea, GC-MS analysis can be used to identify compounds found in these drinks, such as caffeine. Exposing a document to these liquids can mimic some of the characteristics associated with an aged document, such as brownish-yellow discoloration, spotted areas, and, to some extent, fragile edges. The document can also be subjected to intentional tearing and heating, along with the creation of holes to simulate worm holes often found in genuine aged paper. Although the aforementioned features can be present on genuinely aged documents, physical differences are often discernible, permitting a distinction between artificially and genuinely aged documents (Holifield and LaPorte, 2007; LaPorte and Singer, 2012).

8.5 Questioned documents, crime scenes and evidential considerations

The importance of forensic evidence during an investigation and subsequent criminal proceedings cannot be overemphasized. Forensic evidence, in many cases, can ultimately be used to associate a suspect(s) with a victim(s) and/or the crime scene. Crime scenes that involve questioned documents can range from matters of national security to the counterfeiting of a driver's license at the local library. In any case, it is imperative for investigators and forensic examiners to recognize the realm of examinations that can be conducted on documents and the potential information that may be gleaned from their findings. There are some fundamental questions that must be considered in every case, which are discussed in the following paragraphs.

8.5.1 How was the questioned document produced?

The most critical question of any forensic document examination is to determine how the document was produced – handwritten, machine produced, or a combination. If a document is handwritten, then there are materials that do not need to be considered or collected, such as printers, copiers, and typewriters. If the suspect document is machine produced, then it is necessary to ascertain the printing process. A document can be produced via several printing methods. Some of the more common and easily accessible methods for most people include typewriting, inkjet printing, laser printing, photocopying, thermal printing, and faxing. A wealth of forensic information may be acquired in some cases if known office machines are seized from a suspect.

If counterfeit documents are the focus of an investigation, it is crucial to determine how they were printed to ascertain if they are counterfeit or genuine. In order to provide accurate and definitive conclusions, it is often necessary for the forensic document examiner to obtain a genuine specimen for comparison, as well as to have some knowledge about the manufacture of the genuine document. This may include understanding the different printing processes, security features, and procedures used for issuing the authentic document (e.g., passport, social security card, birth certificate). Making a determination of how the counterfeit documents were produced will dictate the type of evidence to search for and could suggest the degree of sophistication of the criminals. For example, if counterfeit materials are produced using an intaglio printing press (very expensive and difficult machinery to obtain), then this could indicate a possible sophisticated criminal syndicate.

8.5.1.1 Case example

This case scenario involves associating counterfeit documents. Two different driver's licenses suspected to be counterfeit were submitted for analysis at different times. Although they contained different biographical information, photographs, and identification numbers, a physical examination revealed the barcodes printed on the back of the driver licenses to be the same. Barcodes on authentic licenses are always unique. A chemical examination of the inks could help establish whether the barcodes were printed from the same source, or if multiple suspects used electronic images of the barcodes. A chemical analysis revealed that the barcodes were created with an inkjet printer, which did not conform to the authentic specimens. Figure 8.13 shows photographs of the barcodes printed on the back of each license (viewed as a top/bottom split).

8.5.2 What evidence can be used to associate a questioned document with the crime scene and/or victim?

The recognition and collection of evidentiary materials is crucial to any investigation, and will likely play an important role during the judiciary process. The prosecution of a case following a criminal investigation can be quite complex and often requires the consideration of an abundance of different types of evidence. The process can be likened

Figure 8.13 The images of the barcodes printed on the back of each license (viewed as a top/ bottom split) reveal that the barcodes are identical. The barcode from the genuine driver license obtained from the suspect was found to be same as those printed on the counterfeit identifications

to piecing together a "jigsaw puzzle," whereby the prosecution presents various types of evidence to the judge and jury so that a conclusion can be rendered based on the collective consideration of all the facts. Indeed, evidence that can help prove an association of the crime with the suspect is invaluable, but often the evidence in questioned document cases is valuable when it can be used to corroborate other evidence presented at trial. Corroborative evidence can be especially effective if there is a plausible explanation for the presence of definitive associative findings, such as DNA, fingerprints, and/or identifiable handwriting. Moreover, additional supporting evidence related to forensic document examinations can be meaningful in cases when a definitive identification of suspect writing cannot be opined due to some type of acceptable limitation.

The collection of suspect materials and/or exemplar specimens will range in complexity and variety depending on the nature of the crime. Most investigators undergo significant training in the processing of seized materials; however, some aspects of document-related crimes can be multifaceted and may require additional assistance. Water soaked, charred, or otherwise tenuous documents may be encountered that will typically require consultation with an examiner. In computer-related crimes, it is often necessary to consult with trained computer forensic examiners. As an example, some printer devices have a limited memory of previously printed materials and document settings, but once they are disconnected some of the information may be lost. Therefore, advice from a qualified individual (e.g., technician, manufacturer representative) may be deemed appropriate.

Consideration should be given to the types of materials used to produce a document. This requires expertise in understanding the resources that could be used by a criminal, especially in cases of counterfeiting. The production process could include printing plates, commercial printing inks, negative and positive films, or other materials that may have been discarded. In cases of counterfeiting, regardless of the printing process, it is highly recommended that any authentic specimens that might have been available to the suspect be seized, including their own identity documents, or possibly those of family members that may have been used as a template. If the authentic specimens have printing defects, which is not uncommon, then they may be associated, or shown to have a common origin, with the counterfeit documents via a forensic examination. Additionally, a scanner or digital camera may have been used, which may impart class and/or individual characteristics into an electronic file that are ultimately transferred to the questioned document. In cases of anonymous letters, the resources used to compile a questioned document could be numerous, ranging from obvious materials like ink and paper to more subtle materials such as staples and hole punching mechanisms. Undeniably, there are limitations to the conclusions that can be rendered in some comparative examinations, but this should not hinder the investigator and/or the forensic document examiner from considering all the possible tests that can be performed.

8.5.2.1 Case example

A female reported to authorities that she was receiving numerous letters of a threatening nature that were becoming progressively more violent. Crayons, markers and an array of pens used on different types of paper were used to write the questioned letters. A suspect was developed, but no fingerprints were identified on the documents and the questioned handwriting was disguised. Therefore, it was concluded that an identification was not feasible based on comparisons with known handwriting. Investigators obtained a search warrant and during the search numerous materials were seized, including crayons, markers, pens, a notepad, and ruled paper. One of the threatening letters was torn and, therefore, was compared with a torn page from a notepad that belonged to the suspect. A comparison was conducted and it was found that the torn edges from the letter and the notepad were exactly aligned. The results from this examination, sometimes referred to as a tear match, are depicted in Figure 8.14. Additional chemical analysis was performed on the paper, writing inks, and crayons.

8.5.3 Are there other forensic examinations that can be performed?

The forensic document examiner must consider the possibility of other types of forensic evidence that may be associated with the questioned document, such as trace evidence, DNA, tape, or any other materials that may be out of the area of their expertise. As an

Figure 8.14 The torn edge from the threat letter received by the victim (Exhibit Q4) is shown to align with a torn page from a notepad that was seized from the suspect (Exhibit K7-1). This method of association is sometimes referred to as a fracture match (*See insert for color representation of the figure.*)

example, DNA analysis may be warranted if an anonymous document contains staining or the envelope has a stamp and/or an adhesive flap that requires moisture such as saliva. If other potential evidentiary materials are discovered (e.g., hairs, fibers) or if requests are made for other analyses, consultation with the appropriate authority to determine the sequence of events should take place. For instance, subjecting a document to ultraviolet radiation may be destructive to DNA. Therefore, the forensic document examiner may have to initially forego this step if it is part of the standard operating procedure.

One of the more common dilemmas that may occur is the decision of whether a questioned document should be expeditiously subjected to destructive fingerprint processing without conducting other possible analyses (e.g., ink and paper analyses). Undeniably, there is tremendous value when fingerprint identification is effected. However, the chemical processing of documents with ninhydrin, 1,8-diazafluoren-9-one (DFO), and physical developer can be very destructive. There are some non-destructive examinations for fingerprints that can be conducted initially, such as visualization with a laser, but the investigator must seriously contemplate other potential evidence and the order in which it should be examined. Consultation with a forensic laboratory and the appropriate examiners should always be considered prior to submitting a request. Furthermore, with the increasing popularity of forensic science in the media, and the general knowledge of laypersons, fingerprints are less likely to be found on anonymous documents. As a consequence, when an expeditious examination is requested, it is highly recommended that non-destructive examinations be fully considered prior to any destructive examinations.

8.5.4 Demonstrating that a suspect altered a document

Not all crime scenes fit the "association" concept. There are a variety of cases whereby establishing an association is not necessary. In theory, this is not viewed as a traditional crime scene but, in practice, should be treated as such by implementing a methodical search. For example, a forensic document examiner may be asked to examine medical records that are purported to be fraudulently prepared. Determining their authenticity by proving or disproving that the suspect fraudulently prepared and/or altered the documents may not require demonstrating an association. This should not preclude the investigator from seeking to obtain known materials. Depending on the circumstances, other undisputed materials could be helpful for testing such as documents of record created around the purported date of the questioned documents or documents of record created around the suspected time of preparation.

8.5.4.1 Case example

A financial agreement of significant value bearing a signature and date was purportedly produced in 1999. Investigators suspected that the agreement was created and signed around 2004. It was concluded that the questioned ink matched an ink formulation produced by the Pilot Corporation that was available prior to 1999, but was still widely

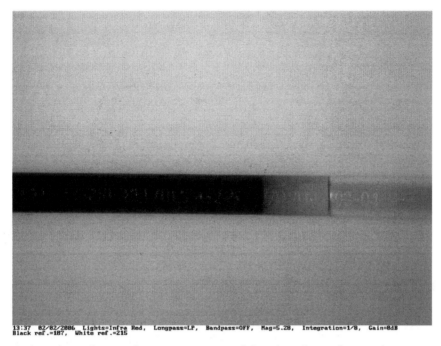

Figure 8.15 The ink used to produce a signature and date found on a financial agreement of substantial value purported to be created in 1999, was found to match an ink formulation produced by the Pilot Corporation. Once the information was provided to the investigator, a Pilot pen was seized from the suspect. The seized pen cartridge contained a date code indicating that the cartridge was filled with ink in April of 2003 (highlighted area). When confronted with the additional information, the suspect proceeded to confess that the financial agreement was created in 2004 (*See insert for color representation of the figure.*)

available. Following the examination, the information regarding the ink analysis was discussed with the investigator and a Pilot pen was subsequently seized from the suspect. The pen was submitted for comparison and it was determined that the ink formulation from the known writing instrument matched the ink used to produce the financial agreement. An examination of the writing instrument revealed that the ink in the cartridge was manufactured in April of 2003 (Figure 8.15). When confronted with the information, the suspect later confessed, stating that the financial agreement was created sometime in 2004 and not in 1999.

8.6 Interpreting results and rendering conclusions

The methodology used to perform chemical testing must be accurate, reliable, and based on sound scientific principles – a basic tenet for the scientific examination of questioned documents – but improper interpretation and the failure to consider limitations of testing results will lead to unsupportable and possibly erroneous conclusions. As stated throughout this chapter, there are often limitations when drawing conclusions from results. Ink can be a challenging substance to analyze because most information pertaining to its manufacturing and chemical composition is considered proprietary information. One important limitation to consider is that ink formulations can be characterized, but chemical analysis is not sufficient to conclude that a questioned ink came from a suspect or known pen to the exclusion of all others.

Not only is chemical analysis not practical to identify a specific writing instrument, it is unlikely to identify a specific ink formulation with certainty. Although it is generally accepted to use the term "match" when two or more ink samples cannot be distinguished at a given level of analysis, it is imperative that the term be defined in all written reports and during court testimony. According to Merriam-Webster, one definition for match is "something or someone that strongly resembles another," while the "same" is defined as "resembling in every relevant respect." That is, the terms – same and match – have different definitions, but there is no doubt that the terms can create ambiguity. Therefore, to avoid any misperception, "match" should always be clearly defined.

Using chemical analysis to conclude that two inks are from the same batch is typically unsupportable. While the primary components of ink are well established, the minor components of ink may help differentiate an ink's manufacturer, as well as similar inks produced at different times by the same manufacturer. As discussed, finding an exact match to an ink to the exclusion of all other formulations of ink is challenging, if not impossible. Thus, not being able to establish the entire universe of inks is already a limiting factor. It is possible for batches to vary in slight ratios of ingredients, and there are no generally accepted methods in ink analysis to quantitatively evaluate every single component in an ink and then measure their respective ratios. Also, it is difficult to conclude with certainty that two or more inks are from different batches. In cases such as these, there is a presumption that the inks are quite similar, but there is some small change to a component as shown in Figure 8.9. However, unlike the ink samples shown in Figure 8.9, where specific information about batch information was provided by the manufacturer, it would be unlikely in most cases to have specific information from a manufacturer. It is very possible that the small difference is due to a new formulation and not a batch variation.

Conducting forensic examinations of evidence for the presence of individual and class characteristics is a fundamental premise used in many forensic science disciplines. As defined by Saferstein (1995), an individual characteristic is "evidence that can be associated with a common source with an extremely high degree of probability" and he goes on to state, "it is not possible to state with mathematical exactness the probability that the specimens are of common origin; it can only be concluded that the probability is so high as to defy mathematical calculations or human comprehension." Class characteristics are general properties of evidence that are associated with a group and not a single source. The results from chemical testing of documents typically fall into class findings.

That is, there are few instances when the results from a single test allow one to make a definitive conclusion with a probability so high as to defy human comprehension. In some cases, using the results from multiple tests can have a cumulative effect when attempting to reach a conclusion.

A forensic examination should begin from a point of neutrality. Once all of the results have been obtained from the testing, and the limitations have been considered, the forensic document examiner must then evaluate and convey the weight of the evidence. Unfortunately, quantitative models to assess evidence based on the results from various types of physical and chemical tests are not fully developed. There are efforts outside of the United States to use Bayes' theorem as part of an approach to statistical inference for ink dating interpretation (Weyermann *et al.*, 2008, 2011). However, these models have not been fully tested nor are they applicable to other questioned document examinations.

SWGDOC, through ASTM International, developed ASTM E1658: Standard Terminology for Expressing Conclusions of Forensic Document Examiners (ASTM, 2008). As stated in Section 3.5 of the standard, "… although the material that follows deals with handwriting, forensic document examiners may apply this terminology to other examinations within the scope of their work …" This standard proposes using a scale of conclusions to express degrees of confidence in the findings to investigators, courts, and other forensic scientists. The scale includes the following terms: Definitive Conclusion (highest degree of confidence and the examiner is certain); Highly Probable (the findings are very persuasive and the examiner is virtually certain, but there is some critical information not available); Probable (there is strong evidence, but the level of confidence of the examiner falls short of virtually certain); Indications (some of the findings are in agreement with and suggestive of a certain proposition); and Inconclusive (the examiner is indeterminate and cannot reach a conclusion due to the limitation of the findings). As used in ASTM E1658, the four gradations of confidence (Definitive, Highly Probable, Probable, and Indications) can be used to include and exclude a proposition, with the Inconclusive level in the middle.

Whatever scale is used, the purpose is to have a standardized method to express conclusions based on quantitative information. Therefore, this set of conclusions is often referred to as a 9-point scale. Some examiners choose to use a modification of the terms such that they have a 7-point or 5-point scale. A good example is how this scale can be applied to results from the 2-PE ink dating method discussed earlier. Consider two different scenarios with the same results. The first scenario is that the document is purported to be 2.5 years old, but it is suspected to have been created in the past 6–9 months. The results from GC-MS testing show that the percentage of 2-PE lost when the unheated samples are compared with the heated samples is 30%. A loss of more than 25% of 2-PE indicates that an ink is less than two years old; however, not every ink has been tested and it is well recognized that storage conditions have an effect on how ink evaporates once it is placed on paper. Now consider the same results, but instead the questioned document is purported to be 10 years old. The differences in the circumstances can change the amount of confidence when reaching a conclusion because the likelihood of a 10-year-old document having such high levels of 2-PE is far less than the likelihood of a document that is 2.5 years old. Having a standardized conclusion scale such as ASTM E1658 allows the examiner to express varying levels of confidence depending on the test results, limiting factors, and case specific circumstances.

Using procedures that are accurate, reliable, and valid for any scientific discipline is a necessary foundation, but practicing examiners should never diminish the importance of interpretation and how conclusions are rendered in written reports and court testimony – one without the other will create an imbalance. On 13 February 2015, the National Institute of Standards and Technology (NIST) held the inaugural meeting for the Organization of Scientific Area Committees (OSAC). The OSAC will coordinate development of standards and guidelines for the forensic science community to improve quality and consistency of work in the forensic science community. Presumably, the OSAC will begin to address current and newly developed guidance for forensic scientists to express accurate and reliable conclusions that reflect the weight of the evidence.

Opinions or points of view expressed in this chapter are those of the author and do not necessarily reflect the official position or policies of the US.

References

Aginsky, V.N. (1993a) Some new ideas for dating ballpoint inks – a feasibility study. *Journal of Forensic Sciences*, **38**(5):134–150.

Aginsky, V.N. (1993b) Forensic examination of "slightly soluble" ink pigments using thin-layer chromatography. *Journal of Forensic Sciences*, **38**:1131–1133.

Aginsky, V.N. (1996) Dating and characterizing writing, stamp pad and jet printer inks by gas chromatography / mass spectrometry. *International Journal of Forensic Document Examiners*, **2**:103–115.

Aginsky, V.N. (1998) Measuring ink extractability as a function of age – why the relative aging approach is unreliable and why it is more correct to measure ink volatile components than dyes. *International Journal of Forensic Document Examiners*, **4**(3):214–230.

Aginsky, V.N. (2002) Current methods for dating ink on documents. Proceedings from the 60th Annual Conference for the American Society of Questioned Document Examiners, 14–18 August 2002, San Diego, CA.

Andermann, T. and Neri, R. (1998) Solvent extraction techniques – possibilities for dating ballpoint pen inks. *International Journal of Forensic Document Examiners*, **4**(3):231–239.

Andrasko, J. (1994) A simple method for sampling photocopy toners for examination by microreflectance FTIR. *Journal of Forensic Sciences*, **39**:226–232.

Andrasko, J. (2001) Changes in composition of ballpoint pen inks on ageing in darkness. *Journal of Forensic Sciences*, **47**(2):324–227.

ASTM (American Society for Testing and Materials) (2005) *ASTM Designation E 1422-05: Standard Guide for Test Methods for Forensic Writing Ink Comparison*. ASTM International, West Conshohocken, PA.

ASTM (American Society for Testing and Materials) (2008) *ASTM Designation E1658-08: Standard Terminology for Expressing Conclusions of Forensic Document Examiners.* ASTM International, West Conshohocken, PA.

Berge, T.G. and Pathak, R. (1998) *Method of ink jet printing using color fortification in black regions.* US Patent No 5,821,957; issued 13 October 1998.

Beshanishvily, G.S., Trosman, E.A., Dallakian, P.B., and Voskerchian, G.P. (1990) Ballpoint ink age – a new approach. Proceedings of the 12th International Forensic Scientists Symposium, 15–19 October 1990, Adelaide, Australia.

Brazeau, L. and Gaudreau, M. (2007) Ballpoint pen inks: the quantitative analysis of ink solvents on paper by solid-phase microextraction. *Journal of Forensic Sciences,* **52**(1):209–215.

Browning, B.L. (1977) *Analysis of Paper,* 2nd edn. Marcel Dekker, Inc., New York, NY.

Brunelle, R.L. and Pro, M.J. (1972) A systematic approach to ink identification. *Journal – Association of Official Analytical Chemists,* **55**:823–826.

Brunelle, R.L. and Reed, R. (1984) *Forensic Examination of Ink and Paper.* Charles C. Thomas, Springfield, IL.

Brunelle, R.L. and Crawford, K.R. (2003) *Advances in the Forensic Analysis and Dating of Writing Ink.* Charles C. Thomas, Springfield, IL.

Bügler, J.H., Buchner, H., and Dallmayer, A. (2005) Characterization of ballpoint pen inks by thermal desorptionand gas chromatography-mass spectrometry. *Journal of Forensic Sciences,* **50**(5):1209–1214.

Bügler, J.H., Buchner, H., and Dallmayer, A. (2008) Age determination of ballpoint ink by thermal desorption and gas chromatography-mass spectrometry. *Journal of Forensic Sciences,* **53**(4): 982–988.

Cantu, A.A. (1995) A sketch of analytical methods for document dating. Part I: the static approach: determining age independent analytical profiles. *International Journal of Forensic Document Examiners,* **1**(1):40–51.

Cantu, A.A. (1996) A sketch of analytical methods for document dating. Part II: the dynamic approach: determining age dependent analytical profiles. *International Journal of Forensic Document Examiners,* **2**(3):192–208.

Conway, J.V.P. (1959) *Evidential Documents.* Charles C. Thomas, Springfield, IL.

Crable, M. (2011) The evolution of tinting dyes and optical brighteners in white paper. Proceedings of the TAPPI PaperCon Conference. Covington, KY.

Ellen, D. (1997) *The Scientific Examination of Documents.* Taylor & Francis, London.

Gaudreau, M. and Brazeau, L. (2002) Ink dating using a solvent loss ratio method. Proceedings of the 60th Annual Conference of the American Society of Questioned Document Examiners, 14–18 August 2002, San Diego, CA.

Green, J. (2012) Reliability of paper brightness in authenticating documents. *Journal of Forensic Sciences,* **57**(4):1003–1007.

Grim, D., Allison, J., and Siegel, J. (2001) Evaluation of desorption/ionization mass spectrometric methods in the forensic applications of the analysis of inks on paper. *Journal of Forensic Sciences,* **52**(1):1411–1420.

Gupta, S.K., Rohillai, D.R., and Jain, M.K. (1981) Paper differentiation by thin layer chromatography of fluorescent brighteners and dyes. *Journal of the Canadian Society of Forensic Science,* **14**(1): 23–31.

Hewlett-Packard Development Company (2014) HP Color Inkjet cp1160 Series Printer – Underprinting. HP Support Center http://h20564.www2.hp.com/hpsc/doc/public/display?docId= emr_na-bpd50083&sp4ts.oid=45899 (last accessed 23 June 2015).

Hilton, O. (1982) *Scientific Examination of Questioned Documents.* Elsevier Science, New York.

Hilton, O. (1993) *Scientific Examination of Questioned Documents* (revised edn). CRC Press, Boca Raton, FL, pp. 31–32.

Holifield, A. and LaPorte, G. (2007) Artificially aged documents. Paper presented at the Mid-Atlantic Association of Forensic Scientists Annual Meeting, 25 2007, Washington, DC.

Houlgrave, S., LaPorte, G., and Stephens, J. (2011) The use of filtered light for the evaluation of writing inks analyzed using thin layer chromatography. *Journal of Forensic Sciences*, **56**(3): 778–782.

Hunter, D. (1974) *Papermaking*. Dover Publications, Inc., New York, NY.

Kelly, J.D. and Cantu, A.A. (1975) Proposed standard methods for ink identification. *Journal – Association of Official Analytical Chemists*, **58**:122–125.

LaPorte, G. (2004) Modern approaches to the forensic analysis of inkjet printing – physical and chemical examinations. *Journal of the American Society of Questioned Document Examiners*, **7**(1):22–36.

LaPorte, G. and Ramotowski, R.S. (2003) The effects of latent print processing on questioned documents produced by office machine systems utilizing inkjet technology and toner. *Journal of Forensic Sciences*, **48**(3):1–6.

LaPorte, G. and Singer, K. (2012) Artificial aging of documents. Proceedings of the American Academy of Forensic Sciences Annual Meeting, 23 February 23, Atlanta, GA.

LaPorte, G., Wilson, J., Cantu, A.A., *et al.* (2004) The identification of 2-phenoxyethanol in ballpoint inks using gas chromatography/mass spectrometry. *Journal of Forensic Sciences*, **49**(1): 155–159.

LaPorte, G., Arredondo, M., and McConnell, T., *et al.* (2006) An evaluation of matching unknown writing inks with the United States International Ink Library. *Journal of Forensic Sciences*, **51**(3):689–692.

Lile, J.E. (1976) Classification and identification photocopiers: a progress report. *Journal of Forensic Sciences*, **21**:923–931.

Lindblom, B.S. (2006) What is forensic document examination? In: J.S. Kelly and L.S. Lindblom (eds) *Scientific Examination of Questioned Documents*. CRC Press, Taylor and Francis Group, Boca Raton, FL.

Mead (1993) *Paper Knowledge*. The Mead Corporation, Richmond, VA.

Meng, P. and Wang, S. (1999) Identification of copy machines through pyrolysis chromatography of toners. *International Journal of Forensic Document Examiners*, **5**:88–90.

Merrill, R.A., Bartick, E.G., and Mazzella, W.D. (1996) Studies of techniques for analysis of photocopy toners by IR. *Journal of Forensic Sciences*, **41**:264–271.

Messier, P. (2005) Notes on dating photographic paper. In: *Topics in Photographic Preservation*, Vol. **11**, Photographic Materials Group of the American Institute for Conservation of Historic & Artistic Works, Washington, DC.

Mitchell, C.A. (1911) *Science and the Criminal*. Little, Brown, and Co., Boston, MA

Mitchell, C.A. (1922) *Documents and their Scientific Examination*. Charles Griffin and Co., Ltd., London.

Munson, T.O. (1999) A simple method for sampling photocopy toners for examination by Py-GC. *International Journal of Forensic Document Examiners*, **5**:303–305.

Osborn, A.S. (1929) *Questioned Documents*, 2nd edn. Boyd Printing, Albany, NY.

Polk, D.E., Attard, A.E., and Giessen, B.C. (1922) Forensic characterization of papers. II: determination of batch differences by scanning electron microscopic elemental analysis of the inorganic components. *Journal of Forensic Sciences*, **22**:524–533.

Roux, C., Novotny, M., Evans, I., and Lennard, C. (1999) A study to investigate the evidential value of blue and black ballpoint pen inks in Australia. *Forensic Science International*, **101**: 167–76.

SDC/AATCC (1971) *Colour Index*, 3rd edn. The Society of Dyers and Colourists/American Association of Textile Chemists and Colorists (SDC/AATCC).

Spence, L.D., Baker, A.T., and Byrne, J.T. (2000) Characterization of document paper using elemental compositions determined by inductively coupled plasma mass spectrometry. *Journal of Analytical Atomic Spectrometry*, **15**:813–819.

Tandon, G., Jasuja, O.P., and Sehgal, V.N. (1995) Thin layer chromatography analysis of photocopy toners. *Forensic Science International*, **73**:149–154.

Tandon G., Jasuja O.P., and Sehgal V.N. (1997) The characterisation of photocopy toners using FTIR. *International Journal of Forensic Document Examiners*, **3**:119–126.

Totty, R. (1999) The examination of photocopy documents. *Forensic Science International*, **46**: 121–126.

Wallace, M.R., Milliken, L.T., and Toner, S.D. (1967) Identification of Dyes in Paper by Extraction and Chromatographic Analysis. *Institute of Materials Research, National Bureau of Standards, Tappi*, **50**(9):121A–124A

Weyermann, C., Kirsch, D., Vera, C., and Spengler, B. (2007) A GC/MS study of the drying of ballpoint pen ink on paper. *Forensic Science International*, **168**:119–127.

Weyermann, C., Schiffer, B., and Margot, P. (2008) A logical approach to ballpoint ink dating. *Science and Justice*, **48**(3):118–125.

Weyermann, C., Kirsch, D., Costa Vera, C., and Spengler, B. (2009) Evaluation of the photodegradation of crystal violet upon light exposure by mass spectrometric and spectroscopic methods, *Journal of Forensic Sciences*, **54**(2):339–345.

Weyermann, C., Almog, J., Bügler, J., and Cantu, A. (2011) Minimum requirements for application of ink dating methods based on solvent analysis in casework. *Forensic Science International*, **210**:52–62.

Wilson, J., LaPorte, G., and Cantu, A. (2004) Differentiation of black gel inks using optical and chemical techniques. *Journal of Forensic Sciences*, **49**(2):364–370.

Witte, A.H. (1963) The examination and identification of inks. In: F. Lundquist (ed.) *Methods of Forensic Science*, Vol. **II**, Interscience, London/New York, pp. 35–77.

CHAPTER 9

Chemical methods for the detection of latent fingermarks

Amanda A. Frick, Patrick Fritz, and Simon W. Lewis

Nanochemistry Research Institute and Department of Chemistry, Curtin University, Australia

"I suppose you won't be able to find one of your famous Clues on the thing?"
"Shouldn't think so, sir. Not with all these fingerprints on it."

Lord Vetinari speaking to Vimes, *Feet of Clay* (Terry Pratchett)

9.1 Introduction

Fingermarks, the impressions left by contact between the skin ridges of the fingertips and a surface, are one of the most easily recognizable forms of identification evidence encountered by forensic investigators. Due to the unique nature of the patterns formed by these ridges, fingermarks have long been considered to provide conclusive proof of an individual's contact with the surface on which their fingermarks are found (Almog, 2000; Champod *et al.*, 2004; Saferstein, 2007). In this chapter, a distinction is made between fingermarks and fingerprints – "fingerprints" are defined here as a deliberately made imprint of the skin ridges of the fingertip (i.e., for the purpose of database entry), while the term "fingermarks" is used to refer to the imperfect (often smudged, distorted or incomplete) ridge impressions left by incidental contact (Wolstenholme *et al.*, 2009; Emerson *et al.*, 2011). The most common of these found at crime scenes are latent or invisible fingermarks, which require some form of treatment to make them visible for subsequent analysis and identification.

This chapter provides an overview of chemical techniques for the detection of latent fingermarks, with a primary focus on methods for the treatment of porous surfaces. This is followed by a discussion of the experimental considerations for research into latent fingermark chemistry before concluding with some comments on current research directions in this field. One of the most crucial aspects of the successful development of

---

Forensic Chemistry: Fundamentals and Applications, First Edition. Edited by Jay A. Siegel.
© 2016 John Wiley & Sons, Ltd. Published 2016 by John Wiley & Sons, Ltd.

latent fingermarks is an understanding of the interaction between a development reagent and its target compound(s). Therefore, it is the chemistry of latent fingermark residue to which attention is first turned.

9.2 Sources of latent fingermark residue

Latent fingermarks are a translucent mixture of aqueous and lipid components (Scruton *et al.*, 1975; Goode and Morris, 1983). This residue is predominantly composed of secretions from the eccrine and sebaceous glands, as well as material from the epidermis and the apocrine glands (Bramble and Brennan, 2000; Jelly *et al.*, 2009) (Table 9.1).

Hundreds of endogenous chemical species may be found in a single fingermark, including water, proteins, amino acids, lipids and salts. Trace amounts of exogenous contaminants, derived from the handling of substances such as food or grease, or from use of cosmetic products on the skin and hair, are also frequently present (Wolstenholme *et al.*, 2009; Koenig *et al.*, 2011; Girod *et al.*, 2012). Some of these substances, such as illicit drugs and explosives residues, are highly important to criminal investigations (Day *et al.*, 2004; Ifa *et al.*, 2008).

Current knowledge of latent fingermark composition is predominantly based on medical literature regarding skin secretions, rather than the unique mixture that is present on the fingertips (Almog, 2000; Bramble and Brennan, 2000; Ramotowski, 2001; Champod *et al.*, 2004; Jelly *et al.*, 2009). Additionally, not all substances on a fingertip are transferred in a single contact; the mechanical transfer of substances from the fingertip to a surface is affected by factors such as fingertip pressure, duration of contact, surface type and temperature, which, therefore, may have an impact on fingermark composition (Scruton *et al.*, 1975; Bramble and Brennan, 2000; Croxton *et al.*, 2006). For this reason, except when stated otherwise, the information below pertains to the composition of skin secretions rather than latent fingermarks.

Table 9.1 Secretory glands of human skin (Champod et al., 2004; Marieb and Hoehn, 2007)

Gland type	Location	Function(s) of gland
Eccrine gland	Entire body, greatest density on forehead, palms and soles of feet	Secrete water and waste products, regulate temperature and blood pH
Sebaceous gland	Entire body except for palms and soles of feet, greatest density on forehead and scalp	Lubricates hair and skin, protect the skin and hair against water, prevents microbial growth
Apocrine gland	Axillary and anogenital regions	Unknown in humans, possible scent glands

9.2.1 Aqueous components

The water-soluble constituents of latent fingermarks are primarily sourced from the secretions of the eccrine glands. More commonly known as sweat glands, they are located all over the body and are found in greatest density on the skin ridges of the palms of the hands and the soles of the feet. They are the sole type of secretory gland on the fingertips; therefore, latent fingermarks generally contain some amount of eccrine sweat, if little else (Champod *et al.*, 2004).

The eccrine glands and their secretory ducts form coiled, tubular structures within the dermis that open directly onto the skin surface (Marieb and Hoehn, 2007). Sweat is produced within the glands and packaged inside vesicles, to be released into the secretory ducts via exocytosis (Marieb and Hoehn, 2007). The main functions of eccrine sweat are to dissipate body heat through evaporation, improve grip on the hands and feet by moistening the skin, and to excrete excess water, electrolytes and waste products such as urea (Dalrymple, 2000; Jelly *et al.*, 2009).

Eccrine sweat is a primarily aqueous secretion consisting of salts and amino acids, as summarized in Table 9.2 (Bramble and Brennan, 2000; Harker *et al.*, 2006).

The exact composition of eccrine sweat varies greatly between individuals and is known to be affected by variables such as health, diet and genetic factors (Jasuja *et al.*, 2009; Croxton *et al.*, 2010). The rate of sweat production is influenced by stress, elevated environmental temperatures and physical activity (Jasuja *et al.*, 2009; Ramotowski, 2001). The response of the eccrine glands to elevated temperatures is gradual and weak, but is much stronger and more immediate when stimulated by stress, which is thought to be an important factor regarding the deposition of latent fingermarks at crime scenes (Ramotowski, 2001).

Table 9.2 Summary of the composition of eccrine sweat (Ramotowski, 2001)

Organic		Inorganic (major)		Inorganic (trace)
Amino acids	0.3–2.59 mg/L	Sodium	34–266 mEq/L	Magnesium
Proteins	15–25 mg/dL	Potassium	4.9–8.8 mEq/L	Zinc
Glucose	0.2–0.5 mg/dL	Calcium	3.4 mEq/L	Copper
Lactate	30–40 mM	Iron	1–70 mg/L	Cobalt
Urea	10–15 mM	Chloride	0.52–7 mg/L	Lead
Pyruvate	0.2–1.6 mM	Fluoride	0.2–1.18 mg/L	Manganese
Fatty acids	0.01–0.1 µg/mL	Bromide	0.2–0.5 mg/L	Molybdenum
Sterols	0.01–0.12 µg/mL	Iodide	5–12 µg/L	Tin
Creatine		Bicarbonate	15–20 mM	Mercury
Creatinine		Phosphate	10–17 mg/L	
Glycogen		Sulphate	7–190 mg/L	
Uric acid		Ammonia	0.5–8 mM	
Vitamins				

Table 9.3 Summary of the composition of apocrine gland secretions (Champod *et al.*, 2004)

Organic	Inorganic
Proteins	Water (>98%)
Carbohydrates	Iron
Sterols	

An infrequent and minor source of aqueous fingermark constituents are the apocrine glands. Apocrine glands are a type of sweat gland associated with hair shaft canals of the armpits and the groin (Ramotowski, 2001). Their function in humans is largely unknown; however, they are thought to act as scent glands as they become functional at puberty, under the influence of androgens. Additionally, apocrine sweat develops a characteristic odor due to bacterial degradation on the surface of the skin (Marieb and Hoehn, 2007).

Apocrine sweat is produced in granules, which are released from the cell together with small amounts of cytoplasm (Marieb and Hoehn, 2007). The secretions of the apocrine glands are otherwise largely similar to eccrine sweat (Table 9.3). Due to their location on the body, apocrine sweat is thought to rarely be a significant contributor to latent fingermark residue, except for cases of sexual assault (Champod *et al.*, 2004).

9.2.2 Lipid components

The lipid material present on the surface of the skin is derived from three sources: the epidermis itself, and the secretions of the sebaceous and the eccrine glands (Drapel *et al.*, 2009). Skin surface lipid composition is not uniform, as the relative contribution of sebaceous and epidermal lipids to total skin surface lipids will vary with anatomical location. Areas of skin rich in sebaceous glands will, unsurprisingly, contain a higher proportion of sebaceous lipids (Boughton *et al.*, 1957; Greene *et al.*, 1970). The constant sloughing of epithelial cells from the stratum corneum, the topmost layer of the epidermis, is a major source of fingermark lipids in areas with low sebaceous gland activity (Nicolaides, 1974; Nikkari, 1974; Asano *et al.*, 2002). Compositional differences between sebaceous and epidermal lipids are outlined in Table 9.4. Eccrine sweat does contain some lipid content, namely in the form of fatty acids and cholesterol, which are derived from the sloughing of epithelial cells within the secretory ducts, but this represents only a small fraction of latent fingermark lipids (Takemura *et al.*, 1989).

The main source of fingermark lipids is sebum, an oily mixture produced by the sebaceous glands that contributes the majority (over 95%) of lipid compounds found on the surface of human skin (Downing and Strauss 1974; Nicolaides, 1974; Nikkari, 1974; Buchanan *et al.*, 1997). Sebum is most commonly incorporated into fingermark residue through the habitual touching of the face and scalp, often referred to as grooming motions (Buchanan *et al.*, 1997; Mong *et al.*, 1999; Asano *et al.*, 2002; Croxton *et al.*, 2010).

Epidermal lipids contribute a minor amount (3–6 %) to total skin surface lipids in these areas (Greene *et al.*, 1970). Therefore, as far as latent fingermarks are concerned, composition of the lipid fraction typically focuses solely on the sebaceous lipids. Sebum is estimated to contribute 5% by weight of latent fingermark residue; however, the actual amount will naturally vary with the extent of grooming and removal of lipid material by washing (Wargacki *et al.*, 2008).

The sebaceous glands consist of one or several lobes, encapsulated by highly vascularised connective tissue (Thody and Shuster, 1989). In humans, they are located almost all over the body, except for the palms of the hands and the soles of the feet; they are found in greatest density on the face and scalp. These glands are associated with hairs on the body, with the secretory ducts of these glands opening directly into the hair shaft canal (Smith and Thiboutot, 2008). The purpose of human sebum is largely unknown; its main functions are thought to include lubrication of the skin and hair, waterproofing of the epidermis, transportation of antioxidants and providing individuals with a unique scent signature (Bonte *et al.*, 1995; Champod *et al.*, 2004; Zouboulis, 2004; Gallagher *et al.*, 2008; Smith and Thiboutot, 2008; Camera *et al.*, 2010; Michael-Jubeli *et al.*, 2011). Most mammals and many birds produce sebum as a means of protecting feathers and fur from water, and to secrete pheromones and other chemical signals. Certain sebaceous fatty acids also have antimicrobial properties and are, therefore, thought to play a role in pathogen defense and the maintenance of skin homeostasis (James and Wheatley, 1956; Nicolaides, 1974; Shalita, 1974).

Sebum is produced within the sebaceous glands via a holocrine mechanism. Lipids are produced within maturing cells and accumulate as these cells (sebocytes) continue to differentiate and migrate towards the center of the gland (Powe, 1972; Smith and Thiboutot, 2008; Ottaviani *et al.*, 2010). During the final stage of differentiation, the mature sebocytes rupture and release their contents into the secretory ducts (Strauss *et al.*, 1962; Thody and Shuster, 1989; Nikkari, 1974; Ramotowski, 2001). The full process of lipid accumulation and secretion onto the skin surface is estimated to take 1–2 weeks (Thody and Shuster, 1989; Zouboulis, 2004).

Table 9.4 Approximate composition of sebum and surface epidermal lipids (Downing and Strauss, 1974)

Constituents	Sebum (wt-%)	Surface epidermal lipid (wt-%)
Glycerides plus free fatty acids	57.5	65
Wax esters	26.0	—
Squalene	12.0	—
Cholesterol esters	3.0	15
Cholesterol	1.5	20

Table 9.5 Composition of sebum (Powe, 1972)

Lipid class	Percentage
Triglycerides	30–50
Free fatty acids	15–30
Wax esters	12–16
Squalene	10–12
Monoglycerides and diglycerides	5–10
Cholesterol	1–3
Cholesteryl esters	1–3
Hydrocarbons	1–3

Sebum is a complex mixture of a number of lipidic compounds, consisting of hundreds of individual molecular species, including wax esters, triglycerides, free fatty acids, cholesterol and squalene (Michael-Jubeli *et al.*, 2011) (Table 9.5).

Sebum composition is species-specific, with marked differences in the types and relative amounts of components observed even between closely related species, such as humans and chimpanzees (Nicolaides, 1974; Downing *et al.*, 1977; Thody and Shuster, 1989; Pappas *et al.*, 2002; Robosky *et al.*, 2008; Smith and Thiboutot, 2008). While many sebaceous components are produced in the sebaceous glands themselves, the presence of some essential fatty acids (such as linoleic acid) indicates that some of these lipids are exogenous in origin, and so are derived from blood circulation rather than *de novo* synthesis in the sebaceous gland. As with eccrine sweat, the composition of sebum can be affected by various factors such as genetics and diet (Ramotowski, 2001).

9.2.3 Sources of compositional variation

The act of depositing a fingermark on a surface is subject to a number of variables that may affect detection. Many factors, including surface temperature, electrostatic forces, fingertip pressure, angle and duration of contact, can also have an effect on the quantity and composition of the material in a latent fingermark, as well as the clarity of the fingermark pattern.

The amounts and composition of skin secretions deposited in a latent fingermark are also subject to variation between individuals, in addition to possible changes over time within a particular individual (Champod *et al.*, 2004; Frick *et al.*, 2013a). This is easily demonstrated through the use of development reagents. By repeatedly collecting and developing fingermarks from a small population, it can be observed that some individuals are consistently "good" or "poor" donors for target compounds such as amino acids or lipids (Almog *et al.*, 2011; Sears *et al.*, 2012; Girod and Weyermann, 2014; IFRG, 2014). This so-called "donor effect" is a well known, but poorly understood phenomenon, and it is often unclear as to why latent fingermarks from certain donors exhibit clear ridge detail, while those from others are barely visible when treated under the same operational conditions (Jones *et al.*, 2001; Almog *et al.*, 2011). It has been established that the increase in sebum production that occurs with the onset of puberty has a dramatic effect on the lipid content of fingermarks deposited by adults compared to young children (Mong *et al.*, 1999). Furthermore, the fingermarks deposited by children contain a higher proportion of volatile constituents, and so evaporate more rapidly than adults' fingermarks (Buchanan *et al.*, 1997). Other contributing factors to fingermark chemistry are thought to include traits such as diet, biological sex and metabolic disorders, which have an influence on skin gland secretion rate and composition (Champod *et al.*, 2004; Frick *et al.*, 2013a), as well as activities such as handling greasy food or washing hands with soap, which can impact on the amount of residue present on the skin ridges of the fingertips (Jones *et al.*, 2001; Archer *et al.*, 2005; Sears *et al.*, 2012).

Latent fingermark composition begins to be altered within a short period of time following deposition (Asano *et al.*, 2002; Lennard, 2007). This adversely affects the efficacy of many latent fingermark development techniques, which are most effective on fingermarks that are less than a few weeks old (Ramotowski, 2001; Archer *et al.*, 2005; Mountfort *et al.*, 2007). There is currently little information, beyond broad trends, on how latent fingermark residue ages, or how it is impacted upon by environmental factors, bacterial activity or possibly even the application of development reagents (Bramble and Brennan, 2000; Archer *et al.*, 2005; Mountfort *et al.*, 2007; Koenig *et al.*, 2011; Girod *et al.*, 2014). Studies into the ageing of latent fingermarks are complicated by difficulties in obtaining homogenous samples, the natural variability between fingermark donors and exogenous contamination, such that a timeframe for latent fingermark degradation processes is difficult to establish (Mong *et al.*, 1999; van Dam *et al.*, 2014; Girod *et al.*, 2014).

One of the first degradation processes that occurs in latent fingermarks is a significant loss of mass, primarily though the evaporation of water, within an hour of deposition (Darke and Wilson, 1977). The remaining residue is a brittle, waxy substance composed of salts, amino acids and non-volatile lipids, that on non-porous surfaces is subject to erosion (Ramotowski, 2001). As such, older fingermarks are less amenable to visualisation with powders, which rely on the mechanical adherence of particles to lipids and moisture, and lipophilic dyes, which partition most readily into liquefied lipids. Further degradation processes involve the breakdown of sebaceous lipids such as squalene, free fatty acids, wax esters and possibly triglycerides into volatile, short-chain compounds (Archer *et al.*, 2005; Mong *et al.*, 1999). Environmental conditions, including light exposure, surface type, temperature, humidity and airflow, are thought to play a significant role in degradation rate (Lennard, 2007; Wargacki *et al.*, 2008; Wolstenholme *et al.*, 2009; Weyermann *et al.*, 2011; Williams *et al.*, 2011). In more extreme cases, such as the immersion of the fingermarks in water, or in cases of arson, successful identification from a fingermark can become much more difficult. Due to the strong, stable hydrogen bonding of amino acids to the cellulose fibers of paper, reagents targeting these compounds can be effective at developing fingermarks that are several decades old (Almog *et al.*, 2004a). However, amino acids are soluble in water, and immersion or exposure to high humidity will dissolve these compounds from a surface. In the case of extreme heat, while amino acids typically deteriorate at these temperatures, some of the new pyrolytic products that can form may be targeted by fingermark reagents instead (Richmond-Aylor *et al.*, 2007).

9.3 Chemical processing of latent fingermarks

An item of evidence examined for latent fingermarks is typically subjected to a series of physical and chemical development methods, ordered so that the success of each technique is not hindered by a preceding one (Almog, 2000). Typically, treatment sequences commence with the simple examination of the surface under ambient and forensic light sources, before proceeding to the use of chemical and physical techniques. As is the norm in forensic analysis, techniques that may cause damage to the item of evidence occur towards the end of the sequence. The development methods utilized in such detection sequences are determined by surface type, which is broadly categorized into porous, non-porous and semi-porous, as well as whether the surface is dry, adhesive, or has been wetted (Ricci *et al.*, 2007). On porous items, for example, fingermark components are not only retained on the surface, but are absorbed into the substrate, resulting in long-lasting impressions that may contain up to three times more amino acid material than non-porous surfaces (Figure 9.1) (Frick *et al.*, 2013a).

It would be impossible to cover the plethora of techniques applied to the detection of latent fingermarks in a single chapter. The overview that follows focuses on reagents used for porous surfaces, although some techniques for non-porous surfaces are included to provide an indication of the full breadth of the methods that can be used.

9.3.1 Amino acid sensitive reagents

Latent fingermark detection on porous surfaces is predominantly carried out using any of a number of reagents that react with amino acids. These methods are highly sensitive, develop fingermarks rapidly, and produce photoluminescent products (or those that can

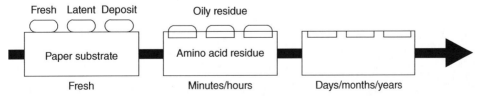

Figure 9.1 Schematic cross-section of a latent fingermark on a paper substrate at various stages after deposition

be treated with metal salts to produce photoluminescent complexes) (Jelly *et al.*, 2009). The last provides a significant advantage when examining dark-coloured or patterned surfaces, on which a developed fingermark cannot be readily seen under conventional lighting.

9.3.1.1 Ninhydrin

Ninhydrin reacts with amino acids to produce "Ruhemann's purple," named after Siegfried Ruhemann, who discovered the reaction in 1910 (Almog *et al.*, 2000; Hansen and Joullie, 2005). Ruhemann's purple is produced by the reaction of ninhydrin with the nitrogen of the amine group in amino acids, as shown in Figure 9.2 (Almog, 2001; McCaldin, 1960).

However, it was not until the 1950s that ninhydrin's potential as a fingermark development reagent became apparent, through the work done by Oden and von Hofsten

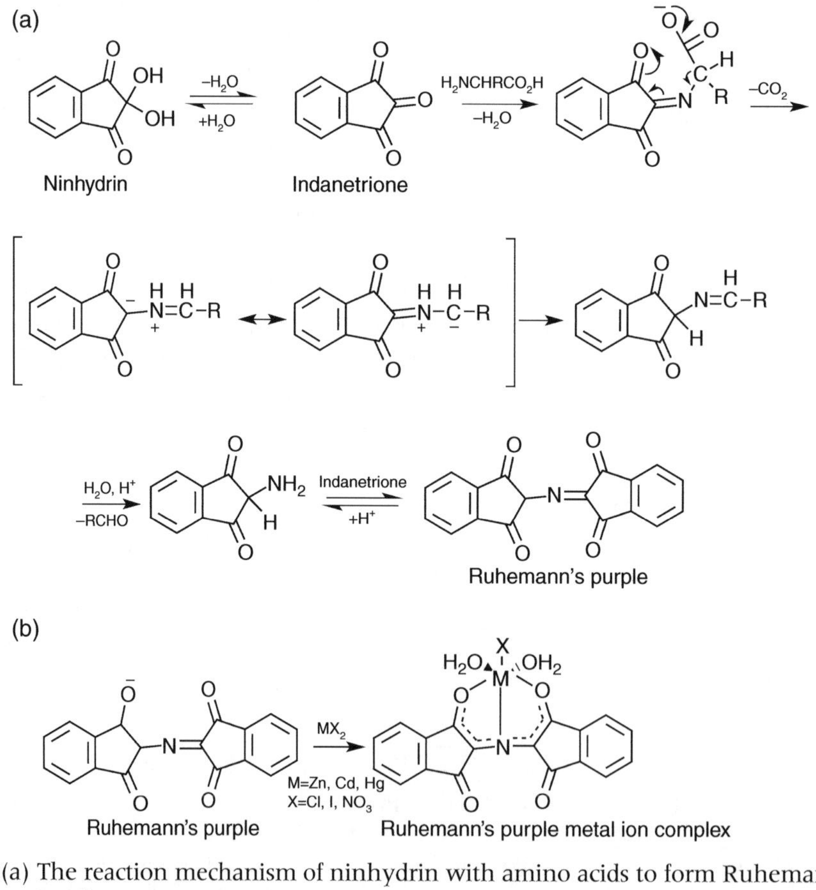

Figure 9.2 (a) The reaction mechanism of ninhydrin with amino acids to form Ruhemann's purple (Friedman and Williams, 1974; Friedman, 2004; Hansen and Joullie, 2005). (b) The reaction of Ruhemann's purple with metal salts to form a complex ion (Davies *et al.*; Lennard *et al.*, 1987)

(Oden and von Hofsten, 1954). Although ninhydrin was shown to successfully react with amino acids in fingermark deposits, it was found that the contrast and sensitivity could be further improved by a secondary metal salt posttreatment (Lennard *et al.*, 1987; Wallace-Kunkel *et al.*, 2007). Ruhemann's purple is an active chelating agent that can form coordination complexes with metal ions, resulting in a luminescent complex that is easy to observe with the right optical filters. Furthermore, the stability of the treated fingermark is also improved (Almog, 2001). Although the relatively slow reaction rate of ninhydrin with amino acids can be accelerated by applying heat, this also increases the unwanted side reaction of ninhydrin with some additives found in paper (Jelly *et al.*, 2009). This side reaction is slower than the desired one, and is not found to cause excessive background staining as long as developed fingermarks are recorded immediately (Stoilovic and Lennard, 2012). Despite the improved development with metal posttreatment, a simpler and more robust method is most commonly used in routine police work. This formulation consists of approximately 0.5% (w/v) ninhydrin in a solvent (mostly alcohol, methylated spirits or HFE-7100), progressing to completion at 50–80% relative humidity over 24–48 hours (Lennard *et al.*, 1987; Almog, 2000). Ninhydrin is commonly dissolved in a polar solvent; however, background staining on certain materials is much higher even with small amounts of polar solvents. To reduce the polarity of the working solution, non-polar cosolvents are commonly used to preserve the exhibits (Hansen and Joullie, 2005).

An important forensic aspect in criminal cases is the possibility of DNA analysis. While DNA analysis of document evidence is not routinely performed, it is an important aspect to consider, especially if other evidence in the case is limited. In the case of ninhydrin, DNA can still be extracted for analysis after the fingermarks have been treated (Presley *et al.*, 1993; Schulz *et al.*, 2004).

To further improve upon the detection of the amino acid fraction of latent fingermarks, efforts have been made into finding analogues of ninhydrin that are still capable of staining similar to Ruhemann's purple formation but with better contrast and sensitivity (Figure 9.3) (Almog, 2001; Hansen and Joullie, 2005). The most successful of these ninhydrin analogues, some of which have found routine use in fingermark detection, are discussed below.

9.3.1.2 1,8-Diazafluoren-9-one

1,8-Diazafluoren-9-one (DFO) was first synthesized in 1950 by Druey and Schmidt (Druey and Schmidt, 1950) and adapted as a fingermark development reagent in 1990 (Grigg *et al.*, 1990; Pounds *et al.*, 1990). DFO is an amino acid sensitive reagent that yields a highly luminescent red product (excitation at 430–580 nm, emission at 560–620 nm) (Champod *et al.*, 2004). Unlike ninhydrin, metal complexes do not increase the sensitivity of DFO, but can cause a change in the color of the product (Conn *et al.*, 2001). The application of heat is required for the reaction to proceed; however, prolonged and high temperatures are detrimental to the luminescent complex (Corson and

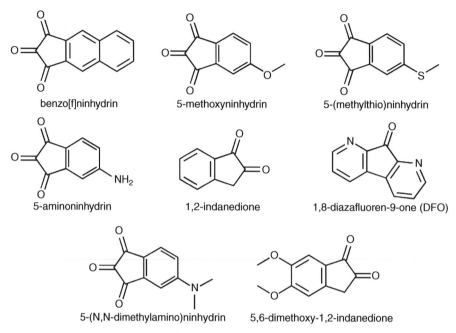

Figure 9.3 Structures of ninhydrin analogues (Jelly *et al.*, 2009)

Lawson, 1991; Masters *et al.*, 1991; Jelly *et al.*, 2009). The reaction mechanism of DFO with amino acids is similar to that of ninhydrin (Wilkinson, 2000). Some analogues of DFO have been synthesized which were found to be superior in luminescence, but the production cost made them financially impractical (Ramotowski, 2012a). A range of different formulations exist depending on the group using the reagent, yet HFE-7100 is widely considered to be the solvent of choice (Jelly *et al.*, 2009; Ramotowski, 2012a). DFO is commonly used prior to ninhydrin in a detection sequence, where ninhydrin may be of use with surfaces that have high background interference in the luminescent mode (Masters *et al.*, 1991; Hardwick *et al.*, 1993). DFO can also be used as a dry contact method, but still requires the application of heat for the reaction to go to completion (Bratton and Juhala, 1995). In addition, several variations exist of the wet contact method for use on thermal papers. Recent recommendations include the use of two-stage processes, such as using an ethanol prewash or a whitening solution to counter the blackening of the background (Anonymous, 2006; Schwarz and Klenke, 2007).

9.3.1.3 1,2-indanedione

In 1997, Joullié *et al.* introduced 1,2-indanedione as a viable latent fingermark development reagent (Hauze *et al.*, 1998). Similar to ninhydrin, 1,2-indanedione converts α–amino acids into aldehydes via an imine intermediate to produce the reaction product "Joullié's pink" (Figure 9.4) (Spindler *et al.*, 2011a).

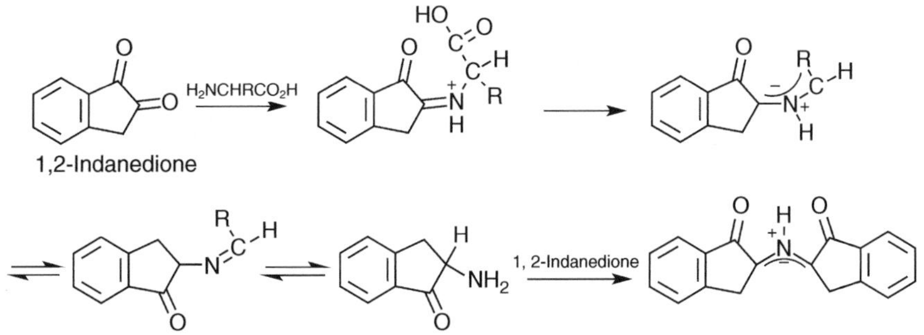

Figure 9.4 Proposed reaction mechanism of 1,2-indanedione and α-amino acids (Petrovskaia *et al.*, 2001)

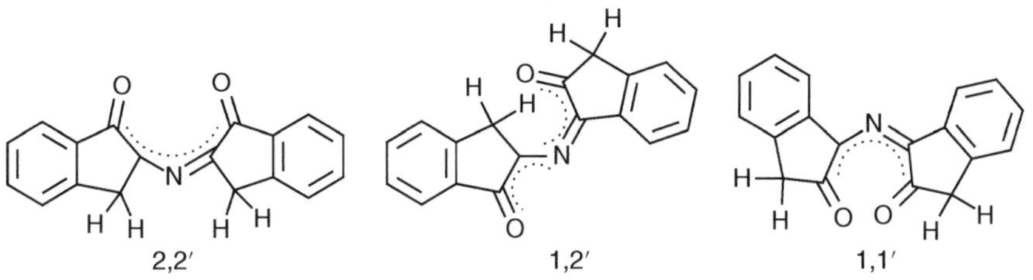

Figure 9.5 The three proposed structural isomers of Joullié's pink that could form from the reaction between 1,2-indanedione and α-amino acids. The enol tautomer of the 2,2′-isomer is hypothesized to be the major structure formed on eccrine-rich latent fingermarks (Spindler *et al.*, 2011)

However, unlike the Ruhemann's purple formation, the complex produced depends on the amino acid, where "R dependence" exists if one considers the general amino acid form of R–NH_2 (Petrovskaia *et al.*, 2001; Alaoui *et al.*, 2005). This general mechanism is referred to as a Stecker degradation. Although the 2′ carbonyl site appears to be the preferential reaction site, recent research indicates that the difference in reactivity between the 1′ and 2′ carbonyl groups is minimal and, therefore, the reaction site is dictated by the strength of the nucleophile (Figure 9.5) (Almog and Glasner, 2010; Spindler *et al.*, 2011a).

The strong nucleophile, together with an active and potentially catalytic substrate, is hypothesized to direct the formation of just the 2,2′- Joullié's pink isomer (Spindler *et al.*, 2011a). A cellulose matrix, such as paper, can further stabilize the reaction intermediates by acting as a surface catalyst, in addition to decreasing the degradation of the reaction product (Spindler *et al.*, 2011a). Furthermore, water molecules can be retained near the reaction site, while trace metals present from the paper production can also add to the catalysis of the reaction. It was previously thought that a complex was formed; however, the two carbonyl sites on the Joullié's pink ligands are required for complexation and, therefore, already occupied (Lennard *et al.*, 1987). This complex can also be formed in solution with amino acids such as alanine and glycine (Alaoui, 2007; Alaoui *et al.*, 2012).

(a) (b)

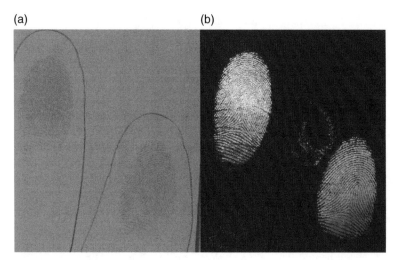

Figure 9.6 Fingermarks on white copy paper developed with 1,2-indanedione, viewed (a) under ambient lighting conditions and (b) through a 550 nm barrier filter with illumination at 505 nm (*See insert for color representation of the figure.*)

While ninhydrin provides superior initial colour under white light, 1,2-indanedione is more sensitive and gives better contrast when viewing fingermarks under luminescent conditions without further treatment (Jelly *et al.*, 2009). Similar to ninhydrin, the addition of a metal to the reagent improves the luminescence intensity and contrast, as well as improving the longevity of the treated fingermarks (Hauze *et al.*, 1998; Roux *et al.*, 2000; Wallace-Kunkel *et al.*, 2007). An image of a fingermark on paper treated with 1,2-indane-dione is presented in Figure 9.6.

Unlike ninhydrin, the metal salt is an integral part of the working solution, rather than an additional posttreatment step. Studies indicate zinc, in the form of zinc chloride ($ZnCl_2$), to be the metal of choice, as it offers the greatest improvement in the luminescence inten-sity. While the exact role that the metal salt has in the 1,2-indanedione reaction is still unclear, apart from acting as a catalyst, there have been several recent studies on this topic (Hauze *et al.*, 1998; Roux *et al.*, 2000; Alaoui *et al.*, 2005; Wallace-Kunkel *et al.*, 2007).

The absorption (visible) band for the Joullié's pink complex is blue–green at 488 and 514.5 nm. Adding $ZnCl_2$ to the formulation causes a blue shift of the complex by 25 nm (Alaoui *et al.*, 2012). The maximum emission occurs between 560 and 570 nm, giving rise to a Stokes shift (difference between absorption and emission maxima) of around 120 nm. Laser-induced fluorescence results in photodegradation of the Joullié's pink complex emission peak at 564 nm. Alaoui used a 1,2-indanedione-glycine solution in methanol to demonstrate this effect, as the emission peak descended to a lower, stable level after about one hour of continuous excitation (Alaoui, 2007). 1,2-indanedione-treated fingermarks were also irradiated for five minutes per day, where the emission maxima decreased in the first week and then remained relatively stable for nearly two weeks. Preliminary results indicate that there is no intermolecular energy transfer from the 1,2-indanedione-glycine ligand to the zinc metal ion (Alaoui, 2007).

To further improve the development of latent fingermark deposits, changes of the reagent formulation have been undertaken since its initial proposal. These include alteration to the solvents used, the pH, as well as the application of heat and dry contact approaches. For the final working solution, two commonly used solvents, HFE-7100 and petroleum ether, were used. The use of HFE-7100 can result in better development than HFC-4310mee, HFE-71de and methanol, and offers lower health and safety risks (Roux *et al.*, 2000; Wallace-Kunkel *et al.*, 2007). Additionally, lipids are not dissolved by HFE-7100, which allows this formulation to be used in sequence with lipid-sensitive reagents (Roux *et al.*, 2000; Wallace-Kunkel *et al.*, 2007). Petroleum ether also offers very good contrast and fluorescence; however, the product is less stable than HFE-7100, and is also highly flammable and toxic. Further solvents have been studied, yet these showed limitations in sensitivity through lower initial color and/or luminescence, expense, toxicity, stability or fingermark preservation (Roux *et al.*, 2000; Wallace-Kunkel *et al.*, 2007).

In addition to the choice of solvent, the acidity of the formulation should be considered. As ninhydrin requires a slightly acidic environment, this was also believed to be necessary for 1,2-indanedione due to the structural similarity (Jelly *et al.*, 2009). The acidified reagent is still widely used; however, Wiesner *et al.* showed that better results could be achieved without the use of acid (Wiesner *et al.*, 2001).

The reaction time can be greatly reduced by applying heat to the treated fingermarks. Ridge development can be observed after 24–48 hours without heating, but only after five days will the results be similar to heat-treated fingermarks (Roux *et al.*, 2000). Heat can be applied using an oven or a laundry press, but the latter results in better luminescence and requires only 10 seconds at 160–165°C compared to 10–20 minutes at 100°C (Roux *et al.*, 2000; Wallace-Kunkel *et al.*, 2007; Stoilovic and Lennard, 2012). Fingermarks developed at a higher humidity result in superior initial color and fluorescence with both the HFE-7100 solvent and petroleum spirits (Roux *et al.*, 2000; Wallace-Kunkel *et al.*, 2007). Studies have found that treated fingermarks are degraded by sunlight, with the intensity decreasing by 80% after 28 days of exposure (Gardner and Hewlett, 2003). As with ninhydrin, DNA can be successfully extracted from 1,2-indanedione-treated fingermarks within the first five days following treatment (Azoury *et al.*, 2002).

Patton *et al.* presented a new technique for the application of 1,2-indanedione called the dry contact method, which consists of sandwiching the porous surface between two preprepared sheets that have been dipped into an alternative 1,2-indanedione reagent formulation and dried prior to use. This method has advantages in developing prints on thermal or very fragile paper and it can be easily transported if travelling is required (Patton *et al.*, 2010). Despite this, an acid-free wet contact formulation, which also does not use the application of heat, is the preferred method for thermal paper, as recommended by the National Centre for Forensic Studies (Stoilovic and Lennard, 2012).

9.3.1.4 *p*-Dimethylaminocinnamaldehyde

p-Dimethylaminocinnamaldehyde (DMAC) was first studied for the purpose of detecting latent fingermarks on paper by Morris and Goode in 1973, and was thought to show promise as an alternative to ninhydrin (Morris, 1976). However, an in-depth study conducted by Sasson and Almog found that DMAC was inferior to ninhydrin on nearly all fingermarks tested (Sasson and Almog, 1978). A significant issue encountered was that fingermarks that were treated more than 72 hours after deposition appeared blurred. This was thought to be due to urea (the proposed target compound) diffusing rapidly through porous substrates (Sasson and Almog, 1978; Ramotowski, 2012a). It has since been established that DMAC does not exclusively target urea in fingermark deposits; it also reacts with primary and secondary amines, including amino acids (Sasson and Almog, 1978).

Brennan *et al.* investigated DMAC as a fuming agent, achieving good results on a wide variety of surfaces (Brennan *et al.*, 1995). Fuming was accomplished by passing the fingermarks over hot DMAC and then conditioning overnight at 200°C and 30% relative humidity. Surfaces that were successfully treated in this study include aluminum cans, thermal paper, cardboard, polythene and glass. It was further found that this method could be used without detriment in sequence with DFO, ninhydrin and physical developer. However, further research concluded that fuming was less effective than using existing processes, except for the treatment of thermal papers (Brennan *et al.*, 1995).

Ramotowski pursued DMAC, initially as a wet contact reagent, then as a dry contact method (Ramotowski, 1996). It was found that, as with the method proposed by Sasson and Almog, red-colored ridge detail was produced with moderate fluorescence using a wet contact approach. However, within several hours the background developed further, giving undefined ridge detail. The dry contact method produced light yellow ridges with significant luminescence. By reducing the concentration and contact time, improved contrast was achieved (Ramotowski, 1996).

Lee *et al.* further investigated the use of DMAC as a dry contact method for thermal paper such as receipts and facsimile paper (Lee *et al.*, 2009). This method differed from the one proposed in the National Centre for Forensic Studies in that methanol was replaced with the less toxic ethanol with no detrimental effect (Stoilovic and Lennard, 2012). It was found that DFO and ninhydrin were vastly superior techniques to DMAC, with 100 % and 50% more ridge detail or fingermarks developed, respectively. In addition, amino acid, urea, and sodium chloride spot tests were prepared, treated with DMAC and their excitation and emission spectra compared to treated latent fingermarks. It was found that the spectra for the amino acids most closely resembled those of a latent fingermark, and it is thought that these are most likely the dominant target compounds for the DMAC reaction (Lee *et al.*, 2009).

9.3.1.5 Alternative reagents

A range of other amino acid sensitive fingermark reagents have been proposed over the last 50 years to overcome some of the limitations of the reagents discussed above (Bleay *et al.*, 2012; Ramotowski, 2012a). Some of the more promising alternatives to ninhydrin

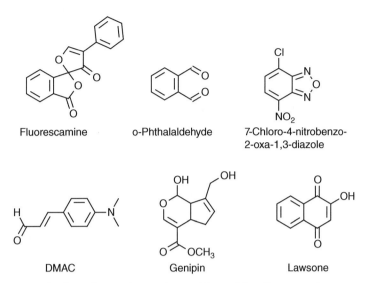

Figure 9.7 Structures of select alternative amino acid sensitive fingermark reagents (Bleay *et al.*, 2012; Jelly *et al.*, 2008)

and its analogues include amino acid assays such as *o*-phthalaldehyde, fluorescamine and NBD-chloride (7-chloro-4-nitrobenzo-2-oxa-1,3-diazole), as well as reagents based on natural compounds such as genipin and lawsone (Figure 9.7) (Jelly *et al.*, 2009).

o-Phthalaldehyde was studied by Mayer *et al.* as an alternative amino acid sensitive fingermark reagent in 1978 (Mayer *et al.*, 1978). This work was based on the amino acid assay developed by Benson and Hare in 1975, which was found to be more sensitive, stable and cheaper than fluorescamine (Benson and Hare 1975). In the presence of 2-mercaptoethanol, it reacts with primary amino acids to yield luminescence at 455 nm when excited with UV radiation at 340 nm (Mayer *et al.*, 1978). It is postulated that the increased use of optical brighteners in paper surfaces, which can be brighter than the luminescence of treated fingermark deposits, is partly to blame for this method's lack of use by the forensic community (Bleay *et al.*, 2012).

Fluorescamine was first investigated in the context of developing fingermarks in 1979 by Lee and Attard, and later re-evaluated by the UK Home Office Centre for Applied Science and Technology (Bleay *et al.*, 2012). In 1972, Weigele reported the use of fluorescamine for the detection of primary amino acids (Weigele *et al.*, 1972). Based upon this information, Lee and Attard proposed a novel fingermark reagent as a ninhydrin alternative (Ramotowski, 2012a). Although the resultant fingermarks were found to be more strongly developed on certain surfaces than ninhydrin, they were less stable. In addition, the use of UV light sources for the excitation was a further disadvantage and this method was not recommended ultimately for routine police work (Bleay *et al.*, 2012).

NBD-chloride was investigated as a viable option for fingermark treatment by Salares *et al.* (Salares *et al.*, 1979). Similar to the development of fluorescamine and *o*-phthalaldehyde, this work was largely based on the results of amino acid detection in the biochemistry field (Fager *et al.*, 1973). Further work by the Forensic Science Research Unit in Canberra, Australia, in the 1980s improved upon the original method by using a cheaper excitation source and compared it against ninhydrin on a range of surfaces (Warrener *et al.*, 1983; Stoilovic *et al.*, 1984). Despite similar development being observed, and improved contrast on older fingermarks, its significant health and safety risks (potent mutagen) ultimately resulted in the rejection of this method (Nelson and Warren, 1981; Stoilovic *et al.*, 1984).

Extracted from the fruit of *Gardenia jasminoides*, genipin has been investigated over the last decade as a novel fingermark reagent (Almog *et al.*, 2004b; Levinton-Shamuilov *et al.*, 2005; Almog *et al.*, 2007). Initially proposed by Almog *et al.* it showed promise due to its luminescence and safety benefits over established methods (Almog *et al.*, 2004b). The clear working solution was shown to provide strong luminescence at an emission wavelength of 620 nm when excited at 590 nm. This research was followed up on with changes to the formulation and more in-depth studies on its applicability on a range of surfaces. It was found that genipin was not as sensitive as existing methods; however, it may offer advantages with surfaces that have strong self-fluorescence and benefit from visualization at longer wavelengths (Almog *et al.*, 2007). This work did, however, prompt the investigation of other natural compounds for the visualisation of latent fingermarks.

Lawsone (2-hydroxy-1,4-naphthoquinone) is the active compound of henna, a natural skin and hair dye derived from *Lawsonia inermis* (Jelly *et al.*, 2008). As a member of a class of compounds called napthoquinones, which are known for their reactions with amino acids, it was thought lawsone might have potential as a novel method for the detection of fingermark deposits on porous surfaces (Jelly *et al.*, 2008). A formulation of 1 g L^{-1} lawsone in 1:4 ethyl acetate:HFE-7100 was found to give the most development in the preliminary study, where emission occurred at 640 nm after excitation 590 nm, similar to genipin. Despite this initial promise, lawsone was found to be less sensitive than existing methods and not further pursued (Jelly *et al.*, 2008).

9.3.2 Reagents based on colloidal metals
9.3.2.1 Physical developer
Physical developer (PD) is a fingermark development method that is based upon a now-defunct photographic development process. It was often observed that the silver nitrate-based physical developer reagent left fingermark impressions on photographic plates that had been handled with bare hands (Goode and Morris, 1983). The Atomic Weapons Research Establishment, working under the Police Scientific Development Branch (PSDB) of the UK Home Office, adapted the physical developer method for the detection of latent fingermarks on porous surfaces in the 1970s (Goode and Morris, 1983; Champod *et al.*, 2004). This original PD formulation is what is referred to as the UK-PD formulation, currently in use by European and Australian forensic laboratories. The UK-PD formulation has remained essentially unchanged since its development in 1975 (Cantu and Johnson, 2001).

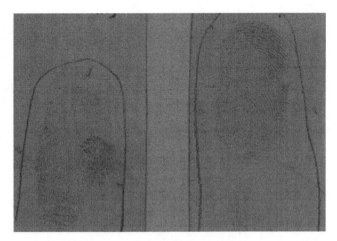

Figure 9.8 Fingermarks on white copy paper developed with physical developer

Essentially, PD treatment involves the reduction of silver ions (Ag^+) to silver particles (Ag^0) in solution by a ferrous/ferric oxidation–reduction system, stabilized by the presence of citrate and a cationic surfactant. The selective accumulation of silver particles on fingermark residue allows the pattern of the ridges to be observed (Figure 9.8).

PD interacts with the water-insoluble fraction of latent fingermark residue, making it one of the few development techniques that can be successfully utilized on porous surfaces that have been exposed to water or high humidity, conditions which wash away the target compounds of amino acid sensitive reagents (Goode and Morris, 1983; Champod *et al.*, 2004). The use of PD on porous surfaces following treatment with amino acid sensitive reagents increases the number of fingermarks detected compared to using the latter reagents alone (de Puit *et al.*, 2011). Though PD is often described as being sensitive to the sebaceous components of latent fingermark residue, evidence suggests that PD also interacts with some water-soluble compounds, such as amino acids, that become trapped in the hydrophobic lipid residue (Champod *et al.*, 2004; Houlgrave *et al.*, 2011).

Despite its effectiveness, the PD working solution remains notorious as an unstable and difficult reagent to work with, making reproducibility an issue (Beaudoin, 2004; Lennard, 2007; Wood and Jame,s 2009; Kupferschmid *et al.*, 2010). Another of the main shortfalls of PD is the need for meticulously clean glassware and reagents. PD utilizes silver nitrate as a source of Ag^+, which tends to stain everything it contacts if not handled carefully, and also leaves PD extremely sensitive to chloride contamination, which causes the formation of silver chloride precipitate and so renders the reagent useless (Rawji and Beaudoin, 2006). Background staining resulting from the deposition of silver on the surface rather than the fingermark is another problem, as the process is irreversible and may obscure the fingermark (Salama *et al.*, 2008). This, together with the time con-suming nature of the treatment process, limits PD from being in widespread routine use unless absolutely required (de Puit *et al.*, 2011).

Table 9.6 Composition of PD stock solutions and working solution (Stoilovic and Lennard, 2012)

Solution	Reagent preparation
Detergent–surfactant solution	2 g n-dodecylamine acetate and 2 g synperonic N dissolved in 500 mL deionized water
Redox solution	30 g ferric nitrate nonahydrate, 80 g ferrous ammonium sulfate hexahydrate, 20 g citric acid and 40 mL detergent–surfactant solution dissolved in 900 mL deionized water in order given
Silver nitrate solution	10 g silver nitrate dissolved in 50 mL deionized water
Maleic acid prewash	25 g maleic acid dissolved in 1 L deionized water
Working solution	12 mL silver nitrate stock solution added to 237 mL redox stock solution

The PD treatment process is a lengthy one that consists of numerous immersion baths and several reagents. Up to two hours may be required to complete the whole procedure. Due to the number of aqueous immersion baths involved in this method, PD is implemented only after documents have been treated with amino acid sensitive reagents, such as ninhydrin, in order to avoid washing away the amino acids before fingermark detection. PD is also employed at the end of a detection sequence because of the destructive and irreversible treatment mechanisms involved.

PD requires the preparation of four stock solutions (surfactant, maleic acid, silver nitrate and redox solution) as well as the working solution, which contains both silver nitrate and the ferric/ferrous redox system (Table 9.6). Commercial preparations are also available as premade redox and silver nitrate stock solutions. It is recommended that glass or plastic trays be used for all PD reagents, as this avoids the silver deposition that occurs with metal trays. Similarly, plastics tweezers are used to handle documents, as any indents left in the paper by metal tweezers will also result in silver deposition.

Emphasis is often placed on the specific order in which the redox solution components must be added (ferric nitrate, followed by ferrous ammonium sulfate and citric acid). The reasoning behind this specific order of addition has been described by Cantu (Cantu, 2000) and is summarized as follows: the main concern regarding the preparation of PD redox solution is the premature oxidation of ferrous ions to ferric ions at high concentrations of the latter, which is overcome by completely dissolving ferric nitrate before adding ferrous ammonium sulfate; citric acid is added last to facilitate the dissolution of any ferric hydroxide that may have formed in solution. However, recent work has found that addition order does not affect the efficacy of the redox solution, indicating that PD is somewhat more robust than previously thought (Sauzier et al., 2013). Earlier documents citing methods for redox solution preparation do not stress any particular order in which the components must be added (Goode and Morris, 1983).

The first step of PD is immersion of the document in deionized water to remove debris, such as might be encountered with documents retrieved from a river or lake, as well as any substances that may interfere with the working solution (Cantu and Johnson, 2001). The next step is immersion in a dilute acid to remove any fillers (namely calcium carbonate) in the paper. These fillers give copy paper a bright white appearance and a smooth finish that prevents pages from jamming or tearing in printers (Mong *et al.*, 1999). Such substances interfere with PD fingermark development, as the carbonate dissolves into the working solution, increasing solution pH, and reacts with silver ions to ultimately form Ag_2O (Cantu and Johnson, 2001; Wilson *et al.*, 2007). This causes irreversible black/brown discoloration across portions or the whole of the document, such that any fingermarks present are likely to be obscured. The acidic prewash removes calcium carbonate via the reaction (Yamashita and French, 2011):

$$CaCO_3 + 2H^+ \rightarrow Ca^{2+} + H_2O + CO_2$$

The main requirement for the pretreatment is that the acid used must not contain chloride (i.e., hydrochloric acid), as this will introduce chloride contamination into the working solution. The most commonly used acid for PD pretreatment is maleic acid, although dilute nitric and acetic acids have also been shown to be effective (Ramotowski, 2000; Cantu and Johnson, 2001; Wilson *et al.*, 2007). Malic acid may be substituted in place of maleic acid in order to reduce reagent costs, and is now the preferred acid pretreatment used by the United States Secret Service (USSS) (Houlgrave *et al.*, 2011). The required time period for this step varies depending on paper type, as the document must not be removed until the neutralization reaction is complete. Therefore, the document must be immersed in the acid until the paper appears translucent and bubbling from the document (carbon dioxide) has ceased (Ramotowski, 2000; de Puit *et al.*, 2011).

Removal of the carbonate fillers weakens paper, leaving it fragile and easily damaged (Rawji and Beaudoin, 2006; Wood and James, 2009). Acid-treated documents must be handled delicately to prevent tearing during subsequent transfers between reagent baths. Blistering and peeling of the paper layers may also occur, depending on paper type and length of pretreatment.

The document is then rinsed in deionized water once again, in order to remove residual acid and carbonates, before being immersed in the working solution (de Puit *et al.*, 2011). The working solution is prepared by mixing a small amount of the silver nitrate stock solution with redox stock solution. It is prepared when needed, as its unstable nature limits its shelf life to only a few days. The redox solution contains the ferrous/ferric system required for silver reduction, as well as citric acid and surfactants that play a role in stabilizing the solution (Houlgrave *et al.*, 2011). Silver nanoparticles are formed by spontaneous reduction in the working solution. The ferrous/ferric redox couple works in conjunction with citric acid to reduce silver ions via the following reactions (Cantu and Johnson, 2001; Yamashita and French, 2011):

$$Ag^+_{(aq)} + Fe^{2+}_{(aq)} \leftrightarrow Ag_{(s)} + Fe^{3+}_{(aq)} \qquad E^o_{redox} = 28.6\,mV \qquad (9.1)$$

$$Fe^{3+}_{(aq)} + H_3Cit_{(aq)} \leftrightarrow FeCit_{(aq)} + 3H^+_{(aq)} \qquad (9.2)$$

$$Ag^+_{(aq)} + Fe^{2+}_{(aq)} + H_3Cit_{(aq)} \leftrightarrow Ag_{(s)} + FeCit_{(aq)} + 3H^+_{(aq)} \qquad E^o_{redox} = 5.0\,mV \qquad (9.3)$$

The formation of the ferric–citrate complex (Reaction 9.2) causes a subsequent decrease in ferric ion concentration of the working solution, thereby shifting the direction of Reaction 9.1 to the right (Ramotowski, 2000; Yamashita and French, 2011). As shown in Reactions 9.1 and 9.3, the addition of citric acid changes the electrical potential of the working solution. These mechanisms shift the overall reaction in favor of silver reduction, but prevent this reaction from progressing too rapidly, allowing better control over fingermark development and reagent stability (Cantu and Johnson, 2001). The concentrations of the redox solution components are finely balanced to achieve this – significant alterations to the redox solution formulation alter the electrical potential of the solution, causing silver reduction to occur either uncontrollably or not at all (Sauzier et al., 2013).

The exact mechanism for the interaction between the silver particles and fingermark residue is unclear; silver deposition will occur on the surface as well as the fingermark under a variety of conditions, to produce dark grey fingermarks on a light grey background on white paper (de Puit et al., 2011). One accepted hypothesis for the interactions between PD and fingermarks is that the deposition of silver on latent fingermark residue is triggered by electrostatic attraction between the negatively charged silver particles and the positive charge conferred to certain fingermark components in the acidic working solution. These are thought to include unsaturated lipids, large, insoluble proteins and possibly amino acids (Cantu and Johnson, 2001; Salama et al., 2008; Becue et al., 2011; Houlgrave et al., 2011). Amine functional groups become protonated in a low pH environment, such as that provided by citric acid (Cantu and Johnson 2001; Champod et al., 2004; Becue et al., 2011). It is though that negatively charged silver particles formed sufficiently close to the fingermark reside become electrostatically attracted, and so adhere to the positively charged residue components, thereby forming nucleating sites for further aggregation of silver particles (Mong et al., 1999; Cantu and Johnson, 2001).

The presence of citrate ions and a cationic surfactant (n-dodecylamine acetate) is intended to stabilize the working solution by preventing the rapid, uncontrolled formation of large silver particles, which results in poorer development contrast (Stoilovic and Lennard, 2012). When silver ions are spontaneously reduced in solution, citrate ions become adsorbed onto the surface of the particles, conferring a negative charge.

This attracts cationic surfactant molecules, thereby forming a protective layer that impedes particle growth, by both preventing the electrostatic attraction of silver ions and acting as a physical barrier. These particles are not thought to play a role in fingermark development (Becue *et al.*, 2011). PD working solutions that contain no surfactants have been devised, but they require a significant reduction in the concentration of redox components (Cantu and Johnson, 2001). Even then, such reagents are not stable for much longer than an hour.

The most commonly used surfactants are Synperonic N and n-dodecylamine acetate. n-Dodecylamine acetate is the cationic surfactant required to stabilize the solution, while Synperonic N, a non-ionic surfactant, is added to facilitate the solubility of n-dodecylamine and prevent it from precipitating out of solution (Goode and Morris, 1983; Ramotowski, 2000). The main problem with using this combination of surfactants is that Synperonic N, once a widely used industrial surfactant, is no longer manufactured in Europe, due to concerns regarding its persistence in the environment and biodegradation to estrogenic compounds (Fields *et al.*, 2004; Soares *et al.*, 2008; Houlgrave *et al.*, 2011). The banning of Synperonic N use in many countries has prompted investigations into alternative non-ionic surfactants to replace it in PD. Work carried out by the USSS indicates that Tween 20 is an effective substitute, and that the USSS working solution is stable for a significantly longer period of time than the UK version (Yamashita and French, 2011; Houlgrave *et al.*, 2011). It has been suggested that the non-ionic surfactant participates in the formation of the surfactant layer surrounding silver particles and that, due to its more complicated molecular structure, Tween 20 may be more effective in this role than Synperonic N (Houlgrave *et al.*, 2011).

The length of PD treatment is left to the discretion of the technician. The document is removed from the working solution when it is felt that adequate development and contrast has been achieved, or after a set period (usually 20–30 minutes) if weak or no fingermark development is observed. Leaving fingermarks in the working solution for longer periods can be detrimental to development quality, as further silver deposition on the surface will cause loss of contrast. Fingermarks may be developed in as little as five minutes, whereas others may require up to an hour. Contrast may be enhanced by using one of several methods, including an additional immersion step in dilute sodium hypochlorite, which converts the silver to Ag_2O, or treatment with a photographic fixing solution (Cantu, 2000; Cantu and Johnson, 2001; LaPorte and Ramotowski, 2003; Stoilovic and Lennard, 2012). Fingermarks that show weak development may be put back into the working solution for further treatment, though this cannot be performed if the document has already been treated with a photographic fixing solution. The last step is to rinse the document in deionized water several times, in order to remove the remaining traces of the working solution trapped in the substrate and, therefore, prevent any further silver reduction (Becue *et al.*, 2011). The document is then left to dry away from direct light, before the developed fingermarks are photographed.

The main differences between UK-PD and that used by the USSS are the use of malic acid rather than maleic acid, and the use of Tween 20 rather than Synperonic N, as described above (Burow *et al.*, 2003). Additionally, the UK method uses a separate glass tray for each stage of PD treatment, whereas the USSS method utilizes a single tray for the full process, disposing of each reagent before progressing to the next step (Cantu and Johnson 2001; Burow *et al.*, 2003). The latter method has the advantage of removing the risk of damaging paper items by excessive handling, although the former allows many items to be treated sequentially, as long as the maleic acid and the working solution still perform adequately (these can be replaced as necessary).

Burow *et al.* observed that working solutions made with deionized water produced by reverse osmosis (RO/DI) can be made with a lower concentration of most components compared the standard formulation made with distilled water (Burow, 2003; Burow *et al.*, 2003). Further modifications proposed by Burow *et al.* include acidifying the working solution with malic acid, such that a separate acid pretreatment is no longer required. Malic acid also forms a complex with ferric ions similar to citrate, which may aid in silver deposition (Burow, 2003). Yapping and Yue reported a PD method that did not require surfactants or maleic acid, due to use of silver ammine (prepared by adding ammoniacal liquor dropwise to silver nitrate solution) rather than silver nitrate in the working solution (Yapping and Yue, 2004). While this formulation was reported to be more stable than the silver nitrate working solution, difficulties in reproducing the silver ammine physical developer have been reported (Cantu, 2005).

9.3.2.2 Multi and single metal deposition

Perhaps the most significant alteration made to PD is the incorporation of a colloidal gold immersion bath, which has given rise to a new series of fingermark development techniques. Multimetal deposition (MMD) was first proposed by Saunders in 1989 (Saunders, 1989). This procedure involves treating documents with a pH 3 solution of negatively charged colloidal gold, which binds to protonated amino acids, peptides and proteins present in fingermark residue via electrostatic attraction (Becue *et al.*, 2007). The gold nanoparticles then serve as nucleation sites for silver deposition via a modified PD working solution, which enhances the relatively weak visual development provided by gold (Champod *et al.*, 2004; Becue *et al.*, 2007). It can be applied to many porous and nonporous surfaces, including banknotes (Schnetz and Margot, 2001), as well as wet surfaces and latent fingermarks that have been previously developed with ninhydrin (Lee and Gaensslen, 2001). The reagents and processes were later optimized by Schnetz and Margot to improve reproducibility and sensitivity (Schnetz and Margot, 2001). The resultant development method is referred to as MMD II (Champod *et al.*, 2004). Despite their high sensitivity (Zhang *et al.*, 2007), both MMD methods suffer many of the same setbacks as PD – the cost of the reagents required, the time consuming nature of the method (Saunders, 1989; Becue *et al.*, 2011), and that it can only be used at the end of a detection sequence (Becue *et al.*, 2007). Additionally, great care is required to prepare and utilize the MMD reagents correctly (Schnetz and Margot, 2001; Becue *et al.*, 2012).

Work by Becue *et al.* into simplifying MMD led to the development of a reagent that can develop latent fingermarks in a single step by functionalizing gold particles to thiolated cyclodextrins. The cyclodextrins are utilized as molecular hosts, thereby enabling an organic dye (Acid Blue 25) to be ferried to fingermarks in aqueous solution, producing blue ridges (Becue *et al.*, 2007). While a less time consuming method than the original MMD, this approach is limited by the need to synthesize the gold-cyclodextrin complexes (Stauffer *et al.*, 2007; Becue *et al.*, 2012).

The single metal deposition (SMD) method substitutes gold chloride for the silver redox reagent. The deposition of gold, rather than silver, onto the colloidal gold particles was found to produce equal or better fingermark development compared to MMD II (Becue *et al.*, 2011). The replacement of the silver enhancement step reduces both processing time and cost compared to the MMD processes (Stauffer *et al.*, 2007). Investigations are ongoing into overcoming remaining issues with the MMD and SMD methods, namely the need for precise control over the pH of the colloidal gold solution, as the quality and performance of the gold nanoparticles are essential to subsequent metal deposition (Becuer *et al.*, 2012).

9.3.3 Lipid-sensitive reagents

There has been some work in recent years focused on the adaptation of histological stains to latent fingermark detection. Dyes used in the demonstration of lipids provide simple and relatively inexpensive fingermark development reagents. Sudan black and gentian violet have been utilized in the detection of fingermarks on porous and non-porous surfaces, and adhesive surfaces, respectively (Lee and Gaensslen, 2001; Ramotowski, 2012b; Garrett and Bleay, 2013). More recently, Oil red O and Nile red have been proposed as development reagents for porous surfaces, as alternatives to PD (Beaudoin, 2004; Braasch *et al.*, 2013). These methods are generally less time consuming, less hazardous, are more cost effective, and provide comparable results to PD with recently deposited latent fingermarks. Additionally, Nile red is a photoluminescent reagent, which provides a significant advantage over other methods in terms of sensitivity and applicability to a wider range of surfaces.

9.3.3.1 Oil red O

Oil red O (ORO) is a lipophilic dye that is structurally related to the Sudan group, a class of dyes employed in histological staining techniques (Figure 9.9). It was first manufactured for use as an industrial dye employed in timber staining (Proescher, 1927). The utilization of ORO as a histological stain was first reported in the late 1920s by French (French, 1926; Proescher, 1927). ORO is used to stain tissue sections, such as adipose tissue, to demonstrate lipid content (Beaudoin, 2004; Becue *et al.*, 2011). Initial solvents for histological preparations of ORO included acetone and ethanol. These have largely

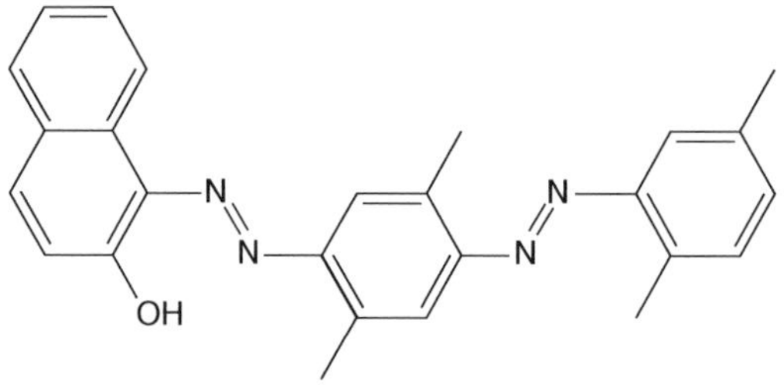

Figure 9.9 Molecular structure of Oil red O

been replaced by isopropanol and propylene glycol, which dissolve out less lipid material from the cell (Fail, 2005).

ORO is a relatively new reagent for latent fingermark development. Its first application to forensic evidence was as a means of detecting latent lip prints by dusting powdered dyes over a surface, similar to dusting for latent fingermarks (Castello *et al.*, 2002; Navarro *et al.*, 2006). In 2004, Beaudoin reported the first adaptation of ORO as a fingermark development reagent, as a possible alternative to PD (Beaudoin, 2004). As PD is currently one of the only conventional fingermark development techniques that can detect fingermarks on porous surfaces that have been exposed to water, a simpler, more effective alternative would be desirable. The ORO method proposed by Beaudoin has several advantages over PD in that it is more stable, easier and less time consuming to use, and produces superior results on some substrates, on both dry and wetted porous surfaces (Rawji and Beaudoin, 2006; Salama *et al.*, 2008; Wood and James 2009).

Beaudoin's ORO formulation is based on that used to stain lipoproteins following cellulose acetate electrophoresis (Salama *et al.*, 2008). Fingermark development involves immersing an item of evidence in the staining solution for 60–90 minutes, producing red-stained fingermarks on a light pink background (Becue *et al.*, 2011). This is followed by rinsing the item, first in a pH 7 buffer to neutralize the alkaline ORO reagent, thus stabilizing and preserving the developed fingermarks, followed by rinsing twice in deionized water (Salama *et al.*, 2008). The carbonate buffer originally proposed as a neutralization reagent has recently been replaced with a more stable phosphate buffer (Beaudoin, 2011).

The development mechanism is the simple diffusion of ORO from the solvent into the lipid fraction of latent fingermark residue (Salama *et al.*, 2008; Beaudoin, 2011). ORO is, therefore, an effective fingermark development reagent as long as the fingermark contains a sufficient amount of sebaceous material. Fingermarks containing a small amount of lipid material may still be visualized by ORO, although the resulting fingermark development is often too faint to enable a comparison. Depending on the surface type, the extent of background coloration may vary, but contrast between the fingermark and the surface is usually sufficient. Results obtained by Salama *et al.* suggest that ORO may not be entirely lipid specific, though it remains to be seen whether or not this is due to the ORO itself, or whether the high concentration of sodium hydroxide in the reagent chemically alters the dye molecule (Salama *et al.*, 2008). Additionally, this high concentration of sodium hydroxide appears to be necessary to improve fingermark contrast (Salama *et al.*, 2008). An alternative formulation utilizes propylene glycol as a solvent, and yields comparable results to the original (Figure 9.10) (Frick *et al.*, 2012).

ORO produces best fingermark development on white or light-colored paper. Treatment of thermal paper receipts with ORO causes the printed text to fade, which is advantageous in revealing fingermark detail, though precautions must be made to photograph the text first (Rawji and Beaudoin, 2006; Frick *et al.*, 2012). Patterned and dark-colored surfaces present the greatest challenge to ORO treatment, as the red fingermark ridges are often obscured on these surface types. ORO cannot develop fingermarks on adhesive surfaces, as interactions between ORO and the glue result in a deep red stain that obscures any fingermark detail (Salama *et al.*, 2008). Very porous paper types, such as newspaper, phone directory pages and brown paper, are also problematic for ORO development, as the fingermark often appears as an indistinct blotch (Rawji and Beaudoin, 2006; Salama *et al.*, 2008; Frick *et al.*, 2012). It has been suggested that surface porosity may affect ORO fingermark development, as lipids may diffuse more rapidly through rough, porous papers, thereby obscuring the ridge pattern of the fingermark (Rawji and Beaudoin, 2006). This is consistent with the observation that fingermarks several weeks old detected with ORO on white copy paper also appear as blotches (Salama *et al.*, 2008).

Comparisons between ORO and PD have found that while ORO produces superior results to PD on fingermarks less than four weeks old, PD is the superior method for detecting older fingermarks (Salama *et al.*, 2008; Frick *et al.*, 2012). ORO performed with decreasing efficacy when treating older fingermarks compared to PD, which can be used to detect latent fingermarks that are up to several decades old (Ramotowski, 2012c).

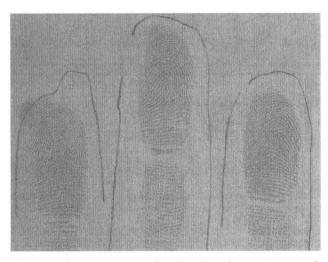

Figure 9.10 Fingermarks on white copy paper developed with ORO in propylene glycol (*See insert for color representation of the figure.*)

It has been hypothesized that the lipids present in latent fingermarks can be divided into two fractions, which goes some way to explaining the differences in performance between PD and ORO (Salama *et al.*, 2008). PD is thought to react with the stable and long-lived "robust fraction", composed of large, water-insoluble proteins and lipoproteins, hence its ability to develop fingermarks that are over several weeks old. The "fragile fraction" is composed of more short-lived compounds such as fatty acids and triglycerides, which may be removed by some solvents of the amino acid sensitive reagents, and is thought to be the target group of ORO (Ramotowski, 2012b). Over time, the "fragile" lipid fraction of latent fingermark residue diffuses through porous surfaces, resulting in the blurred appearance of older fingermarks treated with ORO. Additionally, ORO permeates the entire substrate during treatment, staining any diffused lipids and thus creating a blotch of color, while the accumulation of silver particles is thought to occur only at the surface of porous surfaces, hence the ability of PD to detect fingermarks that are several months old in instances where ORO cannot.

9.3.3.2 Nile red

One of the greatest shortcomings in the detection of latent fingermark lipids on porous surfaces is the lack of photoluminescent methods (Beaudoin, 2012). Neither PD nor ORO are effective at developing fingermarks on dark or patterned papers where the color of the fingermark does not provide sufficient contrast.

Nile red (9-diethylamino-5H-benzo[α]-phenoxazine-5-one) is a neutral phenoxazone dye that is used as a fluorescent probe for the staining of neutral lipids (Fowler and Greenspan, 1985; Greenspan and Fowler, 1985; Bonilla and Prelle, 1987) (Figure 9.11).

Its photoluminescent properties provide increased sensitivity and superior visualization of tissue structures compared to non-photoluminescent lipid stains such as ORO (Fowler and Greenspan, 1985; Bonilla and Prelle, 1987). Nile red exhibits solvatochromism, that is, its absorption and emission maxima vary with solvents of different polarities (Greenspan and Fowler, 1985; Dutta *et al.*, 1996; Golini *et al.*, 1998). It is highly photoluminescent in non-polar media and to a lesser extent in some polar solvents, but photoluminescence is completely quenched in aqueous solution (Fowler and Greenspan 1985; Dutta *et al.*, 1996). This is thought to be due to a twisted intramolecular charge transfer process undergone by the diethylamino group (Dutta *et al.*, 1996; Golini *et al.*, 1998).

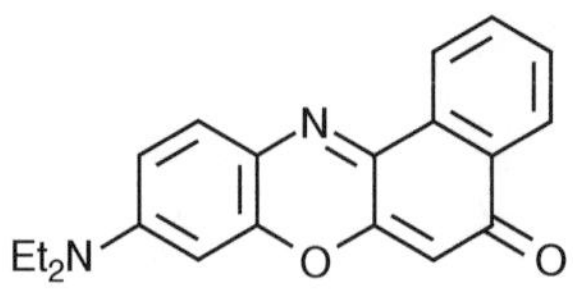

Figure 9.11 Molecular structure of Nile red

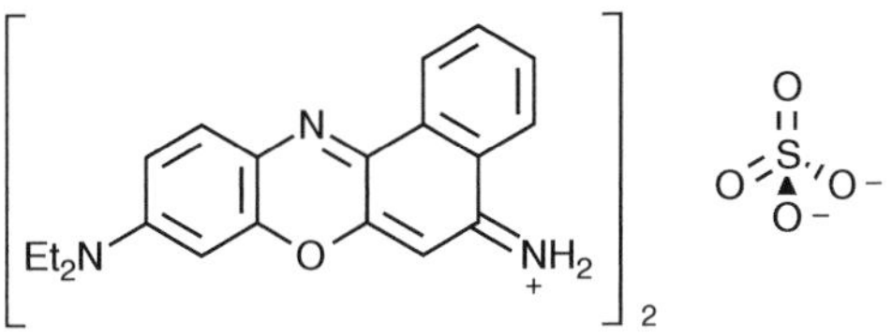

Figure 9.12 Molecular structure of Nile blue A

Nile red has been applied as a novel reagent in forensic applications, including as an enhancement technique for cyanoacrylate-fumed fingermarks and as a potential reagent for the detection of latent lip imprints (Day and Bowker, 1996; Castello *et al.*, 2004). Nile red was first reported as a latent fingermark development reagent by Saunders in 1993 (Ramotowski, 2012b). A stock solution of Nile red in acetone was used to prepare an aqueous working solution. The resultant working solution was noted to have a very short shelf life of less than an hour (Ramotowski, 2012b).

More recent developments by Braasch *et al.* have produced an improved Nile red reagent that imparts both color and photoluminescence to lipid-rich fingermarks, which appear as red on a non-luminescent purple background (Braasch *et al.*, 2013). While this method has been reported to be an effective method of detecting recently deposited fingermarks on porous surfaces that have been wetted, the authors noted concerns regarding the toxicity of the basic, methanolic solvent required due to the poor solubility of Nile red in water, which poses a problem to both personnel exposure and waste disposal (de la Hunty, 2012). Furthermore, the reagent suffers from significant precipitation of Nile red as the methanol portion of the solvent evaporates. A number of Nile red derivatives have been synthesized to overcome these issues, with the aim of producing a photoluminescent, water-soluble, lipophilic reagent (de la Hunty, 2012).

Nile blue A (Nile blue sulfate; Figure 9.12), commonly referred to as Nile blue[1], is a basic phenoxazine dye employed primarily in histology to demonstrate acidic and neutral lipids (Cain, 1947). Nile blue stains acidic components, such as phospholipids and nucleic and fatty acids, a dark blue color, while neutral lipids (i.e., triglycerides) are stained pink or red, and exhibit photoluminescence (Cain, 1947; Dunnigan, 1968; Putt, 1972; Bancroft and Cook, 1994). The dual staining capability of Nile blue is due to the spontaneous hydrolysis of Nile blue A in aqueous media to Nile red, its corresponding phenoxazone (Figure 9.13). The two dyes interact with their respective target compounds by different mechanisms: Nile blue A forms a salt linkage with acidic moieties, while Nile red dissolves preferentially into neutral lipids. Though Nile red is present in the Nile blue histological stain in only trace amounts, this is sufficient to provide coloration to stained tissue sections (Fowler and Greenspan, 1985; Bonilla and Prelle, 1987). It is accepted

---

[1] The term "Nile blue" is often used interchangeably to refer to both the dye Nile blue A, and the histological stain prepared from the same. For clarity, the term "Nile blue" is used here specifically in reference to the aqueous solution, while "Nile blue A" refers to the phenoxazine dye.

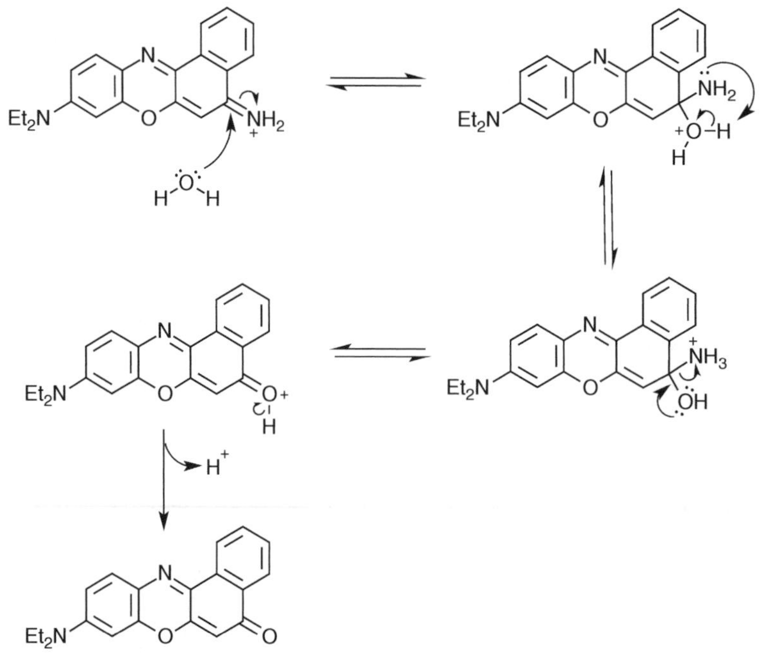

Figure 9.13 Hydrolysis of Nile blue A to Nile red in aqueous solution

Figure 9.14 Fingermarks on white copy paper developed with Nile blue, viewed through a 550 nm barrier filter with illumination at 505 nm (*See insert for color representation of the figure.*)

that the Nile red component is responsible for the photoluminescence emitted by tissues stained with Nile blue reagent (Ostle and Holt, 1982; Canente *et al.*, 1983; Fowler and Greenspan, 1985; Greenspan *et al.*, 1985).

Nile blue has been shown to detect fingermarks in much the same manner as the Nile red reagent on a number of porous surfaces (Figure 9.14) (Frick *et al.*, 2014). In the context of fingermark detection, the utilization of an aqueous Nile blue A solution, rather than being a novel reagent in itself, may be viewed as a simple method of preparing an aqueous Nile red solution, and so negates the need to synthesize a water-soluble Nile red derivative. A more pragmatic advantage is that Nile blue A is significantly less expensive than Nile red. Additionally, the combination of Nile blue A and Nile red in a single solution enables fingermark detection on a wider variety of surface types than is possible with Nile red alone, including non-porous items (Frick *et al.*, 2014). There is further potential for Nile blue as a photoluminescent postcyanoacrylate treatment.

9.3.4 Other techniques
9.3.4.1 Powder techniques
While not strictly a chemical technique, dusting powders are probably the most widely used treatment for the visualization of latent fingermarks on non-porous surfaces, and as such require mention. In the United Kingdom, approximately half of all fingermark identifications are made with dusting powders (Bleay *et al.*, 2012). Dusting for fingermarks is also convenient for surfaces that are not readily transportable, such as cars or window sills (Champod *et al.*, 2004). A large variety of fingermark powders exist and can be composed of fine carbon or metallic particles, luminescent compounds, that is, Rhodamine dyes, or nanoparticles (Stuart, 2013). The application of the powder is performed using fine fiber brushes, or a magnetic wand in the case of powders containing iron particles (Clegg, 2004). The particles adhere to the oils and moisture of the fingermark deposited on the surface, and enable resolution of the ridge pattern. The fingermark can then be photographed or, alternatively, taped, lifted and removed for examination (Clegg, 2004). Powders are generally limited to non-porous surfaces, as the adherence is most effective when the fingermark is on top of the surface (Clegg, 2004). They are not an effective method for developing fingermarks on porous surfaces, as the fingermark deposits are absorbed into the surface (Clegg, 2004). However, some powders are not sensitive, and exhibit a high level of interference from the background, for example, color and inherent luminescence. Dusting powders also become less effective if the fingermark is dried, and some application techniques can damage the latent fingermark (Bandey *et al.*, 2012; Ramotowski, 2012d).

9.3.4.2 Cyanoacrylate fuming
Cyanoacrylate fuming (sometimes referred to as Super Glue® fuming) is a treatment technique whereby the monomer form of cyanoacrylate is vaporized and polymerizes selectively on a fingermark (Ramotowski, 2012e). The formation of polymerized cyanoacrylate enables visualization of ridge detail whilst protecting the fingermark from smearing and degradation. The application of heat is widely used to accelerate the fuming

process, although the use of a vacuum pump instead to reduce the pressure in the fuming chamber and facilitate the evaporation process is also common (Watkin *et al.*, 1994; Lewis *et al.*, 2001). The prevailing humidity has a large effect on the success of fingermark visualization, where higher humidity levels result in better, more visible ridge detail (Watkin *et al.*, 1994). The structure of the polymer, which requires moisture to initiate growth, is affected by the chemical composition of the fingermark (Lewis *et al.*, 2001). Sebum-rich deposits cause "capsule-like" structures to form, whereas eccrine-rich deposits result in a more "noodle-like" shape (Lewis *et al.*, 2001). A range of dyes is commonly used to further enhance visibility of the fingermark, including Basic Yellow 40, Rhodamine 6G and Sudan Black (Chesher *et al.*, 1992; Kempton and Rowe, 1992; Ramotowski, 2012e). Cyanoacrylate fuming has found widespread use for the visualization of fingermark deposits on non-porous surfaces (Ramotowski, 2012e).

9.3.4.3 Vacuum metal deposition

Vacuum metal deposition (VMD) creates a negative fingermark image through the deposition of metals around and between the ridges of latent fingermarks. Gold is evaporated under vacuum and deposits onto the surface on which the latent fingermark has been deposited. The evaporated gold is deposited as clusters, rather than a continuous layer, which increase in size and density with the amount of gold evaporated (Dai *et al.*, 2007). Rather than binding to the latent fingermark, the gold is absorbed into the fingermark residue, and coats the surface as well, providing nucleation sites for the deposition of either zinc or cadmium. The second metal is evaporated and binds preferentially to the gold, but does not deposit on the fingermark ridges, thus creating a negative image of the fingermark (i.e., light fingermark ridges on a dark background) (Champod *et al.*, 2004). VMD is a highly sensitive technique that is effective on a variety of non-porous and semi-porous surfaces, including polymers, which are a difficult surface for fingermark visualization (Dai *et al.*, 2007), and can be used in conjunction with cyanoacrylate fuming (Champod *et al.*, 2004). Though more expensive to apply than cyanoacrylate fuming, VMD has been found to be more sensitive on latent fingermarks over 24 months old (Champod *et al.*, 2004), and can also be applied to surfaces that have been exposed to conditions that render cyanoacrylate fuming ineffective (Dai *et al.*, 2007).

9.4 Experimental considerations for latent fingermark chemistry research

The continued development of fingermark development reagents aims to improve not only current fingermark detection capabilities, but also address factors such as simplicity of the method, cost effectiveness and operational safety, which must be considered in the context of routine, operational use (Lee and Gaensslen, 2001; Champod *et al.*, 2004; Jelly *et al.*, 2009; IFRG, 2014). Although existing methods address most of these issues, current research is focusing on developing the next generation of fingermark treatments with the above borne specifically in mind. The lack of standardization in sample collection and experimental design, in addition to the natural compositional variability of latent fingermark deposits, can hinder meaningful comparisons of results between studies of latent fingermark development methods (Sears *et al.*, 2012). As a result, there have been calls in recent years for more rigorous standards in regards to the number of fingermark donors required, sample collection and treatment, to minimize experimental variation where possible (Jones *et al.*, 2001; Kent, 2010; IFRG, 2014).

Ideally, the number of donors and the number of fingermarks collected should be sufficient to derive valid statistical data, but practical constraints often restrict the number of donors to a handful of individuals; often those working in the immediate vicinity of the researchers (Kent, 2010; Sears *et al.*, 2012). This can have a bottleneck affect, where the sample population is not representative of the general population in terms of the quality of the fingermarks obtained (Jones *et al.*, 2001; Kent, 2010). Guidelines recently proposed by the International Fingerprint Research Group (IFRG) divide experimental approaches into four phases (IFRG, 2014). Phase 1, which encompasses basic, proof-of-concept studies, requires a minimum of 3–5 donors, who provide a range of fingermarks of good to poor development quality. Phases 2–4 encompass more detailed optimization, comparison and evaluation studies, utilizing increasing numbers of fingermark donors and substrates, and even actual items of evidence.

A common approach in studies focused on the lipid fraction of latent fingermarks is to have donors rub the tips of their fingers on areas of skin that are dense in sebaceous glands, namely the forehead and nose, prior to fingermark deposition (Archer *et al.*, 2005; Salama *et al.*, 2008; Houlgrave *et al.*, 2011; Koenig *et al.*, 2011). Such actions are often referred to as "grooming" or "charging" of the fingertips and are intended to deliberately incorporate lipid material into the deposited fingermarks (Kent, 2010). This results in a significantly greater amount of material to be deposited, as well as overrepresenting the lipid fraction of fingermark residue, which may lead to incorrect conclusions regarding the performance of lipid-sensitive development methods (Jones *et al.*, 2001; Croxton *et al.*, 2010; Kent, 2010; Sears *et al.*, 2012). Frequently, cosmetics and other such products present on the skin surface will also be transferred to fingermark samples (Mong *et al.*, 1999). It has been suggested that charged fingermarks are not realistic of latent fingermarks encountered in forensic investigations, and that uncharged fingermarks should be used instead, or as a comparison (Croxton *et al.*, 2010; Sears *et al.*, 2012; IFRG, 2014). Another procedure is to clean donors' hands of any exogenous contamination before allowing time for skin secretions to replenish (Jones *et al.*, 2001). Acetone or alcohols have been recommended as cleaning agents, as soaps may leave fatty acid residues on the skin (Thomas, 1978; Archer *et al.*, 2005).

Fingermark research is complicated further by the inherent difficulties in obtaining reproducible samples (Mong *et al.*, 1999; Sears *et al.*, 2012). Samples collected from the same person at the same time may show significant variation in composition (Koenig *et al.*, 2011).

Latent fingermarks are difficult to deposit in a reproducible and homogenous manner, because of the uneven distribution of eccrine and sebaceous components on the ridge skin (Thomas, 1978). There is evidence to suggest that fingermark composition varies with digit and handedness, although it is not clear whether this has any significant effect of on the detectability of fingermarks from different fingers (Cuthbertson, 1969; Darke and Wilson, 1977). Sample deposition may range from asking donors to touch the surface briefly (Mong *et al.*, 1999), to more controlled procedures that regulate length of contact between the fingertip and the surface, as well as the amount of pressure used in depositing the fingermark (Girod and Weyermann, 2014; de la Hunty *et al.*, 2014; Sutton *et al.*, 2014). It is not specified what basis exists for the precise amount of pressure used (aside from simple control over this variable or to produce a clear fingermark pattern), and as a result it is unclear whether this is representative of "real" fingermark deposition. There is a risk that such strict deposition parameters may divorce the experimental approach from the "reality" of incidental fingermark deposition.

For comparisons between development methods, a generally accepted approach is the "split fingermark" (Figure 9.15), whereby fingermarks are cut in half, and each half is treated separately to enable a comparison (Kent, 2010; IFRG, 2014). Another approach is the "depletion series" method, where the donor is asked to deposit several fingermarks sequentially, without re-charging or allowing eccrine secretions to re-accumulate on the fingertips. In this way, a sequence of fingermarks containing diminishing amounts of material is obtained, which enables assessment of the sensitivity of a development method (Kent, 2010; Sears *et al.*, 2012).

Many studies into development reagents use relatively "fresh" latent fingermarks, which are treated within a short period of collection (usually hours). In an operational

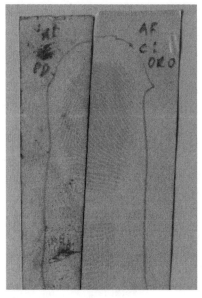

Figure 9.15 Example of a split fingermark on white copy paper, treated with PD (left) and ORO (right) (*See insert for color representation of the figure.*)

context, items may not be examined for latent fingermarks until several days to weeks after deposition (Kent, 2010; IFRG, 2014). Humidity, temperature and light can affect fingermark composition over prolonged storage periods (Jones *et al.*, 2001; Kent 2010). If samples are to be stored in the laboratory, there is a risk that some fingermark development may occur during storage if samples are kept in close proximity to reagents such as 1,2-indanedione. These factors must, therefore, be taken into account if fingermarks are intended to be developed after a prolonged period of time.

The quality of developed fingermark ridge detail is usually assessed visually. There are several fingermark grading schemes currently in use by researchers and industry professionals alike, tailored to suit specific purposes, such as comparisons between two development methods, or overall assessment of the quality of fingermark development (Almog *et al.*, 1999; Bandey, 2004; Becue *et al.*, 2009; Jasuja *et al.*, 2009; Wood and James, 2009; Kent, 2010; McLaren *et al.*, 2010; Sears *et al.*, 2012). Generally, such assessment methods consist of the categorization of developed fingermarks along a scale ranging from "good" to "poor" ridge detail. For example, a grading scale proposed by McLaren *et al.* is widely used to compare the performance of one development reagent against another, along a range of –2 (less effective) to +2 (more effective) (McLaren *et al.*, 2010). Another commonly employed system is that devised by the UK's Home Office Police Scientific Development Branch (HOPSDB) (Bandey, 2004). This system assigns absolute values to fingermark development, using a scale of 0 (no development) to 4 (full development with continuous ridges and excellent contrast). A third scale, reported by Becue *et al.*, is designed to evaluate the usefulness of a developed fingermark to identification, assigning fingermarks a grade of – (no ridge development), ± (some visible development) and + (sufficient development to enable identification) (Becue *et al.*, 2009).

Developed fingermarks must be recorded photographically. If fingermarks are treated with several development methods in a detection sequence, photographing any development following the application of each method is a necessary step to maintain a record of ridge detail, in the event that further treatment impairs any fingermark development produced by a preceding technique (Frick *et al.*, 2013b). Additionally, some development reagents are not stable posttreatment, and deterioration of ridge detail quality may begin to occur within hours in some cases (Braasch *et al.*, 2013).

9.5 Conclusions and future directions

The detection of latent fingermarks remains an essential task for forensic examiners. While existing techniques have provided many years of good service there is a constant search to improve existing techniques and develop novel methods of detection. Some of the more unusual approaches proposed include: the application of nanoparticles, such as fluorescent quantum dots or functionalized nanoparticles (Becue *et al.*, 2009; Moret *et al.*, 2014); disulfur dinitride, which polymerizes selectively on both fingermarks and trace amounts of inkjet inks (Kelly *et al.*, 2008); and immunodetection techniques, whereby antibody conjugates target proteins, amino acids or drug metabolites (Drapel *et al.*, 2009; Spindler *et al.*, 2011b; Boddis and Russell, 2012).

Though successful fingermark detection relies heavily on differences in chemical composition between the latent fingermark residue and its surface (Sears *et al.*, 2012), there remain large gaps in the current understanding of latent fingermark composition. One consequence of this knowledge gap is that the mechanisms behind several of the more commonly used fingermark development reagents are yet to be fully characterized, which presents difficulties for the optimization of these methods. A more detailed understanding of fingermark chemistry, particularly the effects of degradation processes, is vital in order to develop new, more effective development methods, as well as to optimize existing ones. There has been increased interest in recent years into obtaining a greater understanding of latent fingermark composition towards these goals (Mong *et al.*, 1999; Archer *et al.*, 2005; Wolstenholme *et al.*, 2009). For the most part, such research has focused on the groups of compounds currently most pertinent to latent fingermark detection, that is, amino acids and lipids.

A more complete understanding of latent fingermark composition may enable fingermark evidence to provide more information than just the ridge details. It is thought that traits such as age, sex, diet or ethnic background could be inferred from fingermark composition (Williams *et al.*, 2011; Bailey *et al.*, 2012). Such information would be of significant assistance to criminal investigations, if a fingermark found at a crime scene could not be matched to any in a database, or if the clarity of the ridge pattern was in some way unsuitable for comparative purposes (Benton *et al.*, 2010; Bailey *et al.*, 2012; Francese *et al.*, 2013). As fingermark composition changes with degradation, it has also been proposed that a method for estimating the age of a fingermark could be developed, as a means of supporting or discrediting a testimony (Emerson *et al.*, 2011; Weyermann *et al.*, 2011).

Acknowledgements

This chapter contains material drawn from the PhD theses of Amanda Frick and Patrick Fritz who were both supported by Australian Postgraduate Awards. Reece Crocker (Curtin University) is thanked for assistance with the literature search and Georgina Sauzier (Curtin University) is thanked for reviewing the final draft of the chapter.

References

Alaoui, I.M. 2007. Photodegradation studies by laser-induced fluorescence of the reaction product of 1,2-indanedione and glycine. *Spectroscopy Letters*, **40**: 715–721.

Alaoui, I.M., E.R. Menzel, M. Farag, *et al.* 2005. Mass spectra and time-resolved fluorescence spectroscopy of the reaction product of glycine with 1,2-indanedione in methanol. *Forensic Science International*, **152** (2–3): 215–219.

Alaoui, I.M., T. Troxler, and M.M. Joullie. 2012. Fingerprint visualization and spectroscopic properties of 1,2-indanedione-alanine followed by zinc chloride or europium chloride. *Journal of Forensic Identification*, **62** (1): 1–13.

Almog, J. 2000. FINGERPRINTS (DACTYLOSCOPY): Visualization. In: *Encyclopedia of Forensic Sciences* (eds J. Siegel , P. Saukko and G. Knupfer). Academic Press, San Diego, CA, pp. 890–900.

Almog, J. 2001. Fingerprint development by ninhydrin and its analogues. In: *Advances in Fingerprint Technology*, 2nd edn (eds H.C. Lee and R.E. Gaensslen). CRC Press, Boca Raton, FL, pp. 177–209.

Almog, J., and H. Glasner. 2010. Ninhydrin thiohemiketals: basic research towards improved fingermark detection techniques employing nano-technology. *Journal of Forensic Sciences*, **55**: 215–220.

Almog, J., E. Springer, S. Wiesner, *et al.* 1999. Latent fingerprint visualization by 1,2-indanedione and related compounds: preliminary results. *Journal of Forensic Sciences*, **44**: 114–118.

Almog, J., V.G. Sears, E. Springer, *et al.* 2000. Reagents for the chemical development of latent fingerprints: scope and limitations of benzo[f]ninhydrin in comparison to ninhydrin. *Journal of Forensic Sciences*, **45** (3): 538–544.

Almog, J., M. Azoury, Y. Elmaliah, *et al.* 2004a. Fingerprints' third dimension: the depth and shape of fingerprints penetration into paper-cross section examination by fluorescence microscopy *Journal of Forensic Sciences*, **49** (5): 981–985.

Almog, J., Y. Cohen, M. Azoury, and T.-R. Hahn. 2004b. Genipin – a novel fingerprint reagent with colorimetric and fluorogenic activity. *Journal of Forensic Sciences*, **49**: 255–257.

Almog, J., G. Levinton-Shamuilov, Y. Cohen, and M. Azoury. 2007. Fingerprint reagents with dual action: color and fluorescence. *Journal of Forensic Sciences*, **52**: 330–334.

Almog, J., H. Sheratzki, M. Elad-Levin, *et al.* 2011. Moistened hands do not necessarily allude to high quality fingerprints: The relationship betwen palmar moisture and fingerprint donorship. *Journal of Forensic Sciences*, **56** (S1): S162–S165.

Anonymous. 2006. *Use of DMAC on thermal papers*. Fingerprint and Footwear Forensics Newsletter, Home Office Scientific Development Branch.

Archer, N.E., Y. Charles, J.A. Elliott, and S. Jickells. 2005. Changes in the lipid composition of latent fingerprint residue with time after deposition on a surface. *Forensic Science International*, **154** (2–3): 224–239.

Asano, K., C. Bayne, K. Horsman, and M. Buchanan. 2002. Chemical composition of fingerprints for gender determination. *Journal of Forensic Sciences*, **47** (4): 805–807.

Azoury, M., A. Zamir, C. Oz, and S. Wiesner. 2002. The effect of 1,2-indanedione, a latent fingerprint reagent on subsequent DNA profiling. *Journal of Forensic Sciences*, **47**: 586–588.

Bailey, M.J., N.J. Bright, R.S. Croxton, *et al.* 2012. Chromatography/mass spectrometry, X-ray photoelectron spectroscopy, and attenuated total reflection Fourier transform infrared spectroscopic imaging: An intercomparison. *Analytical Chemistry*, **84** (20): 8514–8523.

Bancroft, J.D. and H.C. Cook. 1994. *Manual of Histological Techniques and their Diagnostic Application*, 2nd edn. Churchill Livingstone, Edinburgh.

Bandey, H.L. 2004. Evaluation of fingerprint brushes for use with aluminium powder. *Fingerprint Development and Imaging Newsletter: Special Edition* **54** (4): 1–12.

Bandey, H.L., S.M. Bleay, and A.P. Gibson. 2012. Powders for fingerprint development. In: *Lee and Gaensslen's Advances in Fingerprint Technology*, 3rd edn (ed. R. Ramotowski), CRC Press, Boca Raton, FL, pp. 191–218.

Beaudoin, A. 2004. New technique for revealing latent fingerprints on wet, porous surfaces: Oil red O. *Journal of Forensic Identification*, **54** (4): 413–421.

Beaudoin, A. 2011. Oil red O: Fingerprint development on a 21-year-old cold case. *Journal of Forensic Identification*, **61** (1): 50–59.

Beaudoin, A. 2012. Fingerprint staining technique on dark and wetted porous surfaces: Oil red O and rhodamine. *Journal of Forensic Identification*, **62** (4): 315–324.

Becue, A., C. Champod, and P. Margot. 2007. Use of gold nanoparticles as molecular intermediates for the detection of fingermarks. *Forensic Science International*, **168**: 169–176.

Becue, A., S. Moret, C. Champod, and P. Margot. 2009. Use of quantum dots in aqueous solution to detect blood fingermarks on non-porous surfaces. *Forensic Science International*, **191** (1): 36–41.

Becue, A., S. Moret, C. Champod, and P. Margot. 2011. Use of stains to detect fingermarks. *Biotechnic & Histochemistry*, **86** (3): 140–160.

Becue, A., A. Scoundrianos, and S. Moret. 2012. Detection of fingermarks by colloidal gold (MMD/SMD) – beyond the pH 3 limit. *Forensic Science International*, **219** (1–3): 39–49.

Benson, J.R. and P.E. Hare. 1975. o-Phthalaldehyde: Fluorogenic detection of primary amines in the picomole range. Comparison with fluorescamine and ninhydrin. *Proceedings of the National Academy of Sciences of the United States of America* **72** (2): 619–622.

Benton, M., F. Rowell, L. Sundar, and M. Jan. 2010. Direct detection of nicotine and cotinine in dusted latent fingermarks of smokers by using hydrophobic silica particles and MS. *Surface and Interface Analysis*, **42** (5): 378–385.

Bleay, S.M., V.G. Sears, H.L. Bandey, *et al.* 2012. *Fingerprint Source Book*. The Home Office, LOndon, United Kingdom.

Boddis, A.M. and D.A. Russell. 2012. Development of aged fingermarks using antibody-magnetic particle conjugates. *Analytical Methods*, **4** (3): 637–641.

Bonilla, E. and A. Prelle. 1987. Application of Nile blue and Nile red, two fluorescent probes, for detection of lipid droplets in human skeletal muscle. *Journal of Histochemistry and Cytochemistry*, **35** (5): 619–621.

Bonte, F., P. Pinguet, J.M. Chevalier, and A. Meybeck. 1995. Analysis of all stratum corneum lipids by automated multiple development high-performance thin-layer chromatography. *Journal of Chromatography B*, **664** (2): 311–316.

Boughton, B., R.M.B. MacKenna, V.R. Wheatley, and A. Wormall. 1957. Studies of sebum. 8. Observations on the squalene and cholesterol content and the possible functions of squalene in human sebum. *Biochemical Journal*, **66** (1): 32–38.

Braasch, K., M. de la Hunty, J. Deppe, *et al.* 2013. Nile red: Alternative to physical developer for the detection of latent fingermarks on wet porous surfaces? *Forensic Science International*, **230** (1–3): 74–80.

Bramble, S.K., and J.S. Brennan. 2000. FINGERPRINTS (DACTYLOSCOPY): Chemistry of print residue. In: *Encyclopedia of Forensic Sciences* (eds J. Siegel, P. Saukko and G. Knupfer). Academic Press, San Diego, CA, pp. 862–869.

Bratton, R.M. and J.A. Juhala. 1995. DFO-Dry. *Journal of Forensic Identification*, **45** (2): 169–172.

Brennan, J.S., S.K. Bramble, S. Crabtree, and G. Wright. 1995. Fuming of latent fingerprints using dimethylaminocinnamaldehyde. *Journal of Forensic Identification*, **45** (4): 373.

Buchanan, M.V., K. Asano, and A. Bohanon. 1997. *Chemical characterisation of fingerprints from adults and children*. SPIE (International Society for Optical Engineering), Boston, MA.

Burow, D. 2003. An improved silver physical developer. *Journal of Forensic Identification*, **53** (3): 304–314.

Burow, D., D. Seifert, and A. A. Cantu. 2003. Modifications to the silver physical developer. *Journal of Forensic Sciences*, **48** (5): 1–7.

Cain, A.J. 1947. The use of Nile blue in the examination of lipoids. *Quarterly Journal of Microscopical Science*, **88** (3): 383–392.

Camera, E., M. Ludovici, M. Galante, *et al.* 2010. Comprehensive analysis of the major lipid classes in sebum by rapid resolution high-performance liquid chromatography and electrospray mass spectrometry. *Journal of Lipid Research*, **51**: 3377–3388.

Canente, M., M.J. Hazen, and J.C. Stockert. 1983. Nile blue sulfate staining for demonstration of lipids in fluorescence microscopy. *Acta Histochemica et Cytochemica*, **16** (3): 286–288.

Cantu, A.A. 2000. Silver physical developers for the visualization of latent prints on paper. *Forensic Science Review*, **13** (1): 29–64.

Cantu, A.A. 2005. Letter to the editor. Re: A new silver physical developer. *Journal of Forensic Identification*, **55** (3): 289–290.

Cantu, A.A. and A. Johnson. 2001. Silver physical development of latent prints. In: *Advances in Fingerprint Technology*, 2nd edn (eds H.C. Lee and R.E. Gaensslen). CRC Press, Boca Raton, FL, pp. 241–274.

Castello, A., M. Alvarez, M. Miquel, and F. Verdu. 2002. Long-lasting lipsticks and latent prints. *Forensic Science Communications*, **4** (2). http://www.fbi.gov/about-us/lab/forensic-science-communications/fsc/april2002/verdu.htm/ (last accessed 25 June 2015).

Castello, A., M. Alvarez-Segui, and F. Verdu. 2004. Use of fluorescent dyes for developing latent lip prints. *Coloration Technology*, **120** (4): 184–187.

Champod, C., C. Lennard, P. Margot, and M. Stoilovic. 2004. *Fingerprints and Other Ridge Skin Impressions*. CRC Press, Boca Raton, FL.

Chesher, B.K., J.M. Stone, and W.F. Rowe. 1992. Use of the Omniprint™ 1000 alternate light source to produce fluorescence in cyanoacrylate-developed latent fingerprints stained with biological stains and commercial fabric dyes. *Forensic Science International*, **57** (2):163–8.

Clegg, D. 2004. Fingerprint identification. In: *The Practice Of Crime Scene Investigation* (ed. J. Horswell), CRC Press, Boca Raton, FL, pp. 161–180.

Conn, C., G. Ramsay, C. Roux, and C. Lennard. 2001. The effect of metal salt treatment on the photoluminescence of DFO-treated fingerprints. *Forensic Science International*, **116** (2–3): 117–123.

Corson, W.B. and J.E. Lawson. 1991. Alternate applications of DFO for non-fluorescent visualization. *Journal of Forensic Identification*, **41** (6): 437–445.

Croxton, R.S., M.G. Baron, D. Butler, *et al.* 2006. Development of a GC-MS method for the simultaneous analysis of latent fingerprint components. *Journal of Forensic Sciences*, **51** (6): 1329–1333.

Croxton, R.S., M.G. Baron, D. Butler, *et al.* 2010. Variation in amino acid and lipid composition of latent fingerprints. *Forensic Science International*, **199** (1–3): 93–102.

Cuthbertson, F. 1969. *The Chemistry of Fingerprints*. Atomic Weapons Research Establishment, Aldermaston, UK.

Dai, X., M. Stoilovic, C. Lennard, and N. Speers. 2007. Vacuum metal deposition: Visualisation of gold agglomerates using TEM imaging. *Forensic Science International*, **168**: 219–222.

Dalrymple, B. 2000. FINGERPRINTS (DACTYLOSCOPY): Identification and classification. In: *Encyclopedia of Forensic Sciences* (eds J. Siegel, P. Saukko and G. Knupfer), Academic Press, San Diego, CA, pp. 869–877.

Darke, D.J. and J.D. Wilson. 1977. *The analysis of the free fatty acid component of fingerprints*. Atomic Energy Research Establishment, Harwell, UK.

Day, J.S., H.G.M. Edwards, S.A. Dobrowski, and A.M. Voice. 2004. The detection of drugs of abuse in fingerprints using Raman spectroscopy I: latent fingerprints. *Spectrochimica Acta Part A*, **60** (3): 563–568.

Day, K. and W. Bowker. 1996. Enhancement of cyanoacrylate developed latent prints using Nile red. *Journal of Forensic Identification*, **46** (2): 183–187.

Downing, D.T. and J.S. Strauss. 1974. Synthesis and composition of surface lipids of human skin. *Journal of Investigative Dermatology*, **62** (3): 228–244.

Downing, D.T., J.S. Strauss, L.A. Norton, *et al.* 1977. The time course of lipid formation in human sebaceous glands. *Journal of Investigative Dermatology*, **69** (4): 407–412.

Drapel, V., A. Becue, C. Champod, and P. Margot. 2009. Identification of promising antigenic components in latent fingermark residues. *Forensic Science International*, **184**: 47–53.

Druey, J. and P. Schmidt. 1950. Phenanthrolinchinone und diazafluorene. *Helvetica Chimica Acta*, **33** (4): 1080–1087.

Dunnigan, M.G. 1968. The use of Nile blue sulphate in the histochemical identification of phospholipids. *Stain Technology* **43**, (3): 249–256.

Dutta, A.K., K. Kamada, and K. Ohta. 1996. Spectroscopic studies of nile red in organic solvents and polymers. *Journal of Photochemistry and Photobiology A: Chemistry*, **93** (1): 57–64.

Emerson, B., J. Gidden, J.O. Lay, and B. Durham. 2011. Laser desorption/ionization time-of-flight mass spectrometry of triacylglycerols and other components in fingermark samples. *Journal of Forensic Sciences*, **56** (2): 381–389.

Fager, R.S., C.B. Kutina, and E.W. Abrahamson. 1973. The use of NBD chloride (7 chloro-4-nitro-benzo-2-oxa-1,3-diazole) in detecting amino acids and as an N-terminal reagent. *Analytical Biochemistry*, **53** (1): 290–294.

Fail, R. 2005. Rapid detection of lipid in livers for transplantation. *HistoLogic*, **38** (1): 11–13.

Fields, J.A., A. Wingham, F. Hartog, and V. Daniels. 2004. Finding substitute surfactants for Synperonic N. *Journal of the American Institute for Conservation*, **43** (1): 55–73.

Fowler, S.D. and P. Greenspan. 1985. Application of Nile red, a fluorescent hydrophobic probe, for the detection of neutral lipid deposits in tissue sections: comparison with Oil red O. *Journal of Histochemistry and Cytochemistry*, **33** (8): 833–836.

Francese, S., R. Bradshaw, L.S. Ferguson, *et al.* 2013. Beyond the ridge pattern: multi-informative analysis of latent fingermarks by MALDI mass spectrometry. *Analyst*, **138** (15): 4215–4228.

French, R.W. 1926. Fat stains. *Stain Technology*, **1** (2): 79.

Frick, A.A., P. Fritz, S.W. Lewis, and W. van Bronswijk. 2012. A modified Oil Red O reagent for the detection of latent fingermarks on porous substrates. *Journal of Forensic Identification*, **62** (6): 623–641.

Frick, A.A., P. Fritz, and S.W. Lewis. 2013a. Chemistry of print residue. In: *Encyclopedia of Forensic Sciences* (eds J.A. Siegel, P.J. Saukko and M.M. Houck), Academic Press, Waltham, pp. 92–97.

Frick, A.A., P. Fritz, S.W. Lewis, and W. van Bronswijk. 2013b. Sequencing of a modified Oil Red O development technique for the detection of latent fingermarks on paper surfaces. *Journal of Forensic Identification*, **63** (4): 369–385.

Frick, A.A., F. Busetti, A. Cross, and S.W. Lewis. 2014. Aqueous Nile blue: a simple, versatile and safe reagent for the detection of latent fingermarks. *Chemical Communications*, **50** (25): 3341–3343.

Gallagher, M., C.J. Wysocki, J. Leyden, *et al.* 2008. Analyses of volatile organic compounds from human skin. *British Journal of Dermatology*, **159**: 780–791.

Gardner, S.J. and D.F. Hewlett. 2003. Optimization and initial evaluation of 1,2-indandione as a reagent for fingerprint detection. *Journal of Forensic Sciences*, **48**: 1288–1292.

Garrett, H.J. and S. Bleay. 2013. Evaluation of the solvent black 3 fingermark enhancement reagent: Part 1 – Investigation of fundamental interactions and comparisons with other lipid-specific reagents. *Science and Justice*, **53** (2): 121–130.

Girod, A., and C. Weyermann. 2014. Lipid composition of fingermark residue and donor classification using GC/MS. *Forensic Science International*, **238**: 68–82.

Girod, A., R. Ramotowski, and C. Weyermann. 2012. Composition of fingermark residue: A qualitative and quantitative review. *Forensic Science International*, **223** (1): 10–24.

Girod, A., C. Roux, and C. Weyermann. 2014. La datation des traces digitales (partie II): proposition d'une approche formelle. *Revue Internationale de Criminologie et Police Technique et Scientifique*, **67** (2): 226–249.

Golini, C.M., B.W. Williams, and J.B. Foresman. 1998. Further solvatochromic, thermochromic, and theoretical studies on Nile Red. *Journal of Fluorescence*, **8** (4): 395–404.

Goode, G.C. and J.R. Morris. 1983. *Latent fingerprints: A review of their origin, composition and methods for detection*. Atomic Weapons Research Establishment, Aldermaston, UK.

Greene, R.S., D.T. Downing, P.E. Pochi, and J.S. Strauss. 1970. Anatomical variation in the amount and composition of human skin surface lipid. *Journal of Investigative Dermatology*, **54** (3): 240–247.

Greenspan, P. and S.D. Fowler. 1985. Spectrofluorometric studies of the lipid probe, nile red. *Journal of Lipid Research*, **26** (7): 781–789.

Greenspan, P., E.P. Mayer, and S.D. Fowler. 1985. Nile red: A selective fluorescent stain for intracellular lipid droplets. *Journal of Cell Biology*, **100** (3): 965–973.

Grigg, R., T. Mongkolaussavaratana, C.A. Pounds, and S. Sivagnanam. 1990. 1,8-diazafluorenone and related compounds. A new reagent for the detection of (α-amino acids and latent fingerprints. *Tetrahedron Letters*, **31** (49): 7215–7218.

Hansen, D.B. and M.M. Joullie. 2005. The development of novel ninhydrin analogues. *Chemical Society Reviews*, **34**: 408–417.

Hardwick, S., T. Kent, V. Sears, and P. Winfield. 1993. Improvements to the formulation of DFO and the effects of heat on the reaction with latent fingerprints. *Fingerprint Whorld*, **19**: 65–69.

Harker, M., H. Coulson, I. Fairweather, *et al.* 2006. Study of metabolic composition of eccrine sweat from healthy male and female human subjects by ^{1}H NMR spectroscopy. *Metabolomics*, **2** (3): 105–112.

Hauze, D.B., O. Petrovskaia, B. Taylor, *et al.* 1998. 1,2-Indanediones: New reagents for visualizing the amino acid components of latent prints. *Journal of Forensic Sciences*, **43**: 744–747.

Houlgrave, S., M. Andress, and R. Ramotowski. 2011. Comparison of different physical developer working solutions – Part I: Longevity studies. *Journal of Forensic Identification*, **61** (6): 621–639.

de la Hunty, M. 2012. An investigation of techniques for the development of latent fingermarks on porous surfaces that have been wet: Nile red in sequence with physical developer, and the synthesis and novel application of nile red derivatives, University of Technology, Sydney.

de la Hunty, M., X. Spindler, S. Chadwick, *et al.* 2014. Synthesis and application of an aqueous nile red microemulsion for the development of fingermarks on porous surfaces. *Forensic Science International*, **244**: e48–e55.

Ifa, D.R., N.E. Manicke, A.L. Dill, and R.G. Cooks. 2008. Latent fingerprint chemical imaging by mass spectrometry. *Science*, **321** (5890): 805.

IFRG. 2014. Guidelines for the assessment of fingermark detection techniques. *Journal of Forensic Identification*, **64** (2): 174–200.

James, A.T. and V.R. Wheatley. 1956. Studies of sebum. 6. The determination of the component fatty acids of human forearm sebum by gas-liquid chromatography. *Biochemical Journal*, **63** (2): 269–273.

Jasuja, O.P., M.A. Toofany, G. Singh, and G.S. Sodhi. 2009. Dynamics of latent fingerprints: The effect of physical factors on quality of ninhydrin developed prints – A preliminary study. *Science & Justice*, **49** (1): 8–11.

Jelly, R., S.W. Lewis, C. Lennard, *et al.* 2008. Lawsone: a novel reagent for the detection of latent fingermarks on paper surfaces. *Chemical Communications*: 3513–3515.

Jelly, R., E.L.T. Patton, C. Lennard, *et al.* 2009. The detection of latent fingermarks on porous surfaces using amino acid sensitive reagents: a review. *Analytica Chimica Acta,* **652**: 128–142.

Jones, N.E., L.M. Davies, C.A.L. Russell, *et al.* 2001. A systematic approach to latent fingerprint sample preparation for comparative chemical studies. *Journal of Forensic Identification,* **52** (5): 504–515.

Kelly, P.F., R.S.P. King, and R.J. Mortimer. 2008. Fingerprint and inkjet-trace imaging using disulfur dinitride. *Chemical Communications,* (**46**): 6111–6113.

Kempton, J.B., and W.F. Rowe. 1992. Contrast enhancement of cyanoacrylate-developed latent fingerprints using biological stains and commercial fabric dyes. *Journal of Forensic Sciences,* **37** (1): 99–105.

Kent, T. 2010. Standardizing protocols for fingerprint reagent testing. *Journal of Forensic Identification,* **60** (3): 371–379.

Koenig, A., A. Girod, and C. Weyermann. 2011. Identification of wax esters in latent print residues by gas chromatography-mass spectrometry and their potential use as ageing parameters. *Journal of Forensic Identification,* **61** (6): 652–676.

Kupferschmid, E., L. Schwarz, and C. Champod. 2010. Development of standardised test strips as a process control for the detection of latent fingermarks using physical developers. *Journal of Forensic Identification,* **60** (6): 639–655.

LaPorte, G.M. and R. Ramotowski. 2003. The effects of latent print processing on questioned documents produced by office machine systems utilizing inkjet technology and toner. *Journal of Forensic Sciences,* **48** (3): 1–6.

Lee, H. and R. Gaensslen. 2001. Methods of latent print development. In: *Advances in Fingerprint Technology,* 2nd edn (eds H.C. Lee and R.E. Gaensslen), CRC Press, Boca Raton, FL, pp. 105–175.

Lee, J.L., S.M. Bleay, V.G. Sears, *et al.* 2009. Evaluation of the dimethylaminocinnemaldehyde contact transfer process and its application to fingerprint development on thermal papers. *Journal of Forensic Identification,* **59** (5): 545–568.

Lennard, C. 2007. Fingerprint detection: Current capabilities. *Australian Journal of Forensic Sciences,* **39** (2): 55–71.

Lennard, C.J., P.A. Margot, M. Sterns, and R.N. Warrener. 1987. Photoluminescent enhancement of ninhydrin developed fingerprints by metal complexation: structural studies of complexes formed between Ruhemann's purple and group IIb metal salts. *Journal of Forensic Sciences,* **32**: 597–605.

Levinton-Shamuilov, G., Y. Cohen, M. Azoury, *et al.* 2005. Genipin, a novel fingerprint reagent with colorimetric and flurogenic activity, part II: Optimization, scope and limitations. *Journal of Forensic Sciences,* **50**: 1367–1371.

Lewis, L.A., R.W. Smithwick, III, G.L. Devault, *et al.* 2001. Processes involved in the development of latent fingerprints using the cyanoacrylate fuming method. *Journal of Forensic Sciences,* **46**: 241–246.

Marieb, E.N. and K. Hoehn. 2007. *Human Anatomy & Physiology,* 7th edn. Pearson Benjamin Cummings, San Francisco, CA.

Masters, N., R. Morgan, and E. Shipp. 1991. DFO, its usage and results. *Journal of Forensic Identification,* **41** (1): 3–10.

Mayer, S.W., C.P. Meilleur, and P.F. Jones. 1978. The use of ortho-phthalaldehyde for superior fluorescent visualization of latent fingerprints. *Journal of the Forensic Science Society,* **18** (3–4): 233–235.

McCaldin, D.J. 1960. The chemistry of ninhydrin. *Chemical Reviews,* **60**: 39–51.

McLaren, C., C. Lennard, and M. Stoilovic. 2010. Methylamine pretreatment of dry latent finger-marks on polyethylene for enhanced detection by cyanoacrylate fuming. *Journal of Forensic Identification*, **60** (2): 199–222.

Michael-Jubeli, R., J. Bleton, and A. Baillet-Gufroy. 2011. High-temperature gas chromatography-mass spectrometry for skin surface lipids profiling. *Journal of Lipid Research*, **52** (143–155):

Mong, G.M., C.E. Petersen, and T.R.W. Clauss. 1999. *Advanced Fingerprint Analysis Project: Fingerprint Constituents*. Pacific Northwest National Laboratory, Richland, WA.

Moret, S., A. Becue, and C. Champod. 2014. Nanoparticles for fingermark detection: an insight into the reaction mechanism. *Nanotechnology* **25** (42). doi:10.1088/0957-4484/25/42/425502.

Morris, J.R. 1976. *Processes for Developing Latent Fingerprints*. UK Patent 1428025.

Mountfort, K.A., H. Bronstein, N. Archer, and S.M. Jickells. 2007. Identification of oxidation products of squalene in solution and in latent fingerprints by ESI-MS and LC/APCI-MS. *Analytical Chemistry*, **79** (7): 2650–2657.

Navarro, E., A. Castello, J.L. Lopez, and F. Verdu. 2006. Criminalystic: Effectiveness of lyso-chromes on the developing of invisible lipstick-contaminated lipmarks on human skin. A pre-liminary study. *Forensic Science International*, **158** (1): 9–13.

Nelson, J.O. and P.F. Warren. 1981. NBD-chloride is a potent mutagen in the Salmonella mutage-nicity assay. *Mutation Research/Genetic Toxicology*, **88** (4): 351–354.

Nicolaides, N. 1974. Skin lipids: their biochemical uniqueness. *Science*, **186** (4158): 19–26.

Nikkari, T. 1974. Comparative chemistry of sebum. *Journal of Investigative Dermatology*, **62** (3): 257–267.

Oden, S., and B. von Hofsten 1954. Detection of fingerprints by the ninhydrin reaction. *Nature*, **173**: 449–50.

Ostle, A.G. and J.G. Holt. 1982. Nile blue A as a fluorescent stain for poly-beta-hydroxybutyrate. *Applied and Environmental Microbiology*, **44** (1): 238–241.

Ottaviani, M., E. Camera, and M. Picardo. 2010. Lipid mediators in acne. *Mediators of Inflammation* (Article ID 858176). doi: 10.1155/2010/858176.

Pappas, A., M. Anthonavage, and J.S. Gordon. 2002. Metabolomic fate and selective utilization of major fatty acids in human sebaceous gland. *Journal of Investigative Dermatology*, **118**: 164–171.

Patton, E.L.T., D.H. Brown, and S.W. Lewis. 2010. Detection of latent fingermarks on thermal printer paper by dry contact with 1,2-indanedione. *Analytical Methods*, **2** (6): 631–637.

Petrovskaia, O., B.M. Taylor, D.B. Hauze, *et al.* 2001. Investigations of the reaction mechanisms of 1,2-indanediones with amino acids. *Journal of Organic Chemistry*, **66**: 7666–7675.

Pounds, C.A., R. Grigg, and T. Mongkolaussavaratana. 1990. The use of 1,8-diazafluoren-9-one (DFO) for the fluorescent detection of latent fingerprints on paper. A preliminary evaluation. *Journal of Forensic Sciences*, **35** (1): 169–175.

Powe, W.C. 1972. Laundry soils. In: *Detergency: Theory and Test Methods Part I* (eds W.C. Cutler and R.C. Davis). Marcell Dekker, Inc., New York, pp. 31–63.

Presley, L.A., A.L. Baumstark, and A. Dixon. 1993. The effects of specific latent fingerprint and questioned document examinations on the amplification and typing of the HLA DQ alpha gene region in forensic casework. *Journal of Forensic Sciences*, **38**: 1028–36.

Proescher, F. 1927. Oil red pyridin, a rapid fat stain. *Stain Technology*, **2** (2): 60 – 61.

de Puit, M., L. Koomen, M. Bouwmeester, *et al.* 2011. Use of physical developer for the visualisa-tion of latent fingerprints. *Journal of Forensic Identification*, **61** (2): 166–170.

Putt, F.A. 1972. *Manual of Histopathological Staining Methods*. John Wiley & Sons, Inc., New York.

Ramotowski, R. 1996. *International Symposium on Fingerprint Detection and Identification, Fluorescence visualisations of latent fingerprints on paper using p-dimethylaminocinnamaldehyde (PDMAC)*. Israel National Police, Jerusalem:

Ramotowski, R. 2000. A comparison of different physical developer systems and acid pre-treatments and their effects on developing latent prints. *Journal of Forensic Identification*, **50** (4): 363–384.

Ramotowski, R. 2001. Composition of latent print residue. In: *Advances in Fingerprint Technology*, 2nd edn (eds H.C. Lee and R.E. Gaensslen), CRC Press, Boca Raton, FL, pp. 63–104.

Ramotowski, R. 2012a. Amino acid reagents. In: *Lee and Gaensslen's Advances in Fingerprint Technology*, 3rd edn (ed. R. Ramotowski), CRC Press, Boca Raton, FL, pp. 17–54.

Ramotowski, R. 2012b. Lipid reagents. In: *Lee and Gaensslen's Advances in Fingerprint Technology*, 3rd edn (ed. R. Ramotowski), CRC Press, Boca Raton, FL, pp. 83–96.

Ramotowski, R. 2012c. Metal deposition methods. In: *Lee and Gaensslen's Advances in Fingerprint Technology*, 3rd edn (ed. R. Ramotowski), CRC Press, Boca Raton, FL, pp. 55–82.

Ramotowski, R. 2012d. Powder methods. In: *Lee and Gaensslen's Advances in Fingerprint Technology*, 3rd edn (ed. R. Ramotowski), CRC Press, Boca Raton, FL, pp. 1–16.

Ramotowski, R. 2012e. Vapor/fuming methods. In: *Lee and Gaensslen's Advances in Fingerprint Technology*, 3rd edn (ed. R. Ramotowski), CRC Press, Boca Raton, FL, pp. 97–128.

Rawji, A. and A. Beaudoin. 2006. Oil red O versus physical developer on wet papers: A comparative study. *Journal of Forensic Identification*, **56** (1): 33–54.

Ricci, C., S. Bleay, and S.G. Kazarian. 2007. Spectroscopic imaging of latent fingermarks collected with the aid of a gelatin tape. *Analytical Chemistry*, **79** (15): 5771–5776.

Richmond-Aylor, A., S. Bell, P. Callery, and K. Morris. 2007. Thermal degradation analysis of amino acids in fingerprint residue by pyrolysis GC-MS to develop new latent fingerprint developing reagents. *Journal of Forensic Sciences*, **52** (2): 380–382.

Robosky, L.C., K. Wade, D. Woolson, *et al.* 2008. Quantitative evaluation of sebum lipid components with nuclear magnetic resonance. *Journal of Lipid Research*, **49**: 686–692.

Roux, C., N. Jones, C. Lennard, and M. Stoilovic. 2000. Evaluation of 1,2-indanedione and 5,6-dimethoxy-1,2-indanedione for the detection of latent fingerprints on porous surfaces. *Journal of Forensic Sciences*, **45**: 761–769.

Saferstein, R. 2007. *Criminalistics: An Introduction to Forensic Science*. 9th edn. Prentice Hall, Upper Saddle River, NJ.

Salama, J., S. Aumeer-Donovan, C. Lennard, and C. Roux. 2008. Evaluation of the fingermark reagent Oil red O as a possible replacement for physical developer. *Journal of Forensic Identification*, **58** (2): 203–237.

Salares, V.R., C.R. Eves, and P.R. Carey. 1979. On the detection of fingerprints by laser excited luminescence. *Forensic Science International*, **14** (3): 229–237.

Sasson, Y. and J. Almog. 1978. Chemical reagents for the development of latent fingerprints. I: Scope and limitations of the reagent 4-dimethylamino-cinnamaldehyde. *Journal of Forensic Sciences*, **23** (4): 852–855.

Saunders, G.C. 1989. Multimetal deposition method for latent fingerprint development. *Paper presented at the 74th Conference of the International Association for Identification*. Pensacola, FL.

Sauzier, G., A.A. Frick, and S.W. Lewis. 2013. Investigation into the performance of modified silver physical developers for visualizing latent fingermarks on paper. *Journal of Forensic Identification*, **63** (1): 70–89.

Schnetz, B. and P. Margot. 2001. Technical note: latent fingermarks, colloidal gold and multimetal deposition (MMD) – Optimisation of the method. *Forensic Science International*, **118**: 21–28.

Schulz, M.M., H.D. Wehner, W. Reichert, and M. Graw. 2004. Ninhydrin-dyed latent fingerprints as a DNA source in a murder case. *Journal of Clinical Forensic Medicine*, **11** (4): 202–204.

Schwarz, L. and I. Klenke. 2007. Enhancement of ninhydrin- or DFO-treated latent fingerprints on thermal paper. *Journal of Forensic Sciences*, **52** (3): 649–655.

Scruton, B., B.W. Robins, and B.H. Blott. 1975. The deposition of fingerprint films. *Journal of Physics, D: Applied Physics*, **8**: 714–723.

Sears, V.G., S.M. Bleay, H.L. Bandey, and V.J. Bowman. 2012. A methodology for finger mark research. *Science and Justice*, **52** (3): 145–160.

Shalita, A.R. 1974. Genesis of free fatty acids. *Journal of Investigative Dermatology*, **62** (3): 332–335.

Smith, K.R. and D.M. Thiboutot. 2008. Sebeceous gland lipids: friend or foe? *Journal of Lipid Research*, **49**: 271–281.

Soares, A., B. Guieysse, B. Jefferson, *et al.* 2008. Nonylphenol in the environment: A critical review on occurrence, fate, toxicity and treatment in wastewaters. *Environmental International*, **34** (7): 1033–1049.

Spindler, X., R. Shimmon, C. Roux, and C. Lennard. 2011a. The effect of zinc chloride, humidity and the substrate on the reaction of 1,2-indanedione-zinc with amino acids in latent fingermark secretions. *Forensic Science International*, **212** (1–3): 150–157.

Spindler, X., O. Hofstetter, A.M. McDonagh, *et al.* 2011b. Enhancement of latent fingermarks on non-porous surfaces using anti-L-amino acid antibodies conjugated to gold nanoparticles. *Chemical Communications*, **47** (19): 5602–5604.

Stauffer, E., A. Becue, K.V. Singh, *et al.* 2007. Single-metal deposition (SMD) as a latent fingermark enhancement technique: An alternative to multimetal deposition (MMD). *Forensic Science International*, **168** (1): e3–e9.

Stoilovic, M. and C. Lennard. 2012. *NCFS Workshop Manual: Fingermark Detection & Enhancement*, 6th edn. National Centre for Forensic Studies, Canberra, Australia.

Stoilovic, M., R.N. Warrener, and H.J. Kobus. 1984. An evaluation of the reagent NBD chloride for the production of luminescent fingerprints on paper: II. A comparison with ninhydrin. *Forensic Science International*, **24** (4): 279–284.

Strauss, J.S., A.M. Kligman, and P.E. Pochi. 1962. The effect of androgens and estrogens on human sebaceous glands. *Journal of Investigative Dermatology*, **39** (2): 139–155.

Stuart, B.H. 2013. *Forensic Analytical Techniques*: John Wiley & Sons, Ltd.

Sutton, R., C. Grenci, and L. Hrubesova. 2014. A comparison on the longevity of submerged marks in field and laboratory conditions. *Journal of Forensic Identification*, **64** (2): 143–156.

Takemura, T., P. Wertz, and K. Sato. 1989. Free fatty acids and sterols on human eccrine sweat. *British Journal of Dermatology*, **120**: 43–47.

Thody, A.J. and S. Shuster. 1989. Control and function of sebaceous glands. *Phisiological Reviews*, **69** (2): 383–416.

Thomas, G.L. 1978. The physics of fingerprints and their detection. *Journal of Physics E: Scientific Instruments*, **11**: 722–731.

van Dam, A., J.C.V. Schwarz, J. de Vos, *et al.* 2014. Oxidation monitoring by fluorescence spectroscopy reveals the age of fingermarks. *Angewandte Chemie* **53** (24): 6272–6275.

Wallace-Kunkel, C., C. Lennard, M. Stoilovic, and C. Roux. 2007. Optimisation and evaluation of 1,2-indanedione for use as a fingermark reagent and its application to real samples. *Forensic Science International*, **168** (1): 14–26.

Wargacki, S.P., L.A. Lewis, and M.D. Dadmun. 2008. Enhancing the quality of aged latent fingerprints developed by superglue fuming: Loss and replenishment of initiator. *Journal of Forensic Sciences*, **53** (5): 1138–1144.

Warrener, R.N., H.J. Kobus, and M. Stoilovic. 1983. An evaluation of the reagent NBD chloride for the production of luminescent fingerprints on paper: I. Support for a xenon arc lamp being a cheaper and valuable alternative to an argon ion laser as an excitation source. *Forensic Science International*, **23** (2–3): 179–188.

Watkin, J.E., D.A. Wilkinson, A.H. Misner, and A.B. Yamashita. 1994. Cyanoacrylate fuming of latent prints: vacuum versus heat/humidity. *Journal of Forensic Identification*, **44** (5): 545–556.

Weigele, M., S.L. DeBernardo, J.P. Tengi, and W. Leimgruber. 1972. Novel reagent for the fluorometric assay of primary amines. *Journal of the American Chemical Society*, **94** (16): 5927–5928.

Weyermann, C., C. Roux, and C. Champod. 2011. Initial results on the composition of fingerprints and its evolution as a function of time by GC/MS analysis. *Journal of Forensic Sciences*, **56** (1): 102–108.

Wiesner, S., E. Springer, Y. Sasson, and J. Almog. 2001. Chemical development of latent fingerprints: 1,2-indanedione has come of age. *Journal of Forensic Sciences*, **46**: 1082–1084.

Wilkinson, D. 2000. Study of the reaction mechanism of 1,8-diazafluoren-9-one with the amino acid, l-alanine. *Forensic Science International*, **109** (2): 87–103.

Williams, D.K., C.J. Brown, and J. Bruker. 2011. Characterization of children's latent fingerprint residues by infrared microspectroscopy: Forensic implications. *Forensic Science International*, **206** (1–3): 161–165.

Wilson, J.D., A.A. Cantu, G.A. Antonopoulos, and M.J. Surrency. 2007. Examination of the steps leading up to the physical developer proces for developing fingerprints. *Journal of Forensic Sciences*, **52** (2): 320–329.

Wolstenholme, R., R. Bradshaw, M.R. Clench, and S. Francese. 2009. Study of latent fingermarks by matrix-assisted laser desorption/ionisation mass spectrometry imaging of endogenous lipids. *Rapid Communications in Mass Spectrometry*, **23** (19): 3031–3039.

Wood, M.A. and T. James. 2009. ORO. *The Physical Developer replacement? Science and Justice*, **49** (4): 272–276.

Yamashita, B. and M. French. 2011. Latent print development. In: *The Fingerprint Sourcebook*. Washington: National Institute of Justice, Washington, DC, pp. 7-1–7-67.

Yapping, L. and W. Yue. 2004. A new silver physical developer. *Journal of Forensic Identification*, **54** (4): 422–427.

Zhang, M., A. Becue, M. Prudent, *et al*. 2007. SECM imaging of MMD-enhanced latent fingermarks. *Chemical Communications*, (**38**): 3948–3950.

Zouboulis, C.C. 2004. Acne and sebaceous gland function. *Clinics in Dermatology*, **22**: 360–366.

Further reading

As indicated earlier, a single chapter cannot provide a full coverage of all aspects of the detection of latent fingermarks. For more in-depth study of all aspects of fingermark chemistry, the sources shown here are highly recommended:

Becue, A., S. Moret, C. Champod, and P. Margot. 2011. Use of stains to detect fingermarks. *Biotechnic & Histochemistry*, **86** (3): 140–160.

Champod, C., C. Lennard, P. Margot, and M. Stoilovic. 2004. *Fingerprints and Other Ridge Skin Impressions*. CRC Press, Boca Raton, FL.

Girod, A., R. Ramotowski, and C. Weyermann. 2012. Composition of fingermark residue: A qualitative and quantitative review. *Forensic Science International*, **223** (1): 10–24.

Jelly, R., E.L.T. Patton, C. Lennard, *et al*. 2009. The detection of latent fingermarks on porous surfaces using amino acid sensitive reagents: a review. *Analytica Chimica Acta*, **652**: 128–142.

National Institute of Justice Fingerprint Sourcebook. Available online: https://www.ncjrs.gov/pdffiles1/ nij/225320.pdf; last accessed 24 June 2015.

Nicolaides, N. 1974. Skin lipids: their biochemical uniqueness. *Science*, **186** (4158): 19–26.

Ramotowski, R. (ed.). 2012. *Lee and Gaensslen's Advances in Fingerprint Technology*. CRC Press, Boca Raton, FL.

UK Home Office, *Fingerprint Source Book*. Available online: https://www.gov.uk/government/publications/fingerprint-source-book; last accessed 24 June 2015.

In addition, the following references are highly recommended for those interested in carrying out research into any aspects of latent fingermark chemistry research, to provide guidance on experimental aspects:

Bailey, M.J., N.J. Bright, R.S. Croxton, *et al.* 2012. Chromatography/mass spectrometry, X-ray photoelectron spectroscopy, and attenuated total reflection fourier transform infrared spectroscopic imaging: an intercomparison. *Analytical Chemistry*, **84** (20): 8514–8523.

International Fingerprint Research Group (IFRG) 2014. Guidelines for the assessment of fingermark detection techniques. *Journal of Forensic Identification*, **64** (2): 174–200.

Jones, N.E., L.M. Davies, C.A.L. Russell, et al. 2001. A systematic approach to latent fingerprint sample preparation for comparative chemical studies. *Journal of Forensic Identification*, **52** (5): 504–515.

Kent, T. 2010. Standardizing protocols for fingerprint reagent testing. *Journal of Forensic Identification*, **60** (3): 371–379.

CHAPTER 10

Chemical methods in firearms analysis

Walter F. Rowe

Department of Forensic Sciences, The George Washington University, USA

10.1 Introduction

Firearms examination is the forensic science discipline that attempts to associate fired ammunition components such as bullets, cartridge cases or shot shells with the weapon from which they were fired (AFTE, 2014a). Chemistry aids firearm examiners in a variety of ways. Chemistry provides insight into corrosion of fired bullets and cartridge cases. Chemistry also provides processes for cleaning ammunition components and firearms. Serial number restoration usually employs chemical or electrochemical etching. Reconstruction of shooting incidents may involve the use of chemical tests to identify bullet holes. Determination of muzzle-to-target distances requires an understanding of the chemistry of propellants and of the chemistry of gunshot residues. Powder patterns on substrates such as clothing are often visualized by chemical means. Detection of gunshot residue on the hands of shooters makes use of a variety of instrumental methods of chemical analysis.

10.2 Basic firearms examination

When fired bullets or cartridges are recovered in the course of a criminal investigation a fundamental question that must be answered is what type of firearm fired them. Answering this question requires identifying the class characteristics of a firearm from a visual examination of fired bullets and cartridge cases. These class characteristics include: the caliber; number of lands and grooves; widths of the lands and grooves; direction of twist of the rifling; shape of the firing chamber; size, shape and location of the firing pin; and size, shape and location of extractors and ejectors (Rowe, 2005; Heard, 2008;

Forensic Chemistry: Fundamentals and Applications, First Edition. Edited by Jay A. Siegel.
© 2016 John Wiley & Sons, Ltd. Published 2016 by John Wiley & Sons, Ltd.

Warlow, 2011). The firearms examiner can then use this information to search a computer database to find makes and models of firearm with matching class characteristics (Doyle, 2014).

If a suspect firearm is also found in the course of the investigation, the firearms examiner will attempt associate the fired bullets and cartridges with it. The firearms examiner first verifies that the class characteristics of the firearm match those of the weapon that fired the bullet or cartridge. The examiner test fires the weapon in order to obtain exemplar bullets and/or cartridge cases. The exemplars are compared microscopically with the evidentiary bullets or cases. These examinations are conducted with a comparison microscope, which is really two compound microscopes linked by an optical bridge so that two specimens can be viewed side by side. In the case of bullets, striation patterns in the land and groove impressions on the bullet surfaces left by minute imperfections in the rifling of the weapon's barrel are compared. If these patterns match, the examiner can conclude that it is highly likely that the bullets were fired from the same weapon. The striation patterns produced even by sequentially manufactured barrels have been found to differ. This is the result of the rifling tools wearing during the rifling process. Once attached to a weapon each barrel will experience a unique history of erosion (wear) and corrosion (Rowe, 2005; Heard, 2008; Warlow, 2011).

For cartridge cases, the following features are compared: firing pin impressions, breechblock impressions (sometimes referred to as bolt-face signatures), and extractor and ejector marks. The first two types of marks are the most significant because they are produced by actually firing cartridges in a firearm. Firing pin impressions are made by the firing pin or striker in the firing mechanism of a firearm. These components are typically finished by hand during the manufacturing process. Firing pins are either filed or turned on a lathe. In either case, the hand operation will result in patterns of fine striations that would be virtually impossible to duplicate, even with the same file being used by the same operator. Extractors and ejectors are also finished by hand filing (Rowe, 2005; Heard, 2008; Warlow, 2011).

Firearms comparisons can result in four possible conclusions (AFTE, 2014b). An examiner can conclude that the evidence submitted is not suitable for microscopic examination. The surfaces of the bullets or cartridge cases may be too corroded to preserve microscopic markings. The suspect firearm may also be severely corroded or have undergone significant modifications (e.g., reboring of the barrel). An inconclusive opinion would be reached if the class characteristics of the bullet or cartridge match the weapon but individual characteristics do not. This is not an elimination, because corrosion or wear between the time firearm was discharged at the crime scene and its test firing in the forensic laboratory could have altered the microscopic individual characteristics. An elimination results when the class characteristics of the bullets or cartridges differ from those of the firearm. A firearms examiner makes an identification when all class characteristics that can be determined from the crime scene bullets and/or cartridges agree with the class characteristics of the firearm and the extent of the agreement between individual characteristics observed on the crime scene bullets and/or cartridges and those on the test fired bullets and/or cartridges exceeds the level of agreement of individual characteristics obtained with bullets or cartridge cases fired in different firearms.

10.2.1 Cleaning bullets and cartridges

Bullets and cartridge cases may require cleaning before they can be examined microscopically. Often gently brushing away of adhering dirt will suffice. Vigorous mechanical cleaning is undesirable because vigorous scrubbing can efface striation patterns and produce new ones. Prior to any cleaning, fired bullets should be examined for trace evidence and such evidence should be removed for appropriate analysis (e.g., DNA profiling).

Bullets and cartridges may acquire films or undergo corrosion as the result of exposure to the environment. In arid and semi-arid environments, evidentiary materials on or in surficial soil horizons may acquire calcareous concretions that obscure striation patterns on bullets or microscopic details of firing pin impressions or breechblock markings. Archaeologists and museum conservators have developed procedures for removing calcareous concretions and other films from metal artifacts without damaging them. Solutions of hexametaphosphate and sodium sesquicarbonate [$Na_3H(CO_3)_2$] have been used to clean copper artifacts (Guldbeck, 1972). Commercial products intended for the removal of hard water stains can also be used. The result of the application of such a product to a percussion slide impression on a copper rimfire cartridge recovered from a battlefield in northwestern Wyoming is shown in Figure 10.1.

If a bullet has become badly corroded due to environmental exposure, it may require chemical cleaning before it can be examined microscopically. Lead and lead alloy bullets can react with atmospheric gases such as oxygen, water and carbon dioxide to produce lead oxide (massicot), lead carbonate (cerussite), basic lead carbonate (hydrocerussite) and various salts (Chen and Daroub, 2002; Larson *et al.*, 2011). Caley (1955) developed

Figure 10.1 A .52 caliber rimfire cartridge from the Fetterman Battlefield (1866) before and after removal of calcareous deposits

a procedure for removing corrosion from lead artifacts: the lead objects were soaked in a large volume of cold, dilute hydrochloric acid, washed three times in hot, distilled water, soaked in 10% (w/v) ammonium acetate solution and, finally, washed four times in cold, air-free distilled water. Booker (1980) proposed the following chemical cleaning procedure for lead bullets: (i) ultrasonicate the bullet in water for 30 seconds; (ii) transfer the bullet to a 10% (w/v) aqueous solution of perchloric acid for two minutes (or until effervescence stops); (iii) ultrasonicate for 10 seconds; and (iv) rinse with water. Booker showed that such a regimen will remove corrosion products without destroying the striations on the bullet's surface.

Other metals used in components of ammunition are also subject to corrosion. In particular, many bullet jackets, cartridge cases and primer caps are made of copper–zinc alloys. Copper–zinc alloys are susceptible to a corrosion process called *dezincification* (Anonymous, 2003). In this type of dealloying, zinc, the more active of the two metals, is lost from the alloys. Several hypotheses have been advanced to explain the mechanism of dezincification: (i) zinc preferentially dissolves from the alloy, (ii) copper and zinc dissolve simultaneously with copper redeposited in the form of spongy copper and (iii) both mechanisms operate simultaneously (Jinturkar *et al.*, 1998). For corrosion on copper and copper alloy jackets of bullets, Booker (1980) recommended placing the bullets in a freshly prepared hot aqueous solution consisting of 5% (w/v) oleic acid and 5% (w/v) ammonium hydroxide. The solution should be heated to 80°C. A twenty-minute immersion is usually sufficient to remove all but the most stubborn corrosion; however, the bullets can remain in the hot solution for several hours. Removal of corrosion from a plated lead bullet requires great care to avoid damage to the delicate striation patterns. While the lead core of the bullet provides the copper alloy plating with cathodic protection from oxidation, the plating may flake off during cleaning. Randich *et al.* (2000) developed a cleaning process for jacketed bullets and cartridges in the course of the investigation of a triple homicide. From the crime scene and the victims' bodies investigators recovered a variety of firearms-related evidence, including bullet fragments, full-metal jacket bullets, hollow-point (semi-jacketed) bullets and both fired and unfired cartridges. Further investigation identified potential suspects. Execution of a search warrant for their residence led to investigators recovering expended cartridges of different calibers and fired bullets embedded in a palm tree. Using corroded non-evidentiary items composed of cartridge brass (nominally 70% Cu, 30% Zn) and oxidized non-evidentiary bullets and cartridges, Randich and his coworkers developed the following cleaning procedure:

- The evidentiary item was sonicated in 75°C ultrapure 18 MΩ deionized water for two minutes.
- The item was removed from the deionized water, dried and then sonicated in reagent-grade trichloroethylene at room temperature for two minutes.
- The item was next sonicated in HPLC-grade acetone at room temperature for two minutes.

- The item was then sonicated at 50°C for two minutes in an aqueous cleaning solution composed of 10% (w/v) sodium metasilicate and 10% (v/v) ethylene glycol monobutyl ether at a pH of about 12.5.
- The item was subjected to one or more immersions at room temperature in an acid cleaning solution composed of 3% concentrated sulfuric acid, 5% (w/v) thiourea, and 0.1% (w/v) sodium lauryl sulfate at a pH of about 1.5.
- Finally, the item was sonicated at room temperature for two minutes in ultrapure 18 MΩ deionized water.

The procedure was varied for the bullets removed from the palm tree. These specimens were soaked for several hours in 75°C trichloroethylene. They were also sonicated for varying periods of time in the acid cleaning solution. Organic debris and corrosion were successfully removed from the palm tree bullets; however, the fine striations required for microscopic comparisons were not recovered. Cleaning of the cartridges yielded better results. The microscopic details of the firing pin, breechblock, extractor and ejector marks could be seen. Four fired cartridges from the suspects' residence were matched to fired cartridges from the murder scene. Randich and his coworkers compared their cleaning process for copper and copper alloys with that proposed by Booker (1980). While Booker's procedure adequately cleaned corroded items, the cleaning solution had to be heated and mixed to prevent the oleic acid from separating out. A minimum immersion of a quarter of an hour in Booker's solution was equivalent to thirty seconds in the sulfuric acid–thiourea solution.

Firearms may also be submitted for examination in a corroded condition. Rust must be removed from the gun barrel to prevent rust particles from marking bullets that are test fired. A series of shots may clear the barrel of rust. Soaking and scrubbing the barrel with a detergent solution may also dislodge the rust. However, if most of the rust cannot be removed from the gun barrel, successful striation comparisons may prove impossible. Each successive shot through the barrel will result in a different striation pattern, because each bullet will encounter an altered rust surface (Rowe, 2005).

10.2.2 Analysis of bullet lead

On occasion, the bullets submitted for examination are not suitable for microscopic comparison to test-fired bullets: they are too corroded or too fragmented. Shotgun pellets, which are only rarely marked by the shotgun barrel when they are fired, are also submitted for examination. It would clearly be desirable to associate the ballistic evidence with a suspected shooter, if not to a particular firearm. One approach that has been used in this area is the comparison of the trace elements in the lead of the bullet fragments or shotgun pellets with the trace elements bullets or shotgun pellets from unused ammunition found in a suspect's possession. Guinn (1979, 1982) analyzed the trace elements in bullet lead using neutron activation analysis (NAA). In NAA, samples are irradiated with neutrons in the core of a nuclear reactor. Isotopes of certain elements capture the neutrons and become radioactive. The irradiated samples are removed from the reactor and their gamma-ray emissions measured with a gamma-ray spectrometer. The energies of the emitted gamma rays identify the elements present and their intensities are used to determine how much of each detected element is present in the samples.

Guinn's most famous application of NAA analysis of bullet lead was in the re-investigation of the John F. Kennedy Assassination (Guinn, 1979) by the House Select Committee on Assassinations. Guinn examined five bullet fragments in order to establish how many bullets this collection of evidence represented. At issue was the number of shots fired at President Kennedy and the number of shooters. Based on the antimony and silver concentrations, Guinn concluded that the bullet fragments represented only two bullets. Rahn and Sturdivan (2004) re-examined Guinn's data and data from the Federal Bureau of Investigation's (FBI) unpublished NAA analyses. These researchers concluded that Guinn's original determination that the fragments represented only two bullets was correct. Randich and Grant (2006), however, pointed out that lead in bullets crystalizes in dendrites and that trace metals tend to aggregate at grain boundaries, rendering bullet lead heterogeneous at a microscopic level. Based on this observation, these researchers argued that the bullet fragments analyzed by the Guinn and the FBI could represent between two and five bullets. This leaves the issue of the number of shooters unresolved and probably unresolvable by any chemical or instrumental tests.

For many years the FBI laboratory analyzed trace elements in bullet lead in order to associate bullet fragments from crimes to ammunition found in the possession of suspects. FBI examiners would testify that if the trace elemental profile of the crime scene bullets or fragments matched that of ammunition found in possession in a defendant, then the crime scene bullets or fragments were likely from the same batch of ammunition. Judges and juries were invited to conclude that the crime scene projectiles came from the box of ammunition found in possession of the defendant (Committee on Scientific Assessment of Bullet Lead Elemental Composition Comparison, 2004). The tenuous nature of this reasoning was revealed by the work of Koons and Grant (2002), who analyzed trace elements in bullets produced by one manufacturer using inductively coupled plasma-atomic emission spectrometry. Ten compositionally indistinguishable groups of bullet wire were produced during an almost twenty-hour period; however, the largest group could have produced 1.3 million 40-grain bullets. The FBI asked the National Research Council to convene a committee to examine the validity of the comparison of bullet leads using trace elemental composition. The resulting report (Committee on Scientific Assessment of Bullet Lead Elemental Composition Comparison, 2004) includes the following statements:

- The available data do not support any statement that a crime bullet came from a particular box of ammunition. In particular, references to boxes of ammunition in any form should be avoided as misleading under Federal Rule of Evidence 403.
- Compositional analysis of bullet lead data alone also does not permit any definitive statement concerning the date of bullet manufacture.

- Detailed patterns of the distribution of ammunition are unknown and, as a result, experts should not testify as to the probability that the crime scene bullet came from the defendant. Geographic distribution data on bullets and ammunition are needed before such testimony can be given (p. 7).

In 2005, the FBI terminated bullet lead examinations (Querna, 2005).

10.2.3 Serial number restoration

Serial numbers on firearms provide investigators with the ability to trace the original purchaser of a firearm recovered at the scene of a crime. In the United States, the Bureau of Alcohol, Tobacco, Firearms and Explosives enforces the statutory requirement that gun shops record identifying information of firearms purchasers along with the serial numbers of any firearms purchased. If the original purchaser uses a firearm in the commission of a crime and leaves the weapon at the scene, police investigators can use the firearm's serial number to identify him or her. If firearm has passed through a number of owners, it may be possible to track the possession of the weapon through a string of owners starting with the original owner. For this reason, criminals may attempt to remove or deface a firearm's serial number. Filing or grinding may be used to remove serial numbers; metal punches may also be used to deface serial numbers.

Firearm serial numbers are usually stamped; however, mechanical or laser engraving may also be used. Serial numbers usually appear on the frame of the weapon just above the trigger. Partial serial numbers may also be placed on components such as the removable barrels of semiautomatic pistols. Knowledge of the partial serial number can aid the restoration process.

There are a number of approaches to the restoration of stamped serial numbers: chemical etching (Mathews, 1962; Hatcher *et al.*, 1977; Maehly and Stromberg, 1981), electrochemical etching (Ceccaldi, 1962; Mathews, 1962) and magnetic particle restoration (Wolfer and Lee, 1960). All of these methods depend on the disruption of the crystal structure of metal at the site where the serial number was stamped or engraved. The disrupted crystal structure underlying the serial number can be thought of as a sort of latent image that will be revealed during the restoration process. The disrupted metal is eaten away faster than the surrounding undisrupted metal. Thornton and Cashman (1976) attributed this accelerated dissolution to "change in grain" in the metal structure as the result of the stamping of the serial number. "Change in grain" involves breaking and reforming chemical bonds between adjacent crystalline zones in the metal. Powell (1976) attributed the faster dissolution of metal underlying a stamped serial number to the storage of energy in the deformed metal. Usually when a serial number is ground away, the grinding is stopped when the serial number is no longer visible; regions of strained metal may remain. However, if the grinding has proceeded too far, the serial number may only be partially recoverable or not recoverable at all.

Serial number restoration begins with the firearms examiner photographing the area to be treated. Then the area is then polished with progressively finer grades of sand paper or emery cloth. With simple chemical etching the etching solution is applied to the polished area with a cotton swab. Alternatively, the area to be etched can be surrounded with a clay or wax dam and a volume of etching solution pipetted into the resulting receptacle. Electrochemical etching will speed up the serial number restoration process. The swab soaked in etching solution can be connected by means of an alligator clip to the negative terminal of a battery or regulated DC power supply. Both chemical and electrochemical etching must be carefully monitored with the digits of the serial number being recorded as they appear. It is possible to etch through the serial number, after which the serial number would be unrecoverable. Different etching solutions are used for different metals. Two etching solutions have been widely used for steel: (i) 40 mL concentrated hydrochloric acid, 30 mL distilled water, 25 mL ethanol and 5 gm of cupric chloride; or (ii) 120 mL concentrated hydrochloric acid, 100 mL distilled water and 90 gm cupric chloride. Twenty percent (w/v) aqueous sodium hydroxide can be used to etch aluminum. In another procedure for aluminum, the polished surface is rinsed with 1 N aqueous sodium hydroxide and then treated with 0.1 M mercuric chloride in 0.1 N hydrochloric acid (Rowe, 2005).

The magnetic particle method relies on the fact that when a magnetic field is induced in a ferromagnetic material, discontinuities such as cracks and regions of disrupted metal create variations in the induced field. These variations can be visualized with fine magnetic particles (e.g., magnetite, magnetic iron oxide). The weapon is first magnetized with a strong permanent magnetic or an electromagnet and then the magnetic particles are dusted over the polished area where the serial number is to be recovered. The magnetic particles can also be applied in an oil suspension. When an oil suspension is used, the surface is usually vibrated to move the magnetic particles around (Rowe, 2005).

Mechanically engraving serial numbers does not cause as much disruption of the metal as stamping. Consequently, mechanically engraved serial numbers have proved to be much more difficult to recover. Zaili *et al.* (2007) tested eight different etching solutions on mechanically etched serial numbers on steel. These researchers found that a reagent consisting of 5 gm of copper sulfate, 60 mL distilled water, 30 mL concentrated ammonium hydroxide and 60 mL concentrated hydrochloric acid had the greatest sensitivity. This reagent proved to be capable of recovering engraved serial numbers that had been erased to a depth of 0.04 mm below the original engraving depth. This reagent was also found to recover mechanically engraved serial numbers on aluminum that had been erased to a depth of 0.06 mm (Baharum *et al.*, 2008).

Laser engraving of serial numbers also does not produce as much disruption of the crystal structure of the metal as stamping. Chemical etching methods have not proven successful in recovering laser engraved serial numbers. However, polishing the metal surface to a mirror-like finish does allow the examiner to recover the serial number using reflected light stereomicroscopy (da Silva and dos Santos, 2008).

10.3 Shooting incident reconstruction

Reconstruction of crimes is the ultimate objective of criminalistics (Haag, 2006). In the case of shooting incidents, complete reconstruction would require identifying the weapons that fired all bullets, shotgun pellets and cartridges found at the scene, determining who fired each weapon, establishing the trajectories of all fired bullets and shotgun pellets and estimating the ranges from which the shots were fired. Answering all these questions may involve testing by a number of forensic experts. For example, determining who fired a particular weapon may require development of latent fingerprints by a latent print examiner or the profiling of traces of DNA by a forensic molecular biologist. A forensic chemist may also be called upon the detect gunshot residue on the hands of a suspect shooter. As discussed earlier, firearms examinations attempt to match bullets and cartridge cases to the weapons that fired them. Firearms examiners routinely examine items of clothing and other surfaces for patterns of gunshot residue from which estimations of muzzle to target distances may be made. Such examinations involve chemical and/or instrumental tests.

Determination of bullet trajectories is most commonly done by crime scene investigators, although some firearms examiners undertake such examinations (Garrison, 2003; Haag, 2006). Reconstructing trajectories at a crime scene requires the examiner to identify bullet holes in a variety of materials (e.g., panes of glass, wallboard, wood framing, wooden floors and steel automobile components) and to determine the direction the bullets were traveling. Shooting scene reconstructionists refer to the hole that results when a bullet enters a solid object as the *inshoot* defect; the hole produced when the bullet exits the object is the *outshoot* defect (Garrison, 2003). Chemical tests for lead, copper and nickel can be very useful in identifying inshoot defects, as well as graze or ricochet marks. Bullets may be made of lead or a lead alloy (with either antimony or tin); they can also have brass, nickel or steel jackets. Bullets can have total metal jackets (where the lead core of the bullet is completely enclosed by the jacket), full metal jackets (where the nose and sides but not the base of the bullet are enclosed in the jacket) or semi-jackets (usually where only the sides of the bullet are jacketed) (Rowe, 2005; Heard, 2008; Warlow, 2011). Because of the presence of lead in most firearm primers nearly all marks made by bullet impacts will have traces of lead, even those made by jacketed bullets (Haag, 2006). Jacketed bullets will also leave metallic traces from their jackets.

The colors of the surfaces being examined for metallic traces may interfere with the chemical color tests employed. Consequently, the color tests for lead, copper and nickel should conducted on filter paper or Benchkote™ lifts. Benchkote™ is filter paper with an impermeable polyethylene film backing (GE Healthcare Life Sciences, 2014). Haag (2006) recommends applying the dithiooxamide (DTO) test for copper first, followed by the sodium rhodizonate test for lead. The 2-nitroso-1-naphthol (2-NN) test can also be used to confirm the results of the dithiooxamide test.

Steinberg *et al.* (1984) developed a field test kit for bullet holes that included a DTO test for traces of copper around bullet holes. In their procedure, copper residues were dissolved in ammonium hydroxide and then reacted with DTO reagent. With copper DTO forms a complex whose color has been variously described as green-black (Steinberg *et al.*, 1984), dark green (Lekstrom and Koons, 1986), mossy gray-green to charcoal green (Haag, 2006) or just green (Wheeler and Wilson, 2008). The reaction is shown in Figure 10.2. Recommended concentrations of ammonium hydroxide range from 10% (Lekstrom and Koons, 1986) up to 30% (Haag, 2006). Recommended concentrations of DTO range from 1% (w/v) in ethanol down to 0.1% (w/v) (Lekstrom and Koons, 1986). Lekstrom and Koons (1986) also noted that DTO also produces pink or blue complexes with DTO. The DTO test procedure described by Haag (2006) is:

- Filter paper or Benchkote™ lift is moistened with ammonium hydroxide solution (2:5 dilution of concentrated ammonium hydroxide solution). The lift should be visibly wet.
- The moistened side of the lift is then pressed against the suspected bullet hole.
- The lift should be examined visually and any transferred material photographed. If there is transferred material with a color similar to that of the copper-DTO complex, the 2-NN test should be considered as an alternative test.
- The lift is sprayed lightly with DTO solution. If the dark green copper-DTO complex or the pink or blue nickel-DTO complexes appear the lift should be photographed.

The 2-NN procedure for the detection of copper residues is similar to that for DTO. The only difference is that in the last step the lift is sprayed with a 0.2% (w/v) solution of 2-nitroso-1-naphthol in ethanol. Cupric ion forms a pink complex with 2-NN (Figure 10.3), while ferric ion and zinc ion give green and orange, respectively. This

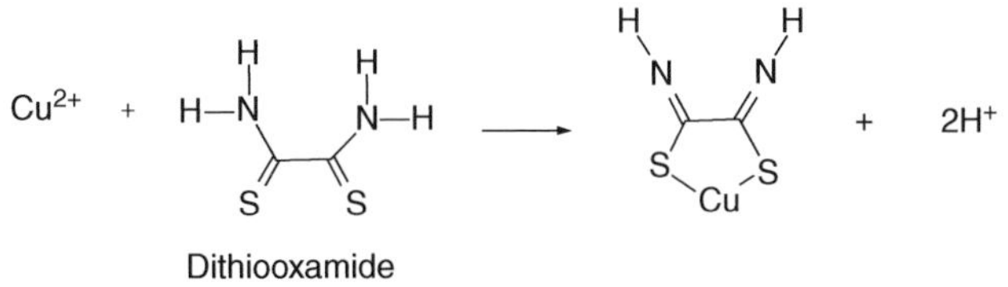

Figure 10.2 Reaction of dithiooxamide (DTO) with cupric ion

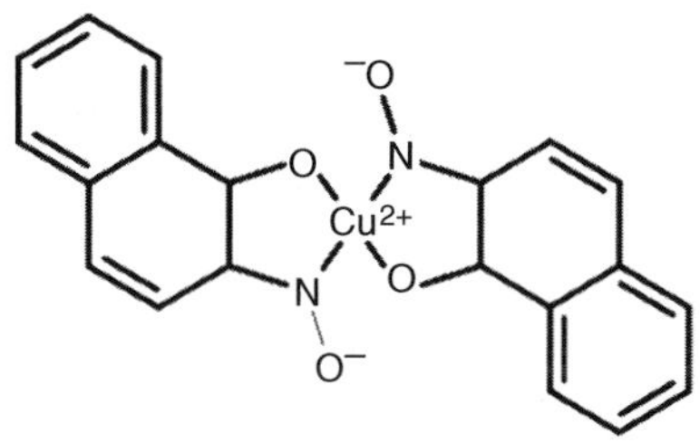

Figure 10.3 Structure of the 2-nitroso-1-naphthol (2-NN)-copper complex

reagent is prone to some false positives. These can be dealt with by allowing the 2-NN-treated lift to become almost dry and then spraying it with the DTO reagent. If the pink color produced by the 2-NN reagent fades and is replaced by the dark green of the copper-DTO complex, copper residues are indeed present. The results of the test should be photographically documented (Haag, 2006).

The most widely used test for traces of lead is the sodium rhodizonate test. Sodium rhodizonate forms a pink complex with lead (Figure 10.4). The test can be performed directly on the object or the traces of lead can be lifted with filter paper or Benchkote™. Haag's (2006) lifting procedure for the sodium rhodizonate test is:

- The surface of the object, filter paper or the Benchkote™ is moistened with pH 2.8 tartrate buffer (1.9 gm sodium bitartrate and 1.5 gm tartaric acid in 100 mL distilled water). The filter paper or Benchkote™ should be moistened until translucent.
- The lifting medium is pressed firmly over the suspected bullet hole.
- The lift is removed from the object's surface and sprayed with freshly made saturated aqueous sodium rhodizonate solution. Lead residues will produce a pink color. The results should be documented photographically.
- Because barium and strontium produce colors somewhat similar to lead, the pink area can be sprayed with 5% hydrochloric acid. The pink color of lead changes to blue and the colors of the barium and strontium complexes fade. The blue color should be photographically documented.

If some of the bullets fired in a shooting incident are nickel plated, holes made by these rounds may be identified using the dimethylglyoxime (DMG) test. DMG forms a bright red complex with nickel (Figure 10.5). Haag (2006) recommends preparing small sections of filter paper or Benchkote™ by pretreating them with an alcoholic solution of DMG. The sections of filter paper or Benchkote™ can be dried and stored. DMG

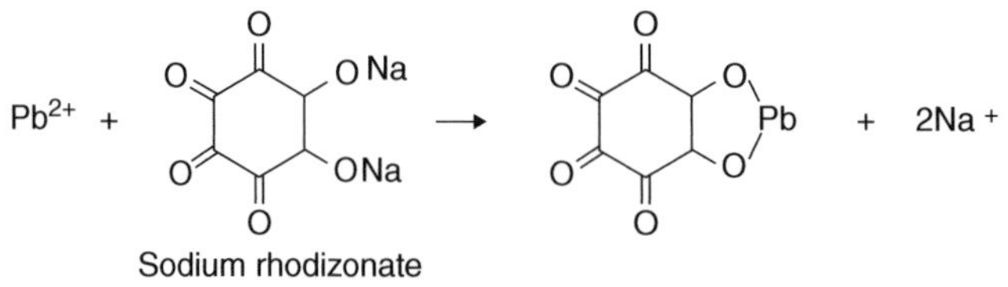

Figure 10.4 Structure of the rhodizonate-lead complex

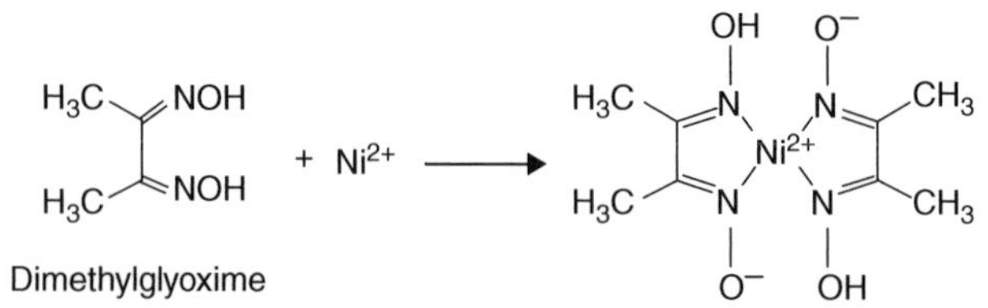

Figure 10.5 Structure of the dimethylglyoxime (DMG)-nickel complex

concentrations ranging from 0.2% (w/v) to 1% (w/v) have been used with success. The suspected bullet hole is treated with the same ammoniacal solution as is used for the DTO test. A pretreated DMG lifter is pressed against the hole and then removed. The red color of the nickel DMG complex should be immediately apparent. If the hole was made by a nickel-plated bullet, the lift should also be tested for copper with the DTO reagent. Nickel-plated lead shot will leave substantial quantities of lead, which can be detected with the sodium rhodizonate test.

10.3.1 Muzzle-to-target determinations

Chemistry plays a significant role in the determination of the range from which a shot was fired. Such a determination may be crucial in the disposition of a shooting case. Suppose the shooter claims that he shot the victim in self-defense at close range when the victim ran toward him in a threatening manner. Gunshot residue found on the victim's skin or clothing would tend to support this version of the event; the failure to find such residue would contradict the claim of self-defense. Understanding generation and deposition of gunshot residue requires an understanding of the chemistry of firearm propellants and cartridge primers. Knowledge of the chemistry of propellant combustion is also very important.

Firearms can be regarded as heat engines that convert chemical energy stored in propellants into kinetic energy of projectiles. The first firearm propellant was black powder, a mixture of saltpeter (potassium nitrate), charcoal and sulfur. This energetic mixture was discovered by the Chinese and knowledge of it reached Europe via the Islamic world by the 13th century. The initial formulations of black powder correspond to the compositions of weak blasting agents. It took many years for black powders suitable for use as propellants to be developed. To attain the fastest burning rate (and hence the fastest release of propellant gases) the components of the mixture must be intimately mixed. One early recipe for black powder called for grinding the saltpeter, charcoal and sulfur together for twelve hours, until a mixture as fine as face powder was obtained. For black powder to function effectively as a propellant its constituents must be interdispersed almost on the molecular level. Black powders used as propellants are almost solid solutions. Because black powders were used in both short-barreled firearms such as pistols and long-barreled weapons such as muskets, it was important to control the burning rate of the powder. This was achieved by controlling the surface areas of the particles. The burning of both black and smokeless powders is a surface phenomenon. All other factors being equal, the higher the surface area of a given weight of propellant the faster it burns and releases gases. The surface area of a propellant is controlled by controlling the size of its grains. The process of corning black powder was developed to control grain size. In the final stage of grinding black powder, water is added to the mixture and the wet cake of black powder is forced through a metal screen with openings of a specified size. This yields a product of uniform size and texture. There is also the added benefit of having the saltpeter (potassium nitrate) partially dissolve and penetrate the pores in the charcoal particles, achieving an even more intimate mixture (Partington, 1998).

Black powder has a number of deficiencies as a propellant. It produces voluminous clouds of smoke that could almost totally obscure battlefields, interfering with tactical movements and diminishing the effectiveness of musketry and artillery fire. Of greater importance was the deficiency of black powder as a propellant. Typically, the combustion products of black powder consist of roughly 44% solids (Lowry, 1968). These solids are in the form of finely divided particles (smoke) and cannot exert any pressure in a gun barrel to propel a projectile.

Despite its deficiencies as a firearm propellant black powder is still manufactured. The ingredients (approximately 75% potassium nitrate, 15% charcoal and 10% sulfur) are crushed and mixed together in heavy wheel mills (Committee on Smokeless and Black Powder, 1998). The pressure exerted by the heavy wheels causes the sulfur to flow and bind together the charcoal and potassium nitrate. The mixture is pressed into blocks that are broken up by passage through a series of rollers. The resulting grains of black powder are sorted according to size by vibrating screens and then coated with graphite.

A number of black powder substitutes are currently on the market. Pyrodex® was introduced by Hodgdon Powder Company in 1976. Like black powder, it contains potassium nitrate, sulfur and charcoal. However, Pyrodex® has less sulfur than black powder and also has potassium perchlorate, sodium benzoate and dicyandiamide (Routon et al., 2011). Hodgdon later introduced Triple Seven® as a sulfur-free alternative to black powder; this powder has no sulfur and contains 3-nitrobenzoic acid (Routon et al., 2011). Many brands use ascorbic acid as the fuel, with potassium nitrate and/or perchlorate as oxidizers (Goodpaster and Keto, 2004; Lang and Boyle, 2009; Bottegal et al., 2010). Ascorbic acid-containing black powder substitutes have been marketed under such names as Golden Powder, Black Mag Powder, Black Mag 3 and Clean Shot (Goodpaster and Keto, 2004).

Nitrocellulose (the main energetic ingredient in smokeless powers) proved to be a more efficient propellant than black powder: upon ignition 100% of nitrocellulose is converted to gases (Lowry, 1968). Nitrocellulose or cellulose nitrate was first prepared in the early 19th century by Braconnot (1833), who reacted wood fibers with nitric acid. Schönbein (1846) was able produce nitrocellulose from cotton fibers (hence the alternative name for nitrocellulose, *guncotton*). The production of nitrocellulose is straightforward: a suitable source of cellulose (e.g., cotton lint) is treated with nitric acid in a sulfuric acid bath (Lowry, 1968). In addition to gun propellants, nitrocellulose was used to manufacture a variety of materials, including shirt fronts and photographic film. Despite its highly flammable nature, it is still used in some automobile primers and fingernail polishes.

There are three types of smokeless powder: single base, double base and triple base (Rowe, 2005; Heard, 2008; Warlow, 2011). Nitrocellulose is the only energetic material in single-base smokeless powder. In double-base powder, nitroglycerine is added to the nitrocellulose. Although nitroglycerine is a primary high explosive, its most important function in smokeless powder is as a plasticizer that "waterproofs" the smokeless powder (Committee on Smokeless and Black Powders 1998). Water acts as a heat sink when powders burn: water has a high heat capacity and a high heat of vaporization. Triple-base powders contain nitroguanidine in addition to nitrocellulose and nitroglycerine. Triple-base powders are not commercially available in the United States (Meng and Caddy, 1997).

Nitrocellulose is highly susceptible to acid-catalyzed breakdown. The breakdown products consist of gaseous nitrogen oxides (Lowry, 1968; Meng and Caddy, 1997). These nitrogen oxides can react with moisture in the smokeless powder grains or with atmospheric moisture to produce nitric and nitrous acids. These acids produce further degradation of the nitrocellulose. Eventually, the nitrocellulose becomes dangerously unstable, prone to spontaneous ignition. To slow down the degradation of nitrocellulose, antacids may be added to the powder formulation. More importantly, stabilizers such as diphenylamine, methyl centralite and ethyl centralite are incorporated in smokeless powders (Meng and Caddy, 1997). Diphenylamine is the most widely used stabilizer in smokeless powder. Ethyl centralite is mainly used in double-base powders. These stabilizers react with nitrogen oxides to produce nitrated daughter compounds (e.g., 2-nitrodiphenylamine, 4-nitrodiphenylamine and N-nitrosodiphenylamine). These daughter compounds can react further with nitrogen oxides. However, eventually the capacity of the stabilizer to absorb additional acidic breakdown products is exhausted, at which point the smokeless powder should be destroyed.

From the perspective of the manufacturer of smokeless powder, the most important feature of a lot of smokeless powder is the burning rate. The powder's burning rate determines the shape of the pressure curve of the propellant gases inside the firearm. Different types of firearms require different pressure curves. A short-barreled handgun, such as a snub-nosed revolver, requires a powder that delivers most of its energy in the time it takes for the bullet to traverse the short barrel. Any gases produced after the bullet exits the barrel are wasted; they impart no further kinetic energy to the projectile. On the other hand, a long-barreled weapon, such as a hunting rifle, requires a powder that burns longer and generates propellant gases throughout most of the bullet's travel down the barrel. As already discussed, the burning of powder grains is a surface phenomenon. To prevent radiant heat from penetrating smokeless powder grains and initiating burning within the grains, opacifiers (e.g., carbon black) are added to smokeless powder (Heramb and McCord, 2002). The burning rate of black powder is mainly controlled by the size of its grains. Very fine particles have a larger surface area for reaction than do coarse particles. Burning rates of smokeless powder grains are controlled by surface area per unit mass (Committee on Smokeless and Black Powder, 1998). Smokeless powder can be manufactured in a wide range of shapes (such as flakes, rods, tubes, balls and flattened balls) and can have perforations that affect burning rates in desirable ways. Take, for example, the burning of a powder grain in the shape of a rod. As the grain burns, the rod's length and diameter decrease, diminishing the surface area of the grain and reducing its burning rate. If the rod is perforated, the interior surface of the perforation will also burn; as it does so it will increase the diameter of the perforation and its surface area. This will tend to balance the decrease in surface area resulting from the burning of the exterior surface. Deterrents are surface coatings that control the initial burning rate of smokeless powder grains (Heramb and McCord, 2002). There are two types of deterrents: penetrating deterrents, compounds that are soaked into the surface of the grains (producing a concentration gradient of deterrent from the grain surface to its interior) and inhibiting or "candy shell" resin coatings (rather like M&M candies). Penetrating deterrents include Herkote® (Paraplex® G-54 Polyester), ethyl and methyl centralite, dibutyl phthalate and dinitrotoluenes; the most common inhibiting deterrent is Vinsol® resin.

Two processes are used in the manufacture of smokeless powder: the extrusion process and the ball powder process (Committee on Smokeless and Black Powders, 1998). In the extrusion process, a block of nitrocellulose containing solvent (typically acetone) and desired additives is forced by hydraulic pressure against a perforated steel plate. A rapidly rotating steel blade below the plate cuts off lengths of extruded nitrocellulose. The extruded powder grains fall on to a moving belt. As the grains move along the belt, the residual solvent evaporates from the grains and the oversized and undersized grains are removed (and returned to the beginning of the process). The grains are next treated with deterrent (if such is called for) and coated with graphite glaze. Graphite glaze prevents the smokeless powder grains from accumulating surface electrical charges during subsequent handling. Cuts of product are taken from the production line and placed in a rest house (where residual solvent evaporates). The cuts undergo ballistic testing to determine their burning rates. Finally, batches of powder are combined to yield a final product with the target burning rate.

In the ball powder manufacturing process the nitrocellulose mixture is extruded through a pelletizing plate and formed into spheres (Committee on Smokeless and Black Powders, 1998). As with the extrusion process, oversized and undersized powder grains are removed from the product stream and recycled back to the beginning of the process. The spherical powder grains may be passed between rollers to produce disks or flakes. Cuts are taken from the process stream and placed in rest houses. The burning rates of each cut from the production line are determined and then cuts are blended to obtain a specified burn rate.

Smokeless powders may contain a number of other additives in addition to stabilizers, deterrents and opacifiers. Smokeless powder grains undergo much tumbling in the course of their manufacture. Fracturing of a powder grain during manufacture would expose a chemically reactive surface and, moreover, one devoid of graphite glaze coating. For these reasons, plasticizers are added to the nitrocellulose "dough" at the beginning of the extrusion and ball processes. Nitroglycerine is an example of an energetic plasticizer. Other plasticizers used in smokeless powder include ethyl centralite, dibutyl phthalate, dinitrotoluenes and triacetin (Meng and Caddy, 1997; Heramb and McCord, 2002). Potassium nitrate and potassium sulfate may be added as flash suppressants (Heramb and McCord, 2002). Firearm muzzle gases contain more carbon monoxide than carbon dioxide; when these gases encounter the outside atmosphere they immediately burst into flame. Flash suppressants break up the free-radical chain reaction of carbon monoxide with atmospheric oxygen. Ignition aid coatings may be added to improve the surface oxygen balance of the grains (Heramb and McCord, 2002). The chemical structures of many of the organic additives found in smokeless powder are shown in Figure 10.6.

Smokeless powders are shipped both to ammunition manufacturers and to retailers who serve the reload market. Many shooters like to load their own ammunition. This may be a cost-cutting measure or a way of producing ammunition with better shot-to-shot consistency or with certain preferred performance characteristics. Smokeless

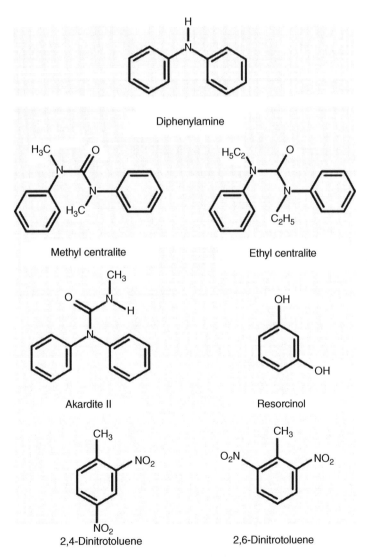

Figure 10.6 Chemical structures of smokeless powder additives

powders for reloading are a potential source of energetic material to fill explosive devices such as pipe bombs (Committee on Smokeless and Black Powders, 1998).

10.3.2 Firearm primers
10.3.2.1 History

Firearms, to be effective weapons, require a means to ignite the propellant. The very earliest handheld firearms were fired by applying a heated wire to hole in the breech end of the barrel. The heated wire was eventually replaced by a "slow match," a cord that had been soaked in a potassium nitrate solution and then dried. The burning "match" could be held in the hand or in a lever-actuated clamp mounted on the side of the firearm. The burning end of the match was thrust into a small amount of fine black powder in a pan mounted on the side of the firearm near the breech end of the barrel. A hole in the barrel allowed the flame from the burning powder to reach the main propellant charge. This ignition system is referred to as a "matchlock action." Such an ignition system had many obvious flaws (e.g., burning "matches" could be extinguished by rain and could ignite the open powder barrels from which artillerymen loaded their guns). Nevertheless, it held on almost to the end of the 17th century.

The "wheel-lock" ignition system was developed in the 16th century and replaced the "matchlock" mechanism on some types of weapons. In the "wheel-lock" system a piece of iron pyrite held in a clamp was forced against a rapidly spinning serrated wheel. The weapon was prepared for firing by winding up a spring attached to the wheel. Pulling the trigger released the wheel and simultaneously moved aside the lid of the pan mounted on the side of the firearm. Sparks struck from the iron pyrite by the spinning wheel ignited the exposed priming charge of black powder in the pan. "Wheel-lock" handguns were widely used by cavalry troopers (for whom management of "matchlock" weapons was particularly difficult). They were also easily concealed on the person yet could almost instantly be fired. "Wheel-lock" mechanisms never supplanted the "matchlock" firing mechanisms: they required close manufacturing tolerances and were consequently much more expensive. Eventually both of these ignition systems were replaced by the snaphaunce and flintlock ignition systems. In the flintlock mechanism, a sharpened piece of flint struck sparks from a curved piece of steel called the frizzen. The frizzen was part of a hinged cover on the priming pan that opened to expose the priming charge to the sparks. Flintlock firing mechanisms were in use from the end of the 17th century well into the 19th century (Rowe, 2005; Heard, 2008; Warlow, 2011).

The development of the caplock or percussion lock ignition system that eventually replaced the flintlock system sprang from research on a class of chemical compounds called fulminates (Greener, 1910). Fulminates are energetic materials that can be exploded by percussion. Mercury fulminate $[Hg(ONC)_2]$ and potassium chlorate ($KClO_3$) are two of the more widely known members of this family. Edward Howard (1800) developed a priming powder composed of mercury fulminate and potassium nitrate (KNO_3). A.J. Forsyth patented the idea of using fulminates to ignite propellants in firearms in 1807 (Greener, 1910). In 1816, Joshua Shaw produced a practical firearm ignition system based on fulminates when he patented a percussion cap containing fulminates (Greener, 1910). In one of the more common versions of the caplock or percussion lock system, the cock, frizzen and pan of the flintlock system were replaced by a hammer, a nipple and a percussion cap. The percussion cap was placed on the nipple, a tube that was mounted on the side of the firearm beneath the hammer. When the cap was crushed by the hammer, the fulminate in the cap ignited and its flame is communicated to the propellant through a small hole in the barrel.

The final development in ignition systems was the self-contained metallic cartridge – a cartridge containing bullet, propellant and primer. Paper cartridges containing bullet and propellant had been in use since the 16th century (Greener, 1910). These required the shooter to tear open the paper with his teeth, pour the propellant charge down the barrel and ram the bullet down on top of the powder. The discovery of the explosive properties of fulminates allowed the ignition source to be incorporated into the cartridge itself. In 1831, Dreyse developed a paper cartridge containing bullet, black powder and mercury fulminate primer (Greener, 1910). The mercury fulminate was placed immediately behind the bullet so that a very long firing pin capable of penetrating the length of the powder-filled cartridge was required. The Dreyse "needle" gun became the standard infantry rifle of the Prussian army. The Dreyse system was further developed by placing the primer at the base of a cardboard tube, so that a short firing pin could be used (Wallace, 2008). Self-contained metallic cartridges eventually replaced cartridges made of paper, cardboard or linen. Pinfire cartridges were the first widely used metallic cartridges. These copper cartridges had a striker pin projecting from the side of the case near its base; when struck by the weapon's hammer, the pin struck a small amount of fulminate, igniting it (Wallace, 2008). Pinfire cartridges could only be loaded in one orientation, precluding their use in magazine-loading firearms. Nevertheless pinfire cartridges were manufactured into the 1930s (Wallace, 2008). In 1854, Smith and Wesson patented two types of metallic cartridge: the rimfire cartridge, in which the primer compound is placed inside the rim of the case, and the centerfire cartridge, in which the primer is incorporated in a cap centrally placed in the cartridge base (Wallace, 2008). Rimfire cartridges were used in a variety of significant 19th century firearms: the Spencer repeating rifle and carbine, the Henry and Winchester repeating rifles and the 1873 Colt Peacemaker revolver. Now, the rimfire ignition system is used only in .22 caliber and .17 caliber cartridges (Wallace, 2008).

10.3.2.2 Modern primers

By the beginning of the 20th century the compositions of the energetic mixtures used in firearm primers had a limited range of ingredients: mercury fulminate, potassium chlorate, antimony sulfide, sulfur and powdered glass. However, residues produced by mercury fulminate and potassium chlorate primers proved to be corrosive and other primer compositions were introduced (Harrison and Gilroy, 1959; Wallace, 2008). Most

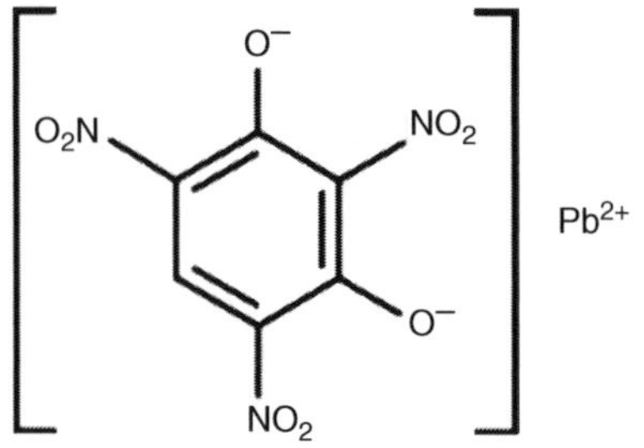

Figure 10.7 Structure of lead styphnate

modern firearm primers consist of four components: initiating explosive, oxidizing agent, fuel and sensitizer (Meng and Caddy, 1997). The most commonly used initiating explosive is lead styphnate. The structure of lead styphnate is shown in Figure 10.7. Barium nitrate [$Ba(NO_3)_2$] is the most common oxidizer but barium peroxide, lead nitrate or lead peroxide may also be used. Most primers use antimony sulfide as the fuel. However, calcium silicide, lead thiocyanate or powdered metals such as aluminum, zirconium, magnesium or titanium may also be encountered. Sensitizers include tetracene (1-(5-tetrolyl)-4-guanyltetracene hydrate), which is most common, pentaerythritol tetranitrate, trinitrotoluene and tetryl. The structures of these compounds are shown in Figure 10.8.

In the 1970s, public health professionals found that law enforcement personnel in the United States who worked in indoor firing ranges could have high blood levels of lead and exhibit neurological and gastrointestinal symptoms of lead poisoning (Fischbein *et al.*, 1979). Eventually the problem was found to be worldwide (Ozonoff, 1994). To mitigate this hazard, ammunition manufacturers began manufacturing total metal jacket bullets in order to reduce ablation of lead particles from the exposed bases of the bullets (Wallace, 2008). The manufacturers also started to replace lead-based primers with lead-free primers. Sintox primers (produced by Dynamit Nobel AG) were the first lead-free primers; they contain tetracene, 2-diazo-4,6-dinitrophenol (diazole), titanium powder, zinc peroxide and nitrocellulose ball powder (Wallace, 2008). Lead-free primers made by US ammunition manufacturers have somewhat similar compositions with some interesting variations. Table 10.1 shows lead-free primer compositions based on material safety data sheets (MSDSs).

When a firearm is fired, the energetic compounds in the primer explode and ignite the propellant in the firearm's cartridge. The propellant burns, generating high-pressure, high-temperature gases that push the bullet or shot mass down the barrel. What emerges from the muzzle of the weapon is a complex mixture of gases and particles. The gases include carbon dioxide and carbon monoxide. Some of the gases are hot enough to be incandescent. Once outside the barrel, the pressure and temperature of the muzzle gases drop quickly. The cloud of particles issuing from the firearm's muzzle contains metallic

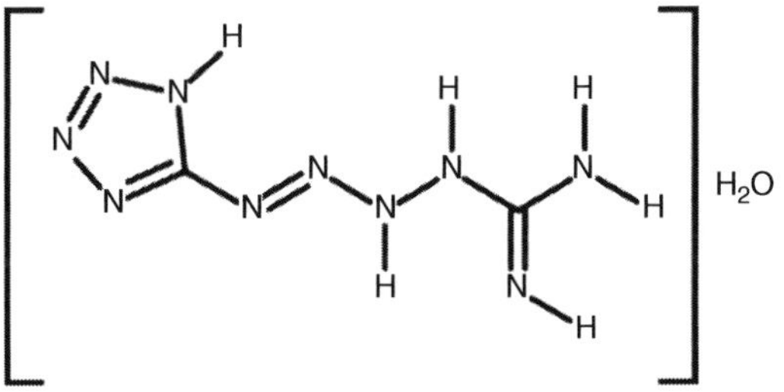

Tetracene (1-(5-Tetrolyl)-4-guanyl tetracene hydrate

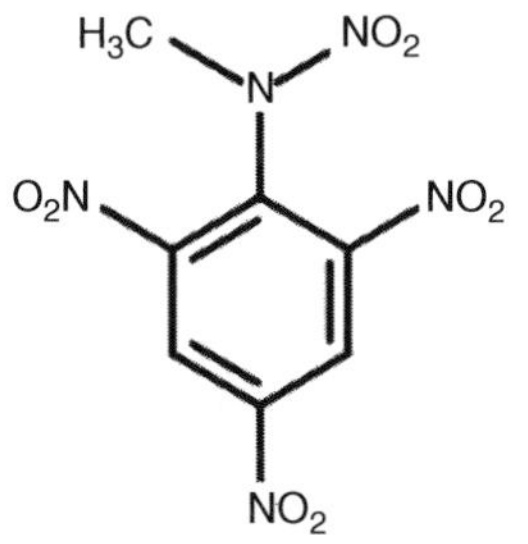

Tetryl

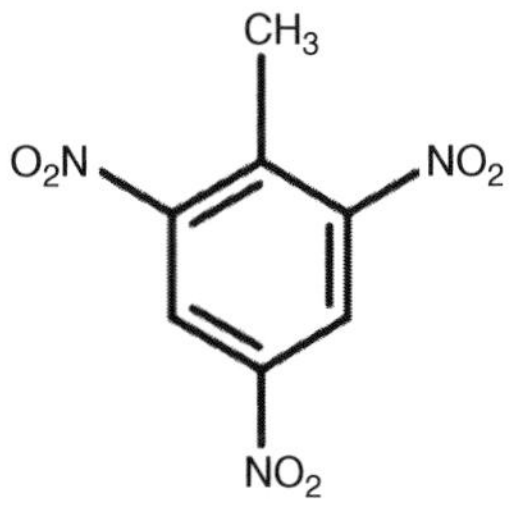

2,4,6-Trinitrotoluene

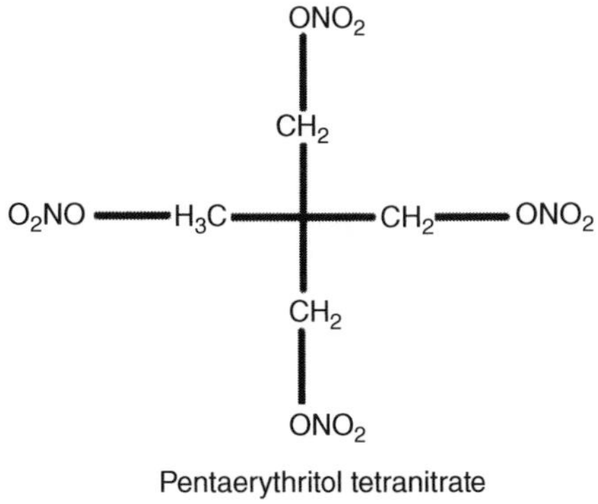

Pentaerythritol tetranitrate

Figure 10.8 Structures of tetracene, pentaerythritol tetranitrate, trinitrotoluene and tetryl

particles from the primer compounds, the bullet and the weapon's barrel. There will also be unburned propellant grains, partially burned propellant grains and solid products of combustion (soot). Air is a viscous medium and atmospheric drag decelerates these particles so that they travel only a few feet. For purposes of later discussion, it is important to distinguish inorganic gunshot residue (IGSR) and organic gunshot residue (OGSR). IGSR consists of metallic particles from the bullet and the primer and various inorganic salts (with nitrates and nitrites predominating). OGSR consists of organic compounds derived from the additives in smokeless powder.

Table 10.1 Compositions of lead-free primers produced in the United States

Brand	Manufacturer	Composition	Reference
BallistiClean	Federal Cartridge Company	diazole, tetracene, potassium nitrate, aluminum, nitrocellulose, nitroglycerin	Federal Cartridge Company, 2010
Clean Fire	CCI/SPEER	diazole, strontium nitrate, nitrocellulose	CCI/SPEER, 2008
Remington Lead-Less	Remington Arms	diazole, tetracene, barium nitrate, nitrocellulose, nitroglycerin	Remington Arms, 2001
WinClean	Olin Corporation-Winchester	diazole, potassium nitrate, boron, nitrocellulose	Winchester, 2013

Muzzle-to-target determinations in shooting incident reconstructions typically involve bullet wounds and GSR patterns on clothing. While determination of muzzle-to-target distances from the appearance of gunshot wounds does not usually involve any chemical or instrumental tests, the features of bullet wounds inflicted at different distances merit discussion because these features highlight the dynamics of the interaction of muzzle gases and GSR with targets. Forensic pathologists distinguish distance shots, intermediate-range shots, near-contact shots and contact shots. Each of these has its own characteristic features. A distant shot is one inflicted from such a distance that no muzzle gases or particulates reach the skin or clothing. For handguns, a distant gunshot wound would be one inflicted from a range greater than 2–3 feet. For more powerful long arms such as rifles or shotguns, this range would be several feet. The main feature of a distant gunshot wound would be a circular or elliptical defect, with an abraded margin (also called a contusion ring) and a grey ring (also called bullet wipe). The abraded margin results from the stretching and tearing of the skin as the bullet or shotgun pellet enters. As it is basically a type of bruise the abraded margin has a purplish discoloration. However, this color may not be readily observable because the abraded margin is usually covered up by the grey ring, which is comprised of material wiped off the surface of the bullet. Bullet wipe is comprised of lubricants (gun oils and lubricants applied to bullets by ammunition manufacturers), metallic traces from the bullet's surface (lead or copper) and fine combustion products from primer and propellants (DiMaio, 1999).

An intermediate-range gunshot wound will be surrounded by a halo of deposited particles. The larger particles dotting the area around the wound are called stippling or tattooing. Technically, stippling consists of large particles on the surface of the skin or clothing; tattooing results from particles of GSR being driven into the skin. There is also a vital reaction of skin (reddening) due to the impact of hot powder grains. The finer particles of GSR are referred to as soot or smudging. If the range of fire is just short enough for the larger particles of GSR (mostly unburned propellant grains) to reach the target, the GSR pattern will consist of stippling; the fine particles will be slowed down by air resistance and not reach the target surface. As the range of fire decreases, the halo of GSR will become smaller, but denser: more stippling (with tattooing) and increasing smudging. The wound will still be circular or elliptical with an abraded margin and grey ring (DiMaio, 1999).

At near-contact range (1–2 inches for handguns) the muzzle flash interacts with the skin or clothing. Hair and skin around the wound will be burned. Natural fibers in clothing will be burned; man-made fibers will be burned or melted. Woven fabrics usually split apart along the warp and weft directions, creating what is often called a cruciate defect. Knitted fabrics typically show an irregular circular hole with burned or melted yarns (Rowe, 2005).

Forensic pathologists distinguish two types of contact wounds: loose contact and tight contact wounds. In a loose contact gunshot the muzzle is in contact with the target, but GSR can escape from the muzzle along the target surface. If the shot is to an area covered by clothing the GSR may spread out between layers of clothing or between the skin and the overlying clothing. Muzzle gases and GSR will also follow the track of the bullet into the body. In a tight contact wound hot muzzle gases and GSR also follow the bullet track, resulting in little deposition on the skin or clothing. When the tight contact wound is inflicted in an area where soft tissue is backed by a boney plate (such as the sternum or the vault of the skull) the high pressure muzzle gases will separate the soft tissue from the underlying bone. This produces a trapped bubble of high pressure muzzle gas. If the muzzle of the weapon is in direct contact with bare skin, the high pressure gas will press the soft tissue against the muzzle and produce a muzzle stamp or impression: a bruise showing the outline of the weapon's muzzle. If the pressure of the muzzle gases is high enough, the soft tissue will split, allowing the release of the gases and producing a stellate (or star-like) defect. The splitting of the soft tissue can also produce blowback of blood and other tissue onto the hands of the shooter and onto the firearm (DiMaio, 1999).

The muzzle-to-target distance of a distant gunshot cannot be determined from a powder pattern because there is not one. The muzzle-to-target distances for near-contact and contact wounds are self-evident. In order to determine the muzzle-to-target of intermediate-range gunshots, firearms examiners fire test shots using the firearm that was used to fire the original powder pattern. The test shots must also be fired using the same ammunition as that used to fire the original pattern (using cartridges from the shooting suspect's remaining box of ammunition, if needed). The same weapon and ammunition must be used to test fire patterns because the condition of the weapon will affect how much GSR it will release. The composition of the smokeless powder in the cartridge affects how completely it burns and how much particulate matter it produces. The test shots are fired at various ranges into cloth targets (usually referred to as witness panels). The muzzle-to-target distance of the original gunshot is determined by comparing its powder pattern to the witness panels (Rowe, 2005; Heard, 2008; Warlow, 2011).

Powder patterns may be on the skin of dark-skinned individuals or on skin discolored by decomposition. Very often the patterns are on dark or bloodstained clothing. In these circumstances, the powder patterns must be visualized in some manner before they can be compared to witness panels. Special infrared photographic techniques may be used. Soot from the discharge of a firearm absorbs infrared radiation very strongly. The hemoglobin in blood absorbs infrared radiation to only a limited degree. Consequently, infrared photography can reveal GSR patterns beneath bloodstains (Eastman Kodak Company, 1972). However, chemical techniques are most widely used for visualization of GSR patterns for purposes of estimation of muzzle-to-target distances.

In addition to lead deposited immediately around the edges of the bullet hole due to contact with the bullet's surface, lead residues may be found in a halo surrounding the bullet hole. These residues may consist of lead particles from the bullet's surface, lead particles left in the weapon's barrel by previous shots, lead particles from lead-based primers and vaporous lead (lead smoke). Vaporous lead is typically deposited by intermediate-range, near-contact and contact shots. Vaporous lead deposits are used for determinations of muzzle-to-target distances (SWGGUN, 2013). These deposits are visualized with the sodium rhodizonate test. The mechanism of this test has been previously discussed.

In addition to metallic traces from the cartridge primers, IGSR contains nitrates and nitrites. The diphenylamine test can used to test for organic nitrates in particles comprising a GSR pattern: a suspected smokeless powder grain is removed from the pattern and tested with a solution of diphenylamine in concentrated sulfuric acid. Acid hydrolysis of nitrocellulose and nitroglycerine will release nitric acid that will oxidize diphenylamine to produce a blue color. The fact that the diphenylamine reagent contains concentrated sulfuric acid makes it unsuitable for the visualization of powder patterns. The diphenylamine test is also not specific for nitrates or nitrites: any strong oxidizing agent will produce the same result. As early as 1928 a test for nitrites was proposed as a test for GSR. Originally, material was removed from the suspected GSR deposit for testing. Walker (1940) proposed a technique for the visualization of GSR patterns on fabrics. Photographic paper was desensitized in a hypo bath, then washed and dried. For visualization of the GSR pattern the photographic paper was first placed in a warm 5% solution of 2-naphthylamine-4, 8-disulfonic acid for ten minutes. The paper is removed from this bath and allowed to dry. The treated paper was placed on a layer of toweling, gelatin side up. The fabric bearing the GSR pattern was placed pattern side down on top of the treated paper, followed by a layer of dry toweling, a layer of toweling soaked in 20% acetic acid and a final layer of dry toweling. The entire stack was pressed for five to ten minutes with a warm iron. The partially burned particles of smokeless powder produced dark red spots on the photographic paper. Nitrites in the GSR pattern reacted with 2-naphthylamine-4, 8-disulfonic acid to form an azo dye. The solution of 2-naphthylamine-4, 8-disulfonic acid could be replaced with

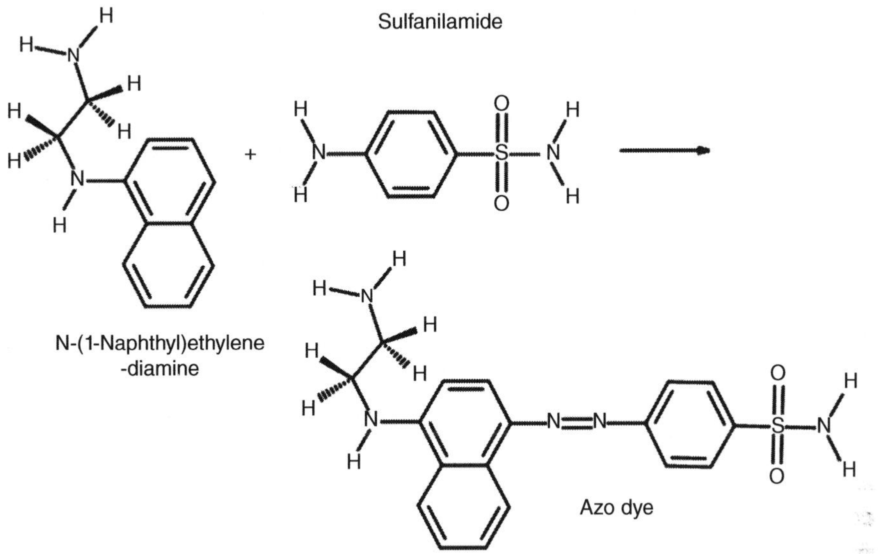

Figure 10.9 An example of the Griess reaction

a solution of 1-naphthylamine and sulfanilic acid. 1-Naphthylamine was eventually recognized as a carcinogen and the Walker test fell into disuse.

Tests to replace the Walker test were based on the diazotization reaction discovered by industrial chemist Peter Griess in the mid-19th century (Figure 10.9). The FBI Laboratory adopted the use of a solution of N-(1-naphthyl)-ethylenediamine dihydrochloride (also known as Marshall's Reagent) and sulfanilic acid (Dillon, 1990). Equal volumes of a methanolic solution of N-(1-naphthyl)-ethylenediamine dihydrochloride and an aqueous solution of sulfanilic acid are combined just before the test is to be conducted. Desensitized photographic paper is soaked in this mixture and used to visualize the GSR pattern. This test is sometimes referred to as the modified Griess test (MGT). Commercial Griess reagent solutions contain phosphoric acid. Nowadays, the MGT reagent would consist of 3% sulfanilamide and 0.3% N-(1-naphthyl)-ethylenediamine dihydrochloride in 5% phosphoric acid. Unfortunately, N-(1-naphthyl)-ethylenediamine dihydrochloride has also been identified as a carcinogen. The RCMP Laboratory replaced N-(1-naphthyl)-ethylenediamine dihydrochloride with 1-naphthol or naphthoresorcinol. Use of 1-naphthol results in the formation of an orange azo dye, while use of naphthoresorcinol gives a yellow azo dye (National Forensic Science Training Center, 2015).

The Israeli National Police have developed a lifting technique that uses a commercially available adhesive lifter (Glattstein *et al.*, 2000a). The lifter is pressed down on the garment bearing the powder pattern using a heated press. After the lifter is removed from the garment, it is treated with a 2% (w/v) ethanol solution of potassium hydroxide solution and heated at 100°C for an hour. Alkaline hydrolysis of nitrocellulose releases more nitrites. Desensitized photographic paper impregnated with the modified Griess reagent is then pressed against the lift for about one minute at 70°C. In this case, the modified Griess reagent consists of 3% (w/v) sulfanilamide and 0.3 % (w/v) N-(1-naphthyl) ethylenediamine dihydrochloride in 5% (v/v) phosphoric acid. This lifting technique with alkaline hydrolysis can also be applied at crime scenes (Glattstein *et al.*, 2000b) and to dead bodies (Glattstein *et al.*, 2000c). Laundering and brushing of clothing can reduce the density and extent of a powder pattern but will not completely remove GSR (Vinokurov *et al.*, 2001).

Haag (2006) recommends using a modified Griess reagent prepared by combining equal volumes of 0.5 % (w/v) sulfanilic acid in distilled water and 0.3 % 1-naphthol in methanol. Desensitized photographic paper or inkjet photopaper is soaked in freshly prepared reagent. The treated paper is then placed on a hard surface, emulsion side up, and the garment bearing the powder pattern is placed on top of it, powder pattern down. A layer of cheesecloth soaked in 15% (v/v) acetic acid is placed on top of the garment and ironed with a hot iron set for "Cotton." GSRs transferred to photographic papers appear as orange spots.

Maiti (1973) noted that bloodstained articles of clothing can present problems with the visualization of GSR patterns: Blood can be transferred to the desensitized photographic paper, obscuring the red azo dye produced in the Griess test. He recommended the following changes to the procedure: 1-naphthol replaces 1-naphthylamine and 4-nitroaniline replaces sulfanilic acid. Finally, a magnesium salt is added to the reagent. Under the alkaline conditions of the reaction, white magnesium hydroxide is produced, which will absorb the azo dye produced by the reaction of 1-naphthol and 4-nitroaniline and mask any transferred blood. To prepare for the test, desensitized photographic paper is soaked in a 0.25% (w/v) solution of 1-naphthol, 4-nitroaniline and magnesium sulfate in 1:1 aqueous ethanol. This paper is dried (and may be set aside for use at a later time). In order to perform the test, the prepared photographic paper is placed on a hard surface; the textile bearing the GSR pattern is placed pattern side down over the photographic paper and covered with a cloth moistened with 10% (v/v) acetic acid. A hot iron is used to press the stack of fabric and transfer nitrite residues to the treated photographic paper. To visualize the GSR pattern the photographic paper is treated with a 10% (w/v) aqueous sodium hydroxide solution. The nitrite pattern appears as blue flecks on a light yellow background.

The nitrite and vaporous lead deposits on the witness panels must also be visualized before comparisons are made with the crime scene pattern. The same procedures (modified Griess test, Maiti test or sodium rhodizonate test) must be applied to the witness panels as were applied to the questioned GSR pattern.

Attempts have been made to apply instrumental methods to the determination of muzzle-to-target distances. Atomic absorption (AA) spectrophotometry (Krishnan, 1974), neutron activation analysis (NAA) (Capannesi *et al.*, 1993), inductively coupled plasma-mass spectrometry (ICP-MS) (Santos *et al.*, 2007), the combination of confocal laser microscopy and inductively coupled plasma-atomic emission spectrometry (ICP-AES)

(Turillazzi *et al.*, 2013) and x-ray fluorescence (XRF) spectrometry (Fonseca *et al.*, 2014) have all been used to detect IGSR and estimate muzzle-to-target distances. Sharma and Lahiri (2009) reported the results of a study in which micro-FTIR was used to measure the percentage transmittance of nitroglycerine in the OGSR. The OGSR was extracted from 2 × 1 cm cuttings from cloth targets, evaporated to dryness and then incorporated in a potassium bromide pellet. The cuttings were obtained at specified distances from the bullet hole. The percentage transmittance of the asymmetric stretching band of the $-NO_2$ group of nitroglycerine at 1650 cm^{-1} decreased with increasing distance from the bullet hole and decreased with increasing muzzle-to-target distance. Zeng *et al.* (2014) used multiple vibration bands of ethyl and methyl centralite to visualize patterns of organic GSRs. Such chemical mapping could be used to visualize the distribution of organic residues around a bullet hole.

10.3.3 Collection of gunshot residue

GSR consists of deposits of both inorganic components and organic compounds. Little is known about the degradation of OGSR. Most IGSR particles are indefinitely stable; however, like other trace evidence they are subject to loss through transfer to other surfaces. Consequently, time is of the essence in the collection of GSR. Some law enforcement agencies have established time limits for collection of GSR. Once the time limits have passed, crime scene investigators will not collect GSR (Schwoeble and Exline, 2000).

GSR can be found in a variety of locations. Handguns leave deposits on the back and palms of the hands. Rifles and shotguns can leave GSR on the face and in head hair. A shooter wiping his hands on his clothing or putting his hands into his pockets can transfer GSR to his garments. GSR can be left on surfaces in close proximity to the weapon's discharge. In particular, in drive-by shootings in which a firearm is discharged inside the vehicle, GSR may be recovered from headliner, seats, headrests, dashboards, windows and doors (Schwoeble and Exline, 2000).

Law enforcement agencies differ in their GSR collection protocols. All agree, however, on the use of sterile gloves by collecting personnel to avoid contamination of the collection media. In many cases, the crime scene investigators who collect GSR are sworn officers who carry firearms; unless special precautions are taken the possibility exists that GSR from the officers' service weapons could contaminate the samples collected at the crime scene. Sampling materials and sampling protocols are, in part, dictated by the method for analyzing GSR that will be used in the forensic laboratory. At the present time, laboratory analyses of GSR focus on IGSR. If neutron activation analysis (NAA), atomic absorption spectrometry or inductively coupled plasma-mass spectrometry (ICP-MS) are to be used, swabs moistened with a 5% nitric acid solution would be used. Because IGSR dissolves in nitric acid, this method of collection cannot be used if scanning electron microscopy coupled with an energy-dispersive X-ray spectrometer (SEM-EDS) is to be used. Samples for SEM-EDS can be obtained using adhesive aluminum stubs, double-sided carbon tape or adhesive lifters. These sampling methods can be used on skin or on fabrics (Schwoeble and Exline, 2000).

10.3.3.1 Analysis of IGSR

The first test for the presence of GSR was the dermal nitrate test (also called the paraffin test, diphenylamine test or the Gonzales test). The use of diphenylamine in sulfuric acid to detect nitrates and nitrates had long been known and paraffin wax had been used to lift GSR on hands and clothing. In 1931,Tomas Gonzales modified the test by using melted paraffin to coat the shooter's hands (Wallace, 2008). Melted paraffin was applied with a brush to the shooter's hand. The paraffin was allowed to cool and then peeled off. The test reagent was applied to the interior surface of the paraffin cast. Dark blue specks were supposedly indicative of the presence of nitrates and nitrites. Unfortunately, a variety of common materials could give false positives: tobacco, tobacco ash, fertilizers (which usually contain sodium or ammonium nitrates), some pharmaceuticals, some paints (e.g., nitrocellulose-based primers), and urine (which contains nitrates) (Wallace, 2008). The test could also give equivocal results even when a person was known to have discharged a firearm (Turkel and Lipman, 1955). In 1968, Interpol recommended that the paraffin test no longer be used (Wallace, 2008).

Because of the large number of possible interferences with the paraffin test, Harrison and Gilroy (1959) developed a method using colorimetric spot tests for the lead, antimony and barium in IGSR. This combination of elements is common in firearm primers but is rare otherwise. Lead and antimony are also found together in bullet lead. The Harrison–Gilroy test begins with the swabbing of the shooter's hands with small squares of cotton cloth that have been moistened with 0.1 M hydrochloric acid. The cloth swab is allowed to dry and 1–2 drops of triphenylarsonium iodide are applied. If antimony is present an orange ring appears on the swab. The swab is dried and then two drops of a freshly prepared 5% (w/v) sodium rhodizonate solution are placed in the center of the orange ring. Both lead and barium will produce a red color with sodium rhodizonate. The swab is dried a third time under reduced illumination (intense illumination will bleach the red color). One to two drops of 1:20 hydrochloric acid are added to the red area. If lead is present the area wetted by the hydrochloric acid will turn blue; if barium is also present a red color will remain in the center of the swab. The Harrison–Gilroy method had the virtue that it focused on an uncommon combination of elements (lead, antimony and barium). However, its detection limits for the elements are not adequate for the reliable detection of GSR from actual firing (Wallace, 2008).

Until the advent of lead-free primers, instrumental analyses of GSR had focused on detection and quantitation of lead, antimony and barium because these elements are present in the majority of cartridge primers. NAA was the first instrumental method employed for the analysis of IGSR (Ruch *et al.*, 1964). In NAA the analytical samples are exposed by bombardment with slow neutrons in the core of a nuclear reactor. Certain isotopes of antimony and barium capture neutrons and are converted into radioactive isotopes (activated). After the samples are removed from the reactor, the gamma-ray emissions of the activated isotopes (in the case of IGSR, ^{131}Ba and ^{122}Sb) are measured with a gamma-ray spectrometer; the energies of the emitted gamma rays identify the activated isotope and the number of emitted gamma rays indicates the amount of activated isotope present in the sample. NAA has been widely used for the analysis of IGSR, despite the fact that it cannot detect lead and requires specially trained personnel and access to a nuclear reactor (Romolo and Margot, 2001).

AA permits the identification and quantitation of lead, as well as antimony and barium. Beer's Law, the fundamental law connecting the absorption of light with the concentration of the absorbing species, requires that the absorptivity or extinction coefficient of the absorbing species be constant over the range of wavelengths passing through the sample. Because the absorbing species are gas phase atoms, this condition can only be met by using a hollow cathode lamp. A separate light source is required for each element; lead, antimony and barium require separate analyses. Flame AA has detection limits for lead that are adequate for detection of that element on hand swabs; however, flame AA detection limits for antimony and barium are too high for this atomization technique to be useful for GSR analysis. Development of flameless atom- izers (carbon rods, tantalum strips and graphite furnaces) made AA usable for IGSR analysis (Romolo and Margot, 2001).

ICP-MS is another bulk analysis method for IGSR. Samples are dissolved in a solvent; the solution is aspirated into a plasma torch where it is atomized and ionized. The ions are skimmed from the plasma torch into a mass spectrometer (usually a quadrupole mass spectrometer). ICP-MS can be used to detect lead, antimony, barium and other elements in IGSR (Santos *et al.*, 2007; Steffen *et al.*, 2007; Souza Sarkis *et al.*, 2007; Abrego *et al.*, 2012; Freitas *et al.*, 2012); it can also be used to differentiate different types of bullets (Udey *et al.*, 2011).

If lead or the combination of lead, antimony and barium are detected on the hands of a suspect shooter, the interpretation of this finding as indicative of the presence of GSR is by no means straightforward. Lead is used in plumbing and is still found in older architectural paints. Antimony and barium can be found on the hands of electricians and automobile mechanics (Havekost *et al.*, 1990). Intensive research on the examina- tion of GSR using SEM-EDS showed that many potential environmental interferences could be avoided by paying attention to the shape and surface texture of the particles as well as their elemental composition (Romolo and Margot, 2001). A scanning electron microscope directs an electron beam at a sample. An image can be produced on a cathode ray tube by capturing secondary electrons (electrons ejected from the sample surface by the electron beam) or backscattered electrons (electrons from the beam deflected by high atomic number elements). If the energies of the electrons in the beam are high enough, they will eject inner shell electrons from atoms in the sample; electrons in higher energy electron shells will fall into the resulting holes with the emission of X-rays. Characteristic particles in IGSR are spheroidal with smooth surfaces (Schwoeble and Exline, 2000;

Table 10.2 Compositions of inorganic gunshot residue particles

Characteristic of gunshot residue	Consistent with gunshot residue	Reference
Pb-Sb-Ba	Pb-Sb	Wolten *et al.*, 1979a
Ba-Ca-Si (with traces of S)	Pb-Ba	
Ba-Ca-Si (with traces of Pb in the absence of Cu and Zn)	Pb	
	Ba (S at trace levels or below	
Sb-Ba	Sb	
Pb-Sb-Ba	Ba-Ca-Si (with and without traces of S)	ASTM, 2010
(may also contain Al, Si, P, S (trace), Cl, K, Ca, Fe (trace), Ni, Cu, Zn, Zr, Sn)	Sb-Ba (with no more than a trace of Fe or S	
	Pb-Sb	
	Ba-Al (with and without a trace of S)	
	Pb-Ba	
	Pb or Sb (only in the presence of particles having elemental compositions in the two columns of the table)	
	Ba (with or without a trace of S)	

Romolo and Margot, 2001). Table 10.2 lists the elemental compositions of individual particles that may be considered characteristic of IGSR or consistent with IGSR. Spheroidal particles containing elements heavier than sodium have been found to rare in the environment (Wolten *et al.*, 1979b). Blank cartridges, cartridge-actuated industrial tools and fireworks have all been considered as possible sources of inorganic-containing particles that could be confused with IGSR (Romolo and Margot 2001). Residue produced by blank cartridges is very similar to IGSR; almost all of the residues from cartridge-actuated devices contained iron; and fireworks produce residues rich in magnesium and/or copper. Lead-free ammunition can produce particles of GSR with the following elemental compositions: strontium, gadolinium-titanium-zinc or gallium-copper-tin (ASTM 2014). Other elements that can be present in residue produced by lead-free ammunition include titanium, zinc, aluminum, silicon, calcium, copper or tin.

10.3.3.2 Analysis of OGSR

Even before the advent of lead-free primers, researchers had begun to explore the feasibility of analyzing the organic components of GSR. Identifying these organic constituents along with lead, antimony and barium could place the identification of GSR on a firmer basis. Any analytical methods applied to the analysis of the organic component of GSR must have both high discriminating power as well as low detection limits. Some of the possible constituents of OGSR have similar chemical structures (e.g., ethyl and methyl centralite) and the quantities deposited on the hands of a shooter may be very small.

Mach *et al.* (1978a) developed a gas chromatography-chemical ionization mass spectrometry method for the analysis of volatile organic compounds (VOCs). These authors wanted to develop a method that was rapid and inexpensive but capable of definitively identifying GSR. Methane was used as the reagent gas for chemical ionization. Chemical ionization with methane results in the transfer of H^+ to the analyte molecules. This type of soft ionization produces limited fragmentation of the analyte molecules, resulting in lower detection limits. Nitroglycerin, in particular, is prone to complete fragmentation of the molecular ion when electron ionization is used. The authors then tested the feasibility of detecting VOCs on the hands of shooters (Mach *et al.*, 1978b), firing a variety of brands of ammunition in a number of firearms. Both acetone swabs and individual flakes deposited on shooters' hands were tested. Analysis of the acetone swabs for VOCs was not successful in this study. On the other hand, individual powder flakes were found to contain nitroglycerin, dinitrotoluene, diphenylamine, ethyl centralite and dibutyl phthalate. Dibutyl phthalate is a common plasticizer and, therefore, not diagnostic of GSR. On the other hand, ethyl centralite is used almost exclusively as a stabilizer in smokeless powders. It is important to note that the compounds detected derived from the cartridge propellant, not from the primer. Other researchers have also employed gas chromatography-mass spectrometry for the detection of VOCs in GSR (Speers *et al.*, 1994; Muller *et al.*, 2007; Weyermann *et al.*, 2009).

The instability of some of the compounds found in OGSR (e.g., nitroglycerin) makes gas chromatography and gas chromatography-mass spectrometry less than ideal methods of analysis because gas chromatography generally requires heating of samples to render them volatile. Room temperature separation techniques, such as liquid chromatography and capillary electrophoresis, have been used to detect VOCs in GSR. Liquid chromatography with ultraviolet detection has been applied to smokeless powders for many years. For example, Espinoza and Thornton (1994) used liquid chromatography with ultraviolet detection to identify diphenylamine and its family of nitrated derivatives in smokeless powders. They were able to detect diphenylamine of the hands of shooters only when shots were fired with larger caliber handguns (e.g., .38 caliber or higher). Espinoza and Thornton found that the detection limit of their method for diphenylamine was too high and suggested electrochemical detection or liquid chromatography-mass spectrometry would be better alternatives. Bratin and Kissinger (1981) compared ultraviolet and electrochemical detection in the liquid chromatographic analysis of VOCs in smokeless powders and in GSR. They used both oxidative and reductive electrochemical detection using a mercury film electrode and were successful in detecting a number of VOCs in smokeless powders. However, nitroglycerin and 2,4-dinitrotoluene were the only VOCs detected on shooters' hands. Their detection limits for oxidative electrochemical detection for diphenylamine and two of its nitrated derivatives (2-nitrodiphenylamine and 4-nitrodiphenylamine) were almost 1–2 orders of magnitude lower than the detection limits for ultraviolet detection. Speers *et al.* (1994) used liquid chromatography with a hanging mercury drop electrode as the detector. The electrode was set to a negative voltage so that the separated VOCs were reduced at the surface of the mercury drop.

Espinoza and Thornton (1994) suggested that liquid chromatography coupled to a mass spectrometer detector would provide lower detection limits for VOCs in GSRs. Yinon (2004) pointed out that LC-MS has become the method of choice for the analysis of trace levels of explosives in forensic laboratories. However, the explosive samples analyzed have been traces on bomb components or traces on the hands of persons who assembled bombs or handled them; VOCs in GSR have rarely been analyzed. Wu *et al.* (2001) developed a liquid chromatography-tandem mass spectrometry method for the analysis of GSR; their method focused on the identification of methyl centralite because of the rarity of this compound in the environment and because methyl centralite is more widely used in Chinese-made ammunition than ethyl centralite. In order to remove most interferences with the detection of methyl centralite, the analysis was performed in multiple reaction mode, in which a precursor ion is selected by the first quadrupole mass spectrometer and a product ion is selected by the second quadrupole mass spectrometer. The precursor ion used had a m/z of 241 and the product ion had a m/z of 134. The separation capabilities of the liquid chromatograph were not actually used. Laza *et al.* (2007) developed a liquid chromatograph tandem mass spectrometry method for the common propellant stabilizers (Arkadite II, diphenylamine, and ethyl and methyl centralite), as well as three degradation products of diphenylamine. The multiple reaction mode was used, with the first mass spectrometer scanning six different precursor ions and the second mass spectrometer scanning seven different product ions. When swabs from the hands of a shooter were analyzed by this method, diphenylamine was detected for all the brands of ammunition test fired, but in many cases at levels below the limit of quantitation (LOQ). Ethyl centralite was detected on swabs for only four of fifteen brands of ammunition test fired. Benito *et al.* (2015) recently reported the development of a polytetrafluoroethylene (PTFE) lifter for lifting GSR from the hands of shooters. Half of the lifter can be reserved for SEM-EDS analysis of IGSR; OGSR is extracted from the remaining half. The PTFE lifter was compared with swabbing as a means of GSR recovery. Shooters fired single rounds of four European brands of lead-free ammunition and the recovered OGSR was analyzed by liquid chromatography with a quadrupole time-of-flight mass spectrometer detector. In the OGSR collected with the PTFE lifter the researchers were able to detect diethyl phthalate plasticizer, the stabilizers ethyl and methyl centralite and 4-nitrosodiphenylamine, a degradation product of diphenylamine.

In capillary electrophoresis a high voltage is applied to a capillary column containing a buffered electrolyte solution. Electroosmotic flow is established in the capillary so that the electrolyte solution flows from the injector end to the detector end. For charged species, separation is based on electrophoretic mobilities. Uncharged species can be separated by capillary electrophoresis by addition of a low concentration of a detergent such as sodium dodecyl sulfate (SDS) to the buffered electrolyte solution. The detergent molecules will form micelles, spherical clusters in which the hydrophobic ends of the detergent molecules are in the interior and the hydrophilic ends are at the surface.

Neutral molecules can dissolve in the interior of the micelles and be carried along with the micelles. In micellar electrokinetic chromatography (MEKC) experimental conditions are arranged so that the micelles are negatively charged and migrate against the electroosmotic flow toward the injector end of the capillary. Neutral molecular species that have low solubility in the detergent micelles will travel toward the detector at the velocity of the electroosmotic flow. Molecules with higher solubility in the micelles will spend less time in the flowing buffer and so will take longer to reach the detector. MEKC is also called micellar electrokinetic capillary electrophoresis (MECE).

Ultraviolet or ultraviolet-visible absorption is the most common means of detecting analytes; however, laser-induced fluorescence is used in capillary gel electrophoresis of DNA. Capillary electrophoresis can also be coupled with electrospray ionization mass spectrometry or with surface-enhanced Raman spectroscopy (SERS). These detection modes provide additional structural information about molecular analytes. Capillary electrophoresis has much higher resolution than either gas chromatography or liquid chromatography. Northrup *et al.* (1991) developed a MEKC method for nitroglycerin, diphenylamine, nitrated diphenylamine derivatives, ethyl centralite, dinitrotoluene isomers and dibutyl phthalate plasticizers and demonstrated its effectiveness in the analysis of residues from expended cartridges. Subsequently, Northrup and MacCrehan (1992) developed a tape lift procedure for the recovery of GSR from hands for MEKC analysis of OGSR. Northrup (2001a) reported a double-sided tape lifting procedure for GSR. The lift could be analyzed by SEM-EDS for the inorganic components of GSR followed by extraction and MEKC analysis of OGSR. In this article Northrup also defined what he termed characteristic organic gunpowder components (COGC); these were the dinitrotoluene isomers, nitrated diphenylamine degradation products, dibutyl phthalate, diethyl phthalate, diphenylamine, ethyl and methyl centralite, and nitroglycerin. Tape lifts taken from one hundred individuals in the general population were analyzed for COGC. Those sampled represented a wide variety of occupations. No COGCs were found on any of these tape lifts. Northrup (2001b) described the application of the double-sided tape lift procedure to casework. Lifts were taken from both hands and clothing. Lifting OGSR from clothing was more successful than lifts from hands. Bernal Morales and Ravilla Vázquez (2004) developed a MEKC method with ultraviolet detection for the simultaneous determination of both IGSR and OGSR. The metal ions were complexed with diaminocyclohexane tetraacetic acid.

Capillary electrochromatography (CEC) combines features of capillary electrophoresis and liquid chromatography. The capillary is filled with stationary phase and the sample is propelled through the capillary by electroosmotic flow. De Perre *et al.* (2012) created a hexyl acrylate monolithic capillary column and used it to separate the organic components of smokeless powders. Both ultraviolet absorption and time-of-flight mass spectrometry were used for detection. This method has not yet been applied to OGSR lifted from skin or clothing.

Rarely have analytical methods for VOCs in smokeless powders and GSR been directly compared. Cascio *et al.* (2004) compared reversed phase liquid chromatography and MEKC using a standard mixture of smokeless powder VOCs. The two techniques were found to resolve all compounds in the standard mixture; they also had comparable elution/migration time reproducibility. MEKC did show better reproducibility of peak areas.

Direct analysis methods for mass spectrometry allow analysts to dispense with preliminary separation of analytes via gas chromatography, liquid chromatography or capillary electrophoresis. Direct analysis methods also do not require elaborate sample preparation. In desorption electrospray ionization (DESI) a mist of ionized solvent droplets is directed at a sample surface. Analyte molecules are ionized, ejected from the sample surface and swept into the inlet of a mass spectrometer. Zhao *et al.* (2008) applied DESI to the hands of shooters and analyzed the sampled ions with tandem mass spectrometry. They successfully detected ethyl and methyl centralite on the hands of shooters. Morelato *et al.* (2012) applied DESI tandem mass spectrometry to adhesive tape stubs used to collect IGSR samples for SEM-EDS analysis. GSR from thirteen different brands of ammunition was analyzed and six of the stubs showed the presence of ethyl centralite. Methyl centralite, diphenylamine and nitrated diphenylamine were not detected. DESI did not interfere with the examination of the stubs with SEM-EDS.

Vibrational spectroscopies have also been applied to the analysis of OGSR. López-López *et al.* (2012) used Raman scattering to examine individual flakes of smokeless powders removed from cloth targets. The powder flakes were dissolved in methyl ethyl ketone and the resulting solution was dried on a microscope slide for Raman analysis. The Raman spectrum of one of these flakes is basically the Raman spectrum of nitrocellulose, which is not surprising given that nitrocellulose comprises the bulk of a grain of smokeless powder and that other organic constituents are present in low concentrations. Other nitrate-ester explosives have similar Raman spectra. Abrego *et al.* (2014) combined Raman microspectroscopy with scanning laser ablation-inductively coupled plasma mass spectrometry (SLA-ICPMS) to analyze GSR particles on tape lifts from skin surfaces for constituents of GSR from lead-free ammunition. SLA-ICPMS successfully identified elemental combinations characteristic of the lead-free primers in the various brands of ammunition used: Al–Ti, Cu–Zn, Cu–Zn–Sn and Sr–Zr. The elemental combination Pb–Sb–Ba was detected in all samples, indicating a "memory" effect due to previously firing ammunition with lead-based primers in the firearm used to fire the lead-free ammunition. The Raman spectra of the GSR particles showed various combinations of organic compounds, such as diphenylamine-centralites, diphenylamine-nitrated diphenylamine and diphenylamine-nitrated diphenylamine-centralites. Bueno and Lednev (2014) examined GSR particles on clothing using attenuated total reflectance -Fourier transform infrared spectrometry (ATR-FTIR). Micro ATR-FTIR image maps were created showing the distribution of OGSR. Because nitrocellulose predominates in grains of smokeless powders, the ATR-FTIR spectra of particles of GSR shows principally the infrared absorption bands of nitrate esters, which do not uniquely identify GSR.

10.4 Conclusion

Although firearms examinations rely primarily on microscopical examinations of bullets and cartridges, chemistry plays an important role in this forensic science discipline. Chemistry provides the means to clean bullets and cartridges, restore serial numbers, detect bullet holes at crime scenes, visualize powder patterns and detect both the inorganic and organic constituents of gunshot residue.

References

Abrego, Z., Ugarte, A., Unceta, N., *et al.* (2012) Unambiguous characterization of gunshot residue particles using scanning laser ablation and inductively coupled plasma-mass spectrometry. *Analytical Chemistry*, **84**, 2402–2409.

Abrego, Z., Grijalba, N., Unceta, N., *et al.* (2014) A novel method for the identification of inorganic and organic gunshot residue particles of lead-free ammunitions from the hands of shooters using scanning laser ablation-ICPMS and Raman micro-spectroscopy. *Analyst*, **139**, 6232–6241.

Anonymous (2003) Dealloying. *Materials Performance*, **42**(5), 66.

AFTE (Association of Firearm and Tool Mark Examiners) (2014a) *What is Firearm and Tool Mark Examination?* http://www.afte.org/AssociationInfo/comm%20&%20info/what%20is.htm (last accessed 27 June 2015)

AFTE (Association of Firearm and Tool Mark Examiners) (2014b) *Range of Conclusions*, http://www.afte.org/ (last accessed 30 June 2015).

ASTM (2010) ASTM E1588-10e1, *Standard Guide for Gunshot Residue Analysis by Scanning Electron Microscopy/ Energy Dispersive X-ray Spectrometry*. ASTM International, West Conshohocken, PA.

Baharum, M.I.M., Kuppuswamy, R, and Rahman, A.A. (2008) Recovering obliterated engraved marks on aluminium surfaces by etching technique. *Forensic Science International*, **177**, 221–227.

Benito, S., Abrego, Z., Sanchez, A., *et al.* (2015) Characterization of organic gunshot residues in lead-free ammunition using a new sample collection device for liquid chromatography–quadrupole time-of-flight mass spectrometry. *Forensic Science International*, **246**, 79–85.

Bernal Morales, E. and Ravilla Vázque, A.L. (2004) Simultaneous determination of inorganic and organic gunshot residues by capillary electrophoresis. *Journal of Chromatography A*, **1061**, 225–233.

Booker, J.L. (1980) Examination of the badly damaged bullet. *Journal of the Forensic Science Society*, **20**, 153–162.

Bottegal, M., Lang, L., Miller, M. and McCord, B. (2010) Analysis of ascorbic acid based black powder substitutes by high-performance liquid chromatography/electrospray ionization quadrupole time-of-flight mass spectrometry. *Rapid Communications in Mass Spectrometry*, **24**, 1377–1386.

Braconnot, H. (1833) De la transformation de plusieurs substances végétales en un principe nouveau. *Annales de Chimie et de Physique*, **52**, 290–294.

Bratin, K. and Kissinger, P.T. (1981) Determination of nitro aromatic, nitramine, and nitrate ester explosive compounds in explosive mixtures and gunshot residue by liquid chromatography and reductive electrochemical detection. *Analytica Chimica Acta*, **130**, 295–311.

Bueno, J. and Lednev, I.K. (2014) Attenuated total reflectance-FT-IR imaging for rapid and automated detection of gunshot residue. *Analytical Chemistry*, **86**, 3389–3396.

Caley, E.R. (1955) Coatings and incrustations on lead objects from the Agora and the method used for their removal. *Studies in Conservation*, **2**(2), 49–54.

Capannesi, G., Ciavola, C. and Sedda, A.F. (1993) Determination of firing distance and firing angle by neutron activation analysis in a case involving gunshot wounds. *Forensic Science International*, **61**, 75–84.

Cascio, O., Trettene, M., Bortolotti, F., *et al.* (2004) Analysis of organic components of smokeless gunpowders: High-performance liquid chromatography vs. micellar electrokinetic capillary chromatography, *Electrophoresis*, **25**, 1543–1547.

CCI/SPEER (2008) CCI/SPEER Rifle, Pistol Primers – CLEAN-FIRE: Material Safety Data Sheet. http://glarp.atk.com/2008/msds/Primers_Clean-Fire.pdf (last accessed 26 June 2015).

Ceccaldi, P.F. (1962) Examination of firearms and ammunition. In: *Methods of Forensic Science*, Vol. **1** (ed. F. Lunquist). Interscience Publishers, New York, pp. 593–637.

Chen, M. and Daroub, S.H. (2002) Characterization of lead in soils of a rifle/pistol shooting range in Central Florida, USA. *Soil and Sediment Contamination*, **11**(1), 1–17.

Committee on Scientific Assessment of Bullet Lead Elemental Composition Comparison (2004) *Forensic Analysis: Weighing Bullet Lead Evidence*. The National Academies Press, Washington, DC.

Committee on Smokeless and Black Powder (1998) *Black and Smokeless Powders: Technologies for Finding Bombs and Bomb Makers*. The National Academies Press, Washington, DC.

da Silva, L. and dos Santos, P.A.M. (2008) Recovering obliterated laser engraved serial numbers in firearms. *Forensic Science International*, **179**, e63–e66.

de Perre, C., Corbin, I., Blas, M. and McCord, B.R. (2012) Separation and identification of smokeless gunpowder additives by capillary electrochromatography. *Journal of Chromatography A*, **1267**, 259–265.

Dillon, J.H. (1990) The modified Griess test: A chemically specific chromophoric test for nitrite compounds in gunshot residues. *AFTE Journal*, **22**, 243–250.

DiMaio, V.J.M. (1999) *Gunshot Wounds: Practical Aspects of Firearms, Ballistics, and Forensic Techniques*, 2nd edn. CRC Press, Boca Raton, FL.

Doyle, J.S. (2014) An Introduction to Firearms Identification. http://www.firearmsid.com/ (last accessed 27 June 2015).

Eastman Kodak Company (1972) *Applied Infrared Photography*. Eastman Kodak Company, Rochester, NY.

Espinoza, E.O. and Thornton, J.I. (1994) Characterization of smokeless gunpowder by means of diphenylamine stabilizer and its nitrated derivatives. *Analytica Chimica Acta*, **288**, 57–69.

Federal Cartridge Company (2010) Ballisticlean Centerfire Rifle & Pistol Ammunition: Material Safety Data Sheet. http://federalpremium.com/downloads/msds/F3005_Ballisticlean_223_Rem_Centerfire_Rifle_Ammunition.pdf (last accessed 27 June 2015).

Fischbein, A., Rice, C., Sarkozi, L., *et al.* (1979) Exposure to lead in firing ranges. *Journal of the American Medical Association*, **241**(11), 1141–1144.

Fonseca, J.F., Cruz, M.M. and Carvalho, M.L. (2014) Muzzle-to-target distance determination by X-ray fluorescence spectrometry. *X-Ray Spectrometry*, **43**, 49–55.

Freitas, J.C., Souza Sarkis, J.E., Neto, O.N., and Viebig, S.B. (2012) Identification of gunshot residues in fabric targets using sector field inductively coupled plasma mass spectrometry technique and ternary graphs. *Journal of Forensic Sciences*, **57**(2), 503–508.

Garrison, D.H. (2003) *Practical Shooting Scene Investigation: The Investigation and Reconstruction of Crime Scenes Involving Gunfire*. Universal Publishers, Inc., Boca Raton, FL.

GE Healthcare Life Sciences (2014). Benchkote. http://www.gelifesciences.com/webapp/wcs/stores/servlet/productById/en/GELifeSciences/28418846 last accessed 27 June 2015).

Glattstein, B., Vinokurov, A., Levin, N., and Zeichner, A. (2000a) Improved method for shooting distance estimation. Part 1. Bullet holes in clothing items. *Journal of Forensic Sciences*, **45**(4), 801–806.

Glattstein, B., Zeichner, A., Vinokurov, A., and Shoshani, E. (2000b) Improved method for shooting distance determination. Part 2 – Bullet holes in objects that cannot be processed in the laboratory. *Journal of Forensic Sciences*, **45**(5), 1000–1008.

Glattstein, B., Zeichner, A., Vinokurov, A., *et al.* (2000c) Improved method for shooting distance estimation. *Part III. Bullet holes in cadavers. Journal of Forensic Sciences*, **45**(6), 1243–1249.

Goodpaster, J.V. and Keto, R.O. (2004) Identification of ascorbic acid and its degradation products in black powder substitutes. *Journal of Forensic Sciences*, **49**(3), 1–6.

Greener, W.W. (1910) *The Gun and Its Development*, 9th edn. Cassell and Company, Ltd., London.

Guinn, V.P. (1979) JFK assassination: bullet analyses. *Analytical Chemistry*, **51**, 484A–493A.

Guinn, V.P. (1982) NAA of bullet-lead specimens in criminal cases. *Journal of Radioanalytical Chemistry*, **72**, 645–663.

Guldbeck, P.E. (1972) *The Care of Historical Collections: A Conservation Handbook for the Nonspecialist.* America Association for State and Local History, Nashville, TN.

Haag, L.C. (2006) *Shooting Incident Reconstruction*. Academic Press, Burlington.

Harrison, H.C. and Gilroy, R. (1959) Firearm discharge residues. *Journal of Forensic Sciences*, **4**(2), 184–199.

Hatcher, J.S., Jury, F.J. and Weller, J. (1977) *Firearms Investigation, Identification and Evidence*. Stackpole Books, Harrisburg.

Havekost, D.G., Peters, R.D., and Koons, R.D. (1990) Barium and antimony distributions on the hands of non-shooters. *Journal of Forensic Sciences*, **35**, 1096–1114.

Heard, B.J. (2008) *Handbook of Firearms and Ballistics: Examining and Interpreting Forensic Evidence*. John Wiley & Sons, Ltd, Chichester.

Heramb, R.M. and McCord, B.R. (2002) The manufacture of smokeless powders and their forensic analysis: A brief review. *Forensic Science Communications*, **4**(2). http://www.fbi.gov/about-us/lab/forensic-science-communications/fsc/april2002/mccord.htm (last accessed 27 June 2015).

Howard, E. (1800) On a new fulminating mercury. *Philosophical Transactions of the Royal Society*, **90**(1), 204–238.

Jinturkar, P., Guan, Y.C., and Han, K.N. (1998) Dissolution and corrosion inhibition of copper, zinc and their alloys. *Corrosion*, **54**(2), 106–114.

Koons, R.D. and Grant, D.M. (2002) Compositional variation in bullet lead manufacture. *Journal of Forensic Sciences*, **47**(5), 950–959.

Krishnan, S.S (1974) Firing distance determination by atomic absorption spectrophotometry. *Journal of Forensic Sciences*, **19**(2), 351–356.

Lang, G.L. and Boyle, K.M. (2009) The analysis of black powder substitutes containing ascorbic acid by ion chromatography/mass spectrometry. *Journal of Forensic Sciences*, **54**(6), 1315–1322.

Larson, S. L., Martin, W.A., Griggs, S., *et al.*, (2011) Comparison of lead dissolution from antique and modern ammunition. *Environmental Forensics*, **12**, 149–155.

Laza, D., Nys, B., De Kinder, J., *et al.* (2007) Development of a quantitative LC-MS/MS method for the analysis of common propellant powder stabilizers in gunshot residue. *Journal of Forensic Sciences*, **52**(4), 842–850.

Lekstrom, J.A. and Koons, R.D. (1986) Copper and nickel detection on gunshot targets by dithio-oxamide test. *Journal of Forensic Sciences*, **31**(4), 1283–1291.

López-López, M., Delgado, J.J., and García-Ruiz, C. (2012) Ammunition identification by means of the organic analysis of gunshot residues using Raman spectroscopy. *Analytical Chemistry*, **84**, 3581–3585.

Lowry, E.D. (1968) *Interior Ballistics: How a Gun Converts Chemical Energy into Projectile Motion.* Doubleday & Company, Inc., Garden City.

Mach, M.H., Pallos, A., and Jones, P.F. (1978a) Feasibility of gunshot residue detection via its organic constituents. Part I: analysis of smokeless powders by combined gas chromatography-chemical ionization mass spectrometry. *Journal of Forensic Sciences*, **23**(3), 433–445.

Mach, M.H., Pallos, A., and Jones, P.F. (1978b) Feasibility of gunshot residue detection via its organic constituents. Part II: analysis of smokeless powders by combined gas chromatography-chemical ionization mass spectrometry. *Journal of Forensic Sciences*, **23**(3), 446–455.

Maehly, A. and Stromberg, L. (1981) *Chemical Criminalistics.* Springer-Verlag, New York.

Maiti, P.C. (1973) Powder patterns around bullet holes in bloodstained articles. *Journal of the Forensic Science Society*, **13**, 197.

Mathews, J.H. (1962) *Firearms Identification*, Vol. **1**. Charles C. Thomas, Springfield, IL.

Meng, H.H. and Caddy, B. (1997) Gunshot residue analysis – A review. *Journal of Forensic Sciences*, **42**(4), 553–570.

Morelato, M., Beavis, A., Ogle, A., *et al.* (2012) Screening of gunshot residues using desorption electrospray ionisation–mass spectrometry (DESI–MS). *Forensic Science International*, **217**, 101–106.

Muller, D., Levy, A., Vinokurov, A., *et al.* (2007) A novel method for the analysis of discharged smokeless powder residues. *Journal of Forensic Sciences*, **52**(1), 75–78.

National Forensic Science Training Center (2015) Evolution of the Modified Griess Test. http://projects.nfstc.org/firearms/embed_swf.htm?link=module12/images/fir_m12_t05_03_a.swf (last accessed 27 June 2015).

Northrup, D.M. (2001a) Gunshot residue analysis by micellar electrokinetic capillary electrophoresis: assessment for application to casework. *Part I. Journal of Forensic Sciences*, **46**(3), 549–559.

Northrup, D.M. (2001b) Gunshot residue analysis by micellar electrokinetic capillary electrophoresis: assessment for application to casework. *Part II. Journal of Forensic Sciences*, **46**(3), 560–572.

Northrup, D.M. and MacCrehan, W.A. (1992) Sample collection, preparation, and quantitation in the micellar electrokinetic capillary electrophoresis of gunshot residues. *Journal of Liquid Chromatography*, **15**(6), 1041–1062.

Northrup, D.M., Martire, D.E., and MacCrehan, W.A. (1991) Separation and identification of organic gunshot and explosive constitiuents by micellar electrokinetic capillary electrophoresis. *Analytical Chemistry*, **63**, 1038–1042

Ozonoff, D. (1994) Lead on the range. *The Lancet*, **343**, 6.

Partington, J.R. (1998). *A History of Greek Fire and Gunpowder.* Johns Hopkins University Press, Baltimore, MD.

Powell, G.L.F. (1976) Comments on the mechanism of the restoration of obliterated serial numbers by acid etching. *Journal of the Forensic Science Society*, **16**(4), 349.

Querna, B. (2005) FBI bullet-lead test terminated. *Analytical Chemistry*, **77**(21), A414.

Rahn, K.A. and Sturdivan, L.M. (2004) Neutron activation and the JFK assassination. *Part I. Data and interpretation. Journal Radiological and Nuclear Chemistry*, **262**(1), 205–213.

Randich, E. and Grant, P.M. (2006) Proper assessment of the JFK assassination bullet lead evidence from metallurgical and statistical perspectives. *Journal of Forensic Sciences*, **51**(4), 717–728.

Randich, E., Fickies, T.E., Tulleners, F.A. *et al.* (2000) Restoration tactics for seriously corroded Cu and Cu-alloy firearms evidence. *Journal of Forensic Sciences*, **45**(6), 1316–1319.

Remington Arms (2001) Remington Arms: Material Safety Data Sheets. http://www.remington.com/pages/news-and-resources/safety-center/material-safety-data-sheets.aspx (last accessed 27 June 2015).

Romolo, F.S. and Margot, P. (2001) Identification of gunshot residue: a critical review. *Forensic Science International*, **119**, 195–211.

Routon, B.J., Kocher, B.B., and Goodpaster, J.V. (2011) Discriminating Hodgdon Pyrodex® and Triple Seven® using gas chromatography-mass spectrometry. *Journal of Forensic Sciences*, **56**(1), 194–199.

Rowe, W.F. (2005) Firearms identification. In: *Forensic Science Handbook*, Vol. **II**, 2nd edn (ed. R. Saferstein). Pearson Education, Inc., Upper Saddle River, NJ, pp. 401–486.

Ruch, R.D., Buchanan, J.D., Guinn, V.P., *et al.* (1964) Neutron activation analysis in scientific crime detection. *Journal of Forensic Sciences*, **9**, 119–132.

Santos, A., Magalhàes, T., Vieira, D.N., *et al.* (2007) Firing distance estimation through the analysis of the gunshot residue deposit pattern around the bullet entrance hole by inductively coupled plasma-mass spectrometry an experimental study. *The American Journal of Forensic Medicine and Pathology*, **28**(1), 24–30.

Schönbein, C.F. (1846) Über Schiesswolle. *Bericht über die Verhandlungen der Naturforschenden Gesellschaft in Basel*, **7**, 27.

Schwoeble, A.J. and Exline, D. (2000) *Current Methods in Forensic Gunshot Residue Analysis*. CRC Press, Boca Raton, FL.

Sharma, S.P. and Lahiri, S.C. (2009) A preliminary investigation into the use of FTIR microscopy as a probe for the identification of bullet entrance holes and distance of firing. *Science and Justice*, **49**, 197–204.

Souza Sarkis, J.E., Neto, O.N., Viebig, S., and Durrant, S.F. (2007) Measurements of gunshot residues by sector field inductively coupled plasma mass spectrometry – Further studies with pistols. *Forensic Science International*, **172**, 63–66.

Speers, S.J., Doolan, K., McQuillan, J., and Wallace, J.S. (1994) Evaluation of improved methods for the recovery and detection of organic and inorganic cartridge discharge residues. *Journal of Chromatography A*, **674**, 319–327.

Steffen, S., Otto, M., Niewoehner, L., *et al.* (2007) Chemometric classification of gunshot residues based on energy dispersive X-ray microanalysis and inductively coupled plasma analysis with mass-spectrometric detection. *Spectrochimica Acta Part B*, **62**, 1028–1036.

Steinberg, M., Leist, Y., and Tassa, M. (1984) A new field kit for bullet hole identification. *Journal of Forensic Sciences*, **29**(1), 169–176.

SWGGUN (2013) Guidelines for Gunshot Residue Distance Determinations. http://www.swggun. org/index.php?option=com_content&view=article&id=50:guidelines-for-gunshot-residue-distance-determinations-&catid=10:guidelines-adopted (last accessed 27 June 2015).

Thornton, J.I. and Cashman, P.J. (1976) The mechanism of the restoration of obliterated serial numbers by acid etching. *Journal of the Forensic Science Society*, **16**, 69–71.

Turillazzi, E., Di Peri, G.P., Nieddu, A., *et al.* (2013) Analytical and quantitative concentration of gunshot residues (Pb, Sb, Ba) to estimate entrance hole and shooting-distance using confocal laser microscopy and inductively coupled plasma atomic emission spectrometer analysis: An experimental study. *Forensic Science International*, **231**, 142–149.

Turkel, H.W. and Lipman, J. (1955) Unreliability of dermal nitrate test for gunpowder. *Journal of Criminal Law, Criminology, and Police Science*, **46**(2), 281–284.

Udey, R.N., Hunter, B.C., and Smith, R.W. (2011) Differentiation of bullet type based on the analysis of gunshot residue using inductively coupled plasma mass spectrometry. *Journal of Forensic Sciences*, **56**(5), 1268–1276.

Vinokurov, A., Zeichner, A., Glattstein, B., *et al.* (2001) Machine washing or brushing of clothing and its influence on shooting distance estimation. *Journal of Forensic Sciences*, **46**(4), 928–933.

Walker, J.T. (1940) Bullet holes and chemical residues in shooting cases. *Journal of Criminal Law and Criminology*, **31**(4), 497–521.

Wallace, J.S. (2008) *Chemical Analysis of Firearms, Ammunition, and Gunshot Residue*. CRC Press, Boca Raton, FL.

Warlow, T. (2011) *Firearms, the Law, and Forensic Ballistics*, 3rd edn. CRC Press, Boca Raton, FL.

Weyermann, C., Belaud, V., Riva, F., and Romolo, F.S. (2009) Analysis of organic volatile residues in 9 mm spent cartridges. *Forensic Science International*, **186**, 29–35.

Wheeler, B.P. and Wilson, L.J. (2008) *Practical Forensic Microscopy: A Laboratory Manual*. John Wiley & Sons, Ltd, Chichester.

Winchester (2013) Olin Winchester Ammunition: Material Safety Data Sheet, http://www.winchester.com/learning-center/ballistic-silvertip/Pages/Material-Safety-Data-Sheets.aspx (last accessed 27 June 2015).

Wolfer, D.A. and Lee, W.J. (1960) Application of magnetic principles to the restoration of serial numbers. *Journal of Criminal Law, Criminology and Police Science*, **50**, 519.

Wolten, G. M., Nesbitt, R.S., Calloway, A.R., *et al.* (1979a) Particle analysis for the detection of gunshot residue. *I. Scanning electron microscopy/energy dispersive X-ray characterisation of hand deposits from firing*. *Journal of Forensic Sciences*, **24**, 409–422.

Wolten, G. M., Nesbitt, R.S., Calloway, A.R., *et al.* (1979b) Particle analysis for the detection of gunshot residue. *II. Occupational and environmental particles*. *Journal of Forensic Sciences*, **24**, 423–430.

Wu, Z., Tong, Y., Yu, J., *et al.* (2001) The utilization of MS-MS method in detection of GSRs. *Journal of Forensic Sciences*, **46**(3), 495–501.

Yinon, J. (2004) Analysis of explosives by LC/MS. In: *Advances in Forensic Applications of Mass Spectrometry* (ed J. Yinon), CRC Press, Boca Raton, FL, pp. 231–274.

Zaili, M.A.M., Kuppuswamy, R., and Harun, H. (2007) Restoration of engraved marks on steel surfaces by etching technique. *Forensic Science International*, **171**, 27–32.

Zeng, J., Qi, J., Bai, F., *et al.* (2014) Analysis of ethyl and methyl centralite vibrational spectra for mapping organic gunshot residues. *The Analyst*, **139**, 4270–4278.

Zhao, M., Zhang, S., Yang, C., *et al.* (2008) Desorption electrospray tandem MS (DESI-MSMS) analysis of methyl centralite and ethyl centralite as gunshot residues on skin and other surfaces. *Journal of Forensic Sciences*, **53**(4), 807–811.

CHAPTER 11
Forensic microscopy

Christopher S. Palenik
Microtrace LLC, USA

11.1 The microscope as a tool

A serial rapist is on the loose. The police have DNA but no database matches. In two of the attacks, athletic jerseys from the suspect were recovered. The jerseys were submitted for analysis to a laboratory without comparison samples, a form of evidence most crime laboratories will not accept (Figure 11.1a and 11.1b). The request: determine if anything can be established about the identity a stain on one of the jerseys. Following a complete analysis of the dust recovered from the clothing, investigators were presented with the following leads: one of the crimes occurred during March or April, and the perpetrator is a drywall installer who works on large scale, indoor commercial projects.

Late one night, a short time after this description was released to the public, police were notified of a suspicious man observed by a retired woman through her apartment window. When pulled over a short time later in a commercial drywall truck and confronted by the police about the serial attacks, the man responded, "I knew it was only a matter of time before you found me." Although this confession was later withdrawn, he was convicted of several attacks on the basis of DNA evidence (Grice, 2002; Palenik, 2007a).

This seemingly apocryphal story is entirely true. The investigative leads that lead to the solution of this matter were obtained entirely by one instrument: a polarized light microscope. While this is an advanced application of light microscopy, this example illustrates many of the strengths that this technique can, by itself, provide. The analysis was fast, the particles were identified specifically, no comparative samples were required, it was completed with relatively inexpensive analytical equipment, and it provided information that DNA could not provide. The specific analyses upon which this conclusion was formed are presented throughout the chapter.

Forensic Chemistry: Fundamentals and Applications, First Edition. Edited by Jay A. Siegel.
© 2016 John Wiley & Sons, Ltd. Published 2016 by John Wiley & Sons, Ltd.

(a)   (b)

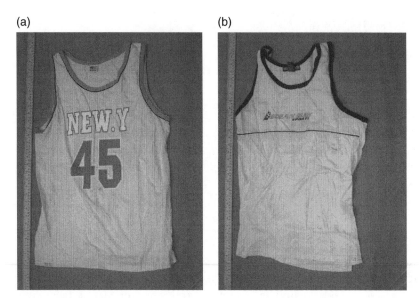

Figure 11.1 Two jerseys recovered from separate crime scenes as they appeared upon receipt at the laboratory (*See insert for color representation of the figure.*)

Although this level of analysis is presently, and unfortunately, outside of the approved scope of most forensic laboratories, the principal concepts of this analysis can be drawn upon and applied to any microscopical examination with only basic training in microscopy. Even the most rudimentary application of light microscopy through the relatively low magnification of a stereo microscope has the potential to provide new insight into many cases and improve the level and quality of information obtained from trace evidence analyses.

Following a few introductory sections, this chapter is divided into five sections: Find, Prepare, Look, Analyze, and Think. These section titles refer to the basic considerations that should be thoughtfully addressed in any analysis, though here they are written as they apply specifically to the application of a microscope to the examination of trace evidence.

11.2 Motivation

The decision to include a chapter on microscopy amidst a book focused entirely on various types of trace evidence (hair, fiber, paint, etc.) may seem odd, particularly when the instrument is one that has been, in many protocols and guidelines, relegated to the second tier of analytical methods. Besides being the central basis for nearly every microanalytical technique utilized in the forensic laboratory, the goal of the chapter is to demonstrate a wide variety of practical and accessible ways in which stereo microscopy, polarized light microscopy (PLM), and scanning electron microscopy (SEM) can be advantageously applied in a modern trace evidence laboratory.

The hope is that this chapter will help the reader begin to develop an appreciation for the ways in which microscopy, even in the hands of a beginner, can provide an immediate and high return on investment in aspects of trace evidence that include: finding and isolating evidence; properly preparing it for further study; characterizing and identifying it, and using the information to formulate and test further hypotheses.

Finally, many responsible for defining forensic policy have arrived at the conclusion that automated sample processing, automated data collection, and automated interpretation are the ultimate goal of forensic analysis (as has effectively become the case in DNA analysis, the so called "gold standard" of forensic evidence). In many areas of forensic science, this hands-off approach (or portions of it) may be relevant and important. In trace evidence, however, the sheer variability of materials and the magnitude of ways in which the significance of these materials may be encountered illustrates why, for the foreseeable future, attempts to remove manual (i.e., human) aspects of trace evidence analysis are not only impractical but will (and do) result in a loss of information and a likely reduction in the evidentiary value obtained from that evidence. This has been shown time and again when reviewing crime laboratory data collected by thoughtless and blind (both mentally and microscopically) application of protocol-based methods. In complete contrast to the prescribed approaches of a protocol, the visual aspect of microscopy requires a more freestyle interaction with a sample. This interaction allows the trained mind to thoughtfully consider the visual significance of the microscopic information in the same way that a good detective might survey a crime scene or the overall scope of a case. The value of this unscripted evaluation in the context of a trace evidence analysis cannot be overstated. For example, the jerseys cited in the introductory story of this chapter were originally submitted to the laboratory for the purpose of identifying a stain that was noted. The hope was that the stain might be indicative of a recently consumed food or restaurant that had been visited. It was only after a more detailed examination of the otherwise invisible background dust lightly covering the entirety of both jerseys that investigative clues began to develop. Unfortunately, paths such as the one that lead to the decision to examine the dust on the jersey are all but eliminated by protocols that require a comparison sample or are limited to the analysis of a select number of items, such as hair, fiber, glass and paint.

It is important to remember that this thoughtful approach is not limited to advanced polarized light microscopy. It can be applied equally to the initial inspection of evidence by stereomicroscopy or to the high resolution of a small particle in the SEM. There is no doubt to the significance that a microscopic examination can provide and the requisite, though often qualitative, human consideration that is necessary to interpret this image. Such an approach must be retained and embraced in all aspects of forensic science, even at the cost of including a method that is not as amenable to the stepwise, turnkey manner of a standard operating procedure. Although this is clearly an opinion amidst a textbook of factual information, it is important to make this point to provide readers with a context in which to consider the information and approach presented in this chapter.

11.2.1 Intimidation

While it is well and good to talk about the benefits of microscopy, many practitioners abandon the microscope because they feel they do not have the theoretical background or experience to properly utilize it. While the full use of any analytical method should be based on a strong theoretical knowledge of light or electron optics, the interaction of photons and electrons with matter, and a strong understanding of the materials and properties that will be studied, a great deal of good can be done with the microscope whilste developing one's knowledge of such concepts. This chapter intends to provide a softer approach towards microscopy, one that illustrates ways in which even a novice can use microscopy to obtain useful information and (hopefully) help them to address some of the commonly encountered pitfalls encountered with its application.

11.2.2 Limitations

This chapter intends to introduce students to and remind practitioners of the various benefits of light and electron microscopy as they exist outside of the prescribed protocols that govern many classes and laboratories. This is not a chapter on theory. There are many excellent references on the theory of microscopy and optical crystallography (McCrone *et al.*, 1978; Bloss, 1999; Goldstein *et al.*, 2007) as well as options for courses (e.g., The McCrone Research Institute). This chapter is meant to serve as an introduction to various ways in which microscopy (stereo, polarized light, and electron microscopy) may be used to advance a forensic investigation and also to illustrate ways in which a practitioner may utilize the various capabilities of microscopy advantageously and practically. While this chapter alone will not make the reader a microscopist, it is hoped that the possibilities discussed here will encourage and help to refine the practical application of microscopy in a crime laboratory setting.

11.3 Scale

Before moving on to the microscope itself, it is important as a microscopist to develop a sense of scale. What is the diameter of a human head hair? (about 70 micrometers). How thick is a typical garbage bag? (about 1.5 mils[1]). How many micrometers in a mil? (25.4). It is helpful to be fluent in the language of scale, as it can help to put the size of small particles and features into context, particularly when discussing observations with

---

[1] 1 mil = one thousandth of an inch

the lay public. Just as importantly, it can benefit sample preparation, analytical method selection, and even interpretation of results.

11.3.1 Scale and magnification

Countless times at meetings, presenters have projected an image on a screen saying that:

> "The projected image is shown at a magnification of 400 X."

This description is incorrect for two reasons. In the same way that a spectroscopist takes care to properly utilize the singular and plural of spectrum (i.e., spectrum and spectra) when referring to data, a microscopist should take care to use the proper vernacular by verbally discussing magnification in terms of "times" rather than "X" (Ford, 1993). While the discussion of "X" versus "times" illustrates a concern for specificity, the true error in the above statement is that the magnification of an image depends on the physical dimensions of that image. The 400 X magnification that the presenter has referenced is the nominal magnification of the image as viewed through the ocular of the microscope (10x ocular * 40x objective = 400x magnification). Since it is unlikely the presenter knows the dimensions of the projected image, discussion of an image in terms of magnification is inappropriate. For example, a 1 cm scale bar projected to fill a 1 m wide display, would be magnified 100 times; however, in a larger meeting room, that same image on a 5 m display would be a magnification of 500 times. Such variables can be easily avoided by simply providing a scale bar (or scale of some type) in an image.

11.3.2 Noting scale

Geologists have the necessary habit of including a common object, typically a rock hammer, coin, ruler or even a person in field images to ensure that the general size of the feature being photographed can be placed into context. The same approach should be used when taking an image of almost anything involving a sample. While the size of an object may be obvious to the examiner, the context is not always apparent in an image. In most instances, a digitally applied scale bar or even a nominal scale, such as an off-the-shelf ruler, placed in a plane of the framed photograph that is in focus will suffice. A typical ruler can be relied on to have a precision that is suitable for most measurements. For example, a study of seven stage micrometers of varying levels of traceability showed that all were within 1% of each other. In instances where more precise measurements are required, a scale with an accuracy and uncertainty appropriate to the measurement needed should be utilized.

11.3.3 Analytical volume and limits of detection

Scale is also important as it relates to the relationship between features in a material and the analytical volume of an instrument. For instance, the resolution of a light microscope is on the order of a micrometer (technically slightly less). In a properly prepared specimen (such as a thin section), this resolution permits visualization of many fine features, including many effect pigments and agglomerates of even finer organic and inorganic pigments (Figure 11.2), yet many organic pigments, such as those responsible for the overall color on the cover of this book, cannot be resolved as individual particles, as their size falls below the resolution limit of the light microscope.

Yet resolution is only part of the picture, as it is also important to understand which instrument is appropriate to the sample or features being studied. Table 11.1 provides some "rule of thumb" values for the range of magnifications anticipated for several microscopes as configured to analyze typical forensic evidence. The values of these ranges should help an examiner to assess the level of information one might expect to obtain using a given technique.

These same considerations are equally important when considering various microscope accessories. The mid-infrared microspectrophotometer (4000–650 cm^{-1}) is practically limited to an aperture size of about 10×10 μm, due to the wavelength of infrared energy in that region of the spectrum (2.5–15 μm). As the thickness of paint layers in automotive samples begins to decrease to thicknesses below ~10 μm, alternative

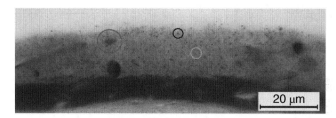

Figure 11.2 Image of a thin paint section (~2 μm thickness) in which effect pigments and some larger agglomerations of pigments can be resolved. However, the finer organic pigments cannot be resolved and are manifest simply as a mass tone color (*See insert for color representation of the figure.*)

Table 11.1 Microscopes commonly encountered in a forensic laboratory and general properties of each instrument

Instrument	Magnification value	Width of field	Practical resolution
Stereo microscope	8–35×	13 mm – 2.9 mm	10 μm
Polarized light microscope	40–630×	2.5 mm to 160 μm	1 μm
Scanning electron microscope (tungsten)	10–20,000×	4.5 mm – 5.6 μm	20 nm
Scanning electron microscope (field emission)	10–100,000×	4.5 mm – 1 μm	5 nm

These values were measured on instruments in the Microtrace LLC laboratory; however, the actual values will vary depending upon the specific configuration. The values are intended to be practical estimates rather than hard, theoretical limitations; for example, the SEM magnifications presented are practical, useful limits for typical forensic samples.

analytical methods and/or alternative sample preparation methods must be utilized in order to collect analytical data about the binders in such layers (such as pressing out a layer in a diamond anvil cell or cutting planar sections). While currently available only in select forensic laboratories, confocal Raman microspectroscopy has an analytical volume on the order of about 10 μm^3, and is thus a useful tool for the characterization of thinner layers and smaller particles (Palenik *et al.*, 2013). While the values discussed above are estimates that vary with the sample and instrument, familiarization with the analytical ranges for the limitations of instruments in each laboratory will help with all aspects of an analysis, from sample preparation through interpretation.

11.4 Finding

The first step in a trace evidence examination often centers on finding appropriate materials to analyze. This may be as simple as lifting a glass particle off of an item of clothing or it may be as complex as searching through a vacuum cleaner bag. In some cases, the target material is known (e.g., hair) while in other instances, the initial goal is to determine what particles, amidst a large assortment of material, may have evidentiary value. While items such as hair, fiber, paint and glass are often obvious candidates, the difficulty in finding these (and other) items depends on the level of detail given to a search as well as the range of particles that are to be considered.

In many protocols, searching is done macroscopically by looking through "scrapings" for an item of evidence hung over a piece of butcher paper. While this is certainly a viable method, it is relatively crude, in that it (i) lacks spatial resolution and (ii) limits the analyst to the recovery of relatively large items of evidence that have been freed from an item (i.e., those that can be observed by the unaided eye). Consideration of this population of finer particles can provide a great deal of additional benefit as shown below.

11.4.1 Spatial resolution

When considering trace evidence sampling, spatial resolution is often considered on a macroscale. Nehse (2011; 2014) has presented findings on the utility of fiber collection based on full body mapping (i.e., collecting and processing tape lifts from an entire person). The spatial information obtained from such investigations can often provide additional information that assists in the reconstruction of an event or it can be used to corroborate a story. For instance, Nehse (2011) shows in one example how this information was used to show that someone was attacked from behind. Due to secondary transfers that can occur during packaging and shipping, particle-mapping samples must generally be collected at a crime scene (with the inspection of the tape lifts conducted at a later date) (Chewning *et al.*, 2008).

The spatial significance evidence need not be limited to the scale of a body. In a matter involving a question of whether or not a hole in the hood of a sweatshirt originated from a bullet or a tear or by some other means, the case came down to the significance of gunshot residue found near the hole versus a lack of particulate evidence on severed fibers in the hole. As can be seen in Figure 11.3, fibers severed by a bullet have a distinct morphology: by stereomicroscopy (in which thermally fused fibers can be seen) (Figure 11.3a), by polarized light microscopy (which show distinct globular ends) (Figure 11.3b), and by scanning electron microscopy (which shows metal particles from the bullet trapped by the molten fiber) (Figure 11.3c and d). It could, therefore, be shown definitively that the severed fibers in this case could not have been formed by a bullet. This also demonstrated that the gunshot residue found on the hood was

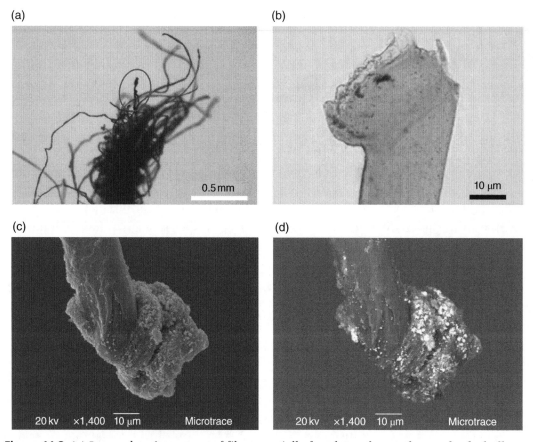

Figure 11.3 (a) Image showing a mass of fibers partially fused together as the result of a bullet severing them. (b) Appearance of a melted, globular end, by transmitted light microscopy. Appearance of a globular end by (c) secondary electron imaging and (d) backscatter electron imaging, showing the appearance of metal particles embedded when the fiber was molten (*See insert for color representation of the figure.*)

unrelated to the process responsible for forming the hole (Palenik and Diaczuk, 2013). Therefore, the definitive evidence in this matter was spatially resolved and came not from around the hole, but rather from within.

11.4.2 Recovery resolution

As illustrated above, significant evidence is not always visible by the unaided eye. Moving beyond the recognition of relatively large (>1 mm) evidence generally requires a more careful approach, which often involves the application of a stereomicroscope. In the attacks discussed in the introduction to this chapter, for example, the outward appearance of the jersey showed little sign of stains or debris (Figure 11.1). It was only by vacuuming the clothing that the fine debris (typically <200 μm) containing the critical particles was recovered. Similarly, in the case of the questioned bullet hole discussed in Section 11.4.1, the area of significance was limited to a few hundred micrometers at the tip of a fiber.

11.4.3 Stereomicroscope

Case after case can be cited in which fine particles, often invisible to the naked eye, have been found and exploited as key evidence. The technology to locate this evidence is not new. Stereomicroscopes, which generally range in magnification from about 8× to 35×, have been around for well over a century. However, the recovery of such particles requires an additional level of consideration, as the significance of many particles found at this level of analysis are not always immediately obvious. In one murder case involving soil, fine metal spheres were observed by stereomicroscopy. The spheres were just barely resolvable by stereomicroscopy (Figure 11.4), yet once found, they could be isolated and analyzed by scanning electron microscopy, which showed that they were composed of iron (Palenik, 2007b). The lesson is that an entire population of accessible but often-unexploited trace evidence exists in this finer fraction of dust.

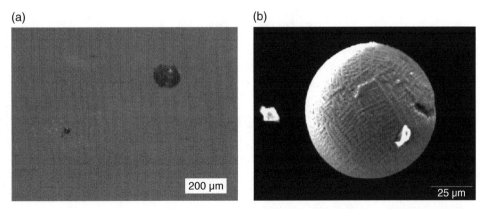

(a)   (b)

200 μm   25 μm

Figure 11.4 (a) Metal sphere recovered from a soil sample as it appeared, barely resolvable, by stereomicroscopy. (b) Metal sphere as it appears by scanning electron microscopy. Elemental analysis shows that it is composed largely of iron

11.5 Preparing

Once items are located, it is necessary to appropriately prepare the items for microanalysis. This preparation often starts with documentation and may include subsampling, particle manipulation, sectioning, and extractions. The goal of sample preparation is to present a sample (or a portion of it) to a micro-analytical method in an optimal manner. Given that these are field samples (as opposed to clean, laboratory generated samples) that are often irregularly shaped, dirty and heterogeneous, a truly optimal preparation may not always be possible. For example, in an academic setting, a metal sample will generally be embedded and polished to a flat surface prior to elemental characterization by energy dispersive X-ray spectroscopy (EDS). In practice, such preparation is often not practical or not possible (as it may be necessary to preserve the evidence for another type of analysis). In the instance of the metal sphere shown in Figure 11.4, polishing it would destroy the morphology of the sphere, which has significance in and of itself. In general, such preparations are generally avoided, as it is entirely reasonable to expect that a sample may require further analysis in the future, either to obtain additional data or to be analyzed by an opposing expert in the course of litigation.

11.5.1 Preservation and documentation

As alluded to in the preceding section, consideration of sample preservation is of prime concern, particularly during the early stages of sample preparation, for it is at this stage of the analysis that exhibits, as well as individual particles within those exhibits, are in their most pristine state. This does not mean that they are pristine but rather that once manipulated the evidence is altered from its initial state. While trace evidence analysis is often considered to be non-destructive, the more appropriate term would often be minimally destructive. Infrared spectroscopy, electron microscopy, and even sampling for DNA analysis require the collection of small samples. While such samples may utilize only a small fraction (<<50%) of a particular sample, some amount of material still needs to be isolated. For this reason, it is critical to document evidence prior to manipulation.

One case, for example, called for the analysis of a stain noted on a piece of facial tissue. The item received contained a large hole in the tissue. After repeated attempts to acquire images of the original item (before it had been sampled), it was eventually learned that no images of the item had been taken in its original state. It was also determined that a portion previously removed had also been consumed during prior testing. This left the entire investigation in the dark with regards to the amount and nature of the residue on the portion of the sample that had been removed. While it was possible, in this case, to conduct further analyses on the minute amounts of remaining residue, it was never possible to address the outstanding question of whether or not the excised portion contained any other information.

Years ago, examiners would fill pages of notebooks with often barely legible handwritten text describing the details of an item of evidence. Today, high-resolution digital cameras and inexpensive digital storage are almost universally employed. With such resources, there is no good reason not to properly photo-document a sample at relevant stages of analysis. Documentation may be conducted with a point-and-shoot camera, a copy stand, a stereomicroscope, a compound microscope, or a scanning electron microscope. Illumination may be transmitted, oblique or coaxial depending on the item being documented. In the case of the jersey shown in Figure 11.1, dark colored spheres were initially observed in the dust; however, with oblique illumination, the spheres were shown to be white. The reason for this difference is that the spheres are opaque to transmitted light (i.e., the spheres are thick enough that light cannot pass through them, so they appear dark in a transmitted light image). When top light (oblique or coaxial reflected illumination) is brought upon the subject, the actual, white color of the particle becomes apparent. This is illustrated in two images of a spray paint sphere shown under different illumination conditions (Figure 11.5a and 11.5b).

Other significant imaging considerations include depth of field (to ensure the area or object of interest is in focus) and color balance. Color balance is the adjustment of red, green and blue color intensities so that the resulting image colors are true (at least qualitatively). In most cases, this is a relatively simple adjustment that can be made through the software of a digital camera. However, this concept may be complicated by the use of LED illumination. In one particularly dramatic instance, it was found that a camera illuminated by a "white" LED source does not properly capture certain shades of green. Figure 11.6a shows the image of a green and yellow tablet, as it would appear to a typical observer. The same tablet is shown as illuminated by a "white" LED collected using a charge-coupled device (CCD) camera mounted to a stereomicroscope (Figure 11.6b). This drastic change in color from green to brown is not an issue of white balance, but rather appears to be due to a difference in illumination. Whereas the macro image was illuminated by the broad spectrum of a tungsten-halogen light, the "white" light LED source utilizes only a few relatively sharp emission lines to produce a white color, which fail to adequately absorb and reflect a broad enough spectrum to accurately reproduce

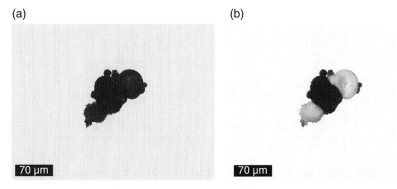

(a)   (b)

70 µm   70 µm

Figure 11.5 (a) White and black spray paint spheres all appear black in transmitted light but in (b) reflected light, the true color of the spheres become apparent (Palenik and Palenik, 2014)

(a)

(b)

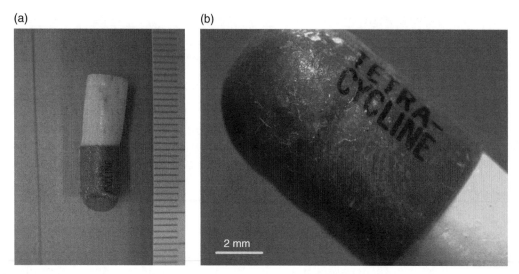

2 mm

Figure 11.6 (a) A macroscopic image of a tablet that is qualitatively consistent with its visually observed color (green and yellow). (b) The same tablet as it appears in an image that was illuminated using a "white" light LED and a digital camera (*See insert for color representation of the figure.*)

the color. Ultimately, care should be taken to ensure that the photos collected are reasonable representations of the item, particle or feature being documented.

11.5.2 Isolation

The concept of isolation was discussed briefly in reference to the act of finding evidence. The idea at this stage of analysis is to separate the portion of the sample to be analyzed from other debris, residues or other adhering materials, so as to present the cleanest, purest possible specimen for analysis. The physical manipulation of evidence can be conducted to a high level of precision under a stereomicroscope using a sharpened tungsten needle. Instructions for preparing needles can be found elsewhere (McCrone *et al.*, 1978). The tungsten needle permits the accurate movement of particles as small as about 10 μm without any additional assistance. The spheres shown in Figure 11.4 have obvious debris adhering to them. With a tungsten needle, a sphere can be quickly transferred to and washed in a drop of water on a microscope slide while being observed through a stereomicroscope.

The goal of this exercise in isolation is to increase the sensitivity of the analytical method being performed. In the elemental analysis of the sphere discussed above, any adhering debris will contribute to the generated X-ray spectrum, which in turn will result in an EDS spectrum that is more difficult to interpret and compare. In contrast, the spectrum from a clean metal sphere will be composed almost exclusively of iron and any alloying elements (e.g., Mg, Cr, etc.). This will permit a more specific identification of the alloy as well as a direct comparison to any known samples that may be available.

In the same way, FTIR and Raman spectroscopy are not particularly sensitive to components at low concentrations, and a spectrum that contains detectable contributions from two or more components is often extremely difficult to interpret. For example, active pharmaceutical ingredients (APIs) are often compounded with excipients, the latter of which often make up the bulk of a tablet by weight. In the analysis of a neat preparation from a suspected pharmaceutical tablet, the resulting spectrum often suggests that the material is composed of cellulose. Microscopic examination often confirms this result. However, through microextraction (*e.g.*, chloroform), which can be conducted on a microscope slide using a micro drop of solvent from a capillary, it is often possible to extract the API, which can then be transferred to a salt plate and identified by vibrational microspectroscopy.

11.5.3 Mounting

Once a sample has been isolated to the point that it is ready for analysis, it is typically necessary to mount the item for the designated analysis. Prior to mounting a sample, however, it is important to consider: (i) what are the goals of analysis; (ii) what techniques(s) will be applied; (iii) how much sample can be used; and (iv) what techniques might be applied in the future? This thoughtful approach will minimize sample consumption and ensure that sample consumption is judicious. With careful particle manipulation, it is often possible to conduct all common analyses while consuming a minute amount of the available material (Bales, 2009). For example, with a single 1 cm length of fiber, it is possible to perform a non-destructive optical characterization by PLM, an examination by fluorescence microscopy, Raman spectroscopy, and microspectrophotometry. A short portion of the fiber (100 μm or 1% of the total fiber length) can then be isolated for infrared spectroscopy and that same fragment can be used for elemental analysis by EDS. If necessary, the length of fiber used for infrared spectroscopy and electron microscopy may be recovered and preserved after analysis. When only smaller amounts of material are present, it may be necessary to move the same preparation from method to method.

11.5.3.1 Polarized light microscopy

In general, mounting of small particles is done under a stereomicroscope, which affords the ability to perform fine manipulations while continuously observing the particle being prepared. Considerations such as sample thickness, opacity, refractive index, particle dispersion, and mounting medium selection are taken into consideration when preparing a sample for polarized light microscopy. A particle that is too thick will prevent light from passing through the sample. A preparation of particles that are not well dispersed will make it difficult to observe and characterize individual particles. When mounting a sample of sand grains, for example, it is helpful to mount grains that have a similar size distribution. Since preparations studied by polarized light microscopy must be covered with a coverslip, grains that are too large or span a large particle size distribution will result in a coverslip that sits at an angle rather than flat. The 90–180 μm size fraction of a soil sample is typically used, which affords several benefits: the grains are large enough to obtain critical optical data (as discussed below), the grains are generally small enough to remain within the depth of focus of most objectives, and by always mounting grains within a similar size range, it becomes possible to estimate birefringence values (which can be important when classifying or identifying substances). While this guidance is taken from analysis of soil, the concepts are applicable to many materials.

The selection of a mounting medium is also an important decision, as a variety of options exist, each with their own pros and cons. Major considerations include the refractive index of the particle being studied and whether or not a mount is intended to be permanent or temporary. Temporary mounting media (often water, xylene or a xylene substitute) are typically used, as particles mounted in this manner can be easily manipulated, the mounting medium can be changed, stains can be introduced, and, when necessary, particles of interest can be picked for analysis by other methods. In contrast, more permanent media make recovery considerably more challenging or, in some cases, impossible.

The practical choice of a mounting liquid comes down to a few main considerations: solubility, contrast and recovery. Obviously, it is important to select a medium in which the sample is not soluble (though if a substance is found to be soluble, that observation in itself can be helpful in the characterization of a particle). The refractive index of a liquid is typically selected so as to achieve a suitable level of contrast between the particle and the mounting medium. When the particle and medium are close in refractive index, the contrast is low. As the difference in refractive index between the particle and medium becomes larger, the contrast observed around the edge of the particles increases, eventually to the point that the interior of the particle being studied is obscured. For this reason, particles are often mounted in a medium with a refractive index that is near, but not necessarily equal to, that of the particle. For example, plant tissue might be mounted in water, while thermoplastic polymers might be mounted in xylene (refractive index ~1.50).

11.5.3.2 Scanning electron microscopy

Major considerations for scanning electron microscopy sample preparation include sample conductivity, geometry, and the mounting substrate. As scanning electron microscopy requires the beam to interact directly with the sample, a covered preparation used for polarized light microscopy is wholly unsuitable. As such, SEM preparations consist of a particle being mounted on some type of sample substrate (e.g., aluminum, carbon, beryllium). The choice of substrate depends on the composition of the particle and the analytical elements of interest. For example, a sample to be analyzed for the presence of carbon should be mounted on a non-carbonaceous substrate (though there are other concerns associated with the analysis of carbon, such as beam contamination (Goldstein *et al.*, 2007)).

Regardless of the material, most SEM sample substrates are conductive. The conductivity of a sample is important, as the electrons (which form the basis of imaging) have a negative charge that can accumulate on a sample if it is not conductive. A build-up of charge, known in the vernacular of electron microscopy as charging, shows a notable effect on imaging; however, it can also impact elemental analysis by reducing the electron beam landing energy (Bean and Kugler, 2007). This latter effect has an impact on both quantitative analytical results and, more importantly for forensic science, can impact peak ratios being compared in overlaid EDS spectra. Samples such as paint, tape, glass, hair, polymers, sand, and seemingly most other forensic samples are not conductive and will (to varying extents) charge. In extreme cases of charging, the specimen will actually act as an electron mirror showing a reflection of the interior of the chamber (Figure 11.7).

Fortunately, there are various ways to counteract charging, which include coating the specimen with a thin (several nanometers thick) layer of carbon or gold or examining an uncoated specimen in a variable pressure SEM. Both methods have their place in electron beam microanalysis. Coating with a metal can often provide the best images, coating with carbon is useful for high resolution imaging and elemental analysis (as there is no interference from a metal in the EDS spectrum), and low vacuum imaging provides the fastest and often most convenient way to analyze a non-conductive specimen. The downside of low vacuum analysis is a significantly expanded spot size. According to Wight (2007), measurements under low vacuum conditions reveal that X-ray production hundreds of micrometers away from the analysis spot can contribute to the collected spectrum.

Another effect that should be taken into consideration when examining forensic samples in the SEM is or geometry (i.e., shape) of the material being analyzed. While

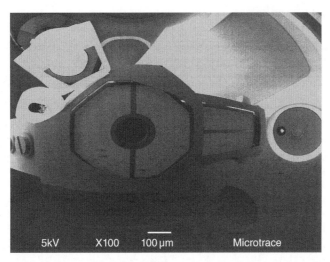

Figure 11.7 SEM image of the interior of the SEM chamber in which an extreme charge buildup on a polyethylene sphere sample turns it into an electron mirror

most textbooks will explain that elemental analysis should ideally be conducted on a flat surface of an infinite depth, such samples (or samples prepared in this manner) are rarely practical for forensic microscopy. As such, it is necessary to make the most of the samples that are available. While there are numerous analytical caveats that must be considered when analyzing EDS data collected from a particle, a few tips can help an analyst acquire the best possible data. First, a sample should be mounted in such a manner that the flattest surface is exposed to the beam. In some cases, this requires a creative mounting method that secures the irregular surface of a sample to the substrate while exposing a flat surface to the beam. Second, the specimen should be oriented in the sample holder and inserted into the SEM so as to ensure that this flat surface has a direct line of sight to the EDS detector. This may mean that the sample needs to be rotated so that elevated areas of the sample are not between the flat surface being analyzed and the X-ray detector. Finally, if attempting to conduct a spectral comparison, utilize questioned and known samples of comparable size. This will reduce differences in peak ratios that can result from the analysis of small particles (as the volume of X-ray emission, which varies by element, may be larger than the actual samples being analyzed). The concept of X-ray emission volume can be understood better by experimenting with a spectral simulation program such as DTSA-II (NIST, 2014).

The above topics touch on a few of the major considerations that should be given attention when preparing samples for analysis by light and electron microscopy. However, it is important to know that there are many additional factors that impact the quality of images and the reliability of data obtained by these methods. Such factors become increasingly important as analyses move from the realm of qualitative to quantitative, from the realm of low magnification to high magnification, and from large to small particle size. While there are a variety of ways to treat such factors, an increased theoretical understanding of light and electron microscopy becomes increasingly important.

11.5.3.3 Specialized sample preparation

In various instances, more involved sample preparation may be required. In the study of paint, fibers, animal hairs, metals, and numerous other materials, cross-sections and polished sections can provide valuable insight into the morphology of such materials and can improve the quality and quantity of analytical information obtainable. There are a multitude of preparation methods, which include the relatively fast method of hand sectioning for fibers (Palenik and Fitzsimons, 1990a, 1990b) and the more involved process of embedding and cutting microtomed sections (Palenik *et al.*, 2013; Groves *et al.*, in press). All methods have their place in the examination of forensic materials; however, a carefully prepared thin section that contains the questioned and known samples can provide the best basis for a comparison of color (by MSP) and composition (by FTIR and EDS). Thin sections also afford one method by which to observe the smallest particles and inclusions resolvable by light microscopy (such as pigments). For high resolution SEM analysis of small particles, it has also been found that ion polished thin sections of small particles and coatings provide useful information (Palenik and Palenik, 2014). Of course, this is not the limit of preparation methods, as various materials-specific methods also exist, such as a method for the preparation of longitudinal sections of hair (i.e., literally splitting a hair) for the characterization of internal hair structure (Palenik, 1983).

11.6 Looking

It is an excellent practice to always look at a sample by light microscopy before moving to a more detailed analytical method (such as FTIR spectroscopy or SEM-EDS). The dust collected from the jersey shown Figure 11.1 appears fairly non-descript when observed by stereomicroscopy. Furthermore, it contains no evidence of the type typically sought after and recovered by a typical trace evidence laboratory. Most of the components that were later identified are not covered in standard operating procedures, and none of the particles in the dust are large enough to recover from typical scrapings or tape lifts. If one were to blindly apply an instrumental protocol such as elemental analysis by EDS, FTIR spectroscopy, or X-ray diffraction, the result would have shown that the recovered dust is composed of calcium carbonate. Alone, this single item of information is of very little use. Calcite is a carbonate mineral found in a wide range of soils and is used in the matrices of a wide number of polymers and other materials as a filler (much like talc or kaolin).

In contrast, a brief look at this same dust by polarized light microscopy provides a great deal of information. First, the sample is composed largely of particles with a fairly uniform particle size distribution. These particles have an angular morphology and consist of relatively freshly fractured surfaces. Looking more carefully reveals a number of other particle types, which would not be identified through a "bulk" analysis due to their relatively low concentration. These particles include small, black spheres, which appear white with transmitted light. Such spheres are typical of spray paint that has assumed a spherical (low surface energy) shape while traveling through the air as a liquid. Another type of crystal, present at a much lower concentration in the dust, is a colorless crystal with a distinct, tabular habit. The shape and size of both the angular and tabular particles are strongly suggestive of an anthropogenic rather than natural origin. Observed floating at the top of the preparation against the coverslip are a few relatively fresh pollen grains identified by morphology as *Quercus* (oak) pollen.

The information that can be obtained by simply looking at a sample by light microscopy covers a broader range of materials and, in many cases, can provide far more practical information than can be obtained many other microanalytical methods. Only by taking the time to observe and ponder the various particles and their microscopic features can one begin to classify, if not identify, particles. Such observations as those discussed in reference to the jersey dust may not always result directly in the identification of a particle; however, the characteristics observed help an analyst to define the analytical path forward and can be used to increase the evidentiary significance of the conclusions.

11.6.1 Light microscopy

The process of looking through a microscope is not something that requires years of training. While experience certainly helps to efficiently prepare samples, recognize potentially significant features, and quickly identify particles, many important characteristics can be observed by a microscopist equipped only with patience and an observant mind. For example, examination of a simple longitudinal preparation of a fiber can show whether it is animal, vegetable, or synthetic. A more detailed look at a synthetic preparation will give indications as to the diameter, colorant (dyed or pigmented), whether it is delustered, and an approximation of the cross-sectional shape (e.g., round, trilobal, bean). A cross-section will give even more detailed information about the cross-sectional shape and the way in which the fiber was dyed. A crenulated cross-section, shown in Figure 11.8, for example, indicates not only the process used to produce the fiber but is also indicative of the fiber composition itself (i.e., rayon). All of these observations are relatively basic to a trained fiber examiner, but each of these features (though perhaps not the significance of them) can be noted simply by anyone with magnification and an observant mind.

These same basic principles of observation can be applied to the characterization of particles or solids of any type using features such as those described in Box 11.1. Even if a particular particle cannot be specifically identified, a classification process such as this will provide information about both the number of components in a sample and their dominant features. In many cases, this can constrain the type or source of material, rule out certain materials, or help to determine the next type of analysis needed. For

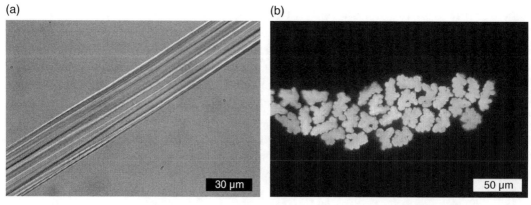

Figure 11.8 (a) Rayon fiber observed in a longitudinal preparation. (b) A preparation showing several rayon fibers in cross-section

Box 11.1 Microscopic characteristics that may be noted in an attempt
to identify a particle

Microscopical Characteristics
Size
Shape (euhedral, subhedral, amorphous)
Texture and cleavage
Color (transmitted and reflected light)
Reflectivity
Inclusions (present or absent)
Magnetism
Hardness

example, a small blue or red flake may suggest a flake of printing ink. Since such flakes are often fairly small and FTIR spectra of printing inks are not particularly useful, this examination might suggest Raman spectroscopy as a next step. The resulting Raman spectrum would likely give information about the blue and red pigments that are present, which could be suggestive of a possible source or origin. In another instance, a particle that is opaque by transmitted light, but reflective by oblique illumination, which is soft and not attracted to a magnet, might suggest some type of soft, non-ferrous metal such as lead. Microchemical or elemental analysis of this particle (or a portion of it) would provide a definitive answer.

11.6.2 Scanning electron microscopy

While the polarized light microscope represents the most versatile tool used for initial characterization, the magnification and contrast provided by a scanning electron microscope affords other information. A typical SEM is equipped with two imaging detectors, one that collects lower energy secondary electrons (which provide topographic information) and a second detector that is sensitive to the higher energy backscattered electrons (which provide contrast that is roughly proportional to the average atomic number of the particle being analyzed). The difference in contrast between these two imaging detectors can be seen in Figures 11.3c and 11.3d, which show the ends of a fiber severed by a bullet. Backscatter electron imaging is used, for example, to find "high-Z" particles of lead, barium and antimony in automated gunshot residue analyses. Secondary electron imaging is more useful for observing surface features such as tool marks or scale patterns on a hair. The examination of such surface features using secondary electrons is often conducted using a reduced accelerating voltage and beam current to reduce beam penetration of the sample and increase resolution (1–5 kV for imaging versus the 20 kV typically used for elemental analysis).

11.7 Analyzing

Most people who have used a microscope in school have done so for the purposes of magnifying a material of interest. However, accessories can be added to both light and electron microscopes that can provide a wealth of quantitative analytical information. While many forensic scientists are familiar with the utility of energy dispersive x-ray spectroscopy to collect elemental information from a sample, fewer understand the way in which two polarizers and the resulting polarized and cross-polarized light can be used as an analytical tool to measure the properties of, and identify, materials by light microscopy. This section of the chapter aims to present some of the ways in which these accessories (as well as a few others) can be used to provide analytical information that extends beyond that which can be simply observed through magnification.

11.7.1 Polarized light microscopy

While the magnification of a polarized light microscope was used to pick out the various components in the dust collected from the jerseys shown in Figure 11.1, polarized light can actually be used to identify the crystalline particles that could not be identified by morphology alone. Using polarized light, the particles from the major component of the dust (that with fresh angular particles of uniform size distribution) were found to have a very high birefringence (relative to other minerals) and are uniaxial with a negative optic sign. Furthermore, one refractive index is equal to approximately 1.660 (a common mounting medium) and the other is considerably lower. These characteristics are consistent with only one common crystalline phase: calcite. The second crystalline phase observed in the dust, present at only a minor level, consisted of tabular particles with a low birefringence and refractive indices closer to, but lower than, 1.540 (another common mounting medium). These crystals are less commonly observed in nature than calcite, but these characteristics are recognizable as gypsum.

The optical properties used to identify these and other crystalline minerals include refractive index, birefringence, optic sign, sign of elongation, interference figures, and pleochroism. Theoretical details for each of these topics is well covered elsewhere (Bloss, 1961; McCrone *et al.*, 1978), as are practical examples of various applications of these concepts as applied to specific materials (fibers: Palenik, 1999; glass: Hamer, 2001; particles: McCrone and Delly, 1973; minerals: Mange and Mauer, 1991). The issue is that intimidation arising from these optical properties appears to remain the single largest hurdle to the more widespread use of polarized light microscopy as an analytical tool in forensic environments. While each of these properties can be measured quantitatively, the remainder of this section is intended illustrate the way in which judicious use of these measurements can be provide useful information quickly and accurately.

11.7.1.1 Polarized light

The study of optical properties requires the use of polarized light, which to this point in this chapter has been referenced only by name. While a full treatment of polarized light and optical properties is beyond the scope of this chapter; a basic understanding of the concept and its significance in terms of microscopy can be achieved through the analogy of polarized sunglasses.

Polarizers (either in sunglasses or on a microscope) act as a filter that permits light waves vibrating in a single plane to pass. This effect can be observed by looking at a reflection on a pond. The light reflected off of the pond surface is polarized in a plane parallel to the surface of the lake. Polarizers in a pair of sunglasses, however, are oriented vertically (they only permit light polarized in a vertical plane to pass). As such, reflections off of the lake surface will not pass through the sunglasses, with the effect that glare from the lake surface is eliminated. Another way to illustrate this effect of polarization is to hold two pairs of sunglasses in front of each other. If the frames are simply held in front of each other, some light will pass through both. However, if one pair of sunglasses is rotated 90 degrees relative to the other, such that the two polarizing lens are perpendicular to each other, the view through the two polarizers will appear black (all light will be blocked).

In the same manner, a polarized light microscope uses two polarizers: one below the specimen and the other above. The two polarizers are typically oriented such that they are perpendicular to each other (i.e., crossed). Therefore, when an empty field of view is observed between cross-polars it appears black. There are also situations where only the lower polarizer is used. This is known as plane polarized light. Both plane and cross-polarized light can be used to probe the optical properties of materials as discussed below.

11.7.1.2 Refractive index

Refractive index is speed at which light travels through a material relative to the speed at which light travels through a vacuum. By comparing the refractive index of two glass particles in a forensic comparison, one is effectively measuring and comparing the speed at which light travels through these samples. For example, light traveling through two windshield glasses ($n_1 = 1.581$ and $n_2 = 1.582$) that vary in refractive index by 0.0001 (the level to which refractive index values are typically measured in forensic glass comparisons) means that light is travelling through the lower index sample at about 13 km/s faster than it does through the higher index fragment.

Light travels through glass and isotropic crystals (such as halite or sodium cyanide) at the same speed regardless of how the particle is oriented. Recall that the atoms within glass have no large-scale periodic arrangement, and so the speed of light passing through glass is indifferent to its orientation. The same holds true for crystals with high symmetry, such as sodium chloride or potassium cyanide. The periodic arrangement of atoms in most crystals, however, is not equivalent in all orientations (i.e., they have lower symmetry), which means that the light passing through such crystals will vary in speed based on the orientation of the crystal relative to the incoming light.

The effect is that, a crystal may have one, two or three optic axes, which result in a crystal having to one (isotropic), two (uniaxial), or three (biaxial) refractive indices.

The measurement of these refractive index values can go a long way towards identifying a crystal. However, exact measurement of these values can be time consuming. That said, a simple assessment of refractive index values of a particle relative to the mounting medium can often provide enough information (along with other easily observable optical properties) to identify many commonly occurring crystalline substances. The contrast of a mineral grain provides the first clue as to the relative value of the refractive index of a material to the medium. The closer the refractive index of a material is to the medium, the lower the contrast. Similarly, the higher the contrast, the more different the refractive index of the material is from the mounting medium. These differences in contrast are illustrated in Figure 11.9, which show grains of quartz mounted in several different media. While contrast can provide an excellent estimate of the difference in refractive index between a sample and mounting medium, it does not tell the analyst if the refractive index of the particle is higher or lower than the medium. However, this can be determined using the relatively straightforward Becke

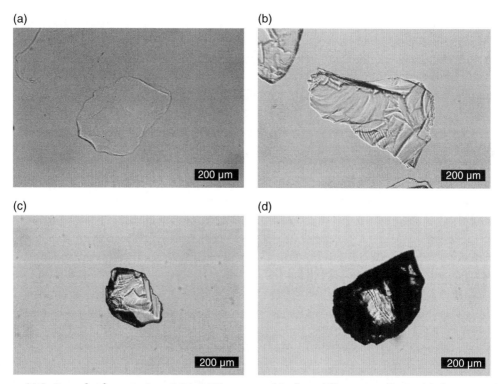

Figure 11.9 Crystals of quartz (n = 1.54–1.55) mounted in four different media. In (a) the mounting medium (n = 1.54) is close to the refractive index of quartz, which results in low contrast around the edge of the grains. In (b) the medium (n = 1.50) is further from that of quartz, which results in a higher contrast. In (c) and (d) the medium (n = 1.33 and n = 1) is increasingly different from that of quartz and the resulting contrast makes observation of the grain almost impossible

line method, the process of which is explained in in detail elsewhere (McCrone *et al.*, 1978). Contrast combined with the Becke line can provide a great deal of insight into the refractive index of a material. For example, all common generic classes of fibers (e.g., nylon, polyester) can be quickly identified through a combination of relative refractive index measurements and birefringence (Palenik, 1999).

11.7.1.3 Birefringence

Birefringence is the difference between two refractive index values in a material. Birefringence can be quantitatively measured in a variety of ways, but the most practical application of birefringence for a particle analyst is through the qualitative visual observation of a crystal as it appears between crossed polarizers. A glass fragment, which has only one refractive index does not have birefringence, and will, therefore, appear black between crossed polarizers. Isotropic crystals (which also have only one refractive index) will also have no visible contrast between crossed polarizers. The birefringence of uniaxial and biaxial crystals is visible between crossed polarizers and a value can be qualitatively estimated based upon the colors observed. Figure 11.10

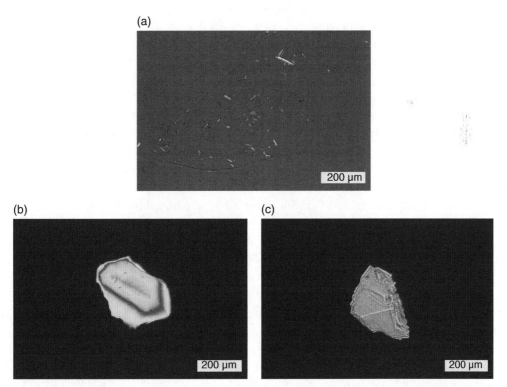

Figure 11.10 Example of crystals observed between crossed polarizers with varying levels of birefringence. (a) Glass is isotropic and has no birefringence, (b) quartz is uniaxial and shows low order interference colors, (c) calcite shows high order interference colors (*See insert for color representation of the figure.*)

shows examples of four crystals spanning a range of birefringence. This method of qualitative estimation of birefringence (none, low, medium or high) is typically of more practical value than a quantitative measurement.

11.7.1.4 Pleochroism

Since *pleo-* refers to many and *-chroism* refers to color, the term pleochroism refers to materials whose color in plane polarized light varies with orientation. Recall that plane polarized light utilizes a single polarizer below the specimen. In the same way that certain crystalline phases (uniaxial and biaxial) cause light to pass through a material at different speeds based on the orientation of the crystal, certain materials also absorb light differently as a function of orientation. This effect is most commonly observed in fibers and minerals, which can have dramatically different colors in different orientations. Figure 11.11 shows a fiber and a crystal as they appear in two different orientations (relative to the polarizer) in plane polarized light. This properly can be used to differentiate certain varieties of minerals, and can be used as a point of comparison when examining fibers.

11.7.2 Energy dispersive X-ray spectroscopy

Energy dispersive X-ray spectroscopy (EDS) is an accessory to the scanning electron microscope whereby the energy of X-rays (photons) is measured and counted. The resulting spectrum provides information about the elemental (as opposed to chemical) composition of a material. Elements from boron upwards on the periodic table can be detected. Detection limits vary based upon the elements, detector and matrix, but can be as low as 0.01 wt%. While EDS is an incredibly powerful method and data collection can be taught to a novice in a short time, the scientific interpretation of any data collected by this method should be considered by someone with a theoretical background in electron–sample and X-ray–sample interaction. This caveat becomes increasingly important when analyses are used for quantitative analysis (or even for spectral overlays). The reason for this caveat is that factors such as excitation voltage, beam current, sample size, sample geometry, sample matrix and vacuum conditions (to name a few) can have a major effect on analytical results. While the EDS software can computationally correct for some of these factors (such as excitation voltage and count rate), other factors such as (sample geometry) are not factored into these calculations or the calculated uncertainty values.

   To illustrate the potential severity of these effects, consider a case that called for a more detailed analysis of two stainless steel fragments that had been identified as such by X-ray fluorescence spectroscopy (a method that uses X-rays instead of electrons to probe the elemental composition of a sample). While the intention was to embed and polish these samples to obtain the best possible quantitative analysis, the analysis started with a simple preparation of the fragments for an initial examination by SEM-EDS. Surprisingly, the fragments were found to be composed of aluminum

(a)

(b)

(c)

(d)

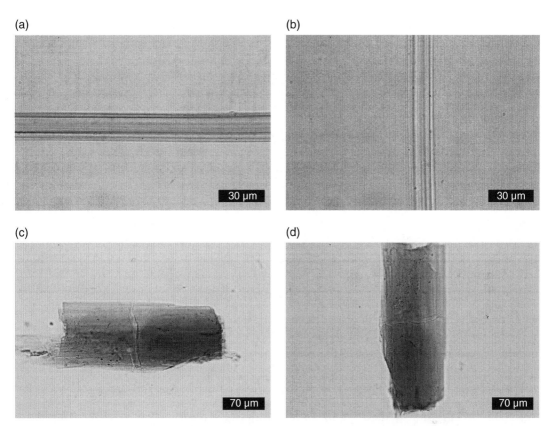

Figure 11.11 A pleochroic fiber changes from colored to colorless when the orientation relative to the polarizer changes from (a) length parallel to (b) length perpendicular. Certain minerals are also pleochroic, as noted by the change in this glaucophane grain when the mineral is rotated from an orientation in which the length is (c) parallel to the polarizer to one that is (d) perpendicular to it (*See insert for color representation of the figure.*)

rather than stainless steel. While a few fine particles of stainless steel were detected that were trapped in the tool marks of the soft aluminum, the samples were not composed of stainless steel. The moral to this story is that XRF, while it is a more sensitive method for studying heavy elements, is not particularly sensitive to lighter elements (such as aluminum). Therefore, the point-and-shoot XRF method was sensitive to the small traces of stainless steel but wholly insensitive to the aluminum that made up the actual particles. This inappropriate analysis coupled with an interpretation that relied entirely upon the instrument-generated results provided a conclusion that was entirely incorrect. While this was not a criminal forensic matter, related concerns over quantitative interpretation of EDS data are significant enough that the use of quantitative EDS is not recommended for forensic glass comparisons (Almirall *et al.*, 2012).

11.7.3 FTIR and Raman spectroscopy

While FTIR and Raman microspectrometers are sold as stand-alone instruments, their value to the forensic community is based largely around the magnification and microscale analytical resolution provided by the microscope to which the spectrometer is attached. In the case of the Figure 11.1 jersey dust, the small paint spheres were analyzed by these methods, which showed they were composed of acrylic paint pigmented with rutile (the light stable polymorph of titanium dioxide). Similar to SEM-EDS data, the instruments themselves are relatively easy to operate. However, the quality of the data depends largely upon the quality of sample isolation and preparation, as both methods provide better data when pure (or relatively pure) compounds are analyzed.

Both techniques have a variety of operating parameters that can be adjusted by a user with some basic understanding of the technique. Infrared spectroscopy can be operated in transmission and reflectance mode, and many systems have an option for a micro- or macro-attenuated total reflectance (ATR) accessory. Transmission spectra generally provide the highest quality data, which are easiest to interpret. ATR and reflectance mode are most useful when analyzing thin films or coatings that cannot necessarily be isolated from their substrate.

Raman spectra can be collected using a variety of different lasers and with various optical configurations. Based on our internal analysis of thousands of samples, the near infrared red (NIR) laser (785 nm) seems to be the most practical choice for general samples analysis. Green lasers (532 nm) are also often included in many systems as a second laser and, at times, this laser can avoid fluorescence issues encountered by the NIR laser. In addition, Buzzini *et al.* (2006) found that in some paint samples, additional pigments not detected by the NIR laser can be detected by the use of a second (green) laser. Higher end Raman microscopes can also operate in a confocal mode, which greatly constrains the analytical volume of the technique, permitting extremely small areas to be analyzed. This also permits samples to be analyzed that are mounted under a cover-slip. This is particularly helpful when studying small particles, such as pigments in paint samples. Palenik *et al.* (2013) showed that confocal Raman spectroscopy can increase the number of pigments identified in a paint sample relative to samples analyzed using a larger (less constrained) laser spot.

In addition to sample preparation and analysis considerations, interpretation of spectra data also requires some fundamental understanding of vibrational spectroscopy. Two common pitfalls of spectral interpretation are (i) a comparison of spectra without identification and (ii) blind reliance on a "hit quality index" generated by a spectral search. In the first case, a comparison without an understanding of the significance of spectral peaks can lead to incorrect interpretation. Without an understanding of what is being compared or why the samples are different, there is no way to know if the similarities or differences originate from the instrument, an artifact, or the sample. In one case, two spectra from a heavily filled black polymer were reviewed after initial analysis. The analyst concluded that the spectra were the same. While it is true that the spectra were the same, a quick look at the spectra showed that they consisted effectively of flat lines with zero transmission. A spectrum with zero transmission means that the detector received none of the infrared light from the source. This can happen if a sample is too thick, or if an IR opaque solid (such as carbon black, the most commonly encountered black pigment) is analyzed. It would be generous to say that these data had any value, but more importantly it illustrates that the value of a comparison is at least partially dependent upon an interpretation of the spectrum.

The second problematic area has to do with automated searching. The search score or hit quality index is a value that ranks the results of an automated database search. While such values are internally useful to the search algorithm, they have no direct bearing on the interpretation of analytical results. There are many instances where the top results from a spectral search have no particular relationship to the sample being analyzed. The point to this entire discussion is that the relative ease of data collection by computer-based analytical tools must not be confused with the more difficult technical and theoretical considerations of sample preparation and data interpretation.

11.7.4 Other methods

A number of other accessories to the light microscope may provide a wealth of additional information: an ultraviolet excitation source (fluorescence microscopy), a controllable hot stage (thermal microscopy), a visible spectrophotometer (microspectrophotometry), two microscopes bridged by a common ocular (comparison microscopy) and a Michelson interferometer (interference microscopy). In addition, there are a number of other contrast methods, such as phase contrast microscopy, which can be configured to help an analyst visualize specific features of a sample.

Once familiar with the basic theoretical and technical concepts of stereo and compound microscopes, the integration of these other accessories and contrast methods is relatively straightforward. While many of these techniques do not see everyday use in a crime laboratory environment, the benefits of such instruments should be familiar to the microscopist, so that they can be called upon when a need arises.

11.8 Thinking

Returning one last time to the jersey shown in Figure 11.1, all of the factual information can be assembled. The nearly invisible dust consisted of a small amount of oak pollen, spheres of a white acrylic paint, fine tabular, anthropogenic crystals of gypsum, and it was dominated by highly angular calcite fragments with a uniform particle size distribution. The presence of fresh oak pollen suggests a particular time – March to April – when oak pollination is at a maximum. The presence of only a relatively small amount of pollen, which was found on only one of the two jerseys, suggests that the dust originated largely from an indoor environment. The presence of gypsum points to drywall operations, which is supported by the spray paint. Furthermore, the spray paint supports an industrial rather than residential environment, as residential painting is often done with rollers as opposed to sprayers. The largest component in the sample, calcite, had a remarkably uniform particle size distribution and fresh surfaces, which suggests it did not originate from a geological source (such as a soil). Given these facts, and examination of reference collections of dust from various industrial operations (Microtrace, 2015), it was hypothesized that this originated from a sanding operation. This was confirmed by the gypsum crystals, which are also part of a drywall compound. Finally, the dust on the two jerseys, recovered from two different attacks, was relatively similar (with the exception of the pollen). This suggests that the perpetrator was a professional tradesman who participated in the same activities from day to day.

In sum, the identified particles provided a great deal of factual information. Independently, these facts do not necessarily have a great deal of value. However, together and in accordance with the scientific method, the factual information could be assembled to suggest the various hypotheses (investigative leads) presented at the start of this chapter. As per the scientific method, the hypotheses can then be tested against other independent facts. For example, investigators later revealed that, indeed, one of the attacks did occur in the spring, which provided further support for the hypothesis drawn from the pollen grains. Similarly, both of the attacks occurred in an industrial park, which was consistent with large-scale operations involving the use of spray paint and a professional tradesman. The factual nature of the actual particle identifications combined with the self-consistency of the investigative hypotheses were sufficient to convince the state police in this jurisdiction to release these leads to the media. In the turn of events described in the introduction to this chapter, which ultimately lead to the confession and eventual conviction of the attacker, it was possible to confirm the validity of each hypothesis.

This example demonstrates the several important lessons. First, the above investigation was performed entirely by light microscopy (though infrared spectroscopy and electron microscopy were used to confirm some of the findings). Second, this lesson illustrates the extent to which a forensic investigation may be taken. In an era where more and more trace evidence laboratories are being closed and trace evidence as a discipline is being reduced to a series of protocols applied to a handful of commonly encountered materials (paint, tape, glass, hair, fibers), it is clear that the trend is to move further away from the origins of trace evidence and the concept of a generalist approach. As a result, investigative analyses such as this are impossible to conduct in most laboratories, which either cannot conduct an analyses without comparison samples, cannot conduct analyses of unknown materials, or are so compartmentalized that there is no one person who has the authority to assemble this information into a cohesive picture. Despite these artificial limitations, this chapter aims to illustrate that microscopy could be used to greater advantage, within existing protocols, in virtually all existing subdisciplines of trace evidence. While the freeform aspects of microscopic observations are not conducive to protocol-based approaches, they do provide a level of information not obtainable by more turnkey methods. It is my hope that in reading this chapter it inspires you to take the time to follow the advice given to me: take the time to prepare, look, and consider each sample through the oculars of a light microscope.

11.9 Thanking

I must first thank Skip Palenik, my father, introduced me to the beauty of logic, the scientific method, and the elegance of practical problem solving through science. My father was also responsible for the analysis of the jersey dust case history that is woven throughout the chapter. Special thanks to the staff at Microtrace, in particular Jason Beckert, Brendan Nytes, Ethan Groves, and Katie White, whose daily hard work and creativity continually uncover new curiosities that lead to fascinating solutions to our clients' problems. Their creativity has also been instrumental in helping us to formalizing the approaches discussed above into a formal process that satisfies ISO 17025 requirements while allowing us to continue our scientifically creative brand of microanalysis. Finally, I would like to thank my wife, Cate, who is always there to support me and my professional hobby while providing a conduit to the world beyond the laboratory.

References

Almirall, J., Naes, B., Cahoon, E., and Trejos, T. (2012) Elemental Analysis of Glass by SEM-EDS, µXRF, LIBS and LA-ICP-MS. Final Technical Report, National Institute of Justice Grant 2005-IJ-CXK069, Florida International University.

Bales, H. (2009) Mounting particles on a TEM grid for analysis by multiple instrumental methods. *Microscope*, **57**(3), 105.

Bean, S.J. and Kugler, V.M. (2007) The effect of the specimen charging on the accuracy of x-ray analysis in the variable pressure SEM. *Microscopy and Microanalysis*, **13**(S2), 1440–1441.

Bloss, F.D. (1961) *Introduction to the Methods of Optical Crystallography*. Holt, Rinehart and Winston, New York.

Bloss, F.D. (1999) *Optical Crystallography*. MSA Monograph Series, Mineralogical Society of America, Chantilly, VA.

Buzzini, P., Massonnet, G., and Sermier, F.M. (2006) The micro Raman analysis of paint evidence in criminalistics: case studies. *Journal of Raman Spectroscopy*, **37**(9), 922–931.

Chewning, D.D., Deaver, K.L., and Christensen, A.M. (2008) Persistence of fibers on ski masks during transit and processing. *Forensic Science Communications*, **10**(3) [online] https://www.fbi.gov/about-us/lab/forensic-science-communications/fsc/july2008/index.htm/research/2008_06_research01.htm (last accessed 1 July 2015).

Ford, B.J. (1993) On correct microscopical English. *The Microscope*, **41**(1), 43.

Goldstein, J., Newbury, D., Joy, D.C., *et al.* (2007) *Scanning Electron Microscopy and X-ray Microanalysis*, 3rd edn. Springer.

Grice, C. (2002) Crime Seen. *Popular Science*, **261**, 60–68.

Groves, E.G. and Palenik, C.S. (in press) Applications of blue light curing acrylic resin to forensic sample preparation and microtomy. *Journal of Forensic Sciences*.

Hamer, P.S. (2001) Microscopic techniques for glass examination. In: *Forensic Examination of Glass and Paint* (ed. B. Caddy), Taylor & Francis, London, pp. 47–73.

Mange, M.A. and Mauer, H. (1991) *Heavy Minerals in Colour*. Springer.

McCrone, W.C. and Delly, J.G. (1973) *The Particle Atlas, Vol.* **I**, Principles and Techniques. Ann Arbor Scientific Publishers, Ann Arbor, MI.

McCrone, W.C., McCrone, L., and Delly, J.G. (1978) *Polarized Light Microscopy*. Ann Arbor Scientific Publishers, Ann Arbor, MI.

Microtrace (2015) *Dusts Reference Collection*. https://www.microtracellc.com/reference-collection/dusts/ (last accessed 27 June 2015).

Nehse, K. (2011) *Fiber Transfer & Distribution*. Trace Evidence Symposium, Kansas City, MO; http://projects.nfstc.org/trace/2011/presentations/Nehse-Fibre-Transfer1.pdf (last accessed 27 June 2015).

Nehse, K. (2014) Fibre evidence and evaluation of findings the benefits of mapping. In: *World Forensic Festival Abstract Book*, World Forensic Festival, 12–18 October 2014, Coex, Seoul, Korea, p. 226.

NIST (2014) *DTSA-II*. http://www.cstl.nist.gov/div837/837.02/epq/dtsa2/ (last accessed 27 June 2015).

Palenik, C.S. and Diaczuk P. (2013) Plumbum microraptus: Microscopic indicators of a bullet hole in a synthetic fabric. *The Microscope*, **61**(2), 51–60.

Palenik, C.S. and Palenik, S.J. (2014) Seeing color: Practical methods in pigment microscopy. *The Microscope*, **62**(2), 71–81.

Palenik, C.S., Palenik, S., Groves, E., and Herb, J. (2013) *Raman Spectroscopy of Automotive and Architectural Paints: In situ Pigment Identification and Evidentiary Significance*. Final Report, National Institute of Justice Award No. 2011-DN-BX-K557, Microtrace, Elgin, IL. https://www.ncjrs.gov/pdffiles1/nij/grants/243162.pdf (last accessed 27 June 2015).

Palenik, S.J. (1983) Light microscopy of medullary micro-structure in hair identification. *The Microscope*, **31**, 129–137.

Palenik, S.J. (1999) Fiber microscopy. In: *Forensic Examination of Fibers*, 2nd edn (ed. M. Grieve). John Wiley & Sons, Ltd., pp. 153–177.

Palenik, S.J. (2007a) The use of microscopic trace evidence to develop investigative leads in criminal investigations. *The Microscope*, **55**(1), 35.

Palenik, S.J. (2007b) Dr Thorndye to the rescue. *The Microscope*, **55**(3), 114.

Palenik, S.J. and Fitzsimons, C. (1990a) Fiber cross-sections: Part I. *The Microscope*, **38**, 187–195.

Palenik, S.J. and Fitzsimons, C. (1990b) Fiber cross-sections: Part 2, A simple method for sectioning single fibers. *The Microscope*, **38**, 313–320.

Wight, S. (2007) Evaluation of the extent of electron scattering in a low vacuum scanning electron microscope. http://www.nist.gov/mml/mmsd/837-02-wight-2008-escattering.cfm (last accessed 1 July 2015).

CHAPTER 12
Chemometrics

Ruth Smith

School of Criminal Justice, Michigan State University, USA

12.1 Introduction

Forensic chemists are primarily employed in the controlled substances, trace evidence, and toxicology sections in a forensic laboratory. Within these sections, instrumental techniques, such as gas chromatography (GC), gas chromatography-mass spectrometry (GC-MS), and infrared (IR) spectroscopy are routinely used for the analysis and characterization of a variety of evidence types, including controlled substances, fire debris, paint, and fibers to name a few. These instrumental techniques generate chromatograms or spectra of the analyzed samples that display the instrument response or signal as a function of retention time (for GC) or wavenumber (for IR spectroscopy). The chromatogram or spectrum of a sample submitted to the laboratory is visually compared to a chromatogram or spectrum of a suitable reference standard for characterization and/or identification purposes. However, the vast number of variables contained in these spectra or chromatograms makes them ideal candidates for further analysis using chemometric procedures. Such procedures can be used to identify patterns in the data, to classify samples based on common data features, or to quantify samples.

The field of chemometrics has been defined as "A chemical discipline that uses statistical and mathematical methods, to design or select optimum procedures and experiments, and to provide maximum chemical information by analyzing chemical data" (Varmuza and Filzmoser, 2009). A variety of procedures are available that are based on pattern recognition and these are potentially of most interest in forensic chemistry. Numerous texts are available in which the chemometric procedures presented in this chapter are discussed in more detail, providing the history of their development, as well as describing the mathematical foundations of the procedures (Beebe *et al.*, 1988; Massart *et al.*, 1988; Lavine and Davidson, 2006; Brereton, 2007, 2009, 2013; Morgan

---

Forensic Chemistry: Fundamentals and Applications, First Edition. Edited by Jay A. Siegel.
© 2016 John Wiley & Sons, Ltd. Published 2016 by John Wiley & Sons, Ltd.

and Bartick, 2007; Varmuza and Filzmoser, 2009). As a result, the focus of this chapter is more on the application of these procedures to forensically relevant data. The more commonly employed unsupervised and supervised pattern recognition methods are presented, providing an overview of each procedure and presenting applications of these procedures to forensically relevant data.

12.2 Chromatograms and spectra as multivariate data

Before beginning a discussion of applications of chemometric procedures in forensic chemistry, it is necessary to consider the data themselves. While this chapter primarily focuses on chromatographic and spectral data as examples of common multivariate data in forensic chemistry, the chemometric procedures described are certainly not limited to only these data types. Consider the chromatogram of a forensic sample (e.g., fire debris evidence, submitted street sample containing controlled substances, etc.), in which abundance is displayed on the y-axis and retention time on the x-axis. Each retention time point represents a variable, meaning that the sample can be described by the p variables contained in the chromatogram. Thus, the chromatogram can be represented as a vector containing p variables. Now consider the n chromatograms for additional samples in the data set: each again can be represented as a vector and the data set can now be represented by the matrix $\mathbf{X}$, which has dimensions $n \times p$, where n is the number of chromatograms (samples) in the data set and p is again the number of variables (Figure 12.1). Spectral data can be considered in a similar manner, representing the data in matrix form prior to performing chemometric procedures.

12.3 Data preprocessing

A critical step prior to applying chemometric procedures is to consider and apply appropriate preprocessing procedures to the data. This is necessary to reduce sources of variance that are inherent in instrumental analysis techniques but that are not related to

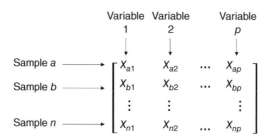

Figure 12.1 Matrix representation of multivariate data

the samples themselves. For example, in gas chromatography, these sources can include noise due to electronic fluctuations, shifts in retention time due to fluctuations in carrier gas flow rate and oven temperature, and differences in peak heights as a result of differences in the volume or mass of sample analyzed. In infrared spectroscopy, differences in optics or in sample thickness change the effective path length, resulting in differences in the intensity of absorption (Miller, 2010).

Data preprocessing procedures are applied to minimize or eliminate these sources of variance that, if left untreated, could be highlighted as actual differences among samples in the subsequent chemometric analysis. In turn, this could lead to incorrect grouping or classification of samples. As a result, the data should be assessed to determine the need for preprocessing and, where necessary, the appropriate procedures should be applied. However, the actual preprocessing procedures applied will vary according to the nature of the data. Baseline correction, smoothing, and normalization are common procedures applied to chromatographic and spectral data. Retention-time alignment procedures are also commonly applied to chromatographic data.

12.3.1 Baseline correction

Baseline correction methods are applied to remove low frequency variation in the baseline signal (Beebe *et al.*, 1998; McIlroy, 2014). The contribution of the baseline in a chromatogram or spectrum can be removed by taking the first derivative of the data (Beebe *et al.*, 1998; Morgan and Bartick, 2007). In doing so, the baseline is no longer meaningful. This method does have a limitation in that it can increase noise in the data. For GC-MS data specifically, a mass spectrum of the baseline region can be subtracted from all total ion chromatograms in the data set (McMaster, 2008). However, care should be taken to ensure that none of the ions present in the subtracted mass spectrum correspond to ions of interest in the sample being analyzed; otherwise, this baseline correction method will result in a reduction in chemical signal. Another option for baseline correction is to mathematically model the baseline, which is then subtracted from each chromatogram or spectrum (Beebe *et al.*, 1998; Morgan and Bartick, 2007). While this method offers a more accurate correction, modeling the baseline in this manner can be time consuming to ensure that no signals from compounds of interest are included in the model.

To illustrate the effect of baseline correction, Figure 12.2a shows the last 20 minutes in the total ion chromatogram (TIC) of a diesel sample analyzed by GC-MS before any preprocessing procedures have been applied (McIlroy, 2014). This region highlights the rise in baseline that results from an increase in stationary phase degradation at the higher oven temperature. The baseline was modeled using an asymmetric sigmoid function and the modeled function then subtracted from the TIC, with the resulting baseline-corrected chromatogram shown in Figure 12.2b. Note the reduction in the magnitude of the baseline after correction.

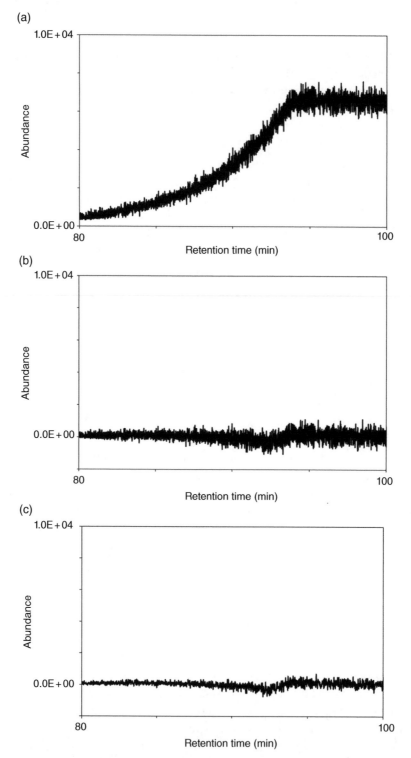

Figure 12.2 Expanded view (last 20 minutes) of the total ion chromatogram of a diesel sample analyzed by GC-MS (a) before pretreatment, (b) after baseline correction using a modeled function, and (c) after smoothing using a fast Fourier-transform smooth with two data points. (Figure based on data used with permission from V.L. McGuffin and J.W. McIlroy.)

12.3.2 Smoothing

Smoothing methods are used to minimize noise in data that occurs as a result of point-to-point fluctuations in signal (McIlroy, 2014). The Savitzky–Golay smooth is a well-known and routinely applied algorithm that fits a polynomial function to a predetermined number of data points (the window), using the least squares method (Savitzky and Golay, 1964; Beebe *et al.*, 1998; Morgan and Bartick, 2007; Pierce *et al.*, 2012). The center data point in the window is replaced by the value calculated on solving the polynomial and the algorithm moves forward by one data point. The process is repeated, moving the window forward by one data point for the remaining data points, each time replacing the center point with the value determined through solving the polynomial. In contrast, the fast Fourier-transform smooth first transforms the data into the frequency domain (Beebe *et al.*, 1998; Pierce *et al.*, 2012; McIlroy, 2014). A low-pass filter is applied to eliminate the high frequency component of the signal, which corresponds to the noise, and then the data are transformed back into the time domain. Figure 12.2c shows the same expanded view of the TIC of a diesel sample, now after applying a fast Fourier-transform smooth with two data points. Note the reduction in noise compared to Figure 12.2b, in which baseline correction was the only preprocessing method applied.

The effect of smoothing on the chromatographic signal should also be assessed to ensure no adverse effects. Figure 12.3 illustrates the effect of smoothing on one peak in the TIC of a diesel sample with no smoothing, after applying a fast Fourier-transform smooth with two data points and after applying a Savitzky–Golay smooth using a 4th order polynomial and 11 data points. Note that there is a reduction in noise in the smoothed signals compared to the unsmoothed but little difference in the peak height. Considering both the reduction in noise and the effect on peak height, the fast-Fourier transform smooth performs slightly better for these data.

12.3.3 Retention-time alignment

Particularly for chromatographic data, retention-time alignment should also be considered as a preprocessing procedure. Fluctuations in carrier gas flow rate and oven temperature, as well as degradation of the stationary phase, can result in differences in retention time over time. Common algorithms to correct such shifts include the correlation optimized warping (COW) algorithm and the peak-match algorithm (Nielsen *et al.*, 1998; Johnson *et al.*, 2003; Pierce *et al.*, 2012; McIlroy, 2014). Both algorithms align peaks in the sample chromatograms to those in a target chromatogram that ideally contains peaks for all compounds represented in the data set. The target chromatogram can be experimentally prepared and analyzed along with the samples or can be generated mathematically after analysis as the average abundance of all compounds represented in the data set.

For alignment using the COW algorithm, the target and sample chromatograms are divided into segments, where the segment size is a user-defined parameter (Nielsen *et al.*, 1998). It should be noted that the definition of segment can vary depending on the software package being used. The segment size may be the number of segments the

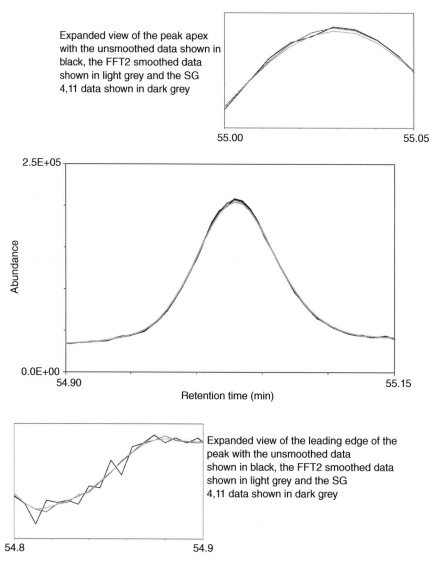

Expanded view of the peak apex with the unsmoothed data shown in black, the FFT2 smoothed data shown in light grey and the SG 4,11 data shown in dark grey

Expanded view of the leading edge of the peak with the unsmoothed data shown in black, the FFT2 smoothed data shown in light grey and the SG 4,11 data shown in dark grey

Figure 12.3 Effect of smoothing on chromatographic signal. The main figure shows the heptadecane (C_{17}) peak in the total ion chromatogram of one diesel replicate and the two insets show expanded views of the signal. In each figure, the unsmoothed signal is shown in black, the Savitzky–Golay smooth with a 4th order polynomial and 11 data points (SG 4,11) is shown in dark grey, and the fast Fourier-transform smooth with two data points (FFT2) is shown in light grey. (Figure based on data used with permission from V.L. McGuffin and J.W. McIlroy.)

chromatograms are divided into or may be the number of data points each segment contains. The algorithm operates by stretching or compressing segments in the sample chromatogram to maximize correlation between peaks in corresponding segments in the target chromatogram. The extent to which the segment can be stretched or compressed is defined as the warp, which is the total number of data points that can be added to or removed from the segment.

The peak-match algorithm firstly identifies the retention times at which peak maxima occur in the target and sample chromatograms (Johnson *et al.*, 2003). The analyst defines a window size (a number of data points) and peak maxima present within this window in both the sample and target chromatogram are considered a match. The retention time axis is then adjusted to shift the apex of the peak in the sample chromatogram to align with the apex of the corresponding peak in the target. This algorithm is effective in aligning chromatograms that contain a different number of peaks, as no adjustment of the retention time axis occurs if a peak is identified in the target but not the sample, or vice versa (McIlroy, 2014).

For alignment to be effective using these algorithms, an appropriate warp (for the COW algorithm) or window size (for the peak-match algorithm) must be selected. If the warp or window is too small, the peaks cannot be aligned whereas, if too large, the peak may be aligned to a neighboring peak, rather than the corresponding peak. Appropriate warps and window sizes can be selected by overlaying chromatograms and assessing the extent of the shift in peaks corresponding to the same compound (McIlroy, 2014).

Figure 12.4a shows an expanded region, corresponding to the tetradecane (C_{14}) peak, in the total ion chromatograms of three diesel replicates (McIlroy, 2014). In this figure, the data have been baseline corrected and smoothed but note that the apex of the C_{14} peak in one replicate is slightly misaligned with respect to the other two replicates. Figure 12.4b shows the same peak in the replicates following retention-time alignment using the COW algorithm with a warp of two data points and a segment size of 75 data points. Note that the apex of the C_{14} peaks in each replicate is now aligned.

12.3.4 Normalization and scaling

Normalization procedures correct for differences in the signal intensity that result from differences in the mass or volume of sample analyzed, as well as differences in instrument response (Beebe *et al.*, 1998; Morgan and Bartick, 2007; Pierce *et al.*, 2012; McIlroy, 2014). A variety of normalization procedures are available, depending on the data and the effect to be corrected. In constant-sum normalization, the value at each data point is divided by the sum of all data points in the sample (Beebe *et al.*, 1998; Pierce *et al.*, 2012; McIlroy, 2014). As an example, a chromatogram can be constant-sum normalized by dividing the abundance at each retention time point by the sum of the abundance at all retention time points in that chromatogram. Figure 12.4b shows the C_{14} peaks in three diesel replicates after retention time alignment. However, there is a slight difference in the peak heights of the replicates as a result of slight differences in injection volume and, hence, the mass of sample analyzed for each replicate (McIlroy, 2014). Figure 12.4c shows the same region of the chromatogram after constant-sum normalization, indicating the decrease in peak height variation and in baseline variation among the replicates.

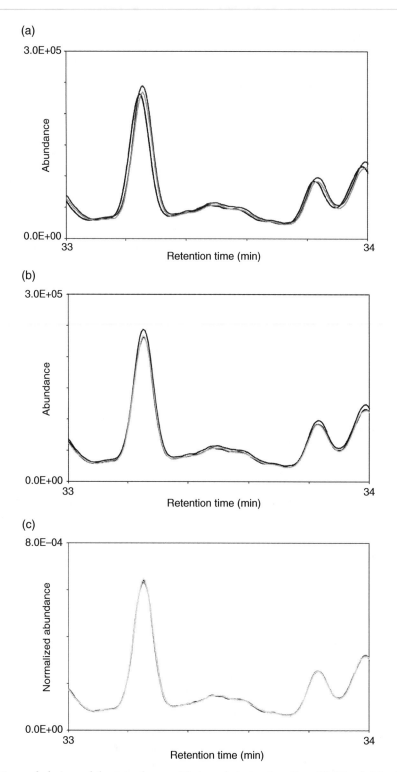

Figure 12.4 Expanded view of the tetradecane (C_{14}) peak (retention time 33.25 min) in the total ion chromatogram of three replicates of one diesel sample: (a) after baseline correction and smoothing; (b) after baseline correction, smoothing, and retention-time alignment to a suitable target chromatogram using the COW algorithm with a warp of two data points and a segment size of 75 data points; and (c) after baseline correction, smoothing, retention-time alignment, and constant-sum normalization. (Figure based on data used with permission from V.L. McGuffin and J.W. McIlroy.)

Chromatographic and spectral data can be normalized to the intensity of a selected peak by dividing each data point by the signal of the selected peak (Beebe *et al.*, 1998; McIlroy, 2014). This peak may be an internal standard, a peak common to all samples, or the peak with maximum intensity. Alternatively, data can be normalized to unit length, in which the value at each data point is divided by the square root of the sum of squares of all data points in the sample (Beebe *et al.*, 1998).

For spectral data, the standard normal variate (SNV) normalization procedure is commonly applied to remove interference effects that are due to radiation scattering and differences in particle size (Barnes *et al.*, 1989; Gemperline, 2006). The mean value of all variables in a sample is subtracted from each individual variable and then divided by the standard deviation of all variables in the sample. This normalization procedure is termed a "weighted normalization," as those variables that show greater deviation from the mean have greater weighting.

In contrast to normalization procedures, which are performed across all variables in each sample, scaling procedures, such as autoscaling or Pareto scaling, are performed on individual variables across all samples in the data set (Morgan and Bartick, 2007). Scaling is often necessary for data sets containing samples that differ in the magnitude of response. The first step in autoscaling involves mean-centering the data; that is, the mean of each variable is subtracted from the corresponding variable in each sample. The mean-centered variables are then divided by the standard deviation of the variable across the data set. Pareto scaling also begins by mean-centering the data; however, in this method, each mean-centered variable is divided by the square root of the standard deviation of the appropriate variable. Autoscaling results in variables with similar variance that are more equally weighted for comparison. However, this scaling method tends to increase the importance of noise. Pareto scaling overcomes this to some extent as the scaled data more closely resemble the original data (van den Berg *et al.*, 2006).

12.4 Unsupervised pattern recognition

Unsupervised pattern recognition procedures require no prior knowledge of the identity of samples in the data set and are used to identify natural groups of samples based on common features. An example of this could be a data set containing IR spectra for different samples in which, following an unsupervised procedure, those samples containing an IR absorption at about 3300 cm^{-1} (corresponding to primary amine) are grouped together. While the exact identity of the samples is unknown, there is now some indication of chemical similarities among the samples. These unsupervised procedures can, therefore, be used to determine similarities and differences among samples in a data set. Further, these procedures are often the first step prior to applying the supervised pattern recognition procedures that are described in Section 12.5. The two more common unsupervised exploratory procedures used in forensic analyses are hierarchical cluster analysis and principal components analysis.

12.4.1 Hierarchical cluster analysis

In hierarchical cluster analysis (HCA), samples are grouped or clustered based on their similarity in multidimensional space, with the goal that samples in the same cluster are more similar to each other than to those in different clusters (Izenman, 2008; Hastie *et al.*, 2009). "Hierarchical" indicates that once a sample becomes a member of a cluster, the sample remains in that cluster for the remainder of the analysis. In agglomerative HCA, all samples are initially considered members of an individual group and the two most similar groups are grouped to form a new cluster (Beebe *et al.*, 1998; Izenman, 2008). The process continues until all samples in the data set are members of the same group. In contrast, for divisive HCA, all samples are initially members of one group and the process proceeds until each sample forms its own cluster (Izenman, 2008). In a forensic context, agglomerative HCA is the most commonly applied procedure and is the focus of the following discussion.

To start, all samples are members of individual groups and the distance between all pairs of samples is calculated. A number of different distance metrics are available but for continuous data the Euclidean distance is often used (Beebe *et al.*, 1998; Lavine and Davidson, 2006; Brereton, 2007, 2009) and is calculated as shown in Equation 12.1:

$$d_{AB} = \sqrt{\left(x_{A1} - x_{B1}\right)^2 + \left(x_{A2} - x_{B2}\right)^2 + \cdots + \left(x_{Ap} - x_{Bp}\right)^2} \tag{12.1}$$

where, d_{AB} is the distance between samples A and B, x_{A1}, x_{A2}, x_{Ap} are the p variables measured for sample A, and x_{B1}, x_{B2}, x_{Bp} are the p variables measured for sample B.

The calculated distances form a distance matrix, which is a symmetrical matrix in which the diagonal entries are zero, representing the distance of a sample to itself. The off-diagonal entries indicate distance between pairs of objects and, as d_{AB} is equivalent to d_{BA}, only the entries above or below the diagonal need to be assessed.

The two samples separated by the shortest distance, which indicates greatest similarity, are identified. These two samples are linked to form a new cluster and the process repeats. That is, the distances are recalculated to generate a new distance matrix, which is again assessed to determine the samples separated by the shortest distance. However, after the first iteration, distances must be calculated between clusters containing more than one sample. A number of different linkage methods are available to calculate the distance between the cluster and all other samples and clusters (Beebe *et al.*, 1998; Lavine and Davidson, 2006; Brereton, 2007, 2009; Izenman, 2008). For example, in the single linkage method (Figure 12.5a), the distance between the sample and the nearest neighbor in the cluster is calculated whereas, in the complete linkage method (Figure 12.5b), the distance between the sample and the farthest neighbor in the cluster is calculated. The data set should be analyzed using different linkage methods and the

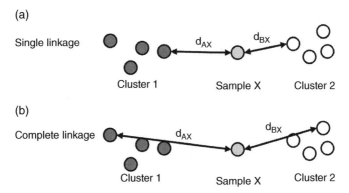

Figure 12.5 Representation of some popular linkage methods in agglomerative hierarchical cluster analysis

clustering obtained should be assessed. For a well-defined data set (i.e., one that contains distinct clusters), the pattern of clusters observed should not vary substantially based on the linkage method although the actual distances or similarity (see below) will vary (Lavine and Davidson, 2006).

The output from HCA is typically displayed as a dendrogram, in which the vertical axis shows the samples in the data set and the horizontal axis shows the distance or the similarity, which is calculated according to Equation 12.2:

$$S_{AB} = 100 * \left(1 - \frac{d_{AB}}{d_{max}} \right)$$

(12.2)

where, S_{AB} is the similarity between samples A and B, d_{AB} is the distance between samples A and B, and d_{max} is the maximum distance between any two samples in the data set. A similarity value of one indicates complete similarity while a similarity value of zero indicates no similarity. From the dendrogram, the samples forming each cluster, the similarity level at which the clusters form, and the order in which the clusters form can all be ascertained.

An exemplar dendrogram obtained following HCA of a data set containing chromatograms of ignitable liquid standards is shown in Figure 12.6. Replicates of each liquid cluster with a high degree of similarity, ranging from 0.865 for replicates of diesel to 0.971 for replicates of torch fuel. The three gasoline samples display high similarity, with Gasoline B and C forming a cluster first at a similarity level of 0.788. Gasoline A subsequently joins the cluster at a similarity level of 0.737. For the petroleum distillates, the torch fuel and kerosene standards first form a cluster, albeit at a similarity level of only 0.022, with diesel joining the cluster at 0.020. While these three liquids are all classed as petroleum distillates, they differ in the range of normal alkanes present; that is, torch fuel contains C_{11}–C_{14}, kerosene contains C_9–C_{17}, and diesel contains C_9–C_{21}. This highlights one of the limitations of HCA; that is, the distances and, subsequently, the similarities are calculated relative to the data set in question, with the result that the two least similar

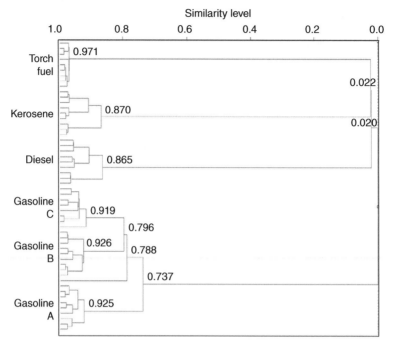

Figure 12.6 Dendrogram generated after agglomerative hierarchical cluster analysis of a data set containing total ion chromatograms of six ignitable liquids, each analyzed in triplicate. Similarity levels are labeled on the dendrogram

samples will have a similarity index close to zero, even if there is some degree of similarity between the two samples in question. While HCA provides a visual representation of the data set, no information is given about the variables that contribute to the cluster patterns observed. As a result, it is often advisable to perform HCA in addition to another exploratory procedure, such as principal components analysis (Lavine and Davidson, 2006).

12.4.2 Principal components analysis

Principal components analysis (PCA) is an unsupervised procedure used to reduce the dimensionality of the data set, so that the underlying patterns are more readily observed (Beebe *et al.*, 1998; Gemperline, 2006; Lavine and Davidson, 2006; Brereton, 2007, 2009; Morgan and Bartick, 2007). Texts are available that give mathematical descriptions of PCA and the algorithms used (Gemperline, 2006; Izenman, 2008; Hastie *et al.*, 2009; Varmuza and Filzmoser, 2009); as a result, only an overview of PCA is given in this section and illustrated with reference to forensically relevant data.

To start, consider the chromatograms of three ignitable liquids –diesel, kerosene, and torch fuel (Figure 12.7a).

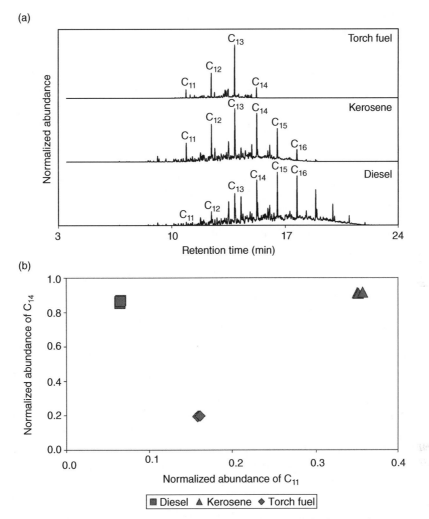

Figure 12.7 (a) Representative total ion chromatograms for torch fuel (top), kerosene (middle), and diesel (bottom) obtained by GC-MS and (b) representation of the three chromatograms based on two variables (normalized abundance of tetradecane (C_{14}) and undecane (C_{11}))

All three contain the normal alkanes, undecane (C_{11}) and tetradecane (C_{14}) and the liquids can be described in two dimensions based on the abundance of these two variables (Figure 12.7b). If an additional variable was included (e.g., dodecane, C_{12}), a third dimension could be added to this plot. Beyond three variables, it is impossible to represent the data in a similar manner, although different combinations of variables could be represented in a series of two- or three-dimensional plots. However, as the chromatograms shown in Figure 12.7a contain 3583 variables, even this is not a practical option.

Herein lies an advantage of PCA, that of dimension reduction. Within multivariate data that contain correlated variables, not all variables will be important or discriminating: some variables will describe chemical information, correlated variables will describe similar information, and other variables will describe noise. PCA defines new axes that are linear combinations of the original variables with an important feature: the new combinations contain uncorrelated variables (Lavine and Davidson, 2006; Morgan and Bartick, 2007). These new axes represent the principal components (PCs) of the data set and, as each PC is a combination of uncorrelated variables, substantially fewer PCs are required to describe the chemical information contained in the data.

In the first step of PCA, the data are mean centered, which results in the samples being centered around the origin of the multidimensional space although their relative position to each other does not change (Beebe *et al.*, 1998). The first principal component (PC1) is positioned to describe the maximum variance (or spread) in the data. The second principal component (PC2) is positioned orthogonally to PC1 and describes the next greatest variance. Subsequent PCs are positioned orthogonally to the preceding PC and describe the next greatest variance not already accounted for.

Samples in the original measurement space are projected perpendicularly onto the PCs and the new coordinates are termed the scores of the sample (Leardi, 2006; Morgan and Bartick, 2007). Thus, each sample is represented in the new space defined by the PCs. Further, information regarding the variables is also obtained in the form of variable weightings or loadings. The weighting, or loading, of a variable on a given PC is a measure of the contribution of that variable to the PC (Beebe *et al.*, 1998; Leardi, 2006).

The scores of the samples and the loadings of the variables are used to generate the two main outputs from PCA. Plotting the sample scores on two or three PCs generates a scores plot in which chemically similar samples are positioned closely and distinctly from chemically different samples. Loadings plots can be generated in a similar manner; that is, plotting the loadings of each variable on two or three PCs. However, for chromatographic or spectral data, the loadings of the variables on a given PC can be plotted against retention time or wavenumber. This way, the variables contributing to the PCs are more readily identified. The loadings plots can also be used to explain the positioning of samples in the scores plot. For example, samples that are positioned positively on PC1 in the scores plot likely contain a high abundance of the variables that have positive loadings on PC1 (Leardi, 2006).

For a given data set, n PCs are defined, where n is the number of samples or variables, whichever is smaller (Lavine and Davidson, 2006; Morgan and Bartick, 2007). However, as the number of PCs increases, a greater proportion of noise is described, such that the last few PCs may only describe noise. Typically, two or three principal components are sufficient to describe the majority of the variance in the data (Lavine and Davidson, 2006). A Scree plot (Figure 12.8) can be used to determine the number of PCs that are necessary to describe the data. In this example, two PCs are sufficient, accounting for 99.96% of the variance in the data.

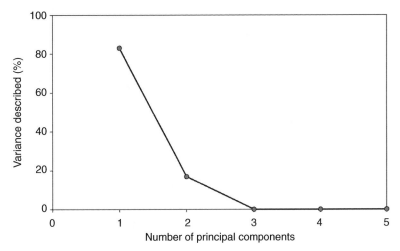

Figure 12.8 Scree plot indicating the variance described by each principal component

The previous discussion focused on a geometric representation of PCA, based on just two variables. In reality, the PCs are identified mathematically by performing eigenanalysis of the covariance matrix generated for the data set (Gemperline, 2006; Morgan and Bartick, 2007). Covariance measures the relationship between two variables, indicating whether a positive or negative relationship exists. The covariance between two chromatograms (or spectra) can be calculated using Equation 12.3:

$$cov\left(x,y\right) = \frac{\sum_{i=1}^{n}\left(x_i - \bar{x}\right)\left(y_i - \bar{y}\right)}{n-1} \tag{12.3}$$

where, cov *(x, y)* is the covariance between chromatograms x and y, x_i and y_i represent the ith variable in chromatogram x and chromatogram y, respectively, $\bar{x}$ and $\bar{y}$ represent the average abundance of all variables in chromatogram x and chromatogram y, respectively, and n is the number of dimensions (or variables). Note that in the process of calculating covariance, the data are mean centered, as shown in the nominator in Equation 12.3. The covariance matrix shows the covariance for pairwise comparisons of all samples in the data set. Entries along the diagonal (e.g., cov *(x, x)*) represent the variance of an individual sample. And, as cov *(x, y)* is equivalent to cov *(y, x)*, the matrix is symmetrical around the diagonal.

Eigenanalysis of the covariance matrix results in a series of eigenvectors and corresponding eigenvalues, which describe the amount of variance described by the eigenvector. The eigenvectors are ranked in order from highest to lowest eigenvalues and the eigenvector with the highest eigenvalue is the first principal component. The eigenvector with the next highest eigenvalue is the second principal component, and so on. For a given PC, the sample scores are calculated as the sum of the product of the mean-centered sample data and the appropriate eigenvector. Different algorithms are available to perform eigenanalysis although in commercial software the singular value decomposition (SVD) algorithm and/or the non-linear iterative partial least squares (NIPALS) algorithm are commonly included (Beebe *et al.*, 1998; Gemperline, 2006; Morgan and Bartick, 2007; Varmuza and Filzmoser, 2009).

It is also worth noting that PCA can be performed through eigenanalysis of the correlation, rather than covariance, matrix (Morgan and Bartick, 2007). Correlation is calculated according to Equation 12.4:

$$corr\left(x, y\right) = \frac{cov\left(x, y\right)}{s_x s_y} \tag{12.4}$$

where, *corr (x, y)* is the correlation between chromatograms *x* and *y*, *cov (x, y)* is the covariance between chromatograms *x* and *y*, and s_x and s_y represent the standard deviations of chromatograms *x* and *y*, respectively. The data are mean centered in calculating covariance (Equation 12.3) and now are divided by the standard deviation. This is equivalent to autoscaling the data. As a result, eigenanalysis of the correlation matrix is used for data sets that contain variables of different magnitude or variables that were measured using different units (Morgan and Bartick, 2007).

To further illustrate PCA, consider again the chromatograms of the three ignitable liquids (diesel, kerosene, and torch fuel) that were introduced in Figure 12.7a. The chromatograms were preprocessed (smoothed, retention-time aligned, and normalized) before PCA, which was performed on the full chromatogram (3583 variables). The scores plot for the first two principal components is shown in Figure 12.9a. These two principal components account for 99.95% of the total variance. Note that the Scree plot for these data was given in Figure 12.8 and indicated that two principal components would be sufficient to describe this particular data set. In the scores plot (Figure 12.9a), replicates of each liquid are positioned closely and the three liquids are distinguished on PC1: diesel is positioned positively, kerosene is close to zero, and torch fuel is positioned negatively on this PC. The second principal component distinguishes kerosene, which is positioned positively on PC2, from the other two liquids, which are positioned negatively.

The loadings plot for the first principal component is shown in Figure 12.9b, in which the loadings are plotted versus retention time. Plots such as these can be used to explain the positioning of the samples on the scores plot. As an example, torch fuel is positioned negatively on PC1 as the more abundant normal alkanes contained in this liquid (C_{11}–C_{13}) have negative loadings on PC1 (Figure 12.9b). The positioning of all samples on each PC can be described in a similar manner, taking into account the loadings of the variables on each PC, as well as the mean-centered data. While often used in combination with HCA (Section 12.4.1), PCA has an advantage in providing

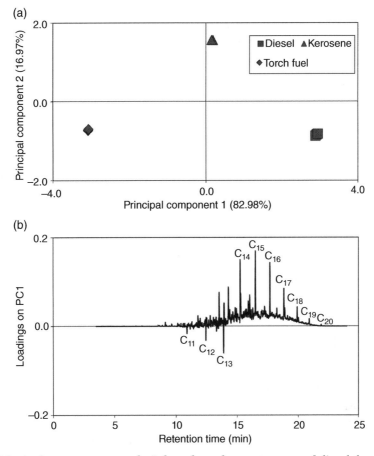

Figure 12.9 Principal components analysis based on chromatograms of diesel, kerosene, and torch fuel (a) scores plot showing PC1 versus PC2 and (b) loadings plot for PC1

information regarding the variables that are most influential and hence, responsible, for the patterns or groups of samples observed in the scores plot.

12.5 Supervised pattern recognition procedures

Supervised pattern recognition procedures build models based on a training set of data and then use these models to determine some property or properties of "new" samples, where "new" indicates samples that were not part of the training set. In forensic chemistry, supervised procedures can be used to "classify" a new sample; that is, to determine the class or identity of the sample. As an example, the training set could contain chromatograms or spectra of reference standards of different controlled substances.

The supervised pattern recognition procedure would be used to develop a model describing class features of these controlled substances. The developed model could then be applied to identify the controlled substance present in a submitted sample. Note that in supervised procedures prior information about the data is needed. In the example above, this is the identity of the reference standards in the training set. This is in contrast to the unsupervised procedures that were described in Section 12.4, for which no prior knowledge of the data was necessary.

The following sections focus on three supervised pattern recognition procedures that are used for classification purposes: *k*-nearest neighbors, discriminant analysis, and soft independent modeling of class analogy. While all three vary in their method of operation, an integral part of any such procedure is to validate the model, which is discussed in Section 12.5.4.

12.5.1 *k*-Nearest neighbors

The *k*-nearest neighbors (*k*-NN) classification method does not involve true model building and is considered relatively simple in mathematical terms (Massart *et al.*, 1988; Adams, 1995; Brereton, 2007; Varmuza and Filzmoser, 2009). Classification is based on the proximity or distance of samples in the multidimensional space (Lavine and Davidson, 2006; Brereton, 2007; Hastie *et al.*, 2009; Varmuza and Filzmoser, 2009; Dougherty, 2013). The distances between a new sample and all members of the training set are determined, with the Euclidean distance being the common distance metric used for this (Equation 12.1). The distances are ranked in order of increasing distance and the *k*-nearest neighbors are identified based on those that have the shortest distance to the new sample. The sample is considered a member of the class to which the majority of its *k*-nearest neighbors belong. Figure 12.10 illustrates a *k*-NN classification. If three nearest neighbors (i.e., *k* = 3) are used for classification, then the new sample is considered a member of class A as two of the three nearest neighbors belong to this class.

Determining the appropriate number of nearest neighbors (*k*) for the classification is integral to the nearest neighbors method. Going back to Figure 12.10, if five nearest

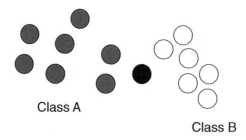

Class A

Class B

Figure 12.10 Classification of new sample (black circle) using three nearest neighbors (*k* = 3). The sample is considered a member of class A, as two of the three nearest neighbors are members of this class

neighbors were used for classification (i.e., $k = 5$), then the new sample would be considered a member of class B, as three out of five nearest neighbors are members of this class. The results and, ultimately, the success of the classification are, therefore, highly dependent on the number of neighbors considered. The number of neighbors can be determined through cross validation of the training set, for which each training set member is treated as an unknown and is classified using a range of k values (Beebe *et al.*, 1998; Varmuza and Filzmoser, 2009). When doing this, the maximum k investigated should be equal to the number of members in the smallest class in the training set. This is necessary to prevent subsequent classification being biased towards classes that have a larger number of members (Beebe *et al.*, 1998; Brereton, 2007).

The k value to retain for subsequent classification is the one that results in a minimum number of misclassifications of the training set members. Note that using only one nearest neighbor can result in the classification being heavily influenced by outliers in the data set. And, as classification is based on a majority vote, an odd number of nearest neighbors is typically used, although "ties" may still occur (Dougherty, 2013). In these cases, methods are available to break any ties that do occur (Beebe *et al.*, 1998).

k-NN is an example of "hard" classification, meaning that every new sample will be classified to one, and only one, class. This has some limitations in that even if the sample does not truly belong to any of the classes represented in the training set, classification will be forced. Beebe describes methods to assess the quality of the classification based on the class membership of the nearest neighbors (Beebe *et al.*, 1998). A more quantitative metric based on the distance of the new sample to the nearest neighbor in the assigned class compared to the spread among members of the class is also described (Beebe *et al.*, 1998).

12.5.2 Discriminant analysis

Discriminant analysis methods can also be used to classify a new sample to a known group. In the first step, the discriminant function(s) that describe separation among classes in the training set is(are) defined and then, in the second step, these functions are used to predict the class of any new sample (Rencher, 1995; Johnson and Witchern, Johnson and Wichern, 1998; Lattin *et al.*, 2003; Koch, 2014). While many methods are available for defining and using these functions, this section focuses on two of the more common approaches: linear discriminant analysis and partial least squares discriminant analysis.

12.5.2.1 Linear discriminant analysis

The premise in linear discriminant analysis (LDA) is to define discriminant functions, which are linear combinations of the original measurement variables that maximize separation of classes within the training set (Cooley and Lohnes, 1971; Tatsuoka and Lohnes, 1998; Rencher, 1995; Lattin *et al.*, 2003; Koch, 2014). This approach was first introduced by Fisher in 1936 as a binary classification method to differentiate iris species

(Fisher, 1936) and then later extended by Rao in 1948 to include multiple classes (Rao, 1948). The discriminant function, y, takes the general form shown in Equation 12.5:

$$y = a_1 x_1 + a_2 x_2 + \cdots + a_p x_p \tag{12.5}$$

where x indicates original measurement variables, p is the number of variables, and a indicates weighting coefficients (Manly, 1986; Tatsuoka and Lohnes, 1998; Rencher, 1995; Lattin *et al.*, 2003). To maximize the ratio of between-class variance and within-class variance, it can be shown (Equation 12.6) that:

$$\frac{SS_B}{SS_W} = \frac{a'Ba}{a'Wa} = \lambda \tag{12.6}$$

where, SS_B and SS_W indicate the between- and within-class sum of squares, respectively, **B** and **W** are the between- and within-class sum of squares and cross-products matrix, respectively, **a** represents the weighting coefficients, and λ is the discriminant criterion, which is also referred to as Fisher's criterion (Cooley and Lohnes, 1971; Tatsuoka and Lohnes, 1998; Rencher, 1995; Lattin *et al.*, 2003).

The next step is to determine the weighting coefficients, **a**, that maximize the discriminant criterion, λ. The solution to this is obtained through eigenanalysis of the $\mathbf{W^{-1}B}$ matrix, resulting in eigenvectors, $a_1, a_2, \ldots, a_r$ and corresponding eigenvalues $\lambda_1, \lambda_2, \ldots, \lambda_r$, where r is the rank of the matrix, **B** (Manly, 1986; Tatsuoka and Lohnes, 1998; Rencher, 1995; Lattin *et al.*, 2003; Morgan and Bartick, 2007). The eigenvalues are ordered from largest to smallest value. Hence, the eigenvalue, λ_1, has the largest value for the discriminant criterion and is obtained when the corresponding eigenvector, a_1, provides the weighting coefficients for the discriminant function. Thus, the first discriminant function, y_1, is defined as (Equation 12.7):

$$y_1 = a_{11} x_1 + a_{12} x_2 + \cdots + a_{1p} x_p \tag{12.7}$$

where a_{1p} indicates the pth element of the first eigenvector, a_1. By definition, the first discriminant function results in greatest separation of the classes, as this results in the largest value for the discriminant criterion. Note the analogy to PCA (Section 12.4.2): both procedures define linear combinations of the original variables, although in PCA these combinations describe variance in the data whereas in LDA these combinations describe group separation (Brereton, 2007; Morgan and Bartick, 2007).

In the case of two classes, only one discriminant function is necessary for separation. However, for c classes, $c–1$ discriminant functions are necessary (Massart *et al.*, 1988; Lattin *et al.*, 2003; Brereton 2007; Morgan and Bartick, 2007). Thus, the second discriminant function, y_2, is obtained when the second eigenvector, a_2, provides the weighting coefficients (Equation 12.8):

$$y_2 = a_{21} x_1 + a_{22} x_2 + \cdots + a_{2p} x_p \tag{12.8}$$

This discriminant function results in the next highest discriminant criterion, λ_2, and offers the next greatest separation of the classes not accounted for by y_1. It follows that subsequent discriminant functions describe the next greatest separation of classes not accounted for by previous functions and so on (Manly, 1986).

As the discriminant functions are linear combinations of the original variables, separation of groups in the training set should be described by the first few discriminant functions (Manly, 1986). Tatsouka and Lattin describe tests that can be used to determine the number of discriminant functions that are necessary to adequately describe group separation (Tatsuoka and Lohnes, 1998; Lattin *et al.*, 2003). These functions can then be used to calculate discriminant scores for samples that are subsequently projected onto the axes defined by the functions. Classification can be achieved by projecting the discriminant score for the "new" sample onto the discriminant axes and then calculating the Mahalanobis distance between the new sample and the centroid of each class (Lattin *et al.*, 2003; Morgan and Bartick, 2007). As the variance–covariance matrix of the data is used to calculate Mahalanobis distance, correlation in the data set is taken into account using this metric (Maesschalck *et al.*, 2000). The new sample is considered a member of the class to which the Mahalanobis distance is shortest (Rencher, 1995; Lattin *et al.*, 2003; Morgan and Bartick, 2007).

To illustrate discriminant analysis, consider a data set of element concentrations detected in different types of document paper. Ten samples of three different paper types (color inkjet, laserjet, and office paper) from the same manufacturer were analyzed using inductively coupled plasma-mass spectrometry. The average concentration of five elements common to each paper type are shown in Figure 12.11a and these data were used as the training set to define the linear discriminant functions. Figure 12.11b plots the discriminant scores of the paper samples for the first two discriminant functions, showing the centroid of each group as a white triangle. The three paper types are clearly distinguished using these two discriminant functions. The functions were then used to calculate discriminant scores for a test set that consisted of ten samples of office paper that were purchased and analyzed at a later date. The test set is projected onto the discriminant function plot and all samples are correctly classified as office paper (Figure 12.11b), based on the shortest Mahalanobis distance to the centroid of this class.

A potential limitation in discriminant analysis is that the data set should contain more samples than variables (Morgan and Bartick, 2007). This is often difficult for chromatographic or spectral data, which typically contain several hundred, if not thousands, of variables. Thus, it is often necessary to select the more important (e.g., more discriminatory) variables to reduce the total number of variables that are included in the analysis. Selection of variables like this can be performed in a number ways, including manual selection, performing PCA first to identify the more important variables, or evaluating the Fisher weight of the variables (Rencher, 1995; Wu *et al.*, 1999; Dixon, 2007; Morgan and Bartick, 2007). This is the ratio of the between-class variance to

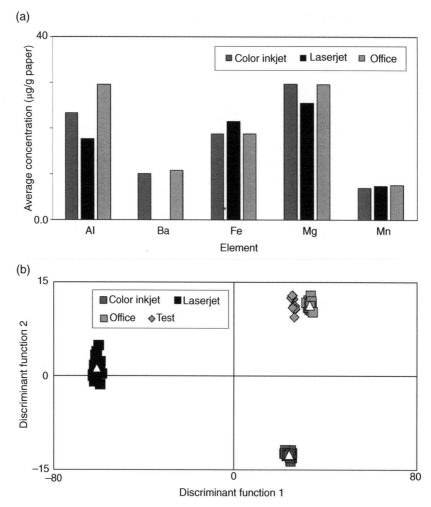

Figure 12.11 Elemental analysis of document paper (a) average concentration of five elements in color inkjet, laserjet, and office paper and (b) linear discriminant analysis based on the element concentrations. The centroid of each paper type is shown as the white triangle and each paper type is distinguished based on the first two discriminant functions. Paper samples in the test set are correctly classified as office paper

within-class variance and variables with larger weights are considered to be more discriminatory (Wu *et al.*, 1999; Dixon *et al.*, 2007; Morgan and Bartick, 2007).

12.5.2.2 Partial least squares discriminant analysis

The partial least squares (PLS) method is commonly used in chemometrics for both classification and regression. The focus in this section is on applications for classification, referred to as partial least squares discriminant analysis (PLS-DA). However, many of the texts listed in the references at the end of this chapter also include sections describing the use of PLS for regression purposes.

The general premise in PLS is to define the relationship between a matrix containing the predictor variables (termed the X block) and the response variables (termed the c block) (Brereton, 2007, 2009; Petraco *et al.*, 2012). In the case of calibration, the c block may be a vector of concentrations whereas, in the case of classification, the c block is a vector containing numerical values that code for membership of a particular class. For example, given two classes A and B, training set samples from class A may be assigned a value of +1 while training set samples from class B may be assigned a value of –1. Thus, in PLS-DA, the c block is a vector containing +1s and –1s that indicate the class the training set members belong to (Brereton, 2007, 2009).

The PLS model can be defined as (Brereton, 2007, 2009; Dixon *et al.*, 2007):

$$X = TP + E$$
$$c = Tq + f$$

where T is a matrix of scores, P is a matrix of loadings, E is the error matrix, q is a vector of "loadings" describing the c block, and f is an error vector for the c block. T and P are analogous to PCA scores and loadings; however, in PLS, the P loadings are neither orthogonal (although the scores are) nor normalized as they are in PCA. Similar to PCA, scores and loadings plots can be generated from the PLS scores and loadings and used to determine those variables that are most important in distinguishing the samples.

A number of different algorithms are available for PLS and are described in the literature; for example, Varmuza and Filzmoser provide a description of many different algorithms while Brereton provides a step-by-step description (Brereton, 2009; Varmuza and Filzmoser, 2009). Briefly, the algorithm is iterative in nature, firstly determining a weight vector (w) associated with the first PLS component. Using w, the PLS scores (t), PLS loadings (p), and the loading for c (q) are calculated and then used to determine the residuals for X and c. The next PLS component is calculated following the same procedure, although this time using the X and c residuals, rather than the original data. This means that beyond the first PLS component, the components relate to the residuals rather than original data; as a result, equations are available to relate the subsequent components to the original data (Varmuza and Filzmoser, 2009). The number of PLS components to retain in the model is an important aspect of model development and is determined using validation procedures, as described in Section 12.5.4 (Brereton, 2009; Varmuza and Filzmoser, 2009).

Once the model is defined and validated, the next step is to predict the class to which a "new" sample belongs. To do this, the relationship between the X block and the c block can be written as shown in Equation 12.9:

$$c = Xb + f = Tq + f \tag{12.9}$$

where *b* corresponds to a vector of regression coefficients that are obtained from Equation 12.10:

$$b = Wq \tag{12.10}$$

where *W* is a matrix containing the weights (*w*) of each PLS component retained in the final model. From this, the class of a new sample, c_{new}, can be predicted from Equation 12.11:

$$c_{new} = xb \tag{12.11}$$

The issue now is to define a suitable threshold to determine which class the new sample belongs to. In the example given above, where Class A was designated +1 and Class B designated −1, the threshold could be set as $c = 0$. If a new sample has a value greater than the threshold (positive value), then it is a member of class A and if the new sample has a value less than the threshold (negative value), then it is a member of class B (Brereton, 2009; Varmuza and Filzmoser, 2009). For more than two classes, however, PLS-DA becomes less straightforward. In these cases, one option is to use a modified version of PLS, which is termed PLS2. Here, the *c* block is a matrix, rather than a vector; Brereton provides a more detailed description of the PLS2 algorithm for the interested reader (Brereton, 2009).

12.5.3 Soft independent modeling of class analogy

The soft independent modeling of class analogy (SIMCA) approach to classification develops models for each class defined in the training set and then uses these models to determine classification of new samples. SIMCA is a "soft" classification procedure, which means that a new sample may be classified as a member of one class, more than one class, or no classes at all. This is in contrast to the previously described "hard" classification procedures in which a sample is classified as a member of one, and only one, class.

In SIMCA, the class models are developed using PCA. The models require some form of optimization, namely, deciding the number of PCs to retain in the final models. This can be done again using a cross-validation procedure wherein the initial model is developed and one PC is retained (Sharaf *et al.*, 1986; Lavine, 1992). This model is used to fit samples in the test set and the residuals for the fit are calculated (Lavine, 1992). Two PCs are then retained and the resulting model is used to fit the test set, again calculating the residuals. This process continues, increasing the number of principal components retained in the model by one each time. The number of principal components that minimizes the residuals is retained in the final model. Models are developed and optimized in a similar manner for each class represented in the training set. As the models are developed independently for each class, each one may retain a different number of principal components (Lavine, 1992). Further, a number of diagnostics are available in SIMCA to assess how well classes in that training set are separated, how well variables describe a particular class, and how well a variable distinguishes between two classes (Lavine, 1992; Brereton, 2007).

Once PCA models are generated and validated for each defined class in the training set, the models can be used to classify "new" samples. Scores for the new sample are calculated and projected onto the model for the first class in the training set. The variance of the residuals for the sample when fit to this class is calculated (Equation 12.12) and compared to the variance of the residuals for members of the class (Equation 12.13) (Sharaf *et al.*, 1986; Lavine, 1992):

$$s_i^2 = \sum\nolimits_{j=1}^{P} \frac{\left(e_{ij}\right)^2}{P-F} \qquad (12.12)$$

$$s_0^2 = \sum\nolimits_{i=1}^{N} \sum\nolimits_{j=1}^{P} \frac{\left(e_{ij}\right)^2}{\left(P-F\right)\left(N-F-1\right)} \qquad (12.13)$$

where s_i^2 is the variance in the residuals for the sample fit to the class, s_0^2 is the variance in the residuals for the class, *eij* are the residuals, P is the dimensionality of the class, F is the number of PCs retained in the model, and N is the number of samples in the class.

The variances are then statistically compared using an F-test, at a specified confidence level (Sharaf, 1986; Lavine, 1992). The null hypothesis (H_0) states that the variance of the sample residuals is less than the variance of the residuals of the class members, which supports the sample being a member of that class. In contrast, the alternate hypothesis (H_a) is that the variance of the residuals of the sample fit to the class is greater than the variance of the residuals of the class members, which supports the conclusion that the sample is not a member of the class.

Once fit to the first class in the training set and deciding on membership (or not), the sample is then fit to the second training set class, following the same procedure. As membership of the sample in each class is tested, the sample may be classified as a member of one class, more than one class, or no classes at all.

12.5.4 Model validation

Model validation is used to test the "success" of the developed model (Beebe *et al.*, 1998; Brereton, 2007, 2009; Morgan and Bartick, 2007; Varmuza and Filzmoser, 2009). For classification models, this translates as how well the model performs in predicting the identity (or class) of a "new" sample. The training set members could be used to assess how well the model performs by determining the predicted class of the training set members. The number of correctly classified samples can then be determined and is usually expressed as a percentage although, the total number of samples (n) in the data set should also be given, so that the success rate of the model can be put into context (Brereton, 2006). The success rate based on the training set members is the "classification ability" of the model but this tends to give an overly optimistic view of the model success as it is based on the data from which the model was originally developed (Massart *et al.*, 1988; Leardi, 2006; Morgan and Bartick, 2007).

Instead, a test set of data should be used to determine the classification success of the model, which is often referred to as the "prediction ability" of the model (Leardi, 2006; Brereton, 2007, 2009; Morgan and Bartick, 2007; Varmuza and Filzmoser, 2009). As before, members of the test set are classified using the model and the percent correct classification is determined. The prediction ability of the model is typically less than the classification ability, as now the model is being used to classify "new" samples that were not present in the original training set. However, a model with a prediction ability substantially lower than the classification ability can indicate that the model is too specific or that the new sample being classified is an outlier (Brereton, 2007).

The test set used to determine the prediction ability of the model can be generated independently or can be generated from the original data set through cross-validation or bootstrap methods. The cross-validation method involves initially omitting a selection of the training set data. This can be a single sample each time (leave-on-out cross-validation) or a subset of data (Massart *et al.*, 1988; de Boves Harrington, 2006; Brereton, 2007; Morgan and Bartick, 2007; Varmuza and Filzmoser, 2009). The model is developed with the remaining training set members, the omitted sample (or subset) is classified using the model, and the percent correct classification of the subset is determined. The cross-validation procedure is repeated many times (in the case of leave-one-out, each sample in the training set is omitted once) and the average correct classification rate determined (Morgan and Bartick, 2007).

Bootstrap methods also involve removing a subset of data from the original data set and using this to validate the model (de Boves Harrington, 2006; Brereton, 2007; Varmuza and Filzmoser, 2009). The required number of samples (equal to the number of samples in the training set) is randomly selected from the training set to become members of the test set. However, the bootstrap method is a resampling method and samples are randomly selected with replacement. This means that, once a sample is selected to become a member of the test set, it is replaced in the training set and can be reselected. As a result, bootstrap methods generate test sets with the same number of samples as the original training set but the same sample can appear more than once in the test set. As before, the bootstrap method can be applied iteratively and the average percent correct classification of the test sets determined to assess the classification success of the model (Brereton, 2007; Varmuza and Filzmoser, 2009).

12.6 Applications of chemometric procedures in forensic science

While chemometric procedures are not routinely used in forensic case work, there is an increasing body of research in which these procedures are being applied in nearly all subdisciplines of forensic chemistry. The following section provides an overview of some of the more recent literature examples (primarily from 2010 onwards) and is certainly not exhaustive. Instead, it aims to highlight the application of many of the procedures discussed in this chapter and demonstrate their potential for future use in case work.

12.6.1 Fire debris and explosives

Several researchers are investigating the use of both unsupervised and supervised pattern recognition procedures for applications in fire debris analysis. In these cases, the procedures are performed on GC-MS data obtained from the debris samples, aiming to identify the presence of any ignitable liquid residues in the samples. Baerncopf *et al.* and Prather *et al.* used PCA to associate simulated fire debris samples to the corresponding ignitable liquid standard (Baerncopf *et al.*, 2011; Prather *et al.*, 2014). In both studies, data analysis was performed on the full TIC, following retention-time alignment and constant-sum normalization.

Turner and Goodpaster considered both weathering and microbial degradation of ignitable liquids in fire debris (Turner and Goodpaster, 2012). Simulated samples containing gasoline were weathered and subjected to microbial degradation before being analyzed by GC-MS. Extracted ion profiles were normalized to unit length and auto-scaled prior to PCA. From the resulting scores plot, the first principal component described the extent of weathering while the second principal component described the extent of degradation. Interestingly, the PCA loadings plots could be used to identify the compounds in gasoline that are susceptible to weathering (i.e., the more volatile compounds) and those compounds more susceptible to microbial degradation (mono-substituted alkylbenzenes and small normal alkanes).

In a series of two papers, Waddell *et al.* described a procedure to first determine the presence of an ignitable liquid in fire debris samples and, second, to classify any liquid according to the chemical classes defined by ASTM International (Waddell *et al.*, 2013, 2014). In the first paper, PCA and discriminant analysis methods were investigated, with a quadratic discriminant model offering greatest classification success (Waddell *et al.*, 2013). For this model, the reported classification success rate was 81.3% based on cross-validation and 70.9% based on an external test set that included samples from laboratory-scale burns and field-test burns. In the second paper, the performance of SIMCA as the classification procedure was investigated (Waddell *et al.*, 2014). Using this "soft" classification, improvements in the classification success rate were reported: 94.2% based on cross-validation and 79.1% based on the external test set.

Both PCA and *k*-NN were used by Kerr *et al.* in their study into the discrimination of six polymers after burning (Kerr *et al.*, 2013). The polymers were selected as those common in household materials and were burned under controlled conditions. The burned samples were analyzed by micro-Raman spectroscopy and the spectra were base-line corrected, smoothed, and normalized prior to PCA. The first two principal components accounted for 54% of the total variance and were sufficient to distinguish the six burned polymers based on their Raman spectra. The *k*-NN approach was then employed, using Euclidean distance and six nearest neighbors for classification. The burned samples were successfully classified according to polymer, with a classification success of 93%, as determined using a bootstrap method.

Banas *et al.* used HCA and PCA to investigate association of postblast debris to the original explosive material (Banas *et al.*, 2010). Simulated samples containing C-4, TNT, and PETN were exploded under controlled conditions and analyzed using FTIR spectroscopy. Reference standards of the same three explosives, along with RDX (which is the major component in C-4), were also analyzed using the same technique. Spectra were preprocessed by taking the first derivative and then normalizing. For HCA, Euclidean distance and single linkage methods were used. In general, postblast debris samples were correctly clustered with the pure explosive, with the exception of some samples containing C-4. Using PCA, the first three principal components accounted for 99.6% of the total variance in the data set and afforded distinction of the four pure explosive materials. Further, the postblast samples were closely positioned to the corresponding pure explosive in the scores plot and the spectral features responsible for the differentiation of the explosives were identified in the loadings plots.

12.6.2 Controlled substances and counterfeit medicines

Rodrigues *et al.* analyzed cocaine samples seized in a region in Brazil using ATR-FTIR then applied PCA and PLS-DA to distinguish and classify samples based on purity and the chemical form of cocaine present (Rodrigues *et al.*, 2013). The IR spectra were mean centered, normalized using the SNV procedure, and truncated to remove spectral regions that did not contain significant absorptions from cocaine or the major adulterants. In the PCA scores plot, the samples were distinguished based on purity and chemical form of cocaine, which were described by PC1 and PC2, respectively. Two separate PLS-DA models were then developed: one to classify the seized cocaine samples based on the proportion of cocaine present ("high" or "low") and the second to classify the samples based on the chemical form of cocaine (salt or base). Cross-validation of the test set indicated successful classification rates of 95% and 97% in classifying the samples according to proportion of cocaine and chemical form, respectively.

Chan *et al.* employed HCA, PCA, and discriminant analysis to associate simulated street samples of heroin cut with a mixture of adulterants to the corresponding samples prior to being cut (Chan *et al.*, 2013). Using the combination of statistical procedures, correct association was possible, with a successful classification rate of 92.4% reported for discriminant analysis with cross validation.

Been *et al.* developed a classification system for the identification of counterfeit pharmaceuticals, based on NIR and Raman spectra of the samples (Been *et al.*, 2011). The NIR spectra of counterfeit samples, as well as the genuine pharmaceutical product, were normalized using the SNV procedure and first-derivative spectra were generated using the Savitzky–Golay method. Raman spectra were truncated to remove spectral contributions from the laser source then baseline corrected. The first derivative of each Raman spectrum was also generated and the spectra were mean centered. Distinction of the counterfeits from the genuine products was possible using PCA and HCA for both the NIR and the Raman spectra. However, for HCA of the Raman spectra, it was not possible to further distinguish among the different counterfeits. For both data sets, k-NN using one nearest neighbor correctly classified all but two spectra in the training and the test sets. For the NIR spectra, the optimal PLS-DA model retained 25 components, with four spectra misclassified in both the training and test sets. For the Raman spectra, the PLS-DA model retained 50 components, with all spectra in the training set successfully classified and five spectra in the test set misclassified. Results from the supervised procedures compared favorably with those of the unsupervised procedures and the latter were then used to further develop a model for the classification of suspected counterfeit samples.

12.6.3 Trace evidence

A number of chemometric procedures have been applied to a variety of different types of trace evidence. Barrett *et al.* employed HCA, PCA, and discriminant analysis to investigate the discrimination of dyed hair samples based on the absorbance spectra obtained using UV-visible microspectrophotometry (Barrett *et al.*, 2011). Spectra were baseline corrected and normalized prior to further data analysis. Three clusters of samples were identified in the HCA dendrogram; these clusters corresponded to the groups of samples that were apparent based on visual assessment of the hairs. The first two principal components accounted for 78.6% of the variance in the data set and the same three groups of samples that were indicated by HCA were observed in the scores plot. The PCA loadings plots provided more information about the variables responsible for the sample groupings, indicating that distinction of the samples was based on differences in the intensity of the absorbance peaks corresponding to the hair and the dye. Discriminant analysis yielded successful classification rates of 89.1% and 75%, based on cross validation and an external test set, respectively.

Bonetti and Quarino focused on multivariate statistical analysis of soil samples collected from various state parks in New Jersey, in both summer and fall (Bonetti and Quarino, 2014). Each soil sample was characterized based on particle size distribution, pH, and organic content and the data were analyzed using PCA and discriminant analysis. PCA was performed initially on the particle-size data for the samples alone, then the pH data were included and, finally, the organic content data were included. By considering all three characteristics, the soil samples were clustered in the scores plot according to state park. Similarly, error rates associated with discriminant analysis were higher based on only on the particle size distribution data (58.3% for summer soil samples and 16.7% for fall soil samples) and decreased as the pH and organic content data were included (33.3% for summer and 3.3% for fall). Differences in the error rates for the summer and fall samples were attributed to greater variability in the summer samples as these samples were collected at 15 meter intervals.

Discrimination of soils was also reported by Jantzi and Almirall, although this time focusing on the elemental composition of the soils (Jantzi and Almirall, 2014). Soils were analyzed using laser-induced breakdown spectroscopy (LIBS) and laser ablation inductively coupled plasma mass spectrometry (LA-ICP-MS) for elemental composition; two different sample introduction methods were used (pellet and tape methods). The elemental profiles generated by each technique were separately subjected to PCA to identify the most discriminatory elements. These elements were then used in discriminant analysis for soil classification and, using leave-one-out cross-validation, correct classification rates ranged from 94.5% for LIBS, using the pellet sample introduction method, to 98.6% for LA-ICP-MS, using the tape method.

Goodpaster *et al.* demonstrated the use of multivariate data analysis procedures to classify electrical tape according to brand (Goodpaster *et al.*, 2009). The FTIR spectra of the backing and the adhesive from 72 rolls of electrical tape were preprocessed by subtracting a constant background, then normalizing to unit length. Association of the tapes according to brand was demonstrated using both HCA and PCA. With discriminant analysis, successful classification rates of 80% and 87% were reported, based on spectra of the backing and adhesive, respectively. The utility of these data analysis procedures was demonstrated using a simulated case example in which a pipe bomb was assembled. The explosive material within the device was wrapped in electrical tape and the outside of the pipe was wrapped in a different-brand tape. The bomb was exploded under controlled conditions and spectra of the tape fragments were successfully classified according to brand.

Association and classification of paint samples using HCA, PCA, and discriminant analysis has also been demonstrated (Liszewski *et al.*, 2010; Muehlethaler *et al.*, 2011; Maric *et al.*, 2014). Maric *et al.* analyzed the primer surface layer in automotive paints using synchrotron FTIR (Maric *et al.*, 2014). Spectra were baseline corrected and normalized prior to PCA. From the subsequent scores plot, 14 groups of samples were apparent, which corresponded to different manufacturers, and from the loadings plots, the spectral features responsible for the groupings were identified. Liszewski *et al.* also analyzed automotive paint samples but focused on the UV absorbers present in the clear coats of the paints (Liszewski *et al.*, 2010). Samples were analyzed using UV-microspectrophotometry, spectra were truncated and normalized to unit length, then subjected to HCA, PCA, and discriminant analysis. The 71 paint samples analyzed formed three distinct clusters using HCA, with the observed clustering being related to the relative absorptions at three different regions in the spectra (at about 250 nm, 300 nm, and 350 nm). Similar grouping of the samples was also observed by PCA using two principal components that accounted for 78.7% of the total variance in the data set. For discriminant analysis, classification success rates were reported as 91.6% and 81.1% based on cross-validation of the training set and an external test set, respectively.

Muehlethaler *et al.* focused their efforts on household paints, analyzing 34 red paints by both IR and Raman spectroscopy (Muehlethaler *et al.*, 2011). Spectra were normalized using the SNV procedure and the IR data were further truncated to remove regions of the spectrum that did not contain useful information. For the IR data, the samples were distinguished in the PCA scores plot based on the resin content (alkyd or acrylic) and the presence, or otherwise, of calcium carbonate. For the Raman data, the samples were distinguished in the scores plot according to pigment composition. Cluster analysis was also performed on each data set and the clusters observed were in agreement with the groupings of the samples observed in the PCA scores plots.

12.6.4 Impression evidence

The applications of chemometric procedures described above primarily focus on spectral or chromatographic data, albeit for different types of evidence. These data types lend themselves to such procedures because of the large number of variables and high correlation among the variables. However, as highlighted in a National Academy of Sciences (NAS) National Research Council report, there is a need for more objective methods of comparison in the majority of forensic disciplines, including impression evidence, which currently is primarily based on visual comparisons of questioned and known samples (National Research Council, 2009).

Petraco *et al.* reported the application of chemometric procedures to both shoe impressions and tool marks (Petraco *et al.*, 2010, 2012). In the first of these studies, five pairs of the same shoe brand and style were worn by the same person for 30-day periods (Petraco *et al.*, 2010). During this time, impressions of the shoes were taken and converted into feature vectors based on the location and number of accidental features. The resulting feature vectors were subjected to PCA for dimension reduction and discriminant analysis for classification. A total of 116 feature vectors were subjected to PCA, determining that 99.5% of the variance in the data set was described by 32 principal components. Discriminant analysis was performed on the reduced data set to classify patterns according to shoe. Using a leave-one-out cross-validation procedure, the rate of successful classification ranged from 77 to 100%, depending on the shoe in question. The authors noted that, as the shoes were subjected to greater wear the accidental characteristics became more distinguishable, resulting in lower rates for successful classification.

A similar methodology was adopted for the statistical analysis of tool mark impressions (Petraco *et al.*, 2012). In this case, the striations left by nine different screwdrivers were converted into binary feature vectors using an image processing program. Various data analysis procedures, including PCA and PLS-DA, were applied to assign the screwdriver to the corresponding striation pattern. Different methods (e.g., bootstrap, leave-one-out cross-validation) were used to assess the classification success, which ranged from 97 to 100% for PLS-DA.

12.7 Conclusions

Although not routinely used for forensic chemistry applications, chemometrics has a long history of use in analytical chemistry. In light of the 2009 NAS report and given the types of data that are typically generated during forensic chemical analyses, chemometric procedures certainly have the potential for increased use by forensic analysts in evidence comparisons. However, caution must be exercised that these procedures do not become "black box" methods. There must be a thorough understanding of the data to be analyzed and the chemometric procedures themselves, as well as their limitations. Without this understanding, there is a real danger of false or misleading conclusions being drawn. Brereton discusses this further in his editorial on the evolution and future of chemometrics (Brereton, 2013) and a whole issue of the journal *Trends in Analytical Chemistry* was dedicated to "Use and Abuse in Chemometrics" (**25**(11), 2006). Thus, while forensic science is moving forward in beginning to apply these procedures, more research, education, and training are necessary before chemometric procedures can become more routinely implemented in forensic laboratories to aid analysts in their work.

Acknowledgements

Thanks to Dr Jay Siegel for the invitation to contribute this chapter and to Dr Victoria McGuffin and Dr John McIlroy for many useful discussions over the years on data pre-processing, data analysis, and chemometric procedures. Thanks also to Emily Riddell who collected the data described in Section 12.5.2. Exemplar data used for illustrative purposes in Sections 12.4.1 and 12.4.2 was based on data collected as part of Award No. 2011-DN-BX-K560 awarded by the National Institute of Justice, Office of Justice Programs, US Department of Justice.

References

Adams, M.J. (1995) *Chemometrics in Analytical Spectroscopy*. Royal Society of Chemistry, Cambridge, UK.

Baerncopf, J.M., McGuffin, V.L., and Waddell Smith, R. (2011) Association of ignitable liquid residues to neat ignitable liquids in the presence of matrix interferences using chemometric procedures. *Journal of Forensic Sciences*, **56**, 70–81.

Banas, K., Banas, A., Moser, H.O., *et al.* (2010) Multivariate analysis techniques in the forensics investigation of the postblast residues by means of Fourier transform-infrared spectroscopy. *Analytical Chemistry*, **82**, 3038–3044.

Barrett, J.A., Siegel, J.A., and Goodpaster, J.V. (2011) Forensic discrimination of dyed hair color: II. *Multivariate statistical analysis. Journal of Forensic Sciences*, **56**, 95–101.

Barnes, R., Dhanoa, M., and Lister, S. (1989) Standard normal variate transformation and de-trending of near-infrared diffuse reflectance spectra. *Applied Spectroscopy*, **43**, 772–777.

Beebe, K.R., Pell, R.J., and Seasholtz, M.B. (1998) *Chemometrics: A Practical Guide*. John Wiley & Sons, Inc., New York, NY.

Been, F., Roggo, Y., Degardin, K., *et al.* (2011) Profiling of counterfeit medicines by vibrational spectroscopy. *Forensic Science International*, **211**, 83–100.

Bonetti, J. and Quarino, L. (2014) Comparative forensic soil analysis of New Jersey state parks using a combination of simple techniques with multivariate statistics. *Journal of Forensic Sciences*, **59**, 627–636.

Brereton, R.G. (2006) Consequences of sample size, variable selection, and model validation and optimisation, for predicting classification ability from analytical data. *Trends in Analytical Chemistry*, **25**, 1103–1111.

Brereton, R.G. (2007) *Applied Chemometrics for Scientists*. John Wiley & Sons, Inc., Hoboken, NJ.

Brereton, R.G. (2009) *Chemometrics for Pattern Recognition*. John Wiley & Sons, Inc., Chichester, UK.

Brereton, R.G. (2013) The evolution of chemometrics. *Analytical Methods*, **5**, 3785–3789.

Chan, K.W., Tan, G.H., and Wong, R.C.S. (2013) Statistical validation for the profiling of heroin by associating simulated postcut samples with the corresponding precut sample. *Journal of Forensic Sciences*, **58**, S199–S207.

Cooley, W.W. and Lohnes, P.R. (1971) *Multivariate Data Analysis*. John Wiley & Sons, Inc., New York, NY.

de Boves Harrington, P. (2006) Statistical validation of classification and calibration models using bootstrapped latin partitions. *Trends in Analytical Chemistry*, **2006**, 11.

Dixon, S.J., Xu, Y., Brereton, R.G., *et al.* (2007) Pattern recognition of gas chromatography mass spectrometry of human volatiles in sweat to distinguish the sex of subjects and determine potential discriminatory marker peaks. *Chemometrics and Intelligent Laboratory Systems*, **87**, 161–172.

Dougherty, G. (2013) *Pattern Recognition and Classification: An Introduction*. Springer, New York.

Fisher, R.A. (1936) The use of multiple measurements in taxonomic problems. *Annals of Eugenics*, **7**, 179–188.

Gemperline, P.J. (2006) Principal components analysis. In: *Practical guide to chemometrics*, 2nd edn (ed P.J. Gemperline), CRC/Taylor & Francis, Boca Raton, FL, pp. 69–104.

Goodpaster, J.V., Sturdevant, A.B., Andrews, K.L., *et al.* (2009) Identification and comparison of electrical tapes using instrumental and statistical techniques: II. *Organic composition of the tape backing and adhesive. Journal of Forensic Sciences*, **54**, 328–338.

Hastie, T., Tibshirani, R., and Friedman, J. (2009) *The Elements of Statistical Learning: Data Mining, Inference, and Prediction*. Springer, New York, NY.

Izenman, A.J. (2008) *Modern Multivariate Statistical Techniques*. Springer, New York.

Jantzi, S.C. and Almirall, J.R. (2014) Elemental analysis of soils using laser ablation inductively coupled plasma mass spectrometry (LA-ICP-MS) and laser-induced breakdown spectroscopy (LIBS) with multivariate discrimination: Tape mounting as an alternative to pellets for small forensic transfer specimens. *Applied Spectroscopy*, **68**, 963–974.

Johnson, K.J., Wright, B.W., Jarman, K.H., and Synovec, R.E. (2003) High-speed peak matching algorithm for retention time alignment of gas chromatographic data for chemometric analysis. *Journal of Chromatography A*, **996**, 141–155.

Johnson, R.A. and Wichern, D.W. (1998) *Applied Multivariate Statistical Analysis*. Prentice-Hall, Inc., Upper Saddle River, NJ.

Kerr, T., Duncan, K., and Myers, L. (2013) Post fire materials identification by micro-Raman spectroscopy and principal components analysis. *Journal of Analytical and Applied Pyrolysis*, **102**, 103–113.

Koch, I. (2014) *Analysis of Multivariate and High-Dimensional Data*. Cambridge University Press, New York, NY.

Lattin, J., Carroll, J.D., and Green, P.E. (2003) *Analyzing Multivariate Data*. Brooks/Cole–Thomson Learning, Pacific Grove, CA.

Lavine, B.K. (1992) Signal processing and data analysis. In: *Practical Guide to Chemometrics*, (ed S.J. Haswell), Marcel Dekker, Inc., New York, NY, pp. 211–238.

Lavine, B.K., and Davidson, C.E. (2006) Classification and pattern recognition. In: *Practical Guide to Chemometrics*, 2nd edn (ed P.J. Gemperline), CRC/Taylor & Francis, Boca Raton, FL, pp. 339–377.

Leardi, R. (2006) Chemometrics in data analysis. In: *Chromatographic Analysis of the Environment*, 3rd edn (ed L.M.L. Nollet), CRC/Taylor & Francis, Boca Raton, FL, pp. 221–241.

Liszewski, E.A., Lewis, S.W., Siegel, J.A., and Goodpaster, J.V. (2010) Characterization of automotive paint clear coats by ultraviolet absorption microspectrophotometry with subsequent chemometric analysis. *Applied Spectroscopy*, **64**, 1122–1125.

Maesschalck, R.D., Jouan-Rimbaud, D., and Massart, D. (2000) The Mahalanobis distance. *Chemometrics and Intelligent Laboratory Systems*, **50**, 1–18.

Manly, B.F. (1986) *Multivariate Statistical Methods. A Primer*. Chapman and Hall Ltd., Bristol, UK.

Maric, M., van Bronswijk, W., Lewis, S.W., and Pitts, K. (2014) Synchrotron FTIR characterisation of automotive primer surfacer paint coatings for forensic purposes. *Talanta*, **118**, 156–161.

Massart, D., Vandeginste, B., Deming, S., *et al.* (1988) *Chemometrics: A Textbook*. Elsevier, New York, NY.

McIlroy, J.W. (2014) *Effects of Data Pretreatment on the Multivariate Statistical Analysis of Chemically Complex Samples*. Michigan State University, East Lansing, MI.

McMaster, M.C. (2008) *GC/MS: A Practical User's Guide*. John Wiley & Sons, Inc., Hoboken, NJ.

Miller, C.E. (2010) Chemometrics in process analytical technology (PAT). In: *Process Analytical Technology: Spectroscopic Tools and Implementation Strategies for the Chemical and Pharmaceutical Industries* (ed. K.A. Bakeev), John Wiley & Sons, Ltd., Chichester, UK, pp. 226–328.

Morgan, S.L., and Bartick, E.G. (2007) Discrimination of forensic analytical chemical data using multivariate statistics. In: *Forensic Analysis on the Cutting Edge: New Methods for Trace Evidence Analysis* (ed. R.D. Blackledge), John Wiley & Sons, Inc., Hoboken, NJ, pp. 333–374.

Muehlethaler, C., Massonnet, G., and Esseiva, P. (2011) The application of chemometrics on infrared and Raman spectra as a tool for the forensic analysis of paints. *Forensic Science International*, **209**, 173–182.

National Research Council (2009) *Strengthening Forensic Science in the United States: A Path Forward*. The National Academies Press, Washington, DC.

Nielsen, N., Carstensen, J.M., and Smedsgaard, J. (1998) Aligning of single and multiple wavelength chromatographic profiles for chemometric data analysis using correlation optimised warping. *Journal of Chromatography A*, **805**, 17–35.

Petraco, N.D.K., Gambino, C., Kubic, T.A., *et al.* (2010) Statistical discrimination of footwear: A method for the comparison of accidentals on shoe outsoles inspired by facial recognition techniques. *Journal of Forensic Sciences*, **55**, 34–41.

Petraco, N.D., Shenkin, P., Speir, J., *et al.* (2012) Addressing the National Academy of Sciences' challenge: A method for statistical pattern comparison of striated tool marks. *Journal of Forensic Sciences*, **57**, 900–911.

Pierce, K.M., Nadeau, J.S., and Synovec, R.E. (2012) Data analysis methods. In: *Gas Chromatography* (ed. C.F. Poole), Elsevier, Waltham, MA, pp. 415–434.

Prather, K.R., Towner, S.E., McGuffin, V.L., and Waddell Smith, R. (2014) Effect of substrate interferences from high-density polyethylene on association of simulated ignitable liquid residues with the corresponding liquid. *Journal of Forensic Sciences*, **59**, 52–60.

Rao, C.R. (1948) The utilization of multiple measurements in problems of biological classification. *Journal of the Royal Statistical Society Series B (Methodological)*, **10**, 159–193.

Rencher, A.C. (1995) *Methods of Multivariate Analysis*. John Wiley & Sons Inc., New York, NY.

Rodrigues, N.V.S., Cardoso, E.M., Andrade, M.V.O., *et al.* (2013) Analysis of seized cocaine samples by using chemometric methods and FTIR spectroscopy. *Journal of the Brazilian Chemical Society*, **24**, 507–517.

Savitzky, A. and Golay, M.J.E. (1964) Smoothing + differentiation of data by simplified least squares procedures. *Analytical Chemistry*, **36**, 1627–1639.

Sharaf, M.A., Illman, D.L., and Kowalski, B.R. (1986) *Chemometrics*. John Wiley & Sons, Inc., New York, NY.

Tatsuoka, M.M. and Lohnes, P.R. (1988) *Multivariate Analysis*. Macmillan Publishing Company, New York, NY.

Turner, D.A. and Goodpaster, J.V. (2012) Comparing the effects of weathering and microbial degradation on gasoline using principal components analysis. *Journal of Forensic Sciences*, **57**, 64–69.

van den Berg, R.A., Hoefsloot, H.C.J., Westerhuis, J.A., *et al.* (2006) Centering, scaling, and transformations: Improving the biological information content of metabolomics data. *BMC Genomics*, **7**, 142–156.

Varmuza, K. and Filzmoser, P. (2009) *Introduction to Multivariate Statistical Analysis in Chemometrics*. CRC Press/Taylor & Francis Group, Boca Raton, FL.

Waddell, E.E., Song, E.T., Rinke, C.N., *et al.* (2013) Progress toward the determination of correct classification rates in fire debris analysis. *Journal of Forensic Sciences*, **58**, 887–896.

Waddell, E.E., Williams, M.R., and Sigman, M.E. (2014) Progress toward the determination of correct classification rates in fire debris analysis II: Utilizing sift independent modeling of class analogy (SIMCA). *Journal of Forensic Sciences*, **59**, 927–935.

Wu, W., Guo, Q., Jouan-Rimbaud, D., and Massart, D. (1999) Using contrasts as data pretreatment method in pattern recognition of multivariate data. *Chemometrics and Intelligent Laboratory Systems*, **45**, 39–53.

Index

Forensic Chemistry: Fundamentals and Applications, First Edition. Edited by Jay A. Siegel.
© 2016 John Wiley & Sons, Ltd. Published 2016 by John Wiley & Sons, Ltd.